SCOTT H. CHANDLER, Ph.D
DEPT. OF PHYSIOLOGICAL SCIENCE, UCLA
2854 SLICHTER HALL
LOS ANGELES, CA 90024-1568

Microelectrode Techniques

The Plymouth Workshop Handbook

Microelectrode Techniques

The Plymouth Workshop Handbook

Edited by

DAVID OGDEN

National Institute for Medical Research,
Mill Hill, London

Published by The Company of Biologists Limited, Cambridge

Published and Printed by The Company of Biologists Limited,
Department of Zoology, University of Cambridge,
Downing Street, Cambridge CB2 3EJ

© The Company of Biologists Limited 1987

ISBN: 0 948601 49 3

First impression 1987

Second impression 1988

Second edition 1994

Contents

Chapter 1

Chapter 2

Chapter 3

Chapter 4

Chapter 5

Chapter 6

Chapter 10

Chapter 11

Chapter 12

Chapter 13

Chapter 14

Chapter 15

Chapter 16

Index

Preface

First edition

The use of microelectrodes as a means of probing the physiological properties of cells has grown enormously since the pioneering work of Ling & Gerard and Hodgkin & Nastuk. Microelectrodes are now used in many areas of biological sciences, to determine not only the membrane properties of cells with voltage clamp techniques, but also to measure the intracellular free ion concentrations of important inorganic ions, to study the overall architecture of cells by injection of markers, to determine the direct connections between cells and, with the advent of the patch clamp approach, to examine membrane properties at the level of single ion channels.

In 1983 a number of us working in the biological sciences became aware of an acute shortage of young research workers with adequate training in such modern electrophysiological techniques. This shortfall was apparent across a wide range of biological sciences as techniques that were originally the province of membrane physiologists interested in excitable cells spread into such diverse areas as developmental biology and plant sciences. Despite such widespread application, opportunities for training were few and diminishing as reduced financial resources bit into University and Research Council funding. Our concern led us to launch a postgraduate Workshop, designed to provide intensive practical and theoretical training in Microelectrode Techniques for Cell Physiology.

It was decided to hold the Workshop at the Laboratory of the Marine Biological Association, Plymouth. The Laboratory has a distinguished record in providing facilities for membrane physiologists to carry out their research and our suggestion was received with enthusiasm by the Director of the Laboratory, Professor Sir Eric Denton FRS, and the Council of the Marine Biological Association. The Natural Environment Research Council, which provides a Grant in Aid to the Laboratory, gave full support to the Director. Generous financial assistance was offered by The Physiological Society, The Nuffield Foundation and the Company of Biologists. The Science and Engineering Research Council, the Medical Research Council and NERC all agreed to support a number of their research students and the Physiological Society, through its Dale Fund, was prepared to give grants to assist those offered places.

The first Workshop took place in April 1984; the demand was very high – 85 applications for 14 places – fully justifying our concerns about the lack of suitable training. Two further Workshops have followed, each drawing more than 50 applicants from all over the United Kingdom, Europe and the rest of the World, including the United States. We hope that it will become as established and prestigious as those at Woods Hole and Cold Spring Harbor. The Research Councils have continued to support their students and the Physiological Society has continued to give grants to assist those offered places through its Dale and Rushton Funds. The Laboratory of the Marine Biological Association at Plymouth has been a generous and supportive host, providing a wealth of marine preparations for our experiments. The Staff of the Laboratory have received our annual invasion with good humour and much practical support. We are especially grateful to the late Dr J. P. Gilpin-Brown, the Bursar of the Laboratory, who has looked after our accounts.

Each Workshop has held firm to the principle of intensive experimental work, together with theoretical sessions, provided by the resident staff and a number of visitors. Each of them is distinguished in their own field and has heavy teaching responsibilities in their own University. The success of the Workshop is a direct consequence of their enthusiasm, dedication and willingness to give up precious research time in order to train others. Each year valuable technical help has been provided by research students or assistants from the laboratories of some of the staff. Our original sponsors have been joined by Smith, Kline & French, Linton Instrumentation and the Sandoz Institute for Medical Research. We have also been fortunate in the number of companies who have generously loaned equipment for use during the Workshop.

This book is the Workshop Handbook. Like the Workshop itself, it is a cooperative venture. All the authors have taught on the Workshop and it has benefitted from the criticisms of the 1986 participants. Three of our number, Peter Gray, Nick Standen and Michael Whitaker have acted as editors of the various contributions and Judy Lewis and Carole Wright, secretaries at the University of Leicester, took our ill-prepared manuscripts and turned them into documents fit for the Press. Our publishers are the Company of Biologists. They, together with the Physiological Society, the Nuffield Foundation and the Research Councils, deserve much credit for making it possible for our tentative plans to materialize into the fully-fledged Workshop. We hope that other groups of scientists will be encouraged by our success and that the Microelectrode Techniques Workshop will prove to be the fore-runner of many other similar ventures.

Anne Warner
Department of Anatomy & Embryology,
University College London.

Second edition

This second edition of 'Microelectrode Techniques' has arisen from the need to extend the set of lecture notes, which is essentially the purpose of this Handbook, to cover techniques that are currently taught on the Workshop but were not in the first edition. In particular, optical techniques – quantitative fluorescence microscopy for ion concentration measurements and flash photolysis – which are often used in conjunction with microelectrode and patch clamp methods nowadays are given a practical treatment in the style of the first edition. The electrophysiological techniques are extended to include channel reconstitution and recording in lipid bilayers and patch clamp recording from cells in tissue slices. The techniques described in the first edition have been updated, notably the chapter on computer analysis, and expanded or rewritten to take account of feedback from the workshop itself and from readers. The aim is to provide an introduction to each technique and its instrumentation, and to extend that to cover the ideas and techniques of analysis and particular aspects of current practice.

The Handbook originates with the Plymouth Workshop and the authors of the second edition are current or past teachers. An updated list of teachers, who keep the Workshops running, is given below.

The Workshop enjoys the continued support and hospitality of the Marine Biological Association of the UK at their laboratories on Citadel Hill, Plymouth. Direct financial support is received from the Medical Research Council, the Science and Engineering Research Council (or its successor), the Company of Biologists and the Physiological Society. There are several Companies listed below who generously loan equipment each year.

David Ogden
National Institute for Medical Research, London.

Companies who have lent equipment

Axon Instruments Inc.
Burleigh Instruments Ltd
Cairn Research Ltd
Cambridge Electronic Design Ltd
Campden Instruments Ltd
Clarke Electromedical Ltd
Devtek Ltd
Digitimer Ltd
Goodfellow Metals Ltd
Linton Instruments Ltd
Medelec Ltd
Newport UK Ltd
Nikon, UK Ltd
NPI Elektronic
Prior Scientific Instruments Ltd
Sandoz Institute for Medical Research, London
Carl Zeiss Ltd

Staff who have taught at the Workshop

David Adams
Claire Aickin
Jeff Allen
Brad Amos
Fran Ashcroft
Jonathon Ashmore
Michael Bate
Boris Barbour
Stuart Bevan
David Becker
Chris Benham
Chas Bountra
Euan Brown
Colin Brownlee
Malcolm Burrows
Armand Cachelin
Mark Cannell
Malcolm Caulfield
Abdul Chrakchri
Graham Collingridge
David Colquhoun
Chris Courtice
Stuart Cull-Candy
Nick Dale
Catherine David
Noel Davies
John Dempster

Sukvinder Dhanjal
Ted Dyett
Frances Edwards
Chris Elliott
Robert Fettiplace
Alistair Forbes
Ian Forsythe
Barbara Fulton
Alasdair Gibb
Jim Gillespie
Peter Gray
Alison Gurney
Jim Hall
James Halliwell
Bert van Heel
Malcolm Irving
Kai Kaila
Bernard Katz
Richard Keynes
Corne Kros
Phil Langton
Luc Leybaert
Ian Macfadzean
Ken MacLeod
Chris Magnus
Alistair Mathie
Robert Meech

Peter Mobbs
Nancy Mulrine
Claire Newland
David Ogden
Machael Pasternack
Voi Piotrowski
Tim Plant
Tim Rink
Jon Robbins
Brian Robertson
David Shepherd
Paul Smith
Nick Standen
Peter Stanfield
Cathy Stansfeld
Alison Taylor
Roger Thomas
Martin Thomas
Richard Vaughan-Jones
Juha Voipio
Ken Wann
Michael Whitaker
Anne Warner
Alan Williams
Roddy Williamson
Mark Yeoman

The front cover is taken from the Workshop poster which was designed by Barbara Fulton

Chapter 1
Using microelectrodes

JAMES. HALLIWELL, MICHAEL WHITAKER and DAVID OGDEN

1. Introduction

Microelectrodes are the basis of the techniques discussed in this book. They are used for: (1) potential recording; (2) current injection; (3) introduction into the cell of ion-selective resins for measuring potential or determining the free concentration of cytosolic constituents.

(1) and (2) are the procedures underlying conventional microelectrode recording, voltage-clamping and patch-clamping. This chapter will be restricted to considering the fundamentals of reliable and accurate measurement of membrane potentials and the experimental manipulation of membrane potential by injection of current. The specialized techniques involved in voltage-clamping and patch-clamping will be treated separately in subsequent chapters, as will the special requirements involved in the use of ion-sensitive electrodes or ionophoresis of drugs.

2. Making microelectrodes

One definition of a microelectrode (ME) might be: 'an electrode constructed with a tip having the dimensions of the order of a micrometre (1 μm)'. Usually this means a glass micropipette of the type pioneered by Ling & Gerard (1949), which is filled with an electrolyte solution to act as a conductor of electricity.

Glass MEs are made by heating a capillary until molten, when it is stretched; while the glass is still plastic but cooling down, the tip draws out, breaks and separates.

(i) By hand: This is not recommended because of lack of reproducibility, although great artists can heat a capillary in a bunsen and pull out a fine tip!

(ii) Vertical puller: This usually has a nichrome filament and a 2 stage pull - the first by gravity, the second electromagnetic. This is fine for MEs up to about 30 to 40 MΩ. In our experience these pullers are less effective for making fine-tipped microelectrodes for use on small cells.

J. V. HALLIWELL, Department of Physiology, The Royal Free School of Medicine, Rowland Hill Street, London NW3 2PF, UK.
M. J. WHITAKER, Department of Physiology, University College London, Gower Street, London WC1E 6BT, UK.
D. OGDEN, National Institute for Medical Research, Mill Hill, London NW7 1AA, UK.

(iii) Horizontal puller 1 (Livingstone type): These pullers are gear driven and have a platinum foil heating element. They are good for fine tips (ME resistance can be 30-300 MΩ) but this type of puller produces rather long wispy shanks.

(iv) Horizontal puller 2 (Brown-Flaming, Ensor, Industrial Science): These pullers usually have a platinum or nichrome heating element with a range of preheat times and a 1 or 2 stage pull. The Brown-Flaming has a gas jet which cools the heater rapidly. Good reproducibility can be achieved with these pullers, but setting them up correctly is time consuming. They are generally good for fine MEs with resistances up to 300-500 MΩ and short shanks.

3. Filling microelectrodes

Glass microelectrodes are usually filled with a salt solution. The composition of the solution can be determined by the individual experimenter and depends on the experimental protocol. Nowadays, the preferred method of filling is to pull electrodes from capillary that has a glass-fibre fused into the lumen. When the ME is pulled the lumen shape is preserved up to the tip. Using fibre-containing capillary, MEs can be backed-filled with small amounts of solution. The solution tracks down the channels formed either side of the fibre right down to the tip. Bubbles *do* form but don't occlude the lumen completely. The exception is when a bubble forms directly in front of an Ag/AgCl sintered pellet in a perspex ME holder. This is easy to remedy.

One should realize, however, that other forms of ME than this exist and that even with what might be considered to be a micropipette there are alternatives to an electrolyte solution as a conductor. Thus, micropipettes have been filled with molten Wood's metal which solidifies to give a continuous metal conductor (in our experience, simultaneously cracking the insulating glass envelope!) (Gesteland *et al.* 1965), or have been drawn over single carbon (graphite) fibres around 7 μm in diameter (Armstrong-James & Millar, 1979) to form a ME with a carbon conductor. These manufacturing techniques produce electrodes with impedances of 200 kΩ to 2 MΩ. Similar values are obtained with glass or varnish-coated tungsten MEs which have been electrolytically etched to a fine tip (Merrill & Ainsworth, 1972); it is possible to electroplate the tips with other metals for positional marking (e.g. iron, which can be visualized by the Prussian Blue reaction) or for lower noise and reduced polarization (e.g. Pt and Pt black). All these MEs which use non-electrolyte filling as the electrical conductor have high DC resistances, however, and are employed in *extracellular* recording and stimulation; in this recording mode they are used for registering the occurrence rather than the accurate wave form of signals (for example, neuronal discharges). An exception to the latter statement is the use of carbon fibre electrodes to measure redox potentials of oxidizable compounds in biological tissues. Metal electrodes behave electrically in moist preparations as if they are a small resistance in series with a larger resistance and parallel capacitance: they are not suitable for measuring standing DC potentials. They do, however, have low noise at the frequencies where most of the power from action potential signals is concentrated. Consequently, a good signal-to-noise ratio is obtained when they are used for extracellular recordings, better than that of electrolyte-filled micropipettes.

Electrolyte-filled (usually with NaCl) micropipettes are also used for extracellular recording of neuronal discharges since they can also faithfully reproduce the potential wave-form down to DC levels. Frequently, these MEs are used to measure potentials set up by synaptic current flow across the resistance of the extracellular space. By considering the potential gradient and its spatial derivative within a tissue, the location of sinks and sources of current can be used to pinpoint synaptic regions within the tissue. This technique is known as current density analysis and the reader is referred to several articles for a complete treatment of the subject (Rall & Shepherd, 1968; Hubbard *et al.* 1969; Llinas & Nicholson, 1974; Nicholson & Freeman, 1975; Nicholson & Llinas, 1975).

4. Connection of microelectrodes to recording circuit

The preparation and the electrode are both wet; electronic circuitry is dry and has metallic conductors. Plain metal/liquid interfaces display junction potentials and can produce gas (hydrogen and oxygen) if current is passed through them. The latter is particularly annoying since the presence of gas bubbles on the electrode simulates the insertion of a capacitance in the circuit at the liquid/ metal junction, thereby limiting DC recording. (Hence the reason that metal MEs are not used for intracellular recording - see above). Connections are therefore made to the recording circuit via non-polarizable reversible electrodes. Silver/silver chloride (Ag/AgCl) electrodes exchange electrons for Cl^- ions in solution. They are usually employed in the form of silver wire coated with silver chloride or a sintered pellet of metallic silver and powdered AgCl pressed around a silver connecting wire. The pellets have the advantages of large current carrying capacity and stability to light. Stability is obviously of prime importance in measuring membrane potentials as is the property of reversibility, i.e. the property whereby the passage of current in either direction through the electrode does not alter the potential difference between the metal and the solution. The Ag/AgCl reference electrode is reversible and of constant potential because AgCl is sparingly soluble and therefore the solution is saturated with respect to AgCl; the concentration of Ag^+ ions in solution is inversely proportional to the $[Cl^-]$ (from the definition of solubility product) and as a result the potential E_{Ag} of metal relative to the solution is given by $E_{Ag} = Const - RT/F \ln[Cl^-]$ (from the Nernst expression for the electrode potential of a metal). Differences in $[Cl^-]$ in the solutions composing these 'half-cells' lead to a standing potential which should not vary unless $[Cl^-]$ does at either electrode. If, in the course of an experiment, $[Cl^-]$ does vary (for instance, as a result of changing the bathing medium in an in vitro preparation) then the AgCl reference electrode should be interfaced to the preparation by means of a salt bridge which will maintain a constant $[Cl^-]$, avoids damage to the AgCl electrode and toxic effects of Ag in the bath. The salt bridge comprises 1-2% agar in 0.15 M NaCl, or with continuous bath perfusion, concentrated KCl may be used to minimise junction potentials (see below). Normally the ME tip itself forms the other salt bridge, with a AgCl wire inserted in the ME barrel, but to equalize the reference electrode potentials exactly an additional salt bridge may be used at the back of the ME, with a AgCl pellet in the ME holder.

AgCl pellet electrodes can be purchased cheaply and are available as discs or pellets, some small enough to insert in the back of wide bore ME glass. There are 3 widely used procedures for chloriding Ag wires; for each the wire must be clean. (1) Electrolytic coating is done by connecting 2 wires immersed in 0.1 M HCl to the poles of a 1.5 V battery and passing current for 20 minutes or so, reversing the polarity at regular intervals, resulting in a uniform but fragile grey coat. (2) Wires may be dipped in molten AgCl (requiring an intense gas/air or gas/oxygen torch and crucible) to produce a tough coating. (3) Ag wires kept in hypochlorite-based bleach become coated with AgCl, and can be changed frequently if necessary.

5. Junction potentials

There is a potential difference set up at the interface between two salt solutions of different ionic composition or concentration by differing diffusional flux of anions and cations across the boundary: this is a liquid junction or diffusion potential. It is described, for a single solute, by equations in the form

$$V = \frac{u - v}{u + v} \frac{RT}{F} \ln \frac{c_1}{c_2}$$

where u is the mobility of the cations in solution and v is the mobility of the anions and c_1 and c_2 are the solute concentrations on opposite sides of the interface. R is the gas constant, T, the temperature and F is Faraday's Constant. Equations for the junction potential in more complex cases are given by Barry & Lynch (1991). Junction potentials develop:

(1) as a result of different anionic and cationic mobilities and

(2) different solute concentrations. They develop between the bath solution and reference salt bridge if these differ, and a modified form of junction potential, termed a tip potential (see below), exists across the ME tip. They cannot be eliminated but the correct electrode configuration ought to aim to stabilize them with respect to experimental solution changes or to reduce them by trying to equalize u and v. A reference salt bridge of 150 mM NaCl and agar is suitable if the composition of the bath is constant.

Experiments in which changes of ionic composition are imposed are often made to determine reversal potentials, and precision is important if permeability ratios are to be derived. The junction potential change at the reference boundary can be minimised by a continuously renewed 2-3 M KCl junction. K^+ and Cl^- have similar mobilities in aqueous solution. Moreover, with high concentrations of KCl, diffusion of KCl from the salt bridge predominates and potential changes due to alterations in salt concentration in the bath are small. A small continuous outflow of KCl solution, generated by 1-2 cm hydrostatic pressure through a 2-5 μm tip, prevents dilution within the bridge by the bath solution and the resulting generation of a concentration gradient and junction potential within the salt bridge and reference electrode itself. However, leakage of KCl into the solution bathing the preparation can lead to unwanted changes of ionic composition, so downstream siting and perfusion of the bath are essential. If this cannot be done, the electrode arrangement can be used to measure junction potentials at the tip of a NaCl bridge when the bath solution is changed in the absence of the preparation, and the results used to correct the membrane potentials recorded when the NaCl bridge is used as reference.

A second problem arises when kinetic measurements of membrane potential response to ionic changes are made, or when the bath solution is changed locally (i.e. not at the reference), for instance with a puffer pipette or similar device. In both cases a junction potential will exist within the bath, even if only transiently, between the recording site and the reference, and will contribute to the potential recorded. As an example, when recording the timecourse of Cl-evoked potential changes by fast perfusion in skeletal muscle fibres, Hodgkin and Horowicz (1960) used differential recording with respect to a blunt ME placed adjacent to the recording site to avoid transient Cl⁻ junction potentials along the bath.

6. Tip potentials

The surface chemistry of the glass is complex: at the tip of a ME it is believed that a restriction of anionic mobility is established. As a result of this, the liquid junction potential is altered by properties of the electrode tip. This is the tip potential (TP). The TP is abolished by breakage of the tip. It can be measured by measuring the potential change when the tip is broken, or by registering the potential change when an electrode of similar manufacture and filling solution, but with its tip broken is added in parallel between the input of the measuring device and the solution bathing the ME tip. A list of TP properties is given in Purves (1981): TPs are (1) of negative sign, up to -70mV with electrodes filled with 3 M KCl and tested in physiological saline; (2) abolished when inside and outside solutions are the same; (3) larger in higher resistance electrodes; (4) may vary widely in magnitude between otherwise similar electrodes; (5) can often be reduced by filling MEs with acidic solutions; (6) can be reduced and even reversed in sign by addition of small concentrations of polyvalent ions (e.g. thorium, Th^{4+}) and high concentrations, 10-100 mM $[Ca^{2+}]$, to the bathing solution.

The TP is the main source of uncertainty in measuring membrane potential because it changes depending on the immediate ionic environment into which the tip is placed. Adrian (1956) studied some properties of TPs of KCl-filled MEs and the change in TP when the tip was in 100 mM NaCl or 100 mM KCl (to simulate cell impalement). The TP is smaller on going from extracellular to intracellular solution, the difference becoming larger the larger the initial potential. For a TP of -5 mV, the error is about 2 mV. Roughly speaking, TP is proportional to ME resistance, and inversely proportional to salt concentration of the external solution. The change in TP in going from one solution to another is proportional to the size of the TP. Thus, if possible, MEs with the smallest resistance, and therefore small TP, should be selected. Adrian used MEs filled by the method of boiling; subsequently it has been shown (Okada & Inouye, 1975) that the method of ME filling and also the time after filling influences TPs. Smaller TPs are observed with the fibre-filling method (e.g. MEs made from filament glass) and with immediate usage.

Clearly, the problem of TP is complex and unpredictable and possibly for this reason is largely ignored in the majority of research reports. Generally speaking, TP is negative and the effect on impalement is to estimate a more positive value of the resting potential as the TP becomes more positive inside the cell. Measurement of the TP requires use of a KCl bridge between AgCl electrodes at both the reference and ME connections. The potential is initially zeroed with KCl connections into the bath solution and the TP recorded as the additional potential when the ME is introduced. A system with differential recording

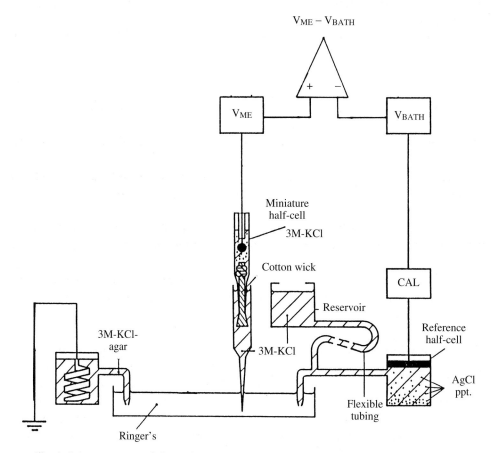

Fig. 1. An arrangement of electrodes permitting measurement of TP or junction potentials in a perfusion bath. The potentials of the reference half electrode and miniature KCl half cell are zeroed with respect to ground with the cotton wick in the bath. Placing the wick into the back of the ME introduces an additional potential due to the TP. During an experiment with changes in the bath composition the flowing KCl reference maintains a constant bath potential. Juction potentials with respect to normal Ringer can be measured by substituting a blunt pipette containing Ringer for ME and changing the bath solution. The calibrator CAL can be used to offset the standing membrane potential to look at small changes e.g. synaptic potentials at high gain without high pass filtering (D. C. Gadsby and D.C.O., unpublished data).

between ME and reference and KCl junctions to permit measurement of TP (or junction potentials of salt bridges introduced in place of the ME) is shown below.

Previous paragraphs have been included as an introduction to the wider range of ME usage; the recording techniques to be described are not dependent on a need to know the absolute value of potential. Even in the case of extracellular field potential analysis it is the difference of potential within the preparation that is of interest and any offsets or error potentials in the recording circuitry are of no consequence provided that they are fixed. It is, however, important to realize that using electrodes in aqueous solution to measure potentials is by no means as theoretically straightforward as putting a voltmeter across the terminals of an electronic component. The theoretical basis of potential measurement in aqueous solution is discussed lucidly by Finkelstein & Mauro (1974).

The rest of this chapter concentrates on using microelectrodes to measure cell membrane potentials.

7. Measuring membrane potentials using microelectrodes

The electronic amplifier used to record potentials via MEs should have the following characeristics:

(i) Input resistance should be at least 100-1000 times the ME resistance in order to measure the full signal at the ME tip and draw negligible current from the signal source.

(ii) Low leakage current. The current flowing into or out of the input terminal of the amplifier should be less than that which would cause a 1 mV drop across the ME or the cell input resistance.

(iii) Response time should be adequate: with a good response time there are ways of overcoming the low-pass filtering properties of a voltage recording set-up.

Capacitance compensation

The low pass filtering properties of the microelectrode and its amplifier are due to inevitable capacitances between the microelectrode and ground (Fig. 1). C_t, the transmural capacitance can be minimized by (i) thick pipette walls (glass pipettes can be made thicker by coating with a layer of Sylgard (Corning) and (ii) a low solution level in the bath to reduce the effective transmural area. C_s, the stray capacitance from the microelectrode and amplifier input to ground (that is, the microscope, bath, stand, etc.) can be reduced by minimizing the length of the ME and connecting wire and by driving the shield of the connecting wire with a low impedance signal from the output of the ME amplifier (Fig. 2). C_a, the input capacitance of the amplifier should be negligible if the amplifier is chosen to have a good response time as suggested. C_{tot} is the total summed capacitance (see Fig. 1 and Chapter 16).

Once C_{tot} is minimized it can be compensated for by using a feedback circuit ofen described as 'negative' capacitance (Fig. 3). This positive feedback circuit provides the current lost through C_{tot}, preventing a potential drop across the electrode

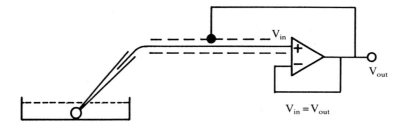

Fig. 2. A low-impedance driven shield to reduce C, the stray capacitance to ground.

resistance. Good compensation clearly depends on the rapidity with which the feedback circuit can supply current. The fully compensated rise time is proportional to the geometric mean of the rise time of the recording amplifier and the rise time of the uncompensated circuit. This suggests that the best strategy is (i) to minimize stray capacitance and (ii) use a head-stage amplifier with a fast rise time. Over-compensation of input capacitance results in damped oscillations at the leading edge of potential steps and finally in continuous oscillation.

8. Manipulating cell membrane potentials

Experimental protocols often require the manipulation of membrane potential (for example, to test passive membrane properties such as input resistance by passing current into the cell). With two MEs in the same cell, one electrode can be dedicated to voltage recording, the other to current injection. This is the preferred method, but simultaneous current injection and voltage recording through a single ME can be achieved - one needs a way of eliminating the potential difference between ME barrel and ME tip caused by the voltage drop due to current flowing through the ME

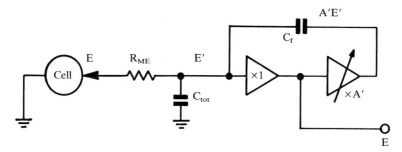

Fig. 3. The potential across C_f is $A'E'-E'=E'(A'-1)$ and the current across C_f is $C_f \times de/dt \times (A'-1)$. To compensate for the current loss across the capacitance to ground when E' is changing with time ($C_{tot} \times dE'/dt$) the gain of amplifier A' is adjusted so that $C_f (A'-1)=C_{tot}$. In this way current is supplied to the input of the amplifier, equal and opposite to the loss through C_{tot}. Another way of regarding this is that of adding *negative capacitance* in parallel with C_{tot}. The effectiveness of a negative capacitance circuit is such that the fully compensated rise time of a circuit is approximately twice the geometric mean of the amplifier's rise time and the uncompensated circuit's rise time (Purves, 1981).

resistance (R_{ME}). These two potentials can be large (i.e. 1 nA through a 50 MΩ electrode = 50 mV). Two methods are available for accomplishing this.

Bridge balance circuits

The modern analogues of the Wheatstone bridge circuitry formerly used for eliminating $E=IR_{ME}$ are still known as bridge balance circuits, though they are not strictly bridge circuits. Nowadays it is usual to subtract electronically from the voltage output a scaled proportion of the input signal driving the current injection circuit (current pump). The circuit in Fig. 4 does just this.

The amplifier gains can be scaled so that the ME resistance can be read off the dial of the BAL potentiometer in MΩ, for example; this is one simple way to measure ME resistance.

Possible artefacts arise from incorrect balancing or the inability to balance the bridge with certainty. First, the method depends on R_{ME} not changing during passage of current through the ME. One should, if possible, use MEs with linear current-voltage (I-V) relationships, or restrict the current to a range of values over which the I-V relation is more or less linear. Second, one cannot compensate completely for the capacitance distributed in the transmural elements comprising the ME tip (C_t). In poor recording conditions, for example deep immersion of the ME in the bath, the frequency response of the recording system is compromised to such an extent that the charging of the cell membrane capacitance cannot be accurately judged (Fig. 5).

Discontinuous current injection method

This is an alternative to bridge balance and eliminates one source of artefact - that of non-linearity of the ME I-V relation. Instead of injecting continuous current, pulses are injected (see Wilson & Goldner, 1975).

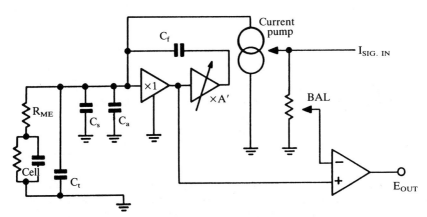

Fig. 4. A circuit to compensate for the potential drop across a ME due to passage of current. The current pump will pass a current proportional to the command voltage and independent of the electrode resistance. The command voltage is therefore scaled and subtracted from the ME potential. The circuit assumes that the ME has a linear current-voltage relation (i.e. obeys Ohm's law). This may not always be true, and should be checked.

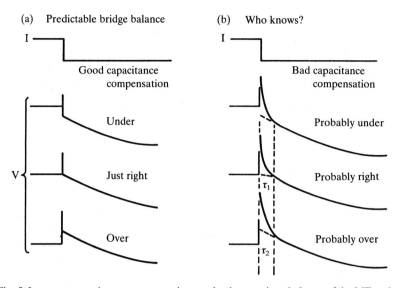

Fig. 5. Incorrect capacitance compensation can lead to spurious balance of the ME resistance.

Fig. 6 shows the principle of the single-electrode switched current clamp. If the membrane time constant is large compared to the electrode time constant, charge will be stored on the membrane capacitance (time constant $C_m \times R_m$,), whereas the potential due to IR_{ME} (time constant $C_{tot} \times R_{ME}$) decays rapidly. If the voltage is sampled and held between time points S ($V_{S\&H}$) at a time when IR_{ME} is zero, only V_m, the true membrane potential is measured and the change in V_m approximates that which would be seen in the normal bridge balance case with a continuous current of one half the amplitude (Fig. 6B). Obviously, the approximation is better the higher the pulse frequency. The voltage due to current passage down the ME does not appear and furthermore, since one monitors voltage after IR_{ME} has decayed, any change in R_{ME} during current passage is of no consequence, provided that the current pump can deliver a truly constant current. Clearly, for rapid switching rates capacity neutralization is important. In addition, the noncompensated C_t must be minimized to allow fast settling. Artefacts encountered with this method of 'balancing' electrode resistance stem largely from inadequate capacity compensation or using too high a pulse frequency. These are illustrated in Fig. 7.

If the capacity compensation is correctly adjusted and the ME is clearly settling within a half cycle time then the $V_{S\&H}$ (sample and hold) should exhibit no step with the onset of current injection. To help settling times asymmetric duty cycles may be employed (e.g. 25 % current injection, 75 % settling time). In these cases the current I, delivered with a duty cycle of D is equivalent to a constant current of $I \times D$.

9. Mechanical stability

The interaction between ME (or patch pipette) and cell needs to be stable on a

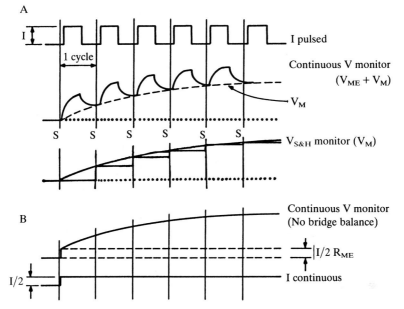

Fig. 6. The principle of the single-electrode switched current clamp is shown. In A, current pulses lead to rapid charging of the ME RC network. The slower charging of the cell membrane RC network leads to a smaller signal which is sampled some time after the end of a current pulse when the charge on RCME has decayed. B illustrates the point that the pulsed current is equivalent to a continuous current of half the pulse amplitude which (a) causes a voltage drop across RME (b) charges the membrane capacitance.

submicron scale to avoid damage to the cell membrane and consequent excessive current leakage. There are two symptoms of mechanical problems, slow drift of the ME tip with respect to cell, and faster vibrations.

Drift can arise from the following causes.

(1) The cell or tissue may move in the bath, a problem that can usually be simply remedied (except when recording from muscle). The bath may move with respect to the microscope stage, remedied by servicing the ratchet drives of the stage movement or by adding locking screws or some other device.

(2) The ME may move in its holder, either axially or laterally, particularly if pressure is applied for injection purposes. Movement of the holder with respect to the micromanipulator can occur with plastic push-in arrangements onto headstage connectors because of 'creep' in the plastic itself. The best arrangement is to secure the ME holder to the micromanipulator directly.

Drift in the micromanipulator can occur between slider assemblies, for example loose rack and pinion drives, screw drives, bearing races or dovetails. If the manipulator is not mounted with flat surfaces then drift can occur due to rocking movements. If the manipulator uses hydraulic drives these are subject to drift because of thermally induced expansion of the fluid, so in this case temperature stability is important.

Vibration of the ME tip with respect to the cell can be minimised by the following

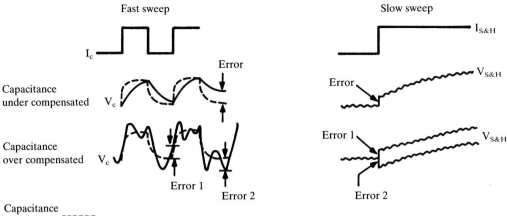

Fig. 7. Artefacts which can arise from inappropriate use of the switched clamp. (I_c, V_c are continuous monitor outputs; $I_{S\&H}$, $V_{S\&H}$ are sampled outputs.)

precautions. The source of vibration is usually floorborne and is eliminated by working on a solid floor, as close as possible to solid ground or at least a supporting wall. Wooden or other unstable flooring can be avoided by sitting the baseplate assembly on brackets bolted to a supporting wall (a cheaper and more effective remedy than an airtable). Vibration reaching the baseplate supports can be filtered by compliant mounts, such as cycle tubes, pneumatic cushions (supplied by optical manufacturers) or regulated compressed air isolation tables, with a heavy table top (total mass 200-500 kg). Generally low frequencies are transmitted more effectively than high through this arrangement; roughly the product of mass and compliance determines the high frequency cut-off. There are two important points to note, that lateral vibration may be as prominent as vertical, and that isolation tables are tuned to a low frequency, 1-2 Hz, to achieve damping and may exagerate inputs in that range.

The combined transmitting microscope stage - baseplate - micromanipulator - ME holder - ME connections make a system of levers with great potential for vibration, or transmitting small movements to the ME-cell impalement. The connections from the fulcrum of the micromanipulator to the cell should be as short, stiff and light as is feasible. Lightweight eg piezo-driven manipulators mounted on the stage may be the best arrangement, otherwise more remote mounting requires manipulators of massive construction to achieve stability and precision, such as the Huxley design. Vibration in steel baseplates should be damped by bonding to laminated or compressed board and flexion avoided by using sufficient thickness or honeycomb designs.

Micromanipulators should provide precision of movement, much less than 1 μm, with stability in the same range. This is achieved by either lightweight design moving short distances mounted close to the cell, on the microscope or stage, or by more substantial units mounted on the baseplate. In the former category piezo drives or hydraulic drives may provide fine movement of limited range mounted on compact optical translation stages. This arrangement has the advantage of remote operation of fine movement. In the second category, micrometer or screw drives acting via reducing levers may provide fine movement in only vertical or, in the Huxley design, all three axes. Baseplate mounting of a coarse unit with a remotely operated fine control is often used.

Detection and diagnosis of vibration and slow drift over μm requires a good experimental microscope with graticule eyepiece or TV system.

10. Microscopy

The experimental microscope is an important part of an ME or patch clamp system, often an essential mechanical component as well as a means for seeing the cell. There are a number of books on the workings of the optical microscope, eg by Bradbury, by Ploem & Tanke, and by Spencer.

Generally, ME (and patch clamp) technique is improved by seeing the tip and cell clearly during impalement or seal formation, and there are a number of examples of significant advances in the application of ME measurements resulting from better optics. The iontophoretic studies of postsynaptic mechanism at the skeletal endplate (McMahan *et al.* 1971; Kuffler & Yoshikama, 1975) and autonomic ganglia (McMahon & Kuffler, 1971; Lichtman, 1977) used Nomarski differential interference contrast (DIC) optics to see synaptic elements clearly. The same optical system with infrared video microscopy can be used to see and record from soma or dendritic elements in brain slices (Stuart *et al.* 1993). DIC and fluorescence microscopy are used to identify cell types and developmental stages when investigating changes in electrophysiological properties (see chapter on dye injection and cell labelling). The use of indicators for intracellular ion concentrations requires good optics for fluorescence microscopy (see chapter on fluorescent indicators).

Some general considerations are given here. Good optical resolution requires a good microscope objective, particularly as high a numerical aperture (NA) as is practicable. 'Resolving power', the separation of 2 distinct points, is proportional to 1/NA. The practical restriction in an upright microscope is mechanical, because high NA is achieved by decreasing the distance between objective lens and specimen. An upright microscope is essential for work in thick specimens such as tissue slices or embryos and the minimum working distance for placing MEs is about 1.5 mm. The best NA currently available with this working distance is 0.75 in a 40× water immersion objective (Zeiss 0.75 W 40×), although water immersion objectives exist with NA 0.9 or 1 but shorter working distance[1]. For single cells dispersed or in culture an inverted microscope can be used. If the chamber base is made of a thin coverslip working distance is no longer a problem (the ME or pipette comes from above) and the maximum NA obtainable with an oil immersion objective, 1.3-1.4,

---

1. Zeiss have recently introduced a 63× 0.9 NA W 1.4 mm w.d. objective in their infinity corrected range, but this has not been evaluated at time of press.

can be used. Illumination of the specimen should also be with as high an NA condenser as possible; the same restriction of working distance means that a condenser of NA 0.9 can be used with upright microscopy (good illumination is important in thick specimens) but only 0.5-0.6 with inverted.

Contrast methods such as phase contrast, DIC or Hoffman greatly improve the visibility of thin cells and small structures on large cells, and one or other of these is really necessary. Phase contrast is inexpensive and adequate for single cells, but for thicker specimens DIC optics are favoured. CCD TV cameras are also inexpensive and convenient.

For fluorescence microscopy the brightness of the specimen increases with the NA^2 and decreases with $1/(\text{magnification})^2$, so high NA low magnification objectives are often used for microspectrofluorimetry.

11. Summary

The main sources of error in measuring potential with intracellular microelectrodes should be clear. They are:

(1) Varying tip potentials of the ME.

(2) Varying junction potentials.

(3) Asymmetry of electrode reference potentials and their dependence on salt concentration in the bath solution.

(4) Inadequate amplifier frequency responses when monitoring fast signals.

(5) Errors in potential measurement when injecting current because of 'bridge balance' or ME resistance change, or due to too high a switching rate when using the discontinuous current injection method.

Most of what we have discussed in this chapter is available in greater detail in Purves (1981). Purves' book is a good source of reference to more advanced texts.

References

ADRIAN, R. H. (1956). The effect of internal and external potassium concentration on the membrane potential of frog muscle. *J. Physiol., Lond.* **133**, 631-658.

ARMSTRONG-JAMES, M. & MILLAR, J. (1979). Carbon fibre microelectrode. *J. Neurosci. Methods* **1**, 279-287.

BARRY, P. B. & LYNCH, J. W. (1991). Liquid junction potentials and small cell effects in patch clamp analysis. *J. Memb. Biol.* **121**, 101-117

BRADBURY, S. *An Introduction to the Optical Microscope* and PLOEM, J. S. & TANKE, H. J. *Introduction to Fluorescence Microscopy.* Both from Oxford University Press and the Royal Microscopical Society. Affordable softcover handbooks.

FINKELSTEIN, A. C. & MAURO, A. (1974). Physical principles and formalisms of electrical excitability. In *Handbook of Physiology - The Nervous System 1.* pp. 161-212. Bethesda: American Physiological Society.

GESTELAND, R. C., LETTVIN, J. Y. & PITTS, W. H. (1965). Chemical transmission in the nose of the frog. *J. Physiol., Lond.* **181**, 525-559.

HODGKIN, A. L. & HOROWICZ, P. (1960). The effects of sudden changes in ionic concentrations on the membrane potential of single muscle fibres. *J. Physiol.* **153**, 370-385.

HUBBARD, J. I., LLINAS, R. & QUASTEL, D. M. J. (1969). *Electrophysiological Analysis of Synaptic Transmission.* pp. 265-293. London: Edward Arnold.

KUFFLER, S. W. & YOSHIKAMA, D. (1975). The distribution of acetylcholine sensitivity at the postsynaptic membrane of vertebrate skeletal twitch muscles: iontophoretic mapping in the micron range. *J. Physiol.* **244**, 703-730.

LICHTMAN, J. W. (1977). The reorganisation of synaptic connexions in the rat submandibular ganglion during postnatal development. *J. Physiol.* **273**, 155-177.

LING, G. & GERARD, R. W. (1949). The normal membrane potential of frog sartorius fibers. *J. Cell Comp. Physiol.* **34**, 383-396.

LLINAS, R. & NICHOLSON, C. (1974). 'Analysis of field potentials in the central nervous system'. In *Handbook of Encephalography and Clinical Neurophysiology.* (ed. A. Redmond), vol. 2. Part B, Section V. pp. 2B 61-92 Amsterdam: Elsevier.

MCMAHON, U. & KUFFLER, S. W. (1971). Visual identification of synaptic boutons on living ganglion cells and of varicosities in postganglionic axons in the heart of the frog. *Proc. R. Soc. B*, **177**, 485-508.

MCMAHON, U., SPITZER, N. C. & PEPER, K. (1972). Visual identification of nerve terminals in living isolated skeletal muscle *Proc. R. Soc. B*, **181**, 421-430.

MERRILL, E. G. & AINSWORTH, A. (1972). Glass-coated platinum-plated tungsten microelectrodes. *Med. Biol. Engng* **10**, 662-672.

NICHOLSON, C. & FREEMAN, J. A. (1975). Experimental optimization of current source-density technique for anuran cerebellum. *J. Neurophysiol.* **38**, 369-382.

NICHOLSON, C. & LLINAS, R. (1975). Real time current source-density analysis using multielectrode array in cat cerebellum. *Brain Res.* **100**, 418-424.

OKADA, Y. & INOUYE, A. (1975). Tip potential and fixed charges on the glass wall of microelectrode. *Experientia* **31**, 545-546.

PURVES, R. D. (1981). *Microelectrode Methods for Intracellular Recording and Ionophoresis.* London: Academic Press.

RALL, W. & SHEPHERD, G. M. (1968). Theoretical reconstruction of field potentials and dendrodendritic synaptic interactions in olfactory bulb. *J. Neurophysiol.* **31**, 884-915.

SPENCER, M. (1982). *Fundamentals of Light Microscopy.* Cambridge University Press.

STUART, G. J., DODT, H-U. & SAKMANN, B. (1993). Patch clamp recordings from the soma and dendrites of neurones in brain slices using infrared video microscopy. *Pflugers Archiv.* **423**, 511-518.

WILSON, W. A. & GOLDNER, M. M. (1975). Voltage-clamping with a single microelectrode. *J. Neurobiol.* **6**, 411-422.

Chapter 2
Voltage clamp techniques

JAMES V. HALLIWELL, TIM D. PLANT, JON ROBBINS and
NICK B. STANDEN

1. Introduction

Much of what we know about the properties of ion channels in cell membranes has
come from experiments using voltage clamp. In general, the method allows ion flow
across a cell membrane to be measured as electric current, whilst the membrane
voltage is held under experimental control with a feedback amplifier. The method
was first developed by Cole (1949) and Hodgkin *et al.* (1952) for use with the squid
giant axon. Since then, many variants of the technique have evolved and voltage
clamp analysis has been extended to a wide range of tissues.

The usefulness of the voltage clamp stems firstly from the fact that it allows the
separation of membrane ionic and capacitative currents. Secondly, it is much easier to
obtain information about channel behaviour using currents measured from an area of
membrane with a uniform, controlled voltage, than when the voltage is changing freely
with time and between different regions of membrane. This is especially so as the
opening and closing (gating) of most ion channels is affected by the membrane potential.

This chapter will concentrate on voltage clamp methods that are used to measure
currents from whole cells or large areas of membrane containing at least a few
hundred channels; such currents are usually called macroscopic currents. We shall
give particular attention to the single and double microelectrode methods. The use of
both these techniques is increasing; double microelectrode clamp is being used for
studies of channels expressed in *Xenopus* oocytes, while the switched clamp circuitry
used for single microelectrode clamp is sometimes used with whole cell patch
pipettes because it provides a neat way of dealing with the series resistance problems
that can arise with such electrodes. The patch clamp technique has extended the
application of voltage clamp methods to the recording of ionic currents flowing
through single channels, but in its whole cell configuration has also become the most

J. V. HALLIWELL, Department of Physiology, Royal Free Hospital Medical School,
Rowland Hill Street, London NW3 2PF, UK
T. D. PLANT, I Physiologisches Institut, Universität des Saarlandes, D-6650
Homburg/Saar, Germany
J. ROBBINS, Department of Pharmacology, University College London, Gower Street,
London WC1E 6BT, UK
N. B. STANDEN, Ion Channel Group, Department of Physiology, University of Leicester,
PO Box 138, Leicester LE1 9HN, UK

widely used method for recording macroscopic currents. Whole cell patch clamp is mostly described in Chapter 5, though we shall mention the application of some of the general principles of voltage clamp to the technique in the present chapter.

2. Theory

The basis of the voltage clamp may be understood by consideration of the simplified equivalent circuit for the cell membrane shown in Fig. 1A. Here, C is the membrane capacity while the channels that allow ionic current, I_i, to flow through the membrane are represented by the variable resistor R. The current I_m flowing through the circuit will be the sum of I_i and a capacity current,

$$I_m = I_i + C \frac{dV}{dt}$$

In voltage clamp experiments the voltage is usually forced to change in a square step fashion, being changed as rapidly as possible from one steady level to another (Fig. 1B). Under these conditions a brief spike of capacity current flows at the edges of the pulse, but when the voltage is steady dV/dt is zero and so the capacity current is zero. The ionic current may therefore be obtained free from capacity current once the change in voltage is over. In most experiments it is ionic current which is measured to give information about the permeability properties of channels and the mechanisms by which they open and close, though some studies concentrate on components of the capacity current which are related to channel gating.

For the voltage clamp of a particular preparation to be adequate the system used should meet certain criteria for the accuracy of voltage control, speed, and isopotentiality.

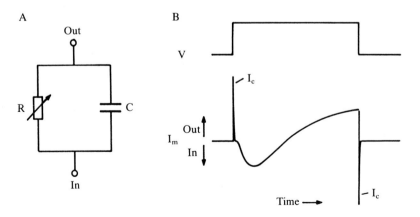

Fig. 1. (A) Simple electrical analogue of the cell membrane. (B) Typical currents recorded from a voltage clamped excitable cell when the membrane potential is stepped in a square fashion to a level at which ionic channels open. Spikes of capacity current occur at the edges of the pulse; during the pulse inward, and then outward ionic currents flow. The time course is typically in the order of a millisecond.

Control of membrane potential

The accuracy with which the membrane voltage is controlled depends on having sufficiently high gain in the clamping amplifier. This can be seen by considering the schematic voltage clamp circuit of Fig. 2, as discussed by Moore (1971). The membrane potential, V_m, is measured by the voltage follower, which has very high input impedance and so draws negligible input current. The clamping amplifier, of gain A, compares V_m with the command potential E, and passes current through the access resistance R_a (which might consist of an electrode and the cytoplasmic resistance) to control V_m. The output of the clamping amplifier, V_o, is given by

$$V_o = eA = A\ (E-V_m)$$

This output is divided between the access resistance and the membrane (for the moment we will assume that R_s, the series resistance, is zero), so for a current I

$$V_o = V_m + R_a I$$

Substituting for V_o and rearranging gives

$$V_m = E\ \frac{A}{1+A} - \frac{R_a I}{I+A}$$

Thus, as the gain A is increased, the membrane potential approaches the command potential more closely, and the effect of the access resistance is reduced.

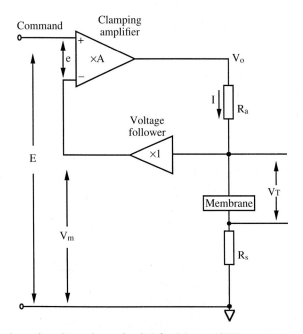

Fig. 2. Simplified schematic voltage clamp circuit (after Moore, 1971).

Series resistance

In practice, there is normally also a resistance in series with the membrane and between the voltage recording electrodes. This is represented by R_s in Fig. 2. When a current I flows across the membrane this resistance leads to a discrepancy between the measured membrane potential V_m (which is what the clamping amplifier controls) and the true potential difference across the membrane, VT. The size of the error is $I{\times}R_s$, so that the problem is most likely to be serious when large membrane currents are flowing. The usual approach is to measure R_s (see below) and so decide whether the error IR_s is significant. If so, compensation for R_s may be achieved by adding a voltage signal proportional to the membrane current, and scaled appropriately, to the command voltage of the clamping amplifier. Usually the level of compensation is set using a potentiometer and compensation for around 80-90% of the measured R_s is possible, with attempts to increase the level further driving the clamp circuit into oscillation. An example of R_s compensation in a two-electrode clamp circuit is given later in this chapter.

Series resistance is usually measured under current clamp conditions by applying a step change in current, I. Theoretically, the voltage response first shows a discontinuous jump of size IR_s, followed by a rise in voltage with initial slope equal to I/C, where C is the membrane capacitance. Under experimental conditions, with a current step of finite risetime, it is unfortunately often quite hard to distinguish the size of the initial voltage jump. This problem, and a method of correcting for the risetime of the current step, have been discussed by Binstock *et al.* (1975).

Series resistance can often be a problem in whole-cell patch clamp experiments. In this technique, described in Chapter 4, the patch pipette is used both to record voltage and as the path for current flow into the cell. This means that the access resistance of the pipette, usually in the order of a few MΩ, contributes to the series resistance. Permeabilized patch whole cell recording methods using nystatin or amphotericin B usually give access resistances 2- to 3-fold higher, with correspondingly greater possible problems due to R_s. For these reasons it is important to measure Rs in whole cell patch clamp experiments, as described in Chapter 4, and to calculate from the maximum size of the currents being studied whether the resulting voltage error will be serious. If so, a good proportion of R_s can usually be compensated using circuitry provided in most commercial patch clamp amplifiers. In extreme cases, it is also sometimes possible to check the true membrane potential of the cell with a separate microelectrode.

Clamp speed

The clamp should be able to change membrane potential sufficiently rapidly for the capacity current transient to be over by the time that ionic current is measured. Clearly, this criterion is most severe when fast ionic currents are of interest.

Voltage clamp systems have at least two, and usually more, time constants in their feedback circuit, and such systems tend to oscillate as gain is increased (see e.g. Moore, 1971). Generally, the experimental aim is for something close to a critically damped response, in other words the fastest rise in voltage which just avoids oscillation, or a slightly underdamped response, which is faster but has a slight voltage overshoot. Lags in the system will be caused by the need to charge the membrane capacity through an access resistance, and by the time constant(s) of the control amplifier itself. There will also be a delay in the measurement of membrane potential caused by capacitance at the input of the voltage follower, and at the microelectrode if one is used. This delay may be reduced by use of a voltage follower with capacity compensation.

The gain of the control system needs to be high to ensure good voltage control. Katz & Schwarz (1974) have given a theoretical analysis of the conditions for a critically damped response with a simplified voltage clamp circuit (see also Smith *et al.* 1980). They point out that with high gain, critical damping is achieved either when τ_L, the time constant of the membrane-electrode-solution load, is considerably greater than τ, the amplifier time constant, or when the converse is true. For systems with low resistance electrodes (for example axial wires) $\tau \gg \tau_L$ will generally apply, whereas $\tau_L \gg \tau$ will apply for microelectrode clamps. These authors also describe the use of current feedback to reduce the effect of access resistance. In summary, the frequency response of the control system needs to be adjusted for the particular preparation to be clamped, and gain and capacity compensation are adjusted to give the best risetime obtainable.

Isopotentiality (space clamp)

Membrane current should be recorded from an area of uniform potential, so that the current comes from a population of channels that are all experiencing the same voltage. How fully this criterion is met depends on the geometry of the preparation and electrode system used. With axial wire clamps an internal wire electrode ensures isopotentiality of a length of cylindrical axon or muscle fibre. Microelectrode clamps, which deliver current at a single point, will give a good clamp in round cell bodies, though attached structures such as axons and dendrites may not be controlled. Whole cell patch clamp is often applied to small cells where very good space clamp is achieved, though again long processes may not be controlled. In some cases a clamp which controls voltage at one point only may be sufficient to examine currents from the limited area of membrane that is clamped. An example is the two electrode point clamp of the motor endplate of skeletal muscle (Takeuchi & Takeuchi, 1960). Currents arising from electrical activity of unclamped or poorly controlled regions of membrane can often be recognized as 'notches' of inward current which may occur a relatively long time after the start of a depolarizing voltage step, and correspond to a very abrupt rise in the current-voltage relation for inward current.

3. Techniques and their applications

A variety of different voltage clamp methods are shown in Fig. 3. The choice of method depends largely on the size and shape of the preparation to be investigated and is discussed below. In general the methods all have a feedback amplifier which

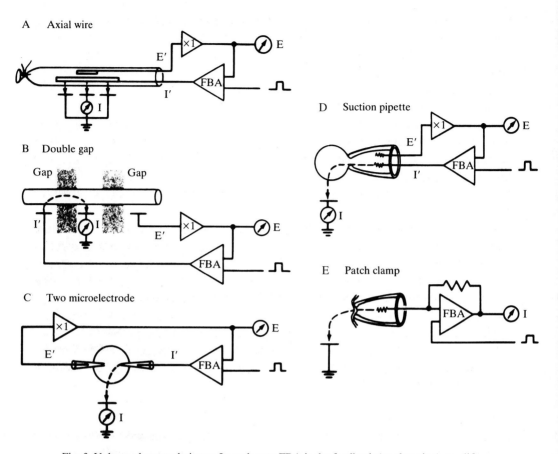

Fig. 3. Voltage clamp techniques. In each case FBA is the feedback (or clamping) amplifier, E′ is the voltage electrode and I′ is the current electrode. (Reproduced with permission from Hille, 1984).

receives a signal from a voltage recording electrode, E′ and compares this with a command potential. The difference between these signals is amplified and applied to the membrane as a current via a current electrode I′. In the circuits illustrated the current I is recorded at the electrode that grounds the bath; it may also be recorded as the voltage drop across a resistor in series with the current electrode, or across the resistance of the external solution. The patch clamp circuit shown operates rather differently, being essentially a current-to-voltage converter.

Axial wire methods

These methods are applicable to large cylindrical preparations such as giant axons of squid and *Myxicola* and barnacle giant muscle fibres. The internal voltage and current electrodes are formed by wires inserted longitudinally down the centre of the preparation, though sometimes a separate measurement of membrane potential is made with a microelectrode. The method gives a very rapid clamp as the access resistance is low, and is often combined with perfusion methods to control the internal solution (see Baker, 1984). Descriptions may be found in Hodgkin *et al.* (1952), Cole & Moore (1960) and Chandler & Meves (1965).

Gap methods

These methods are also applicable to elongated preparations, though their diameter does not have to be large as it does for axial wires. Thus the method has been used, for example, for studies on myelinated axons (Nonner, 1969), and vertebrate muscle fibres (Hille & Campbell, 1976). The preparation runs through a number of pools of solution separated by air, sucrose or vaseline gaps of high electrical resistance. The end pools usually contain intracellular solution and the cut ends of the preparation lie in them. These pools allow measurement of membrane potential and delivery of current.

A rather different type of gap method can be used for large spherical cells. In these techniques the cell is held in a funnel shaped hole in a partition separating two solution-filled chambers, where it seats against the walls to form an electrical seal. The cell membrane is broken or permeabilized on one side so as to give access to the inside of the cell. Thus the cell is clamped between two pools of solution, one in contact with the cell exterior, and the other with the cytoplasmic solution. This method has been used with molluscan neurones (Kostyuk & Krishtal, 1977), and an elegant system of this type has recently been developed to give a rapid low noise clamp of *Xenopus* oocytes for the study of gating currents of channels expressed from mRNA (Tagliatela *et al.* 1992).

Microelectrode clamps

In these techniques, microelectrodes are used both to measure the membrane potential and to deliver the current that controls it, resulting in a rather high access resistance. Since current is delivered at a point, the method can be used to space clamp cells which are approximately spherical or are short cylinders. Alternatively, a point clamp may be obtained in a long cell; for example in the endplate region of a muscle fibre. A special case is the three electrode clamp (Adrian *et al.* 1970; Stanfield, 1970) which was developed for use at one end of a muscle fibre. Two electrode clamps can be applied to cells large and robust enough to tolerate two intracellular electrodes (e.g. Meech & Standen, 1975; Adams & Gage, 1979; Smith *et al.* 1980). Smaller cells may be too fragile to allow the insertion of two electrodes without causing damage. Alternatively, the cell under study may not have been impaled under direct visual control, so that the placement of a second electrode is impracticable. In these circumstances a single electrode voltage clamp may be employed. The first of two available recording strategies was originally employed by Wilson & Goldner (1975) and involves switching the microelectrode rapidly (at 3-20 kHz) between voltage recording and current passing modes with a clamp feedback amplifier controlling the amount of current delivered. Such methods have been applied to study kinetically slow membrane conductances in central neurones (Johnston *et al.* 1980; Halliwell & Adams, 1982) and to analyse faster synaptic currents in the central nervous system (Brown & Johnston, 1983; Finkel & Redman,

1984). Another application of the single electrode clamp has been to control voltage and measure macroscopic current in neurones whilst leaving a second microelectrode free to make intracellular injections (Adams *et al.* 1982). Despite these advantages the switching single electrode clamp is inferior to two electrode configurations in terms of its slow voltage response time and the introduction of noise into the recording of both current and voltage, which is inherent in the manner the clamp system is implemented: for fuller description of the technique and its optimization the reader is referred to Finkel & Redman (1984) and to the last section of this chapter. A second voltage-clamp strategy was elaborated by Blanton *et al.* (1989) for situations where neurones cannot be visualised directly when it became clear that it is possible to use the whole-cell variant of the patch clamp technique beneath the surface of a brain slice preparation. The use of lower resistance patch pipettes in conjunction with switching techniques has improved noise levels and reduced series resistance errors in a variety of experimental situations, one of which is illustrated below.

Suction pipette methods

In these methods a cell is sealed to a pipette that has a fairly large tip orifice (up to 50 μm). The membrane within the opening is destroyed, giving low resistance access to the interior of the cell. The pipette may be used both to pass current and to record potential, though the series resistance provided by the pipette can cause problems in the voltage measurement. Sometimes a microelectrode or a second suction pipette is used to overcome this difficulty. A clamp that switches between voltage recording and current passing, as used for the single microelectrode method, may also be used to overcome this series resistance problem, since with this system no current flows at the instant when voltage is sampled. Suction pipette methods have been used with preparations that can be dissociated to give isolated cells, for example neurones and cardiac myocytes, and are usually used to perfuse the interior of the cell as well as to provide voltage clamp. Accounts of suction pipette techniques may be found in Byerly & Hagiwara (1982) and Kostyuk & Krishtal (1984).

The whole-cell form of the patch clamp technique uses a pipette with a smaller tip (usually 1 μm) and can be applied to many types of cells which are too small for other voltage clamp methods.

4. Practice and experimental details

This section will be restricted to descriptions of the two microelectrode and single electrode voltage clamp methods.

Two electrode voltage clamp

Fig. 4 shows a schematic diagram for a two electrode voltage clamp circuit, and an example of a circuit for a clamping amplifier is described in the appendix to this chapter and illustrated in Fig. 7. The detailed methods described here refer to clamping molluscan giant neurones, though the methods for other preparations such

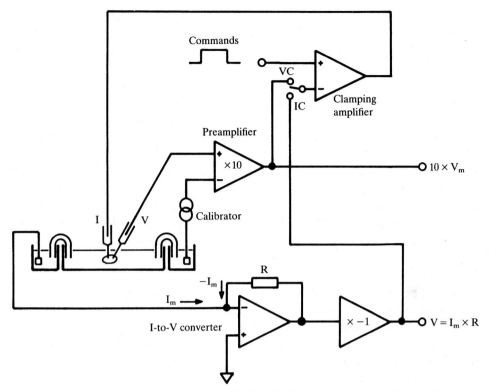

Fig. 4. Two electrode voltage clamp circuit. The circuit may be switched either to clamp voltage (VC) or current (IC).

as oocytes are essentially the same. Several commercial amplifiers are available for two electrode voltage clamp. These are made by, for example, Axon Instruments (Axoclamp or Geneclamp), NPI, or WPI (S7070A or OOC-1).

A schematic diagram of the clamp circuit is shown in Fig. 4. The cell is penetrated with two microelectrodes, one to record voltage and the other to pass current. A preamplifier records membrane potential and the clamping amplifier passes current to control this potential.

Electrodes and voltage recording. The microelectrode resistance is made as low as possible while avoiding excessive damage to the cell. Thin-walled glass may be used to obtain a low resistance for a given tip size; for molluscan neurones voltage electrodes of 5-10 MΩ and current electrodes of 1-2 MΩ are commonly used.

The voltage electrode is mounted in a holder containing an Ag/AgCl pellet, and plugs directly into one headstage of the preamplifier. A second headstage follows the bath potential via an Ag/AgCl pellet and agar bridge, and the preamplifier records differentially between these two inputs with a gain of 10. The electrode is shielded by wrapping in aluminium foil to within 1-2 mm of the tip, and then insulating this foil from the bath with parafilm and vaseline. The shield both reduces capacity coupling between current and voltage electrodes, and, since it is connected to the driven screen, which provides capacity compensation for the preamplifier, also increases the speed of voltage recording. Other shielding methods include coating the electrode with conductive paint, and some workers also use a grounded shield around the current electrode or between the two electrodes. Keeping the level of the bath solution as low as possible also reduces coupling between the electrodes.

The voltage signal from the preamplifier is fed to the clamping amplifier, and also to an oscilloscope, pen recorder, and bleater (voltage-to-frequency converter plus audio amplifier).

Clamping amplifier. This is a differential amplifier with a gain up to 10 000 which may be varied at two different stages. The circuit used (see Appendix) also has a high voltage (~120 V) amplifier in its output stage. Its frequency response may be varied by altering the capacity in the feedback stage of the differential amplifier.

Command pulses and holding potential. Command pulses are derived from a digital pulse generator and are fed into the clamping amplifier input via a summing amplifier, which also receives an offset voltage for zeroing the amplifier. Because of the ×10 gain of the preamplifier, the command voltage must be 10 times the desired voltage step. The command summing amplifier is followed by a variable low-pass filter, which may be used to reduce the risetime of the command voltage step. This often makes it easier to obtain a fast voltage clamp step while avoiding oscillation.

To set the holding potential a steady voltage signal can be applied to the command summing amplifier in the same way as command pulses. An alternative is to use a precision voltage source (the calibrator of Fig. 4) in series with the bath electrode and so provide an accurate holding potential. Commercial clamping amplifiers usually provide an inbuilt potentiometer to set the holding potential.

Measurement of membrane current. Membrane current is measured via a bath electrode which is held at virtual ground by a current-to-voltage converter circuit (Fig. 4). The operational amplifier strives to keep its two inputs at the same potential, in other words to keep its negative input at ground. To do this it must pass a current equal to $-I_m$ around its feedback loop, so that the voltage at its output equals $-I_mR$. Thus with a feedback resistor of 1 MΩ the circuit will give 1 mV.nA^{-1}. The operational amplifier should have a high input impedance ($>10^{12}$ Ω) so that no current flows into its negative input. For this reason an amplifier with an FET input is used. The whole control circuit may be made to clamp current, rather than voltage, by feeding the output of this current-to-voltage converter to the clamping amplifier.

Series resistance compensation. The voltage output from the current-to-voltage converter may also be fed to a potentiometer, which adds a proportion of the signal back to the clamping amplifier command input to compensate for R_s. For example, a series resistance of 10 KΩ will give a voltage error of 10^{-2} mV per nA of current, so that this voltage should be added to the command voltage. In the present circuit, this would be achieved by adding 1/10 of the output of the I-to-V converter (1 mV.nA^{-1}) to the command voltage; this allows for the effective division of command voltages by 10.

Leakage and capacity compensation. It is possible to provide analogue compensation for linear leakage current through the membrane, and for most of the capacity current. For leakage compensation, a suitable proportion of the membrane voltage is subtracted from the current signal before it is fed to the oscilloscope or other recording apparatus. For capacity compensation, the voltage is differentiated, and the resulting signal is subtracted. Often two or three differentiators with different time constants are used.

Experimental procedure. With the control circuit switched to current clamp, or with the clamping amplifier disconnected from the current electrode, a cell is first penetrated with the voltage electrode. The calibrator is used to back off the resting potential, and so provide a holding potential for the voltage clamp. The current electrode is then inserted; this seems to be aided by a current pulse repeated at about 1 Hz. The gain of the clamping amplifier is set low and the voltage clamp switched on. The rise time of the clamp is then increased by increasing the gain and adjusting the capacity compensation of the preamplifier, while observing the response to a small hyperpolarizing command pulse.

Single electrode voltage clamp

In Chapter 1, an alternative to the 'bridge balance' technique is presented for the elimination of the voltage drop across the resistance of an impaling microelectrode, whilst the potential of a cell is being changed by the passage of current through the electrode. In essence, this scheme is the basis of the method proposed by Brennecke & Lindemann (1974a,b) and implemented independently by Wilson & Goldner (1975). Fig. 5A is a schematic diagram of the Wilson and Goldner system, which is the technique elaborated in commercially available amplifiers and most home-built devices. A_1 is a high input-impedance high speed amplifier to which the impaling microelectrode is connected *via* a non-polarizable Ag/AgCl half-cell. A_1 measures the voltage, with respect to ground, of the electrode shank: this potential comprises the membrane potential (V_m) of the impaled cell *plus* the voltage drop (V_e) across the resistance of the electrode caused by current (I_o) injected into the cell through the microelectrode. This current is supplied from a controlled current source or pump (CCS) to the input of A_1 *via* a 2 MΩ resistor. The input resistance of A_1 being very large (>10^{11} Ω), the current from the CCS passes down the electrode into the cell. The output of A_1 is fed to a sample and hold device (SH$_1$) which in turn is connected to A_2, the clamp feedback amplifier, where, after sampling in SH$_1$, the voltage is compared with the clamp command potential (V_c). When operating in voltage-clamp mode, the output of A_2 is led *via* the electronic switches S$_1$ in a feedback loop to control the output of the CCS. The gain of the system is controlled by adjusting the sensitivity of the CCS to its input voltage. It will be noticed that the voltage at the input of A_1 is the sum of V_e+V_m; in order that the cell be clamped at its true potential, V_e has to be eliminated. This is accomplished by electronic control of the state of switch S$_1$ in order to pass current discontinuously. Consider the feedback loop closed by S$_1$ being connected to A_2: the output of $A_1 = V_m + V_e$. If S$_1$ changes state, I_o becomes zero and V_e decays to zero with the time constant $= R_e.C_s$, where R_e is the microelectrode resistance and C_s is the associated stray capacitance (see Chapter 1).

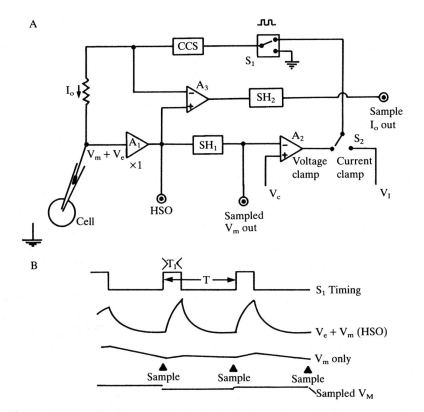

Fig. 5. (A) Schematic diagram of single electrode clamp circuit. (B) Diagram showing the timing of the states of switch S_1 and the corresponding signals recorded at HSO and SH_1 (sampled V_m). Also shown is the varying theoretical potential on the cell membrane alone (V_m only) and the instants at which SH_1 samples the potential at HSO. T is the cycle time and T_1 is the time for which current is passed. Note that duty cycle = $T_1/T \times 100$ %. See text for further details.

If SH_1 samples when the decay of V_e is complete, the output of SH_1 equals the true cell membrane potential, which will have been changed slightly, owing to charge transfer onto the membrane capacitance during current injection. When S_1 changes again, another current injection is made, based on the difference between the new value of V_m and V_c. The timing of the states of S_1 (see Fig. 5B) is such that current is regularly injected for between 25-50% of the total time (25-50% duty cycle) and the sample point of SH_1 is such that V_e has decayed completely before an upgraded current injection is re-imposed. (In practice, the sample point of SH_1 is timed to allow maximum decay of V_e: i.e. it occurs *just* before the current injection period). Thus the cell is clamped close to the value of V_c by reiterating this procedure at high frequency (3-20 kHz).

The main limiting factor determining the maximum frequency at which the system will run is the decay time of the microelectrode voltage signal. To reduce this the amplifiers are furnished with efficient capacity compensation circuitry, but attention must be paid to minimizing both the electrode resistance and the capacitance to ground that cannot be compensated completely (for example, capacitance distributed within the immersed shank of the microelectrode (see Chapter 1)). Other considerations required in operating this clamp system are: (i) that the cell membrane time constant should be long (>10^2 times greater) compared with the time constant of the recording electrode: this ensures that the charge transfer during current injection is effective in changing V_m; (ii) that the switching frequency is at least twice the frequency component of the fastest current that is to be resolved (in accord with Sampling Theory).

Recording current and voltage. Reference to Fig. 5 will indicate that the output of SH$_1$ should approximate true V_m: the faster the switching of cycling rate, the better the approximation will be. This value is read out to a voltage recording device. Current injected into the cell is measured by observing the voltage drop across the 2 MΩ resistor with amplifier A$_3$. Because of the discontinuous current delivery, this voltage is also sampled with a sample and hold device SH$_2$. The sample timing is not critical; any time during the current injection period will suffice. However, because current is passed for only part of the cycle, the signal from SH$_2$ is scaled by a percentage factor equal to the duty cycle (e.g. for 50% duty cycle it is halved, for 25%, divided by four and so on). This arises because the membrane capacitance stores the electrical charge transferred and the current is charge transferred per unit time. The scaled signal from the current monitoring amplifier is also read out to a recording device to measure current.

Use of single electrode clamp in other modes

(i) Discontinuous current clamp. By changing the state of switch S$_2$ (Fig. 5A) the feedback loop of the voltage clamp circuit may be broken and the CCS be brought under control of an external current injection command potential (V_I). Provided, as before, that attention is paid to the frequency of switching such that adequate time is allowed for the decay of the potential V_e then V_m will be registered faithfully at the SH$_1$ output. In this configuration the cell will be under conditions of current clamp.

(ii) Conventional bridge-balance operation. Incorporated in most instruments using the above techniques is the facility to disconnect the circuitry for switched current injection, together with the sample and hold devices, and pass continuous current from the CCS. In this case, an appropriately scaled proportion of the command potential V_I (equal to the voltage drop V_e) is subtracted from the output of A$_1$ to eliminate V_e. This has been discussed in more detail in Chapter 1.

Experimental application of single microelectrode clamp with sharp and other microelectrodes

The input amplifier and the final stage and output resistor of the CCS are contained in a small probe that is mounted close to the preparation and into which is connected the electrode holder. The electrodes used demand the same characteristics and treatment for reducing stray capacitance as mentioned above for 2 electrode clamping (i.e. the lowest resistance electrode compatible with experimental aims and the minimum non-compensatable capacitance). Because sharp electrodes employed with the single

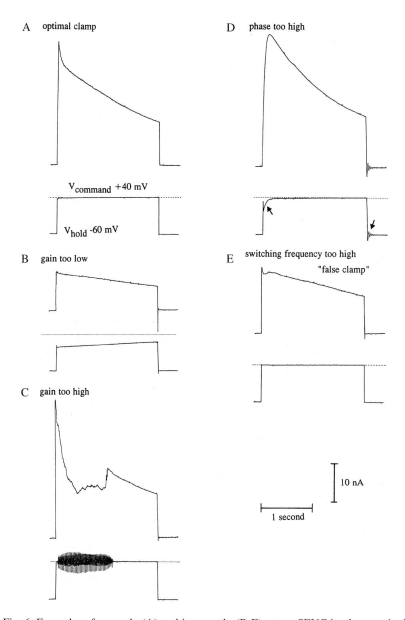

Fig. 6. Examples of correctly (A) and incorrectly (B-E) set-up SEVC implemented with an Axoclamp 2A instrument. Voltage dependent potassium currents were recorded from a neuroblastoma × glioma hybrid cell (NG108-15). Currents (upper traces) were evoked by 100 mV voltage commands from −60 mV for 2s (lower traces). Whole cell variant of the patch-clamp technique was employed; series resistance 8 MΩ, headstage gain X1. (A) Gain 7 nA/mv, phase-lag 0.2 ms and switching frequency 8 kHz. (B) As for A but gain reduced to 0.3 nA/mV. (C) As for A but gain increased 20 nA/mV. (D) As for A but phase-lag increased to 200 ms; note oscillations on the voltage trace (arrows). (E) As for A but switching frequency increased to 21 kHz; note that the voltage trace appears convincing but it is not reflecting the true membrane voltage owing to a systematic voltage error because of insufficient settling time for the voltage transient across the electrode resistance. Inspection of the headstage voltage output would reveal this error.

electrode clamp technique are usually for studying small cells, the large rejection rate on grounds of tip inadequacy (e.g. inability to impale cells) and a large consumption of micropipettes owing to fragility and propensity to tip blockage preclude complicated time-consuming procedures in the production of the electrodes. Often, attention to the recording situation and simple procedures such as coating with silicone oil or rubber will suffice to reduce electrode capacitance.

Impalement of small cells is often best accomplished by oscillating the capacity compensation circuit briefly ('zap', 'buzz' or 'tickle') or by applying a short ~20 ms intense current injection. Once a stable impalement has been achieved in the conventional bridge mode, the preamplifier may be switched to discontinuous current clamp mode and for further adjustment it becomes necessary to monitor the circuitry at three points to adjust the instrument's performance. Outputs from SH_1 (voltage) and SH_2 (current) are observed as is the voltage record just prior to SH_1, usually termed head stage out (HSO) or continuous voltage output. HSO indicates the signal V_m+V_e (Fig. 5) and with regular periodic current injection commands it is possible to observe the decay characteristics of V_e with a rapid oscilloscope sweep. Capacity compensation is adjusted to allow the fastest switching rate compatible with a full decay of V_e. Finally the SH_1 signal is monitored to check that V_m changes smoothly from rest with the onset of a current application and without showing an abrupt step, which could introduce an error in measuring V_m under conditions of current injection, similar to a series resistance error in a conventional clamp system. Errors of this sort (discussed in Chapter 1) are caused by a combination of inadequate capacity compensation (under or overadjustment) and an excessively fast cycling frequency so that the electrode transient contributes a significant component of the sampled voltage. Assuming that these checks have been made, V_c can be adjusted to the indicated resting value of V_m, S_2 can be set to voltage clamp mode and the cell brought under voltage control. Clamp gain is increased until the point of oscillation is almost reached and then reduced slightly. Clamp steps may now be imposed on the cell, as for the two electrode clamp. This may reveal a tendency for the clamp to oscillate owing to excessive loop gain. In this case, adjustment of the phase relation between voltage recording and current injection may negate this tendency. Most single electrode clamp devices possess an adjustment of this sort; in addition to introducing phase shifts, this control also changes the loop gain of the clamp in a frequency dependent manner.

The final parameter which *can* be changed but is often preset on single electrode clamp systems is the duty cycle. A duty cycle <50% allows more time for the V_e signal to decay but requires that a larger current be passed for the shorter period, thereby increasing the amplitude of V_e and also the chances of exceeding the current-passing capability of the (normally) high resistance electrode. Choice of duty cycle is obviously a compromise and the reader is referred to the more rigorous treatment of the subject by Finkel & Redman (1984).

As mentioned above the switching method can be employed with suction electrodes for the whole-cell patch clamp technique where series resistance errors can be minimised. However, from the discussion above it can be seen that inappropriate

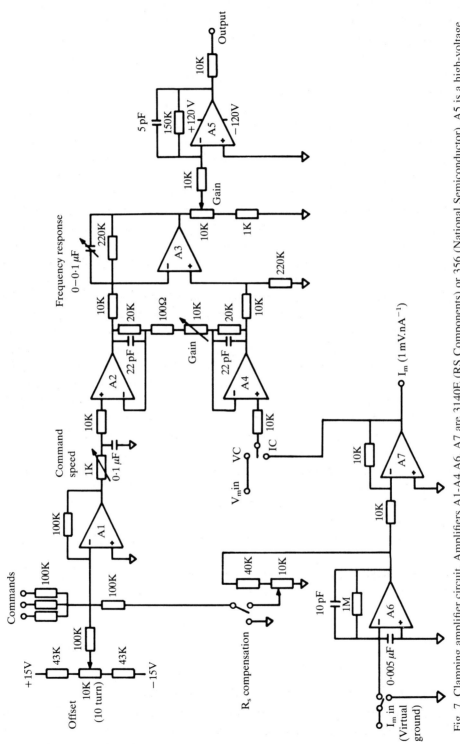

Fig. 7. Clamping amplifier circuit. Amplifiers A1–A4, A6, A7 are 3140E (RS Components) or 356 (National Semiconductor). A5 is a high-voltage operational amplifier - Teledyne Philbrick 1022 or Analog Devices 171J.

adjustment of the voltage clamp instrument can lead to errors that are equally bad. Figure 6 shows some results from a single electrode clamp experiment on a cultured neuroblastoma/glioma hybrid cell. Optimal clamp conditions reveal that a step depolarisation from resting potential (−60 mV) to +40 mV elicits a declining outward current. Inspection of the voltage trace reveals that within 10 ms the voltage attains the command potential. As with all modes of voltage clamp, insufficient loop gain causes a disparity between the potential command and that achieved (Fig. 6B). Because voltage is not constant, this situation, which *does* give a true measure of both current and voltage is not much better from the point of view of assessing membrane currents than an experiment with current clamp; if a full voltage command is assumed by the experimenter, the current measured will be underestimated. Conversely, with the gain too high, the clamp becomes unstable and both current and voltage traces display oscillations, which, in ignorance of the voltage trace, could lead an experimenter to misinterpret an actually smooth elicited current as a fluctuating one. At least, with the switching clamp method, attention to the voltage trace can alert the investigator to these insufficiencies; furthermore, the voltage indicated by the instrument reflects the true potential of the cell membrane subject to geometrical considerations pertaining to space-clamp. More hazardous are errors where too much lead or lag has been applied in the clamp timing; overshoot of the clamp potential with concomitant underestimation of command potential can ensue (Fig. 6D). Only very close attention to the voltage trace can signal this condition. Finally, when switching frequency is too high, the voltage samples the value of membrane potential together with an error due to insufficient decay of the voltage drop across the electrode (Fig. 6E): the command potential, while seeming to have been attained, in fact, falls short of the intended and consequently leads to an underestimate of the elicited current. The latter condition can be assessed and eliminated by careful inspection of the continuous headstage input voltage of the amplifier.

To conclude and recapitulate, the optimization of the single electrode clamp method can be effected by attending to the following points.

(i) Lowest electrode resistance that is feasible.

(ii) Minimize stray capacitance.

(iii) Fastest switching rate compatible with full decay of electrode voltage transient.

(iv) Maximum gain commensurate with clamp stability.

(v) Clamp noise can be reduced by filtering the output appropriately.

Appendix

A clamping amplifier for two microelectrode voltage clamp

Fig. 7 shows a clamping amplifier circuit designed for two microelectrode clamp of relatively large cells, such as molluscan neurones. The circuit is based on a conventional three operational amplifier differential amplifier (A2, A3, A4), with the addition of a high voltage output stage (A5) that enables the amplifier to pass the high currents which may be required to clamp large cells. The frequency response of the amplifier may be altered by means of an 11-way switch which varies the feedback capacitor in the range below 0.1 µF.

Command pulses are fed in through a summing amplifier (A1) and are then passes through a variable low-pass filter so that their rise time may be adjusted. Membrane current is measured using a virtual-ground circuit (A6), while A7 inverts the current signal to the normal convention. The amplifier may be used to clamp membrane voltage (VC mode) or membrane current (IC mode). The membrane potential signal (V_m in) is derived from a differential preamplifier with a gain of 10. Compensation for series resistance (R_s compensation) is achieved by feeding a proportion of the membrane current signal to the summing point of amplifier A1.

References

ADAMS, D. J. & GAGE, P. W. (1979). Ionic currents in response to depolarization in an Aplysia neurone. *J. Physiol., Lond.* **289**, 115-141.

ADAMS, P. R., CONSTANTI, A., BROWN, D. A. & CLARK, R. B. (1982). Intracellular Ca^{2+} activates a fast voltage sensitive K^+ current in vertebrate sympathetic neurones. *Nature* **296**, 746-749.

ADRIAN, R. H., CHANDLER, W. K. & HODGKIN, A. L. (1970). Voltage clamp experiments in striated muscle fibres. *J. Physiol., Lond.* **208**, 607-644.

BAKER, P. F. (1984). Intracellular perfusion and dialysis: Application to large nerve and muscle cells. In *Intracellular Perfusion of Excitable Cells.* (ed. Kostyuk, P. G. & Krishtal, O. A.) pp. 1-17. Chichester: Wiley.

BINSTOCK, L., ADELMAN, W. J., SENFR, J. P. & LECAR, H. (1975). Determination of the resistance in series with the membranes of giant axons. *J. Membr. Biol.* **21**, 25-47.

BLANTON, M. G., LO TURCO, J. J. & KRIEGSTEIN, A. R. (1989). Whole cell recording from neurons in slices of reptilian and mammalian cortex. *J. Neurosci. Meth.* **30**, 203-210.

BRENNECKE, R. & LINDEMANN, B. (1974a). Theory of a membrane-voltage clamp with discontinuous feedback through a pulsed current clamp. *Rev. Sci. Instr.* **45**, 184-188.

BRENNECKE, R. & LINDEMANN, B. (1974b). Design of a fast voltage clamp for biological membranes, using discontinuous feedback. *Rev. Sci. Instr.* **45**, 656-661.

BROWN, T. H. & JOHNSTON, D. (1983). Voltage-clamp analysis of mossy fiber synaptic input to hippocampal neurones. *J. Neurophysiol.* **50**, 487-507.

BYERLY, L. & HAGIWARA, S. (1982). Calcium currents in internally perfused nerve cell bodies of Limnea stagnalis. *J. Physiol., Lond.* **322**, 503-528.

CHANDLER, W. K. & MEVES, H. (1965). Voltage clamp experiments on internally perfused giant axons. *J. Physiol., Lond.* **180**, 788-820.

COLE, K. S. (1949). Dynamic electrical characteristics of the squid axon membrane. *Arch. Sci. Physiol.* **3**, 253-258.

COLE, K. S. & MOORE, J. W. (1960). Ionic current measurements in the squid giant axon. *J. Gen. Physiol.* **44**, 123-167.

FINKEL, A. S. & REDMAN, S. J. (1984). Theory and operation of a single microelectrode voltage clamp. *J. Neurosci. Methods* **11**, 101-127.

HALLIWELL, J. V. & ADAMS, P. R. (1982). Voltage-clamp analysis of muscarinic excitation in hippocampal neurones. *Brain Res.* **250**, 71-92.

HILLE, B. (1984). *Ionic Channels of Excitable Membranes.* Sunderland: Sinauer.

HILLE, B. & CAMPBELL, D. T. (1976). An improved vaseline gap voltage clamp for skeletal muscle fibres. *J. Gen. Physiol.* **67**, 265-293.

HODGKIN, A. L., HUXLEY, A. F. & KATZ, B. (1952). Measurement of current-voltage relations in the membrane of the giant axon of Loligo. *J. Physiol., Lond.* **116**, 424-448.

JOHNSTON, D., HABLITZ, J. J. & WILSON, W. A. (1980). Voltage-clamp discloses slow inward current in hippocampal burst-firing neurones. *Nature* **286**, 391-393.

KATZ, G. M. & SCHWARTZ, T. L. (1974). Temporal control of voltage-clamped membranes: an examination of principles. *J. Membr. Biol.* **17**, 275-291.

KOSTYUK, P. G. & KRISHTAL, O. A. (1977). Separation of sodium and calcium currents in the somatic membrane of mollusc neurones. *J. Physiol., Lond.* **270**, 545-568.

KOSTYUK, P. G. & KRISHTAL, O. A. (1984). Intracellular perfusion of excitable cells. (IBRO handbook series: *Methods in the Neurosciences*; vol. 5). Chichester: Wiley.

MEECH, R. W. & STANDEN, N. B. (1975). Potassium activation in Helix aspersa neurones under voltage clamp: a component mediated by calcium influx. *J. Physiol., Lond.* **249**, 211-239.

MOORE, J. W. (1971). Voltage clamp methods. In *Biophysics and Physiology of Excitable Membranes.* (ed. Adelman, W. J.) pp. 143-167. New York: Van Nostrand.

NONNER, W. (1969). A new voltage clamp method for Ranvier nodes. *Pflügers Arch.* **309**, 176-192.

SMITH, T. G., BARKER, J. L., SMITH, B. M. & COLBURN, T. R. (1980). Voltage clamping with microelectrodes. *J. Neurosci. Methods* **3**, 105-128.

STANFIELD, P. R. (1970). The effect of the tetraethylammonium ion on the delayed currents of frog skeletal muscle. *J. Physiol., Lond.* **209**, 209-229.

TAGLIATELA, M., TORO, L. & STEFANI, E. (1992). Novel voltage clamp to record small fast currents from ion channels expressed in Xenopus oocytes. *Biophys. J.* **61**, 78-82.

TAKEUCHI, A. & TAKEUCHI, N. (1960). On the permeability of end-plate membrane during the action of transmitter. *J. Physiol., Lond.* **154**, 52-67.

WILSON, W. A. & GOLDNER, M. M. (1975). Voltage clamping with a single microelectrode. *J. Neurobiol.* **6**, 411-422.

Chapter 3
Separation and analysis of macroscopic currents

NICK B. STANDEN, NOEL W. DAVIES and PHILIP D. LANGTON

1. Introduction

In voltage clamp experiments, ionic flow is measured as electrical current. Macroscopic currents recorded from whole cells, or from relatively large areas of cell membrane, represent the sum of the ion fluxes through many channels, generally of more than one type. The first step in the interpretation of such currents is usually their separation into constituent components, each associated with a particular sort of ion channel. This chapter will consider some of the means by which this separation may be achieved, and will also give an introduction to methods for the analysis of the currents isolated in this way.

2. Current separation

In some experimental situations, the total membrane current may result almost entirely from ion flow through one type of channel, so that current separation is not necessary. This is often the case when currents flowing through transmitter-activated channels are elicited by the application of an agonist. When voltage clamp experiments involve a change in membrane potential, however, the situation is rarely so simple, since many types of channel are voltage-gated. In this case some means of current separation is needed. The initial process is often the removal of leakage and capacity currents by analogue compensation, by computer, or both, followed by methods of the type outlined below. Several of these methods involve the subtraction of currents recorded under one experimental condition from those under another, for example with and without a particular channel blocking substance. Such subtraction is usually done on the experimental current records themselves, but sometimes on their current-voltage relations. The widespread use of microcomputers together with programs for data acquisition and analysis has made this process straightforward. Subtraction does require, though, that currents should be relatively stable over the time period needed to change the experimental conditions, or at least that any change should be predictable enough to be controlled for.

Ion Channel Group, Department of Physiology, University of Leicester, PO Box 138, Leicester LE1 9HN, UK

Although the various strategies for current separation will be described separately, it is very common to combine two or more methods to isolate the current of interest; for example by using a blocking substance in combination with an appropriate voltage clamp protocol. Similar methods to those described below may also be applied to a novel cell type so as to characterize the channels present in terms of their permeant ionic species, pharmacology and voltage-dependence.

Time of measurement

If the ionic channels that contribute to the total membrane current have sufficiently different kinetics, it may be possible to separate currents merely by measuring them at a suitable time during a voltage clamp pulse. For example, in squid axon, current measured at the end of a 10 ms depolarization will represent nearly pure current through delayed K^+ channels, as nearly all Na^+ channels will have inactivated by that time. Although this method may sometimes be adequate for the routine analysis of records, it needs experimental justification by one of the other methods described below.

Ionic substitution

This method may be used to identify and separate currents in terms of the ionic species that carry charge. A classical example may be found in the study of squid

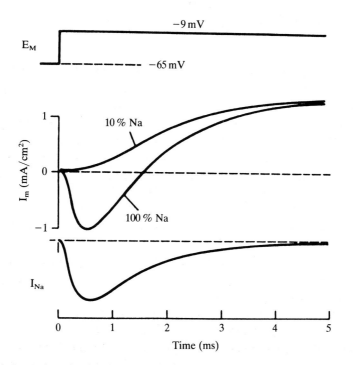

Fig. 1. Separation of Na^+ and K^+ currents in squid axon by ionic substitution. The low-sodium record (10% Na) is subtracted from that measured in sea water (100% Na) to give the Na^+ current shown below. (Reproduced with permission from Hille, 1984; after Hodgkin, 1958).

axon by Hodgkin & Huxley (1952a); see also Hodgkin (1958), shown in Fig. 1. Current was measured for a depolarization from −65 mV to −9 mV, first in sea water and then in artificial saline with only 10% of the Na$^+$ of sea water. In this preparation, where only two sorts of channel predominate, the current in 10% Na$^+$ is flowing through delayed rectifier K$^+$ channels, and the Na$^+$ channel current may be obtained by subtracting this current from the record in sea water (Fig. 1). If it is possible to perfuse the inside of the cell under study, ionic substitution may be applied to intracellular as well as extracellular solutions. Whole cell patch clamp is now the most common technique for recording macroscopic currents and, with this method, the composition of the pipette solution can normally be used to control the intracellular ionic composition, while a variety of perfusion methods for larger cells may be found in Kostyuk & Krishtal (1984).

Obviously ionic substitution alone cannot be used to separate currents through different channels which have the same major permeant ion, for example different types of K$^+$ channel. Ca^{2+} channels also pose special problems, both because Ca^{2+} entry may activate other channels (for example K$^+$ or Cl$^-$ channels) and because replacement of extracellular Ca^{2+} by other ions can shift the voltage dependence of gating of many types of channel. So current records made in Ca^{2+}-free solution, or in the presence of Ca^{2+} channel blockers, may not only lack Ca^{2+} current, but also have altered currents through other channels. For this reason it is often necessary to use a combination of methods to separate currents through Ca^{2+} channels. An example is illustrated in Fig. 3, and is discussed in the following section.

Pharmacological separation

Many substances are known that block ionic channels. These channel blockers form useful tools for the separation of currents, for the identification of channel types and also as ligands for use in the purification of channel proteins. Blockers range from simple inorganic ions through synthetic drugs to complex naturally occurring toxins. The range of blockers available is increasing as new toxins are discovered and new blocking compounds are synthesized, and so pharmacological methods provide increasingly powerful means of isolating currents through particular channel types. Blockers may act from the outside of the membrane, from the inside, or may be membrane permeant. Some block with high affinity, acting at nanomolar concentrations, while for others millimolar concentrations may be needed. Some blockers are specific for one type of channel, others may block several types, often with rather different affinities. An excellent account of blocking mechanisms is given by Hille (1992), and some blocking substances for different channels are given in Table 1 and the accompanying references.

Blocking substances are used in one of two main ways to separate currents. First, it may be possible to block all currents except the one that is of interest. Depending on the range of channels present, this may involve one or more blockers, sometimes in combination with ionic substitution. Secondly, a blocker specific for the channel of interest may be used. Current is recorded in the presence of the blocker and subtracted from that in its absence to yield the current required.

Table 1. *Some blockers for ion channels*

Channel	Blocked by	References
Sodium	TTX, STX, local anaesthetics	Catterall (1980)
Calcium		
Generic calcium	Co^{2+}, Cd^{2+}, Ni^{2+}, La^{3+}	Byerly & Hagiwara (1988)
T-type	No specific ligand	
L-type	Dihydropyridines: e.g. Nifedipine	Triggle & Janis (1987);
	Phenylalkylamines: e.g. Verapamil	Fox *et al.* (1987)
	Benzothiazepines: e.g. Diltiazem	
N-type	ω-Conotoxin GVIA	Plummer *et al.* (1989)
P-type	ω-Agatoxin-IVA	Mintz *et al.* (1992)
Potassium		
Delayed rectifier	TEA^+, Cs^+, Ba^{2+}, Dendrotoxins,	Stanfield (1983); Cook &
	4-Aminopyridine, Noxiustoxin,	Quast (1990); Hille (1992)
	Quinidine, Capsaicin, Phalloidin	
A-current	TEA^+, 4-Aminopyridine, Quinidine,	Cook & Quast (1990);
	Dendrotoxins	Hille (1992)
Inward rectifier	Ba^{2+}, TEA^+, Cs^+, Rb^+	,,
Ca^{2+}-activated	TEA^+, Cs^+, Quinidine,	,,
	Charybdotoxin*, Apamin,	
	Leiurotoxin I†	
ATP-dependent	Cs^+, Ba^{2+}, sulphonylureas e.g.	Ashford (1990)
	Glibenclamide	
Chloride	DIDS‡, IAA-94§, Furosemide,	Cabantchik & Greger (1992)
	A9Cǁ, AABs¶, SITS**	

*Selective for the large conductance Ca^{2+}-activated channel (BK_{Ca})
†Selective for the small conductance Ca^{2+}-activated channel (SK_{Ca})
‡DIDS: 4,4′-diisothiocyanatostilbene-2,2′-disulphonic acid
§IAA-94: indanyloxyacetic acid
ǁA9C: anthracene-9-carboxylate
¶AABs: arylamino-benzoate derivates
**SITS: 4-acetamido-4′-isothiocyanatostilbene-2,2′-disulphonic acid

An example of a straightforward pharmacological separation of currents at the frog node of Ranvier is shown in Fig. 2. In this preparation, the Na^+ channel blocker tetrodotoxin (TTX) leaves only K^+ current flowing through delayed rectifier K^+ channels. Alternatively, the Na^+ current may be studied after block of the delayed K^+ channels with tetraethylammonium ions (TEA^+).

Fig. 3 shows an example of a more complex procedure in which a combination of methods is used to isolate the current flowing through Ca^{2+} channels in a smooth muscle cell from a rat cerebral artery, recorded using whole cell patch clamp. To reduce K^+ currents as far as possible, Cs^+ is substituted for K^+ in the pipette solution that perfuses the cell interior. Cs^+ both acts as a K^+ substitute and as a K^+ channel blocker. The size of the Ca^{2+} current is increased by raising external $[Ca^{2+}]$ to 10 mM. Ca^{2+} current is obtained by subtraction using Co^{2+} as a Ca^{2+} channel blocker. Currents are first recorded in response to a series of voltage clamp pulses in an extracellular solution without Co^{2+} (Fig. 3A). A rapid perfusion system is then used to apply solution to which 2 mM Co^{2+} has been added and the voltage clamp series is

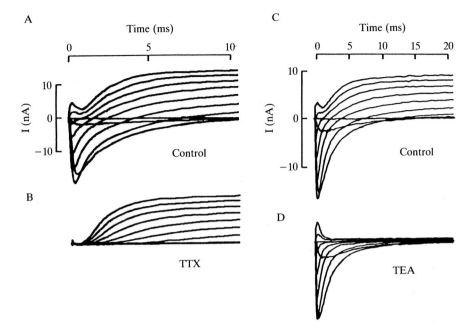

Fig. 2. Block of Na$^+$ and K$^+$ channels in frog node of Ranvier. (A) Control currents recorded when a node was depolarized to a series of potentials ranging from −60 mV to +60 mV in 15 mV steps. (The depolarization was preceded by a 40 ms hyperpolarization to −120 mV from the holding potential of −90 mV). (B) 300 nM TTX blocks Na$^+$ channels, leaving only I_K. (C) Control measurements in another node. (D) 6 mm TEA blocks K$^+$ channels, leaving I_{Na}. (Reproduced with permission from Hille, 1984).

repeated (Fig. 3B). The Ca^{2+} currents (Fig. 3C) are obtained by subtracting the records in the presence of Co^{2+} from those in its absence. Finally the stability of the currents with time may be checked by removing the Co^{2+} again and recording a second set of currents in its absence.

Voltage pulse protocol

When the activation or inactivation of different channels occurs, at least in part, over different ranges of membrane potential, it may be possible to separate currents by choosing a suitable voltage pulse regime.

A good example of a current that may be isolated in this way is the early outward K$^+$ current, or A-current, first described in molluscan neurones, but which also occurs in vertebrate neurones and various other cell types. The channels carrying this current are usually almost fully inactivated at holding potentials close to the normal resting potential. Hyperpolarization removes the inactivation and the channels may then be opened by depolarization. The A-current may therefore be separated by subtraction of the current obtained in response to a depolarizing voltage step from the current recorded when the same step is applied preceded by a hyperpolarizing pulse, as shown in Fig. 4A.

Another method of separation that uses pulse protocol is the removal of all current

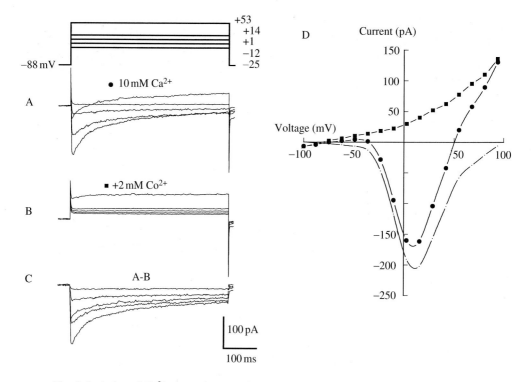

Fig. 3. Isolation of Ca^{2+} channel current in an arterial smooth muscle cell. (A) Whole cell currents recorded in response to the depolarizing voltage steps shown above in external solution containing 10 mM Ca^{2+}. (B) Currents recorded during application of the 10 mM Ca^{2+} solution to which 2 mM $CoCl_2$ was added to block currents through Ca^{2+} channels. (C) Ca^{2+} currents obtained by subtraction of records in B from those in A. (D) Corresponding current-voltage relations obtained in 10 mM Ca^{2+} (●), 10 mM Ca^{2+} + 2 mM Co^{2+} (■), and for the current obtained by subtraction as A–B (broken line). The pipette (intracellular) solution contained (mM): CsCl, 130; $MgCl_2$, 1; EGTA 5, pH 7.2 with CsOH.

carried by a particular ion by making measurements at the equilibrium potential for that ion. For example, 'tails' of Na^+ or Ca^{2+} current may be recorded when the membrane potential is returned to the K^+ equilibrium potential, E_K, after a depolarizing pulse. The method is illustrated for Na^+ current in Fig. 4B. A depolarizing pulse activates Na^+ channels so that an inward Na^+ current develops. The potential is then stepped to E_K and the inward current jumps to a larger value as the driving force on Na^+ ions flowing through open Na^+ channels (= V–E_{Na}) is increased. The tail current then declines rapidly as Na^+ channels close at the hyperpolarized potential. The instantaneous value of the tail current measured immediately after the potential is changed to E_K will be proportional to the number of Na^+ channels open at the end of the preceding pulse, and will be uncontaminated by K^+ current.

Channel expression

Virtually complete isolation of the ionic currents through a particular channel may be possible by expressing the cloned channel in a cell type that contains little or no intrinsic ionic conductance. The most common method is to inject *Xenopus* oocytes with mRNA encoding a particular ion channel (see Dascal, 1987 for review). This mRNA is then translated by the oocyte and the channel protein becomes embedded in

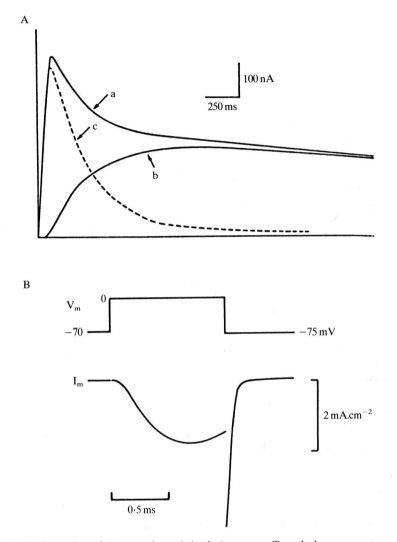

Fig. 4. (A) Separation of A-current in an *Anisodoris* neurone. Trace b shows current recorded for a step from −40 mV to −5 mV. This consists almost entirely of delayed K$^+$ current, I_K. If the holding potential is set to −80 mV, a step to −5 mV now activates both I_K and A-current (trace a). Subtraction of b from a gives the A-current (broken line). For steps to potentials more negative than about −35 mV, I_K is not activated, so that I_A may be recorded directly. (Reproduced with permission from Connor & Stevens, 1971). (B) Sodium tail current in squid axon. The membrane potential is stepped to 0 mV for 1 ms so that Na$^+$ channels open, and is then repolarized to E_K (−75 mV) when a rapidly declining tail of Na$^+$ current is seen. This figure and Fig. 7 have been drawn using a computer model of the voltage-clamped squid axon written by P. R. Stanfield and based on the equations given by Hodgkin & Huxley (1952b).

the membrane within 2 or 3 days. Although oocytes do have some intrinsic currents, the expressed currents are usually many times larger than this and can be studied in isolation. Other expression systems include some insect and mammalian cell lines that have almost no endogenous currents. An example is shown in Fig. 6, which shows currents recorded from a murine erythroleukaemia (MEL) cell transfected with DNA encoding a delayed rectifier current (hPCN1) from a human heart cDNA library (Shelton et $al.$, 1993). MEL cells that have not been transfected do not have measurable voltage-activated currents, so that the currents measured represent pure currents through the expressed channel.

It is important to realise, however, that the current recorded from a cloned channel after expression may not always exactly reflect its behaviour in the native cell. This is because the subunit structure of the native and expressed channels may not always be the same, and because expressed channels may sometimes lack certain regulatory proteins even though they are still functional. This is particularly true of voltage-activated K^+ channels which form as tetramers of individual subunits (Hille, 1992; Pongs, 1992). As such structural details become clearer, however, channel expression should provide an increasingly powerful method for studying the behaviour of a single type of channel in isolation.

3. Analysis of currents

There is only space here to give an outline of some of the basic methods available for the analysis of ionic currents recorded in macroscopic voltage clamp experiments. More detail and a wider variety of approaches may be found in the original literature. The book by Hille (1992) covers many techniques for the analysis of currents and also acts as a useful source of references to research papers.

Current-voltage relations

The first step in the analysis of an ionic current is often to summarize its voltage-dependence in the form of a current-voltage (I-V) relation. For transmitter-activated channels that show little voltage-dependent gating, the current-voltage relation can be obtained by measuring the current induced by application of the transmitter chemical at a series of different voltage-clamped potentials. The potential at which the current reverses direction (the reversal potential) gives information about which ion or ions may be carrying current through the transmitter-activated channel.

Examples of current-voltage relations for voltage-activated channels are shown in Fig. 5, which shows I-V relations for the peak Na^+ current, I_{Na}, and delayed K^+ current, I_K, of squid axon. I_K first becomes measurable at about -50 mV and increases with further depolarization, both because more channels open and because the electrochemical driving force on K^+ (the difference between the membrane potential, V, and the potassium equilibrium potential, E_K) increases. I_{Na} first becomes increasingly negative (inward) as V becomes more positive because depolarization leads to increasing numbers of open Na^+ channels, but then starts to decrease again as

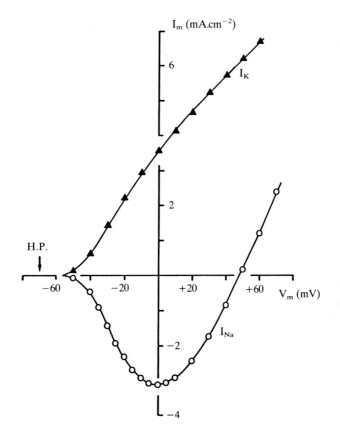

Fig. 5. Current-voltage relations for I_{Na}, and I_K in squid axon. Peak I_{Na} (○) and steady-state I_K (▲) are plotted against the membrane potential during the voltage-clamp step. Holding potential −70 mV.

the reduction in driving force on Na^+ becomes more important. I_{Na} becomes zero as the potential reaches E_{Na}, and at potentials positive to this the Na^+ current is outward. Thus the shape of the current-voltage relations for the current through these voltage-activated channels is determined both by the effect of membrane potential on channel open probability and by its effect on the driving force on the permeant ion.

Activation curves and Boltzmann relations

To describe the voltage-dependence of a voltage-activated channel we normally wish to study the way in which voltage affects channel gating independent of its effect on the electrochemical driving force. A way to extract the effect of voltage on channel gating alone is to measure instantaneous (or tail) currents at a fixed voltage, so that the driving force is constant. The method is very similar to that used to measure Na^+ current tails as described earlier, and an example taken from a cloned delayed rectifier K^+ channel, hPCN1, expressed in a cell line is shown in Fig. 6. Fig. 6A shows the voltage protocol used: a series of pulses in which the voltage is stepped from the holding potential to a variable potential followed by return to a fixed voltage of −40 mV. The first depolarization leads to activation of K^+ channels, giving rise to the K^+ currents seen in Fig. 6B. On stepping to −40 mV the current jumps to an instantaneous value proportional to the number of channels open immediately before the voltage step, and then declines exponentially as channels close (Fig. 6C). Plotting the relative size of the instantaneous current against the voltage during the first depolarization gives a measure of the channel open probability as a function of voltage, usually called the activation curve (Fig. 6D).

The activation curve has a characteristic shape that reflects the nature of channel gating: the movement of charges in the voltage-sensitive region of the channel protein in the membrane voltage field. It may be fit by a Boltzmann relation of the form

$$\frac{I}{I_{max}} = \frac{1}{1 + \exp\left[(V_h - V)/k\right]} \tag{1}$$

where I/I_{max} is the normalized instantaneous current, V_h is the voltage at which half activation occurs, and k is a factor determining how steeply the activation curve

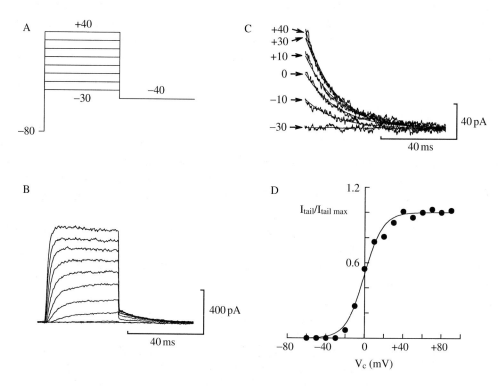

Fig. 6. Measurement of the activation curve of a delayed rectifier K^+ channel using tail currents. The channel (hPCN1) was cloned from a human cDNA library and expressed in a murine erythroleukaemia cell. (A) The voltage-clamp protocol used. (B) The family of currents observed using whole cell patch clamp. Note the outward tail currents that occur after the membrane potential is returned to −40 mV. C shows these tail currents at a higher gain, together with exponential fits to their decline with time. (D) Activation relation obtained as described in the text. The curve is drawn according to the Boltzmann function of eqn (1) with $V_h = -0.1$ mV and $k = 9.3$ mV.

changes with voltage. The Boltzmann relation describes the voltage-dependence of channel gating, the factor k showing the equivalent number of charges on the channel protein that move in the membrane voltage field to give rise to channel opening. One elementary charge will give $k = 24$ mV (channel open probability will change e-fold for a 24 mV change in voltage), while the k of 9.3 mV for the curve of Fig. 6D corresponds to a gating charge of 2.6 elementary charges. Activation curves, and the parameters V_h and k of the Boltzmann fits to them, are used very commonly to characterize voltage-activated ion channels.

Single channel recording methods have provided measurements of the single channel current for many types of ion channel. This sometimes gives another way to construct an activation curve, which can be useful if for some reason it is difficult to measure tail currents, provided that the type of channel involved is known. The mean macroscopic current, $\bar{I}$, through a homogeneous population of ion channels will be given by $\bar{I} = Nip$, where N is the total number of channels, i is the single channel current, and p the probability that a channel is open. The macroscopic current divided by the unitary current measured at the same voltage, $\bar{I}/i$, therefore equals Np, so that dividing the macroscopic I-V relation by the single channel current-voltage relation gives the voltage-dependence of Np. If, as is usually the case, N does not change over the course of the measurement, this is equivalent to the activation curve for the channel.

Channel inactivation

Many ion channels, like Na^+ channels of squid axon, and many Ca^{2+} and K^+ channels show inactivation as well as activation.

Inactivation can be studied using two-pulse experiments. The steady-state dependence of inactivation on membrane potential, also often called the inactivation curve or h_∞ curve, can be measured with the pulse protocol shown in Fig. 7A. The first pulse (prepulse) is made long enough for inactivation to develop fully, and its voltage is varied, while the second (test) pulse of fixed size is used to measure the degree of inactivation. The current flowing during the test pulse is expressed relative to its size in the absence of a prepulse to generate the inactivation curve (Fig. 7B). In most cases channel inactivation is also voltage dependent, and can be fit by a Boltzmann function of the same type as that used to fit activation curves.

The time course of the development of inactivation at a particular potential can be measured using a prepulse of fixed voltage, but whose duration is varied. The membrane potential may be returned briefly to the holding level between the pre- and test pulses to allow channel activation to return to its resting level (e.g. Gillespie & Meves, 1980). Similarly, recovery from inactivation can be studied by varying the duration of the interval between a prepulse and a test pulse.

Fitting the kinetics of currents

The kinetic analysis of current waveforms recorded under voltage-clamp is to some extent dependent on the type of model that is used to interpret channel gating. The initial process in this analysis, however, is usually to fit a mathematical function to

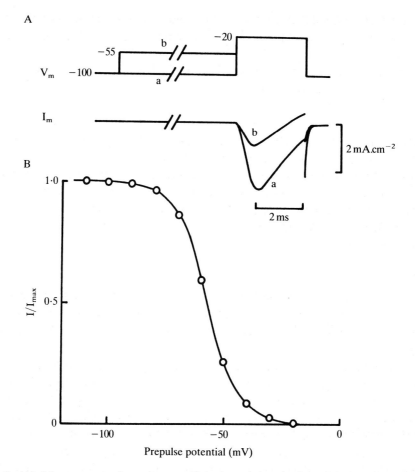

Fig. 7. (A) Measurement of steady-state Na⁺ channel inactivation with a two-pulse experiment. The example traces show I_{Na} for a depolarization to −20 mV with no prepulse and with a 50 ms prepulse to −55 mV. Holding potential −100 mV. (B) Steady-state Na⁺ inactivation curve.

the experimental records of current, or to plots of conductance or of channel open probability against time. These may normally be fit by either an exponentially rising or falling function, or such a function raised to a power greater than one, or by the sum of such functions.

As an example, we may consider the time course of the increase in delayed potassium current in squid axon when the membrane potential is stepped from a value where the channels are shut to a level where they open significantly. The initial value of the current, I_0, will be zero if all channels are shut at the holding potential, and the rise of the current to its final value, I_∞, may be fitted by

$$I_t = I_\infty(1-e^{-t/\tau})^y \tag{2}$$

where I_t is the current at time t and τ is the time constant for activation. The power y is 4 in this case, corresponding in the classical Hodgkin-Huxley formalism to the idea that four particles must move in the membrane to open a channel. A value of y greater than one gives an initial delay in the rise of current, so that it follows a sigmoid time course.

If the membrane is repolarized to a potential where the channels will shut again, so that now $I_\infty = 0$, the current falls as

$$I_t = I_0(e^{-t/\tau})^y \tag{3}$$

where I_0 is the initial value of I immediately after repolarization. This gives an exponential decline in current, with a time constant equal to τ/y.

Many types of channel show both activation and inactivation, so that the current through them first increases and then declines with time. The necessary expression is then the product of a rising and a falling function. For example, assuming a holding potential where $I_0 = 0$, and that inactivation is exponential and will be complete at the step potential used

$$I_t = I'_\infty(1-e^{-t/\tau_1})^y.e^{-t/\tau_2} \tag{4}$$

I'_∞ is now the maximum value which the current would reach in the absence of inactivation, and can be estimated from experimental records by measuring τ_2, the time constant for inactivation, and multiplying the recorded I_t by e^{t/τ_2} (e.g. Adams & Gage, 1979).

Having fit the macroscopic currents with exponential functions as described above, channel kinetics are usually described by some form of model. The models used most widely are still variants of that originally developed by Hodgkin & Huxley to analyse the currents recorded under voltage clamp from squid axon (Hodgkin & Huxley, 1952b; Hodgkin, 1958). The model describes channel gating in terms of the movement of charged particles in the cell membrane, and provides a powerful empirical description of channel kinetics. Hodgkin-Huxley descriptions have been developed for almost every type of voltage-dependent channel so far described, and the model is very clearly described by Hille (1992).

Equivalent models may be drawn in the form of state diagrams, which have become increasingly popular since the advent of single channel recording because they are also useful for interpreting kinetics at the single channel level. Such models envisage the channel as adopting several possible conformational states, each of which may correspond to the channel pore being either open or closed. The simplest model in which there is just one open and one closed state, with voltage dependent rate constants for transitions between them

$$C \rightleftharpoons O$$

will lead to an exponential change in channel open probability, and so current, in response to a step change in voltage. Sequential models, in which the channel passes through a series of closed states before opening, can predict a sigmoid rise in current. An example originally proposed by Armstrong (1969) to describe delayed rectifier K^+ channel kinetics in squid axon is given below:

$$C4 \rightleftharpoons C3 \rightleftharpoons C2 \rightleftharpoons C1 \rightleftharpoons O$$

Here $C4$, $C3$, $C2$ and $C1$ are different closed states, while in state O the channel is open. This type of model is further discussed in Chapter 7.

Noise analysis of macroscopic currents gives information about single channels

The macroscopic current recorded under voltage clamp is the sum of the current through many ionic channels. Even when the average number of channels open is constant, the exact number open will change from instant to instant. This variation will result in fluctuation of the measured current or 'noise'. The techniques of noise analysis exploit this fluctuation to obtain information about the underlying single channel events.

The single channel current, i, can be estimated by measuring the variance of the macrosopic current, I. If a series of measurements of I are taken, I_1, I_2 .... I_n, the variance, σ_I^2 is given by

$$\sigma_I^2 = \frac{1}{n} \sum_{j=0}^{n} (I_j - \bar{I})^2 \qquad (5)$$

and is usually calculated by computer. $\bar{I}$ is the mean current.

We assume that each individual channel that contributes to the macroscopic current passes a single-channel current which is either zero (when the channel is in a closed state) or i (when the channel is open). Since the mean value of the macroscopic current will be given by $\bar{I} = Nip$, where N is the total number of channels and p the channel open-state probability, it can be shown that the variance and the mean macroscopic current will be related by

$$\sigma_I^2 = \bar{I}i(1-p) \qquad (6)$$

Equation 6 shows that the variance will be maximal when $p = 0.5$, so that half the channels are open on average, and will be zero when p is either 0 or 1. Rearranging (6) gives

$$i = \frac{\sigma_I^2}{\bar{I}(1-p)} \qquad (7)$$

so that if $p \ll 1$, the single-channel current i may be measured by dividing the variance by the mean current. If the open-state probability is not so small, an alternative version of equation (7) may be used. Variance and mean current will also be related by

$$\sigma_I^2 = i\bar{I} - (\bar{I}^2/N) \qquad (8)$$

which yields a parabolic relation between variance and mean current. (The variance is maximal when $p = 0.5$ and is zero when p is 0 or 1). Experimental plots of variance against mean may be fitted to equation (8) to give estimates for i, the single-channel current, and N, the total number of channels (e.g. Sigworth, 1980).

Experimentally the variance may be measured when the current is in a steady state, for example in the presence of a constant low concentration of acetylcholine to activate endplate channels (Anderson & Stevens, 1973). Alternatively, the time course of the variance throughout a voltage pulse can be measured by ensemble averaging the currents recorded from many repetitions of the same pulse, and subtracting the average from each individual record (Sigworth, 1980). Finally, noise analysis methods can also be used to give information about channel kinetics (see e.g. Neher & Stevens, 1977).

We thank Professor Peter Stanfield for the computer program for simulation of ionic currents in squid axon.

References

ADAMS, D. J. & GAGE, P. W. (1979). Characteristics of sodium and calcium conductance changes produced by depolarization in an Aplysia neurone. *J. Physiol., Lond.* **289**, 143-161.

ASHFORD, M. L. J. (1990). Potassium channels and modulation of insulin secretion. In *Potassium Channels: Structure, Classification, Function and Therapeutic Potential*, (ed. N. Cook), pp. 181-231. Ellis Horwood.

ANDERSON, C. R. & STEVENS, C. F. (1973). Voltage-clamp analysis of acetylcholine produced end-plate current fluctuations at frog neuromuscular junction. *J. Physiol., Lond.* **235**, 655-691.

ARMSTRONG, C. M. (1969). Inactivation of the potassium conductance and related phenomena caused by quaternary ammonium ion injection in squid axons. *J. Gen. Physiol.* **54**, 553-575.

BYERLY, L. & HAGIWARA, S. (1988) Calcium channel diversity. In *Calcium and Ion Channel Modulation* (ed. A. D. Grinnell, D. Armstrong & M. B. Jackson), pp. 3-18. New York: Plenum.

CABANTCHIK, Z. I. & GREGER, R. (1992). Chemical probes for anion transporters of mammalian cell membranes. *Am. J. Physiol.* **262**, C803-C827.

CATTERALL, W. A. (1980). Neurotoxins that act on voltage-sensitive sodium channels in excitable membranes. *Ann. Rev. Pharmacol. Toxicol.* **20**, 15-43.

CONNOR, J. A. & STEVENS, C. F. (1971). Voltage clamp studies of a transient outward membrane current in gastropod neural somata. *J. Physiol., Lond.* **213**, 21-30.

COOK, N. S. & QUAST, U. (1990). Potassium channel pharmacology. In *Potassium Channels: Structure, Classification, Function and Therapeutic Potential*, (ed. N. Cook), pp. 181-231. Ellis Horwood.

DASCAL, N. (1987). The use of *Xenopus* oocytes for the study of ion channels. *CRC Crit. Rev. Biochem.* **18**, 317-387.

FOX, A. P., NOWYCKY, M. C. & TSIEN, R. W. (1987). Kinetic and pharmacological properties distinguishing three types of calcium currents in chick sensory neurones. *J. Physiol., Lond.* **394**, 149-172.

GILLESPIE, J. I. & MEVES, H. (1980). The time course of sodium inactivation in squid giant axons. *J. Physiol., Lond.* **299**, 289-307.

HILLE, B. (1984). *Ionic Channels of Excitable Membranes.* Sunderland: Sinauer.

HILLE, B. (1992). *Ionic Channels of Excitable Membranes.* 2nd Edition. Sunderland: Sinauer.

HODGKIN, A. L. (1958). Ionic movements and electrical activity in giant nerve fibres. *Proc. R. Soc. Lond. B* **148**, 1-37.

HODGKIN, A. L. & HUXLEY, A. F. (1952a). Currents carried by sodium and potassium ions through the membrane of the giant axon of *Loligo. J. Physiol., Lond.* **116**, 449-472.

HODGKIN, A. L. & HUXLEY, A. F. (1952b). A quantitiative description of membrane current and its application to conduction and excitation in nerve. *J. Physiol., Lond.* **117**, 500-544.

KOSTYUK, P. G. & KRISHTAL, O. A. (1984). Intracellular perfusion of excitable cells. (IBRO handbook series: Methods in the neurosciences; vol. 5). Chichester: Wiley.

MINTZ, M., VENEMA V. J., SWIDEREK, K. M., LEE, T. D., BEAN, B. P. & ADAMS, M. E. (1992). P-type calcium channels blocked by the spider toxin ω-Aga-IVA. *Nature* **355**, 827-830.

NEHER, E. & STEVENS, C. F. (1977). Conductance fluctuations and ionic pores in membranes. *Ann. Rev. Biophys. Bioeng.* **6**, 345-381.

PLUMMER, M. R., LOGOTHESIS, D. E. & HESS, P. (1989). Elementary properties and pharmacological sensitivities of calcium channels in mammalian peripheral neurones. *Neuron* **2**, 1453-1463.

PONGS, O. (1992). Molecular biology of voltage-dependent potassium channels. *Physiol. Rev.* **72**, S69-S88.

SHELTON, P. A., DAVIES, N. W., ANTONIOU, M., GROSVELD, F., NEEDHAM, M., HOLLIS, M., BRAMMAR, W. J. & CONLEY, E. C. (1993). Regulated expression of K^+ channel genes in electrically silent mammalian cells by linkage to β-globin gene-activation elements. *Receptors and Channels* **1**, 25-37.

SIGWORTH, F. J. (1980). The variance of sodium current fluctuations at the node of Ranvier. *J. Physiol., Lond.* **307**, 97-129.

STANFIELD, P. R. (1983). Tetraethylammonium ions and the potassium permeability of excitable cells. *Rev. Physiol. Biochem. Pharmacol.* **97**, 1-67.

TRIGGLE, D. J. & JANIS, R. A. (1987). Calcium channel ligands. *Ann. Rev. Pharmacol. Toxicol.* **27**, 347-369.

Chapter 4

Patch clamp techniques for single channel and whole-cell recording

DAVID OGDEN and PETER STANFIELD

1. Introduction

The *patch clamp* technique was first used by Neher and Sakmann (1976) to resolve currents through single acetylcholine-activated channels in cell-attached patches of membrane of frog skeletal muscle. The method they used (described by Neher, Sakmann & Steinbach, 1978) and subsequent refinements (Hamill, Marty, Neher, Sakmann & Sigworth, 1981) have led to techniques for high resolution recording of current in *excised membrane patches* in addition to those that remain *cell-attached*. Single channel recording yields information about unitary conductance and kinetic behaviour of ionic channels already partly investigated by classical voltage clamping and by noise analysis; it is also leading to the discovery of new classes of ion channel. Together with the method of whole-cell recording, which permits the application of voltage clamping to cells that are too small for microelectrode methods, patch clamp techniques also permit investigations of the physiological role of ionic channels in cells otherwise inaccessible to voltage clamp and to cells that are not electrically excitable.

2. Method for patch clamp recording

The principle of the method is to isolate a patch of membrane electrically from the external solution and to record current flowing into the patch. This is achieved by pressing a fire-polished glass pipette, which has been filled with a suitable electrolyte solution, against the surface of a cell and applying light suction. Providing both glass pipette and cell membrane are clean, a seal whose electrical resistance is more than 10 $G\Omega$ is formed. Under such conditions, the glass pipette and the cell membrane will be less than 1 nm apart.

A high seal resistance is needed for two reasons. First, the higher the seal

DAVID OGDEN, National Institute for Medical Research, Mill Hill, London NW7 1AA, UK
PETER STANFIELD, Department of Physiology, University of Leicester, Leicester LE1 9HN, UK

resistance, the more complete is the electrical isolation of the membrane patch. Secondly, a high seal resistance reduces the current noise of the recording, permitting good time resolution of single channel currents, currents whose amplitude is in the order of 1 pA.

Fig. 1 illustrates an equivalent circuit for the recording set-up. The fraction of the current flowing through the patch that will be collected by the recording pipette is

$$R_{seal}/(R_{seal} + R_{pipette})$$

where R_{seal} is the seal and $R_{pipette}$ the pipette resistance.

As Fig. 1 also illustrates, it is necessary that the patch from which recordings are made be small in area, compared with the area of membrane of the cell as a whole. In terms of Fig. 1, it is necessary that $R_{cell} \ll R_{patch}$ if single channel currents are not to alter the membrane potential of the cell significantly.

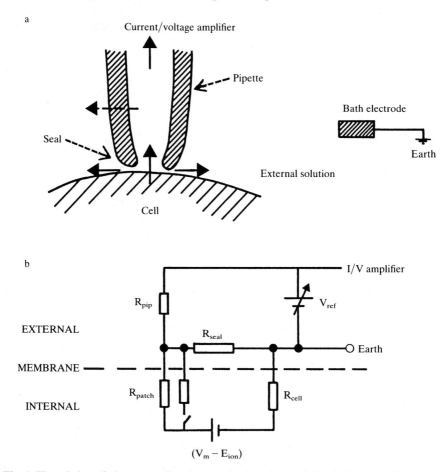

Fig. 1. The relation of pipette to cell and the equivalent electrical circuit during patch clamp recording. The opening of an ion channel is represented as the closing of the switch in Fig. 1b. The pipette resistance and seal resistance are in series between the amplifier and earth (the external solution) and the patch and cell resistances are in parallel with the seal resistance in this path.

The background noise level is also minimized by a high seal resistance. The variance of the current noise (in A^2) through a resistor (R, Ω) is related to the Johnson voltage noise due to the resistance, being given by:

$$s_i^2 = 4kTf_c/R$$

where k is Boltzmann's constant, T is temperature (°Kelvin), and f_c is the bandwidth (Hz) i.e. the low pass filter setting. Thus, for a 10 GΩ resistor at 20°C, the standard deviation of the current noise at 1 kHz will be 0.04 pA, but for a resistor of 100 MΩ it will be 0.4 pA. In the recording situation used in patch clamp (Fig. 1), resistor current noise will depend on all the resistive paths to ground from the amplifier input, decreasing as resistance increases. Since the patch resistance is high (approximately 100 GΩ or more), the low seal resistance predominates and will result in noise that will prevent good resolution of currents smaller than 4 or 5 pA. Such was the situation in the earliest patch clamp experiments, where the seal resistance was less than 100 MΩ. In spite of this difficulty, information was obtained about acetylcholine-activated channels (Neher & Sakmann, 1976), about the blocking effects of local anaesthetics on such channels (Neher & Steinbach, 1978), and about glutamate-activated channels of insect muscle (Patlak, Gration & Usherwood, 1979; Cull-Candy, Miledi & Parker, 1981). The achievement of 'gigaseals' (>1 GΩ, Neher, 1981; Hamill et al. 1981), however, radically improved the quality of recording and made it possible to study channels of lower unitary conductance.

Such seals are generally achieved in an all-or-nothing fashion and result in a dramatic improvement in signal-to-noise ratio. The conditions that appear to be required for the formation of a gigaseal are the following.

First, the surface membrane of the cell used must be clean and free of extracellular matrix and connective tissue. Cells in tissue culture are often preferred; adult cells generally must be cleaned enzymatically or mechanically.

Secondly, solutions should be free of dust and of macromolecules such as the components of serum in tissue culture media. Solutions are filtered using 0.2 μm filters (of the detergent-free type). Cell cultures are washed several times to remove serum.

Thirdly, the pipette tip is clean, often by fire-polishing.

Fourthly, during the period just prior to seal formation, a small positive pressure is applied to the pipette to generate an outflow of solution from the pipette tip and so keep it free of debris.

Loose-patch clamp

In certain situations, however, a low seal resistance can be an advantage. Such a method is the so-called 'loose-patch' clamp (Almers, Stanfield & Stühmer, 1983), developed to allow measurement of the distribution of ionic channels over cell membranes and, making use of the low resistance seal and the resulting gap that will exist between membrane and recording pipette, to permit investigation of the lateral mobility of ionic channels. Here large diameter (approximately 10 μm) pipettes are used to collect current from many ionic channels at the same time. Currents that are collected by the pipette must be multiplied by a 'seal factor' ($\{R_{seal}+R_{pipette}\}/R_{seal}$) to correct for the fraction of the membrane current that flows to ground through the resistance of the seal.

The method has been used to investigate sodium channels of skeletal muscle, channels that are activated by stepping membrane potential. Changing the voltage of the inside of the pipette to achieve this change in membrane potential will result in large currents flowing though the resistance of the seal to the grounded bath solution. This leakage current must be subtracted. Additionally, the voltage applied as V_{ref} (in Fig. 1) must be bigger than the desired change in membrane potential by the seal factor ($\{R_{seal}+R_{pipette}\}/R_{seal}$) to correct for the division that occurs between pipette and seal resistances. These corrections can be applied by a mixture of analogue and digital means. If a single pipette is used, no correction can be made for an additional source of error, activation at an uncontrolled voltage of the ionic channels in the membrane under the rim of the pipette, where the seal is formed. Concentric, double-barrelled pipettes have been used to remove this source of error (Roberts & Almers, 1984).

Gigaseal patch clamp

Cell-attached and excised patches; whole-cell recording.

Fig. 2 summarizes the main modes of gigaseal recording. Much work is done using patches in the cell-attached mode, but the resting potential of the cell is not known and neither intra- nor extracellular ionic concentrations can be changed easily. For these reasons, it is sometimes essential to work using a cell-free mode, with excised or ripped-off patches. There are two kinds:

Inside-out - made by pulling the membrane patch off the cell into the bath solution.

Outside-out - made by applying suction to destroy the membrane isolated by the patch pipette and then pulling the pipette away from the cell. The membrane should reseal to give a patch of membrane whose intracellular face is in contact with the pipette solution.

Whole-cell recording is achieved by destroying the membrane patch using suction so that the cell, whose interior then comes into contact with the solution in the pipette, may be voltage or current-clamped. The cell contents equilibrate over time with the solution within the pipette (Fenwick, Marty & Neher, 1982). Further details of these modes are given below.

Giant patches and measurement of pump currents

A procedure has been described for making gigaseals with large diameter pipettes (10-25 μm) and has been used to study electrogenic pump and exchange transport where the unitary currents are small (Hilgemann 1989, 1990; Collins, Somlyo & Hilgemann, 1993). It uses one of a variety of lipophilic 'glues' (eg. 40/60 w/w Parafilm/mineral oil, α-tocopherol) to improve seal resistance and produces large areas of electrically isolated membrane. Currents generated by ion exchange processes such as the Na^+/Ca^{2+} exchanger, the sodium pump, and many exchangers driven by the electrochemical gradient on Na^+, are too small to resolve as unitary

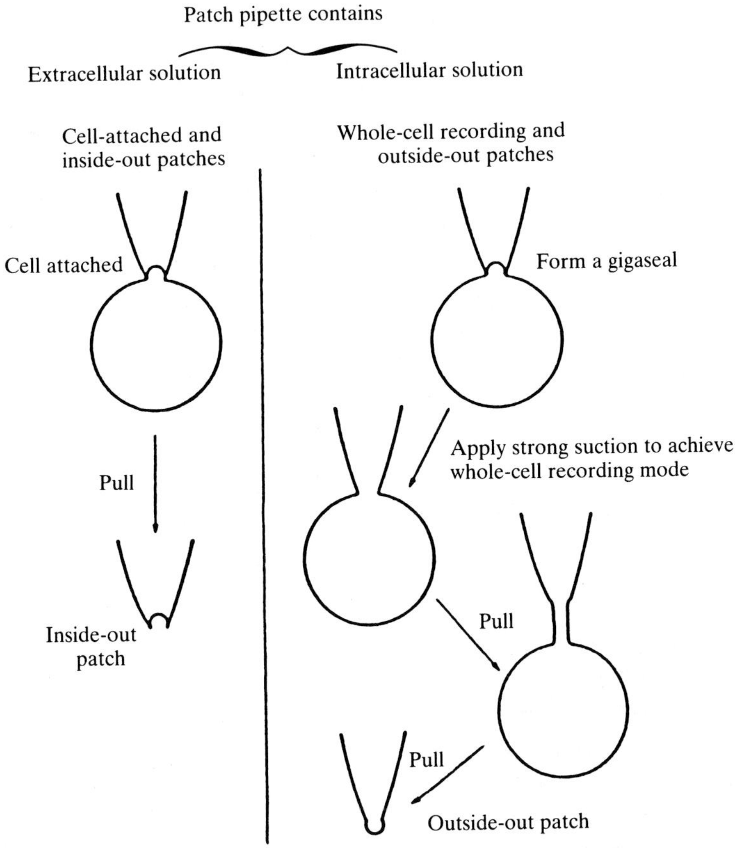

Patch pipette contains

Extracellular solution   Intracellular solution

Cell-attached and   Whole-cell recording and
inside-out patches   outside-out patches

Cell attached   Form a gigaseal

Pull   Apply strong suction to achieve
                                                       whole-cell recording mode

Inside-out   Pull
patch

                                                       Pull

                                          Pull   Outside-out patch

Fig. 2. Diagram illustrating the methods of making cell-attached and inside-out patches (left hand panel) and whole cell and outside-out patches.

currents and in normal size patches are not present in sufficient number to generate a measurable current. In large patches, currents due to exchangers and pumps are well resolved. They are used specifically in the inside-out configuration to permit changes at the cytosolic surface that modify the rate and characteristics of transport measured as the pipette current.

3. Instrumentation of single channel recording

A diagram of the essential features of the amplifier used in the headstage of a patch clamp is shown in Fig. 3. Essentially, this amplifier is a current-to-voltage converter which has a high gain, owing to the large feedback resistor R_f, and which is arranged so that the potential inside the pipette, V_{ref}, may either be held at a steady level or changed in a step-wise fashion. A description of the patch amplifier is given in Chapter 16, but the following points may be considered here.

(i) The input of the amplifier has JFET transistors of low leakage current and noise

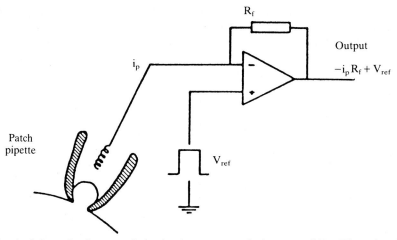

Fig. 3. Schematic diagram of the headstage current/voltage amplifier. The gain (V_o/i_p, mV/pA) is set by the feedback resistor R_f,

$$V_o = -R_f \cdot i_p + V_{ref}$$

V_{ref} is composed of the sum of V_{hold}, V_{null} and $V_{command}$, and is subtracted from the output at a later stage.

level. The experimenter needs to take care to avoid damage to these from static potentials, for instance by simultaneously touching an earth when connecting a pipette.

(ii) The *feedback resistor* determines the sensitivity, the background noise level and range of current measurement. If i_p is the patch membrane current and ($V_{out}-V_{ref}$) the voltage output,

$$-i_p R_f = V_{out} - V_{ref}$$

The feedback resistor contributes current noise which decreases inversely with the resistance. As described above, the variance of current noise at bandwidth f_c is given by

$$s^2 = 4kTf_c/R_f$$

It follows that for low noise, high gain recording, R_f should be high, say 50 GΩ.

However, since V_{out} will swing by a maximum of only ±12 V, the output of the headstage will be saturated if i_p exceeds 240 pA. In whole-cell recording, currents may often exceed this level. The value of R_f must therefore be chosen to suit the experiment, either by prior selection of a suitable headstage or by using an amplifier that enables the experimenter to switch the value of R_f remotely. Several commercial instruments possess the latter facility. Alternatively, a good compromise may be achieved with a fixed 5 or 10 GΩ resistor for R_f.

Capacitor feedback. The noise associated with the feedback resistor is eliminated by using instead a capacitor to feedback current to the inverting input, producing an output that is the integral over time of the pipette current. To correct for this the output is taken via a differentiator stage. The improvement in signal to noise is about

30% when noise from other sources is minimised, a worthwhile improvement for high resolution single channel recording. The disadvantage is the need to discharge the voltage on the capacitor as the output voltage limit of the amplifier is approached, producing a reset transient which, although brief (50 μs) may occur frequently with large standing currents such as those encountered in whole-cell recording.

(iii) The potential in the pipette is equal to V_{ref}. This potential is set for zero current by offsetting electrode potentials at the beginning of the experiment. This can be done either manually or by using the 'search mode' or 'tracking mode' of the patch clamp amplifier, which uses an integrator to keep the current at zero, adjusting V_{ref} accordingly. Once a high resistance seal is obtained, and the amplifier is switched to its voltage clamp mode, V_{ref} may be changed without causing large currents between pipette and bath, (such as those that occur with the 'loose-patch' method). Changing V_{ref} will change the patch membrane potential. Most patch amplifiers possess a 10-turn potentiometer, labelled V_{HOLD} or V_{PIP}, which allows the holding potential of the patch to be altered. Pulses from an external source ($V_{COMMAND}$) may be used to change V_{ref} in a step-wise way. Generally, since noise applied to the headstage with the command signal appears in the current trace, the command voltage (and the noise applied with it) is divided 10- or 50-fold in the headstage, requiring a command pulse 10- or 50-fold larger than the pipette potential but with better signal/noise.

(iv) Fast changing commands, such as the leading edges of rectangular pulses, give rise to large currents due to charging stray capacitance associated with pipette and cell. These may saturate the amplifier and must be offset by adjustment of compensation circuits. Separate compensation is usually provided for 'fast' (primarily stray) and 'slow' (cell) capacitance.

(v) Particular attention needs to be paid to earthing and screening. The signal earth point of the amplifier should attach directly to the principal earth point of the set-up. The earth (or ground) socket on the headstage is connected within the cable to the signal earth in the patch clamp amplifier and is for connexion of the bath electrode. All metal surrounding the bath (baseplate, manipulators, dish holders, etc) and the microscope stage, nosepiece, objectives and condenser should have low resistance ($<0.5\Omega$) connexions to the principal earth point. The case of the headstage amplifier may be connected to V_{ref} to reduce charging transients during voltage steps. The case should therefore be isolated from earth when it is mounted on the micromanipulator. V_{ref} may be available at a socket on the headstage with the intention that it might be used to minimise capacity current transients by connection to screens near the pipette (a 'guard' potential).

(vi) It is important to remember the convention for membrane current, viz. outward flux of cations (or inward flux of anions) is positive. The output polarity of a patch clamp may be $+$ or $- R_f i_p$ as determined by the internal circuitry after the headstage and therefore may be correct for whole-cell recording (and inverted for cell-attached), as is the case in most commercial instruments, or may be correct for cell-attached and incorrect for whole-cell recording (e.g. the Biologic amplifier has this arrangement).

(vii) The following instruments are particularly useful for patch clamp work: a 4- or 8-pole lowpass filter, normally with a Bessel characteristic; an FM or digital (VTR or DAT) tape recorder; a computer interface and software for on-line voltage pulses and data analysis, and a wide bandwidth chart recorder such as a UV, light pen (Medelec) or electrostatic type.

4. Patch pipettes

Choice of glass

The very high resistance of the seal between cell membrane and glass pipette means that some hydrophobic chemical interaction occurs between the two. The chemical composition of the glass may therefore influence the ability to form seals, although the comprehensive review by Rae & Levis (1992) of the properties of glasses of different composition indicates that no firm conclusions can be drawn concerning the ability of different types of glass to form seals. Other considerations that need to be taken into account in choosing electrode glass are as follows:

(i) The ability to form pipettes whose tips have an appropriate size and taper, factors that may influence the area of membrane isolated in a patch or the series resistance in whole-cell recording,

(ii) The degree to which background noise needs to be reduced.

(iii) Glass is doped with heavy metals as minor components to reduce the melting point. Heavy metal ions may leach into the pipette solution and modify channel properties.

The composition and properties of different types of glass are discussed by Rae & Levis (1992).

The following types of glass are commonly used.

Soft (low melting point) glass. This type of glass was initially used for patch clamp recording because it is easily pulled and shaped at relatively low temperatures, producing large aperture pipettes with blunt tips. It is not so commonly used to make recording pipettes but can be useful for making large aperture tips for cleaning tissue slices or drug application. Some soft glasses contain lead or barium, and haematocrit tubes are often made of this type of glass.

'Pyrex' borosilicate glass (Corning type 7740). Standard microelectrode glass is usually of this type, which is also most commonly used for patch pipettes. Pulling and polishing require high temperatures, usually orange heat of a platinum wire. Some argue that the sealing properties are better than those for soft glass (see Sakmann & Neher, 1983) although Hamill *et al.* (1981) state that seals are less stable with glass of this type. Low resistance pipettes (tip diameter 1-3 µm; resistance 1-5 MΩ) are most easily pulled from thin walled glass. Thick walled pipettes generally seal better and have lower noise levels, because of the lower electrode capacitance (see below). The facility to back-fill pipettes made from electrode glass that has a fused internal filament is particularly convenient.

Glasses for low-noise single channel recording. The current noise in high resolution recording is influenced by the type of glass used because of differences in the resistive and capacitative properties of the thin glass wall at the tip. These properties are reviewed by Rae & Levis (1992). The most significant points noted are (i) the large improvement in noise levels produced by Sylgard or other coating of the pipette shank with all types of glass and (ii) the considerably superior properties of quartz pipettes conferred by the low leakage conductance across the glass at the tip. Quartz softens at temperatures too high for conventional electrically heated filaments and pipettes are pulled either with a gas/oxygen flame or by a laser heated puller.

Effects of glass on channel properties. Two studies show that the type of glass used to make pipettes can affect the properties of channel recordings. Cota & Armstrong (1988) showed that two types of low melting point glasses soda or potash glasses produced additional fast inactivation of K channel currents when compared with channel properties determined with a borosilicate glass. This was subsequently shown to be due most likely to Ba ion release from the soda glass (Copello *et al.* 1991). Furman & Tanaka (1988) compared the properties of cGMP-activated channels in photoreceptor membrane recorded with pipettes made from one of six types of glass. They found differences in the current/voltage relation that could be reversed when EDTA or EGTA (which bind heavy metals with high affinity) were included at mM concentration in the pipette solution with all but one of the glasses tested. These and other reports indicate the necessity to avoid those types of glass that have been shown to affect channel properties and to test for contamination with chelators of heavy metals and by comparing different glass types before use on channels of well defined properties.

A second potential source of contamination should be noted. This arises from the use of stainless steel needles for backfilling pipettes, which are thought to be a source of polyvalent ions. It has also been reported that the hardeners in some plastic syringes can block nicotinic receptor/channels.

Manufacture and fire polishing of pipettes

Fig. 4 illustrates the double pull procedure used to make patch pipettes. With a conventional upright puller, the solenoid is disconnected and gravity alone provides the pull. Consistency in pipette formation can be increased if a digital voltmeter (set to V_{rms}) is used to monitor the voltage across the heating element.

Fire polishing of pipettes is done with a heated Pt wire, mounted on the stage of a compound microscope, with stage movements or micromanipulators used to bring pipette tip and wire close together. High magnification (200-400×) good quality optics makes this step easier to carry out.

The tip diameter of pipettes may be estimated before and after fire polishing by measuring the pressure needed to expel air bubbles from the tip into clean methanol. This pressure may be measured as the volume change applied to a 10 ml syringe connected to the pipette. As a guide, starting at 10 and using the scale on the syringe, a bubble number of 5 or less indicates pipettes of >10 MΩ resistance suitable for cell-attached recording, of 6 or more low resistance pipettes suitable for whole cell recording. This procedure is useful particularly when setting up a pipette puller.

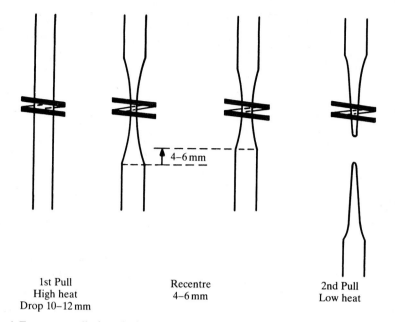

4–6 mm

1st Pull	Recentre	2nd Pull
High heat	4–6 mm	Low heat
Drop 10–12 mm		

Fig. 4. Two-stage pull of patch pipettes.

Coating pipettes to reduce capacitance

Pipettes used in single channel recording are usually coated to thicken their walls and reduce capacitance to the bath solution. There are two reasons for wishing to minimize this stray capacitance. First, high frequency noise levels are reduced and secondly, as mentioned above, capacity currents associated with stepping the potential of the inside of the pipette are also reduced.

The r.m.s. current noise associated with stray input capacitance is given by

$$s_i = 2\pi f_c . C . s_V$$

where s is the r.m.s. noise for current (i) and voltage (V), f_c is the bandwidth, and C the total capacitance from input to earth. Current noise increases in proportion with voltage noise as the recording bandwidth is increased and also with the capacitance.

The current needed to charge the stray capacitance on changing the pipette potential is given by

$$i = C . dV/dt,$$

and will be particularly large for a fast potential step. If at the leading or trailing edge of such a step $dV/dt = 100$ mV/20 µs, the current needed to charge a stray capacity of 10 pF will be 50 nA. Since a headstage with a 50 GΩ feedback resistor will be saturated by a current that exceeds 0.24 nA (see above), saturation would occur and important information about channels that are rapidly activated by the voltage step would be lost. Generally the *fast* capacity compensation circuits of patch amplifiers have a range too small to compensate fully pipette capacities larger than 10 pF.

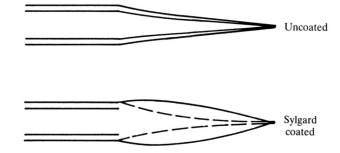

Fig. 5. Patch pipette before and after coating with Sylgard and curing.

Most of the stray pipette capacitance arises across the pipette wall between the pipette and the bath solution. It can be reduced by coating the pipette with a thick layer of Sylgard 184, an inert, hydrophobic, translucent elastomer resin rapidly cured by heating. The coat both thickens the wall of the pipette and reduces the area of capacitance coupling by preventing the bath solution creeping up and wetting the pipette. Coating should run from close to the tip (preferably closer than 250 µm) and continue up the pipette well past the beginning of its taper. The procedure for coating pipettes and curing the Sylgard is illustrated in Fig. 5. As an alternative to Sylgard coating, pipettes can be dipped with backpressure before filling into a molten mix of 'Parafilm' in mineral oil (or some other molten wax) or after filling immersed in a silicone coating such as 'Sigmacote'. Both these procedures prevent fluid contact up the shank of the pipette and thicken the wall at the tip.

It is best to observe two additional measures to reduce stray capacity. First, the depth of bath solution in contact with the pipette wall should be minimized. Secondly, the pipette holder should be kept free of excess fluid; often it is helpful to dry the holder with compressed air or N_2 before inserting a new pipette. A discussion of methods for low-noise recording is given by Rae & Levis (1992).

5. Formation of a gigaseal

As the pipette is advanced through the surface of the bath solution, slight positive pressure is applied to the inside of the pipette to keep the tip free of contamination, either by mouth or from a manometer (a water-filled U-tube that permits fine pressure adjusment by changing the pressure head of water). The current is set to zero, either by altering V_{ref} manually with the V_{null} control or by using the search mode of the amplifier. The pipette resistance is then monitored by measuring the current in response to 0.1 to 0.5 mV step-changes in V_{ref}. Contact with the cell is indicated by a rise in pipette resistance. *Gentle* suction (5-20 cm of water) is now applied to the pipette, when a seal should be formed. Seal formation may occur immediately, or it may take some while to happen. It may be helped by polarisation of the pipette to around −40 mV once a high resistance has developed.

A fresh pipette should be used for each attempt to make a seal.

The process of seal formation can be observed with good microscope optics. Gigaseals may form immediately on contact with the cell surface with very little negative pressure, particularly with high resistance pipettes. In these cases the profile of the membrane is flat within the pipette tip. More often an Ω-shaped region of membrane is drawn into the pipette tip and the seal forms initially slowly and suddenly achieves a high value. Photographs are given of membrane-pipette seals by Sakmann & Neher (1983). Milton & Caldwell (1990) report the formation of a bleb of membrane lipid in the pipette tip based on microscope observations of seal formation and propose a role for the bleb in forming gigaseals. Sokabe & Sachs (1990) give a microscopic analysis of the pressure/area relations of membrane patches.

Giant patches

Hilgemann (1989, 1990, see Collins *et al.* 1992) reports the use of various hydrophoboic mixtures to facilitate the formation of seals with large tipped pipettes. These include viscous Parafilm/oil mixtures and α-tocopherol. With minimal suction, these agents promote the formation of flat stable seals desirable for access to the inner surface of inside-out patches.

Liquid junction potentials

Where differences in ionic composition exist between pipette and bath solution, a correction must be applied for the junction potential across the boundary between the two solutions. This potential arises from differences in concentration and mobility of the ion species in the two solutions and maintains electroneutrality across the boundary between the two solutions. The magnitude depends on the relative mobility and concentration gradient of ions diffusing in the two directions and, depending on the solutions used, may be 2-12 mV in physiological solutions. The error arises because the potential will be present when pipette current is zeroed before sealing, but when the seal is made the concentration gradients form part of the driving potential for ion movement, leaving the applied offset potential still imposed in the pipette.

With reference to Fig. 3, the initial sequence of steps made during patch recording are as follows. (i) Zero current output of the amplifier is checked with the pipette out of the bath - this is usually zero volts. (ii) The pipette is put in the bath solution and an offset current flows as a result of asymmetries in the AgCl half cell potentials, the junction potentials across the pipette tip to bath and bath into the reference electrode salt bridge, and any voltage on the amplifier V_{ref}. For present purposes, suppose that the AgCl potentials are equal and opposite and the reference junction potential is zero, and so do not contribute. The current flowing into the pipette is driven by the difference between bath potential and pipette potential and is set to zero by adjusting V_{ref} (which sets the pipette potential with respect to the bath). In most patch amplifiers, this is done with a 'Null' potential control that changes V_{ref} without registering on the voltage monitor. This value of V_{ref} is taken as zero holding potential and is the zero current potential of the pipette with respect to bath potential. However, by convention the polarity of the junction potential is bath minus pipette, opposite to the polarity of the V_{ref} potential applied in the pipette. (iii) A seal is formed and the liquid junction potential disappears, leaving the pipette potential displaced from true zero by an amount equal to the initial pipette-bath potential difference, opposite in polarity to the junction potential. In a cell-attached or inside-out recording, the potential is on the outside of the membrane and the true transpatch potential is the holding potential minus the initial pipette-bath potential (i.e. plus the conventional junction potential). For whole-cell recording on breaking through, a new junction potential appears between the pipette and cytosol immediately on patch rupture and disappears over the first minute or so as the cell is dialysed with pipette solution, leaving the initial pipette-bath potential added to the holding potential (i.e. the conventional junction potential should be subtracted).

The error due to the junction potential can be demonstrated and quantified fairly easily. However, it is necessary to have a reference junction that remains at zero when the composition of the bath is changed. This can be done with a 2-3 M KCl solution in the reference salt bridge and stable AgCl wire or pellet provided the junction between salt bridge and bath is continuously renewed by slow outflow of KCl; for example from a blunt pipette some distance from the recording site (see Neher, 1992; also Chapter 1). Zero junction potential is set at the recording pipette by having the same solution in the bath as would be in the pipette during an experiment. The potential is measured at zero current, either in voltage clamp or current clamp, and set to zero. The bath is changed to another solution (e.g. external solution) and the change in zero current potential measured. This is the excess potential that would be present in the pipette after seal formation and should be subtracted from the holding potential in cell attached or added in whole cell to give the real membrane potential.

This problem is reviewed, with particular reference to patch clamp recording, by Barry & Lynch (1991) and by Neher (1992). The polarity has been given the convention bath-pipette, opposite to that determined in the experiment above. The junction potential can be calculated from the Henderson equation (given e.g. in Barry & Lynch, 1991) if the activities and mobilities of the ions are known. As an example, in the simple but common case where the bath contains a 150 mM NaCl solution, while the pipette has an equal concentration of KCl, the junction potential can be predicted from:

$$E_{j.p.} = (RT/F).\ln[(u_K + u_{Cl})/(u_{Na} + u_{Cl})]$$

where u_{Na}, u_K, and u_{Cl} are the mobilities of Na^+, K^+, and Cl^- and $E_{j.p.}$ is the junction potential, bath relative to pipette. $E_{j.p.}$ has in this case a calculated value of +4.3 mV, so the pipette will have an excess potential of −4.3mV with respect to the bath. In whole-cell recording, zero mV imposed in the pipette will be −4.3 mV; in cell-attached recording zero will be +4.3 mV. In cases where solutions differ more, particularly with Cl^- ion replacement, the junction potential can be calculated with equations and ion mobilities given by Barry & Lynch (1991) or taken from the measured values given by Neher (1992).

Further problems arise when junction potentials occur at the reference/bath solution junction. It is usual to have a salt bridge of NaCl bath solution between the reference AgCl electrode and bath. This has no junction potential into the bath unless the external solution is changed, when a potential may arise and add to the holding potential. Procedures for measuring liquid junction potentials and discussion of more complex effects seen with external cell perfusion by dissimilar solutions are given by Neher (1992).

A second, but less common problem, is that salts of low solubility product (e.g. calcium or barium phosphates, fluoride or sulphate) may precipitate at the interface of pipette and bath solutions.

6. Choice of preparation

The choice of cell to use for single channel work depends ideally on the experiment that is envisaged. In practice, the choice of preparation can hinge on whether its cells readily make seals with glass and on the density of ionic channels in the membrane. Preparations that are used include cells from adult tissues, cultured cells, channels in membranes of native lipid, and channels reincorporated into artificial bilayers as large vesicles.

Adult cells

These have to be cleaned of connective tissue and extracellular matrix, using enzymes such as collagenase and proteases. When dissociating tissues, the presence of DNAse helps prevent clumping; however, individual protocols for enzymatic dispersal are empirically determined. In CNS tissue slices, the cell surface can be cleaned by irrigation from a 'cleaning' pipette, and this procedure is increasingly applied to other tissues.

The channel density is often very high in adult tissue - sodium channels may have a density of 100 μm^{-2} or more - making it impossible to isolate a single or small number of channels.

Cultured cells

These provide the most widely used preparation for single channel recording. Channel density is often lower than in adult cells, and the membrane may have less extracellular matrix. Commonly used *primary cultures* include myotubes and explant neuronal cultures grown from new-born rats or mice, from *Xenopus* larvae and from invertebrates. Various cell lines which provide convenient models for physiological processes are also widely used.

Channels in native lipid membranes

Large liposomes or vesicles have been developed using freeze/thawing of microsomes (Tank, Miller & Webb, 1982) dehydration/rehydration procedures (Keller & Hedrich, 1992) or KCl and water loading of cells, coupled with protease treatment (Standen, Stanfield, Ward & Wilson, 1984).

Reincorporation

Lipid membranes can be formed on the tips of patch pipettes and channel reincorporated into these bilayers from microsomes (Coronado & Latorre, 1983 see Chapter 5). Such a method is particularly useful in the investigation of channels in specific membrane fractions. For example, this and related methods can be used to investigate channels in membranes of the transverse tubular system of muscle fibres or in intracellular membrane systems such as sarcoplasmic reticulum and endoplasmic reticulum.

7. Cell-attached recording

This mode of recording is used to study single channel currents. The resolution of such unitary currents should be optimized by observing the precautions detailed above for reducing stray capacitance. Good mechanical stability and freedom from vibration are needed to keep cell-attached patches for long enough to record sufficient data.

Activation of channels operated by binding of transmitters may be produced by having a suitable concentration of transmitter or agonist in the patch pipette. Opening of voltage-activated channels may be achieved by stepping the potential inside the pipette. Often the pipette solution will have the same ionic composition as the bath solution; substitution of ions may be made for studies of permeability properties of channels.

The membrane potential of cell-attached patches (V_{patch}) is determined by that of the cell interior (V_{cell}) as well as by the pipette (V_{ref}). The membrane potential is therefore:

$$V_{patch} = V_{cell} - V_{ref}$$

Generally, V_{cell} is not known. Sometimes, for example in large cells such as skeletal muscle fibres, it is possible to record this potential with an impaling microelectrode. Suitable analogue means may be used to subtract it from V_{ref}, so that V_{patch} is the quantity monitored. Breaking the membrane patch at the end of the experiment with strong suction, so that the pipette is in contact with the cell interior, can also be used to obtain an indication of V_{cell}.

Examples of currents through nicotinic receptors, activated by acetylcholine, are shown in Fig. 6. Fig. 7 shows unitary potassium currents through delayed rectifier channels of skeletal muscle, activated by a depolarizing step. In these records, outward membrane currents are shown as upward deflections since they are, by convention, positive quantities. Most commercial patch clamp amplifiers are

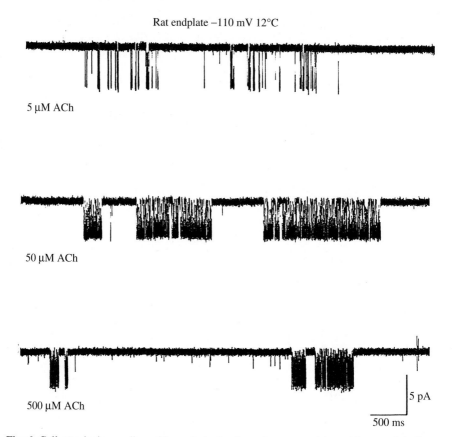

Fig. 6. Cell-attached recording of inward single channel currents activated by acetylcholine in the pipette solution at the rat skeletal neuromuscular junction. Potential −110 mV. 12°C. Bandwidth 4 kHz. Note the grouping of channel openings into bursts and clusters and the increase in open probability within clusters as the ACh concentration was increased. N. K Mulrine and D. Ogden.

arranged so that current flowing into the pipette (outwards across the membrane of cell-attached patches) produces a negative output. For clarity in description of their experiments, patch clampers should be careful to use the correct convention for membrane currents and membrane potentials.

Recording from expressed channels. Ion channels modified by molecular biological procedures and expressed in cell membranes are used to study the relation between structure and function. They can be transfected in cell lines or expressed in *Xenopus* oocytes by injection of mRNA into the cytosol, or of DNA into the nucleus. Procedures for recording from cell lines are the same as those described above. Recording from oocytes is technically more difficult because of the presence of accessory follicular cells and a vitelline membrane which must be removed to gain access to the plasmalemma. Procedures for treating oocytes for patch recording are described by Methfessel *et al.* (1986).

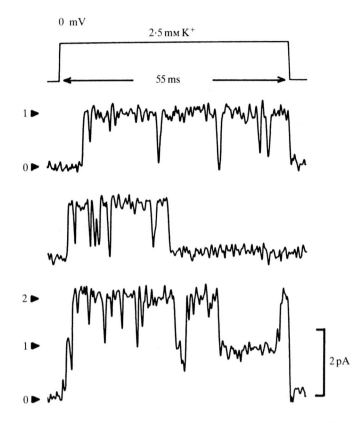

Fig. 7. Single channel records of potassium channel currents in skeletal muscle fibre activated by a 55 ms depolarization of the membrane potential from -100 mV to 0 mV. Cell attached recording. Note three levels of current corresponding to simultaneous opening of 0, 1 and 2 channels. Reproduced from Standen, Stanfield & Ward (1985), with permission.

8. Cell-free, excised patches

Inside-out patches are made from the cell-attached configuration as indicated in Fig. 2. Pulling off a membrane patch often results initially in the formation of a vesicle of membrane in the pipette tip. The outer face must be broken open, which may be done by briefly taking the membrane through the bath solution/air interface; by exposure to a low Ca^{2+} solution; or by momentarily making contact with a droplet of paraffin or a piece of cured Sylgard.

Once the patch is formed, channels activated or modulated by substances applied to the cytoplasmic membrane face may be studied. Such studies have included investigations of the effects of Ca^{2+} (Marty, 1981; Barrett, Magleby & Pallotta, 1981; Colquhoun, Neher, Reuter & Stevens, 1981); of catalytic subunit of protein kinase (Shuster, Camardo, Siegelbaum & Kandel, 1985); and of ATP (Noma, 1983; Spruce, Standen & Stanfield, 1985).

Outside-out patches are formed after breaking into the cell with the procedure indicated in Fig. 2. Such patches have been used by a number of workers for the study of the pharmacology of channels activated by transmitters or hormones (Hamill, Bormann & Sakmann, 1983; Gardner, Ogden & Colquhoun, 1984; Nowak, Bregestovski, Ascher, Herbert & Prochiantz, 1984; Cull-Candy & Ogden, 1985). This configuration of the cell-free mode often has the disadvantage of a higher level of background noise, probably because of a lower resistance of the seals achieved and because of the larger area of membrane in the patch. Further it has been reported that channel kinetics may differ when recorded in outside-out, excised patches from those found in cell-attached recording (Trautmann & Siegelbaum, 1983; Fernandez, Fox & Krasne, 1984).

Fast concentration changes. To study neurotransmitter-activated channels, it is important to be able to apply the neurotransmitter on a timescale similar to that encountered in the synapse, within 1 ms. This can be achieved with outside-out patches because of the very narrow unstirred layer of solution adjacent to a patch when compared to that surrounding a whole cell, where solution changes require tens of milliseconds. Methods for fast, submillisecond, concentration changes have been described by Maconochie & Knight (1989) and Lui & Dilger (1991) utilising electronic valves, and with piezo driven stepping of a stream of solution onto the pipette tip by Dudel, Franke & Hatte (1990).

9. Whole-cell recording

Whole-cell recording is achieved by rupturing the patch of membrane isolated by the patch pipette, which brings the cell interior into contact with the pipette interior. The process of whole-cell recording is as follows.

After forming a gigaseal, the fast capacity transients associated mainly with pipette capacitance to the bath are compensated. Next, the potential of the pipette interior (V_p) is adjusted to a level similar to the anticipated membrane potential of the cell. Rupture of the membrane patch is achieved by applying strong suction or sometimes brief voltage transients, and the process is conveniently monitored by applying test pulses of 1 to 5 mV amplitude. Rupture is indicated by the sudden appearance of large capacity transients at the leading and trailing edges of the pulse. The sequence of changes is shown in Fig. 8.

Series resistance errors in whole cell recording

The equivalent circuit used to represent the whole-cell recording condition is illustrated in Fig. 9. Current in the pipette, defined for whole-cell recording as positive flowing from pipette to cell, flows first through a resistance in series with the cell membrane to represent the pipette tip. If this series resistance is R_s and the pipette current is i_p, a potential difference pipette − cell of amplitude $i_p R_s$ results. This potential needs to be subtracted from the command potential to yield the cell potential. The series resistance compensation present in commercial patch clamps allows a fraction of $i_p R_f$ to be added to the command to partly correct the error.

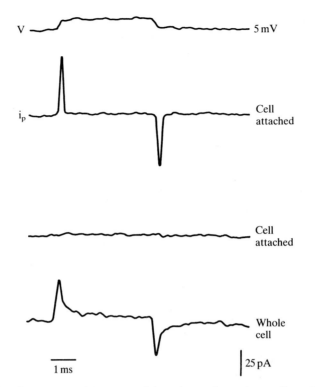

V ⎯⎯⎯⎯⎯⎯⎯⎯ 5 mV

i_p ⎯⎯ Cell attached

⎯⎯⎯⎯⎯⎯ Cell attached

⎯⎯ Whole cell

1 ms | 25 pA

Fig. 8. (a,b) Transient currents due to stray (pipette) capacitance in a cell-attached record before (a) and after (b) cancellation with C_{fast} and τ_{fast}. (c) Transients due to cell capacitance in a whole cell recording with 5 mV rectangular input to V_{ref}. 2 kHz bandwidth. The cell capacitance was 4 pF.

The pipette current has two components, one ionic ($=V_c/R_c$) and the other capacitative ($=C.dV_c/dt$). The time course of the change in the potential of the cell, $V_c(t)$, from its initial value $V_c(0)$ is given by:

$$V_c(t) = \{V_c(0) - V_c(\infty)\}\{1 - e^{-t/\tau}\}.$$

V_c changes on an exponential curve of time constant

$$\tau = C.R_sR_c/(R_s+R_c),$$

usually approximated well by $\tau=C.R_s$ because $R_c \gg R_s$.

This time constant characterizes the response of the voltage clamped cell not only to applied voltage steps, but also to currents originating in the cell membrane, which are therefore effectively low pass filtered with a bandwidth

$$f_c = 1/(2\pi\tau).$$

The time course of the current following a sudden change of V_p, also shown in Fig. 9, has a declining exponential timecourse. Initially current flows mainly into the capacitance. Since $V_c(0)$ is steady at the beginning of the step in V_p, the initial pipette current flowing into the cell discharges the capacitance and is given by $i_p(0)=V_p/R_s$. The pipette current declines exponentially with $\tau=C.R_s$ to a steady level. The size and time constant of the error voltage due to series resistance is minimized simply by keeping R_s as small as possible by using low resistance pipettes. If the last condition is kept (and it may be expected to hold in most conditions), $\tau=R_sC$ and for small currents $V_c(\infty)=V_p$.

When using whole-cell recording, it is important to know the errors that arise from series resistance and cell capacitance. These quantities may be estimated from the current transients shown in Figs 8 and 9. Provided $R_s \ll R_c$, the area of the transient, which gives the charge on C, is equal to $C.V_p$, while the time constant of the decline is given by $\tau=R_sC$. The area may be estimated by digitizing the current record and integrating, the time constant by curve fitting. Estimates of $R_s = V_p/i_p(0)$ from the initial amplitude of the transient are subject to error because the rise of V_p usually has a time course of 10-20 µs and because of low-pass filtering. Typical values of the series resistance are 3-20 MΩ if low resistance (1-5 MΩ) pipettes are used. The resistance may be as high as 50 MΩ, however, and is liable to frequent fluctuation during recording. For a series resistance of 10 MΩ and a cell capacity of 12 pF (corresponding to a spherical cell of 20 µm diameter of membrane capacitance 1 µF.cm^{-2}), $\tau=120$ µs for settling of the clamp and $f_c=1.3$ kHz.

The currents associated with charging the cell capacitance should be compensated, using the 'slow' capacity compensation of the patch clamp amplifier. This system injects current into the pipette to cancel the capacitative currents, avoiding saturation of the headstage during large voltage pulses. In most patch clamp amplifiers, the slow capacity compensation is separately switched, and has controls for the size of the capacitance and the time constant (or series resistance) of the decline of capacity currents. It is important to note that capacity compensation subtracts capacity current from the amplifier output. It does not compensate for the slow change in cell membrane potential due to series resistance. Most commercial amplifiers have separate series resistance compensation which allows a fraction of the pipette current times series resistance to be added to the command voltage as in conventional voltage clamp (see Chapter on voltage clamp techniques). This works best if R_s is small and should be used with caution because of frequent fluctuations of R_s during recording, to the extent that the best procedure is to monitor and adjust the compensation digitally via an interface immediately before each voltage command. A second way of dealing with series resistance in whole-cell recording is to use a switched (discontinuous) single electrode voltage clamp in which the membrane potential is monitored at zero current flow, so eliminating the error voltage due to i_p (see Chapter 2). Switched clamps have inherently greater noise, 10-100 fold, than patch clamps and can only be used with large membrane currents.

Diffusional exchange between pipette solution and cytosol

In conventional whole-cell recording, the contents of the interior of the cell come into diffusional equilibrium with the solution in the patch pipette. It is achieved

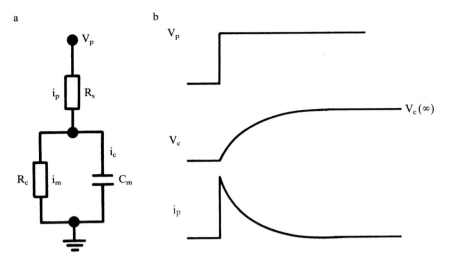

Fig. 9. (a) Equivalent circuit of whole cell recording. Current i_m flows in the cell resistance R_c and i_c in the capacitance. Pipette current, $i_p=i_m+i_c$, flows in the series resistance R_s between pipette and cell and produces a voltage error $V_p-V_c=i_pR_s$. (b) Time course of changes of V_c and i_p following a step of V_p.

$$V_c = V_p \frac{R_c}{R_c + R_s} (1 - e^{-t/\tau})$$

$$i_p = \frac{V_p}{R_c + R_s} + V_p \frac{R_c}{R_s(R_c + R_s)} e^{-t/\tau}$$

$$\tau = C_m . R_sR_c/(R_s + R_c)$$

$$V_c(\infty) = V_pR_c/(R_c + R_s)$$

$$i_p(0) = V_p/R_s \qquad i_p(\infty) = V_p/(R_c + R_s).$$

more rapidly for ions than for large intracellular molecules, which do, however, get washed out of the interior of the cell in time. The time course of exchange depends on the size of the pipette tip and the cell volume, as well as on the molecular size and the degree of buffering by cytoplasmic constituents. The advantages are control of the intracellular ionic composition and the perfusion into the cell of agents such as dyes that indicate ion concentrations, second messengers and 'caged' second messengers, antibodies to intracellular proteins and small proteins generally. The disadvantage is the loss of important cellular components, including small proteins, into the pipette.

The time course of equilibration of Na^+ ions during whole-cell recording was reported by Fenwick, Marty & Neher (1982) to have a time constant of about 5 s in chromaffin cells, judged by changes in the Na current.

Junction potentials due to small ions will exist between pipette solutions and the cytoplasm. This source of error will not be a constant, but will change towards zero as equilibrium between pipette and cellular contents proceeds. Marty & Neher (1983) estimate a maximum error of 12 mV from this source. A displacement of the potential dependence of ionic currents that occurs during whole-cell recording may be related to this changing junction potential.

The time course of equilibration of larger molecules is much slower, particularly if they bind to cellular components or are metabolised. For molecules such as fluorescent dyes of mass number 500-1000 Da equilibration, judged by the fluorescence increase, takes several minutes, and depends very much on how good the access is through the pipette tip. Observation suggests that as well as a low access resistance it is important to minimise the amount of material taken into the pipette tip during seal formation. An analysis of the time course of equilibration of dextran-tagged dyes as a function of size and access resistance has been made by Pusch & Neher (1988). They present evidence that components of upto about 20 kDa may be lost from the cell.

There are well documented instances of changes in the properties of ionic currents, and also cell responses involving second messengers, that occur during whole cell recording and which can be attributed to loss of cytosolic components into the pipette. Best known is the rundown of slowly-inactivating Ca conductance that occurs during intracellular perfusion. Evidence suggests that this is slowed by inclusion of ATP and Mg in the pipette, and may be improved by inclusion of protease inhibitors such as leupeptin or phosphatase inhibitors. Loss of the calcium mobilising hormonal response of exocrine cells during conventional whole cell recording was noted by Horn & Marty (1988) who introduced the method of *patch permeabilisation* to prevent rundown of the responses. Instead of rupture by pressure of the membrane under the pipette they achieved electrical access to control the membrane potential by including the Na-ionophore nystatin in the pipette solution, preventing loss of high molecular weight components. The method has been used with other ionophores, such as amphotericin, and a bacterial toxin (α-toxin from *Staph. aureus*, Khodakhah *et al.* 1992) that makes pores large enough to admit dyes and 'caged' second messengers <1 kDa. Permeabilisation is generally more difficult to implement than patch rupture because the ionophores prevent seal formation and are unstable in solution.

Use has also been made of the exchange of cell and pipette constituents during whole cell recording to change the concentrations of internal ions and of other substances in a systematic way. A perfusion system that takes account of the technical problems encountered has been described by Soejima & Noma (1984).

Measurement of changes in membrane capacitance

The facility of whole cell recording to make precise measurements of membrane capacitance has been developed as a method for measuring changes in membrane area that occur during exocytosis. The procedure is to analyse the admittance of the whole cell equivalent circuit into components due to capacitance, series resistance and membrane conductance by use of a phase-locking amplifier. The resolution is sufficient to see unitary changes associated with the incorporation of single (albeit 'giant') vesicles into the cell membrane, and has given considerable information about the process of vescicle-membrane fusion (reviewed by Almers & Tse, 1990). Technical descriptions of the method are contained in the initial report by Neher & Marty (1982) and in articles by Lindau & Neher (1988) and Fidler & Fernandez (1989).

10. Identification of ionic channels

The nature of ionic channels may be identified by (1) the mode of activation; (2) the permeant ions (under physiological conditions); (3) the unitary conductance; and (4) selective block by drugs, ions or toxins.

(1) Ionic channels may be opened by the following. (a) Neurotransmitters and chemical analogues acting on external receptors coupled directly to the channel opening mechanism. The open probability of these channels depends on transmitter concentration and they are usually involved in fast synaptic transmission. Examples are nicotinic acetylcholine channels and glutamate-activated channels. (b) Membrane potential, for example sodium, potassium and calcium channels opened by depolarisation during the action potential. (c) Intracellular ion concentration (e.g. Ca-activated K-channels) or other cellular constituents (e.g. K-channels closed by internal ATP). (d) The open probability of some channels is modified by intracellular second messengers released by hormone action. An example is the Ca channel of cardiac muscle, after phosphorylation by cAMP-dependent protein kinase. Most of these kinds of channels are also opened by one of the above means.

(2) Once a channel is opened, the ions that permeate can be determined by measuring changes of the reversal potential and unitary conductance when the concentration of ions in the external or internal medium is changed.

(3) The unitary conductance under specified conditions may vary considerably among different channels with the same ion selectivity, for example there are at least 3 types of Ca-activated K-channel distinguishable on this basis.

(4) Some channels are blocked selectively by low concentrations of drugs or toxins, for example block of Na channels by external tetrodotoxin, of delayed rectifier K-channels by TEA derivatives or of nicotinic acetylcholine channel activation at the neuromuscular junction by α-bungarotoxin. Block by other agents may be less selective; for example block of Na or acetylcholine channels by local anaesthetics or K-channels by quinine.

References

ALMERS, W., STANFIELD, P. R. & STÜHMER, W. (1983). Lateral distribution of sodium and potassium channels in frog skeletal muscle: measurements with a patch clamp method. *J. Physiol., Lond.* **336**, 261-284.

ALMERS, W. & TSE, F. W. (1990). Transmitter release from synapses; does a preassembled fusion pore initiate exocytosis. *Neuron* **4**, 813-818

BARRY, P. H. & LYNCH, J. W. (1991). Liquid junction potentials and small cell effects in patch clamp analysis. *J. Memb. Biol.* **121,** 101-118

BARRETT, J., MAGLEBY, K. L. & PALLOTTA, B. S. (1981). Properties of single calcium activated potassium channels in cultured rat muscle. *J. Physiol., Lond.* **331,** 211-230.

BYERLY, L. & HAGIWARA, S. (1982). Calcium currents in internally perfused axons of *Lymnea stagnalis. J. Physiol., Lond.* **322,** 503-529.

BYERLY, L. & YAZEJIAN, B. (1986). Intracellular factors for maintenance of Ca currents in internally perfused neurones of the snail *Lymnea stagnalis. J. Physiol., Lond.* **370,** 631-651.

COLLINS, A., SOMLYO, A. V. & HILGEMANN, D. (1992). The giant cardiac membrane patch method: stimulation of outward Na-Ca exchange current by Mg-ATP. *J. Physiol. Lond.* **454,** 27-58.

COLQUHOUN, D., NEHER, E., REUTER, H. & STEVENS, C. F. (1981). Inward current channels activated by internal Ca in cultured cardiac cells. *Nature, Lond.* **294,** 752-754.

COPELLO, J. SIMON, B. SEGAL, Y., WEHMER, F., SADAGOPA RAMANJAM, V. M., ALCOCK, N. & REUSS, L. (1991). Ba^{2+} release from soda glass modifies single K channel activity in patch clamp experiments. *Biophys. J.* **60,** 931-941.

CORONADO, R. & LATORRE, R. (1983). Phospholipid bilayers made from monolayers on patch pipettes. *Biophys. J.* **43,** 231-236.

COTA, G. & ARMSTRONG, C. M. (1988). K-channel inactivation induced by soft glass pipettes. *Biophys. J.* **53,** 107-109.

CULL-CANDY, S. G., MILEDI, R. & PARKER, I. (1981). Single glutamate activated channels recorded from locust muscle fibres with perfused patch-clamp electrodes. *J. Physiol., Lond.* **321,** 195-210.

CULL-CANDY, S. G. & OGDEN, D. C. (1984). Ion channels activated by 1-glutamate and GABA in cultured cerebellar neurons of the rat. *Proc. R. Soc.* B **224,** 367-373

DUDEL, J. FRANKE Ch. & HATT, H. (1990). Rapid activation, desensitization and resensitization of synaptic channels of crayfish muscle after glutamate pulses. *Biophys. J.* **57,** 533-545

FENWICK, E. H., MARTY, A. & NEHER, E. (1982). A patch clamp study of bovine chromaffin cells and their sensitivity to acetylcholine. *J. Physiol., Lond.* **331,** 577-599 and 599-635.

FERNANDEZ, J. M., FOX, A. C. & KRASNE, S. (1984). Membrane patches and whole cell membranes: a comparison of electrical properties in rat clonal pituitary cells. *J. Physiol., Lond.* **356,** 565-585.

FIDLER, N. & FERNANDEZ, J. (1989). Phase tracking: an improved phase detection technique for cell membrane capacitance measurements. *Biophys. J.* **56,** 1153-1162

FISCHMEISTER, R., AYER,K. & de HAAN, R. L. (1986). Some limitations of patch clamp techniques. *Pflügers Arch.* **406,** 73-85

FURMAN, R. E. & TANAKA, J. C. (1988). Patch electrode glass composition affects ion channel currents *Biophys. J.* **53,** 287-292.

GARDNER, P., OGDEN, D. C. & COLQUHOUN, D. (1984). Conductances of nicotinic ion channels opened by different agonists are indistinguishable. *Nature, Lond.* **309,** 160-162.

HAMILL, O. P., BORMANN, J. & SAKMANN, B. (1983). Activation of multiple conductance state chloride channels in spinal neurons by glycine and GABA. *Nature, Lond.* **305,** 805-808.

HAMILL, O. P., MARTY, A., NEHER, E., SAKMANN, B. & SIGWORTH, F. J. (1981). Improved patchclamp techniques for high-resolution current recording from cells and cell-free membrane patches. *Pflügers Arch.* **391,** 85-100.

HILGEMANN, D. (1989). Giant excised cardiac sarcolemmal membrane patches: Na & Na/Ca exchange currents. *Pflügers Arch.* **415,** 247-249

HILGEMANN, D. (1990). Regulation and deregulation of cardiac Na-Ca exchange current in giant excised sarcolemmal patches. *Nature* **344,** 242-245

HORN, R. (1991). Diffusion of nystatin in plasma membrane is inhibited by a glass-membrane seal. *Biophys. J.* **60,** 329-333

HORN, R. & MARTY, A. (1988). Muscarinic activation of ionic currents measured by a new whole cell recording method. *J. Gen. Physiol.* **92,** 145-159

KELLER, B. U. & HEDRICH, R. (1992). Patch clamp techniques to study ion channels from organelles. *Methods in Enzymology* **207,** 673-680

KHODAKHAH, K., CARTER, T., TOROK, K., SMITH, S. & OGDEN, D. (1992). Patch permeabilisation with *Staphylococcal* α toxin in whole cell recording. *J. Physiol.* **452,** 160P.

LINDAU, M. & NEHER, E. (1988). Patch clamp techniques for time resolved capacitance measurements. *Pflügers Arch.* **411**, 137-

LIU, Y. & DILGER, J. P. (1991). Opening rate of ACh receptor channels. *Biophys. J.* **60**, 424-432

MACONOCHIE, D. & KNIGHT, D. E. (1989). A method for making solution changes in the sub millisecond range at the tip of a patch pipette. *Pflügers Arch.* **414**, 589-596

MARTY, A. (1981). Ca-dependent K channels of large unitary conductance. *Nature, Lond.* **291**, 497-500.

MARTY, A. & NEHER, E. (1983). Tight seal whole cell recording. In *Single Channel Recording* Ed Sakmann, B. & Neher, E. Plenum, N.Y.

METHFESSEL, C., WITZMANN, V., TAKAHASHI, T., MISHIMA, M., NUMA, S. & SAKMANN, B. (1986). Patch clamp experiments on Xenopus oocytes: currents through endogenous channels, and implanted nicotinic and sodium channels. *Pflügers Arch.* **407**, 577-588.

MILTON, R. L. & CALDWELL, J. H. (1990). How do patch clamp seals form? - a lipid bleb model. *Pflügers Arch.* **416**, 758-765

NEHER, E. (1981). Unit conductance studies in biological membranes. In: Techniques in Cellular Physiology (ed. P. F. Baker). Amsterdam: Elsevier, N. Holland.

NEHER, E. (1992). Correction for liquid junction potentials in patch clamp experiments. *Methods in Enzymology* **207**, 123-130.

NEHER, E. & MARTY, A. (1982). Discrete changes of cell membrane capacitance observed under conditions of enhanced secretion in bovine adrenal chromaffin cells *Proc. Natl. Acad. Sci. USA* **79**, 6712-6716.

NEHER, E. & SAKMANN, B. (1976). Single channel currents recorded from membrane of denervated frog muscle fibres. *Nature, Lond.* **260**, 799-802.

NEHER, E., SAKMANN, B. & STEINBACH, J. H. (1978). The extracellular patch clamp: a method for resolving currents through individual open channels in biological membranes. *Pflügers Arch.* **375**, 219-228.

NEHER, E. & STEINBACH, J. H. (1978). Local anaesthetics transiently block currents through single acetylcholine receptor channels. *J. Physiol., Lond.* **277**, 152-176.

NOMA, A. (1983). ATP-regulated K channels in cardiac muscle. *Nature, Lond.* **305**, 147-148.

NOWAK, L., BREGESTOWSKI, P., ASCHER, P., HUBERT, A. & PROCHIANTZ, A. (1984). Magnesium gates glutamate activated channels in mouse central neurons. *Nature, Lond.* **307**, 462-465.

PATLAK, J. B., GRATION, K. A. F. & USHERWOOD, P. N. R. (1979). Single glutamate channels in locust muscle. *Nature, Lond.* **278**, 643-645.

PUSCH, M. & NEHER, E. (1988) Rates of diffusional exchange between small cells & patch pipette. *Pflügers Arch.* **411**, 204

RAE, J. L. & LEVIS, R. A. (1984). Patch clamp recording from the epithelium of the lens with glasses selected for low noise and improved sealing properties. *Biophys. J.* **45**, 144-146.

RAE, J. L. & LEVIS, R. A. (1992). Glass Technology for patch clamp electrodes. *Methods in Enzymology* **207**, 66-91

ROBERTS, W. M. & ALMERS, W. A. (1984). An improved loose patch clamp method using concentric pipettes. *Pflügers Arch.* **402**, 190-196.

SAKMANN, B. & NEHER, E. (1983). Geometric parameters of pipettes and membrane patches. Chap. 2. *Single Channel Recording* (ed. B. Sakmann & E. Neher). New York: Plenum.

SHUSTER, M., CAMARDO, J., SIEGELBAUM, S. & KANDEL, E. R. (1985). cAMP-dependent protein kinase closes serotonin sensitive K-channels in *Aplysia* sensory neurons. *Nature, Lond.* **313**, 392-395.

SOEJIMA, A. & NOMA, A. (1984). Mode of regulation of the acetylcholine sensitive K-channel by the muscarinic receptor in rabbit atrial cells. *Pflügers Arch.* **400**, 424-431.

SOKABE, M. & SACHS, F. (1990). The structure and dynamics of patch clamped membranes: a study using differential interference contrast microscopy. *J. Cell. Biol.* **111**, 599-606.

SPRUCE, A. E., STANDEN, N. B. & STANFIELD, P. R. (1985). Voltage dependent ATP sensitive K channels of skeletal muscle membrane. *Nature, Lond.* **316**, 736.

STANDEN, N. B., STANFIELD, P. R. & WARD, T. A. (1985). Properties of single potassium channels formed from the sarcolemma of frog skeletal muscle. *J. Physiol., Lond.* **364**, 339-358.

STANDEN, N. B., STANFIELD, P. R., WARD, T. A. & WILSON, S. W. (1984). A new preparation for recording single channel currents from skeletal muscle fibres. *Proc. R. Soc.* B **221**, 455-464.

TANK, D. W., MILLER, C. & WEBB, W. W. (1982). Isolated patch recording from liposomes containing functionally reconstituted Cl channels from Torpedo electroplax membrane. *Proc. Natn Acad. Sci. USA* **79,** 7749-7753.

TRAUTMANN, A. & SIEGELBAUM, S. (1983). The influence of membrane patch isolation on single acetylcholine channel current in rat myotubes. In *Single Channel Recording* (ed. B. Sakmann & E. Neher). New York: Plenum.

General

HILLE, B. (1992). *Ionic Channels of Excitable Membranes.* 2nd Ed. Sinauer Associates, Sunderland, Mass.

SAKMANN, B. & NEHER, E. (eds) (1983). *Single Channel Recording.* New York: Plenum.

LEVITAN, I. B. (1985). Phosphorylation of ion channels. *J. Memb. Biol.* **87,** 177-190.

Ion Channels. Vol 207 of Methods in Enzymology (1992).

Chapter 5
An introduction to the methods available for ion channel reconstitution

ALAN J. WILLIAMS

1. Introduction

Reconstitution describes the disassembly of a complex structure; the isolation of one or more of the components of that system and the reassembly of these components into an intelligible, measurable system. When applied to membrane proteins, such as ion-channels, it describes the solubilisation of the membrane, the isolation of the channel protein from the other membrane constituents and the reintroduction of that protein into some form of artificial membrane system which facilitates the measurement of channel function. However, in practice, the term is often applied less rigorously in the study of ion channel function and can be used to describe the incorporation of intact membrane vesicles, including the protein of interest, into artificial membrane systems that allow the properties of the channel to be investigated.

In this chapter I will describe methods that are currently in use for incorporation of both native and purified channel proteins into artificial membranes. The diversity of the subject means that the technical aspects of the various approaches cannot be covered in detail; rather, this chapter is designed to act as an introduction to ion channel reconstitution. Detailed descriptions of experimental protocols can be obtained from the literature cited throughout the chapter.

Ion-channel function can be monitored in a number of ways; by monitoring isotope flux in isolated tissues or membrane vesicles, or by using electrophysiological techniques, such as whole cell voltage-clamp or patch-clamp. What then are the advantages of monitoring channel function following incorporation into some form of artificial membrane?

2. Advantages of ion-channel reconstitution

Not all species of ion-channel are amenable to study by conventional voltage or patch-clamp techniques. Ion-channels found in intracellular membrane systems such

Department of Cardiac Medicine, National Heart and Lung Institute, University of London, Dovehouse St., London SW3 6LY, UK as the endoplasmic or sarcoplasmic reticulum networks are not readily accessible to extracellular patch electrodes and the elucidation of the single-channel properties of a number of types of channel from these membrane systems has relied heavily on the incorporation of isolated membrane vesicles, and more recently purified channel proteins, into artificial membranes where single-channel properties can be investigated under voltage-clamp conditions (Miller, 1982a; Tomlins *et al.* 1984; Lai and Meissner, 1989; Williams, 1992; Rousseau *et al.* 1988; Bezprozvanny *et al.* 1991).

The transfer of a membrane protein to an artificial membrane of defined phospholipid composition can have a number of advantages for the determination of the biophysical properties of the channel. Ion-channel reconstitution makes possible the investigation of the influence of membrane lipid composition on channel function. For example, the influence on channel function of net membrane surface charge can be investigated by varying the phospholipid composition of the bilayer into which the channel is incorporated (Bell and Miller, 1984; Moczydlowski *et al.* 1985; Coronado and Affolter, 1986).

Similarly, several of the techniques described in this chapter allow the investigator to set and alter the ionic composition of the solutions bathing both faces of the channel protein. Such manipulations are essential for the comprehensive characterization of ion-channel conduction and selectivity (Lindsay *et al.* 1991; Tinker and Williams, 1992; Tinker, Lindsay and Williams, 1992a).

The functional state of purified water-soluble proteins such as cytoplasmic enzymes can be readily determined in solution. However, a functional assay for a purified membrane transport protein, such as an ion-channel, is completely dependent upon the reconstitution of the protein into a membrane which provides a suitably hydrophobic environment for the protein and a barrier through which the channel can catalyze the movement of ions.

3. Methods of ion-channel reconstitution

In this section I will describe methods for the formation of artificial lipid bilayers and the incorporation of ion-channels into these membranes. I will not discuss strategies of membrane vesicle isolation or channel protein purification. These topics are covered in a number of recent reviews (Evans, 1990; Levitski, 1985; Catterall *et al.* 1989; Jones *et al.* 1990; Silvius, 1992). Reviews dealing with channel function following reconstitution are also available (Miller, 1983; Coronado, 1986; Montal, 1987).

The aim of ion-channel reconstitution is to incorporate the channel into an artificial membrane in which its function can be investigated. For practical purposes this means a membrane system in which ion flow through the channel can be studied under voltage-clamp conditions; ideally with good enough resolution to permit the measurement of single-channel open and closed lifetimes. The starting point for such studies is the formation of an artificial planar phospholipid bilayer.

4. Planar phospholipid bilayers

The bilayer is formed across an aperture which links two fluid filled chambers. The size of the aperture and the material of which the aperture is made vary in the techniques to be discussed here, however in all cases bilayers are formed using one of two basic techniques. Readers are advised to consult White (1986) for a detailed description of bilayer formation and the physical properties of planar bilayers.

Spreading from dispersions of phospholipid (painted bilayers)

This method was first described by Mueller and colleagues (Mueller *et al.* 1962; Mueller and Rudin, 1969). The apparatus used in our laboratory for the formation of painted bilayers is shown in Fig. 1. It consists of a block (A) into which is cut an oblong chamber (the *trans* chamber) connected to a second circular chamber which holds the bilayer cup (B). The cup contains a well (volume 500 µl - the *cis* chamber). The face of the cup adjacent to the well is machined to form a thin (approximately 200 µm) septum through which is drilled a hole (200 µm diameter) that, once the cup is located in the block, connects the two chambers. The painted bilayer is formed across this hole.

Bilayers are formed from a dispersion of either one or a mixture of purified phospholipids in a non-polar solvent such as n-decane. Pure phospholipids can be obtained from Avanti Polar Lipids, Alabaster, Alabama 35007, USA. The phospholipid dispersions are made from stock solutions of the required phospholipids

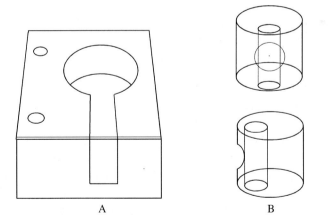

A B

Fig. 1. Experimental chambers for the formation of painted bilayers. (A) Styrene co-polymer block into which is machined the oblong *trans* chamber (volume 1.0 ml) and a well into which the bilayer cup fits. The front face of the block consists of a glass panel through which it is possible to view the hole in the bilayer cup. The two wells at the left of the block are filled with 3 M LiCl and connect the experimental chambers to the head amplifier via agar bridges and silver-silver chloride electrodes. (B) Two views of the styrene co-polymer bilayer cup. The top diagram shows a front view of the cup as it is orientated in the block (A). The lower panel shows the cup from the side and demonstrates the machining of the front wall of the cup to form the septum through which is drilled the hole on which the bilayer is formed. The volume of the *cis* chamber is 0.5 ml.

in chloroform, stored at −80°C. Chloroform is evaporated under a stream of nitrogen and n-decane is added to give the desired final phospholipid concentration (20-50 mM).

Prior to bilayer formation, the hole on which the bilayer is to be formed is "primed" with a small quantity of the phospholipid dispersion. The priming dispersion is allowed to dry, the cup is positioned in the block and both *cis* and *trans* chambers are filled with the desired experimental solution. Additional phospholipid dispersion is then drawn across the hole using a "stick". This implement varies considerably from laboratory to laboratory and may be a small brush, a plastic rod or in some cases an air bubble at the end of a glass capillary.

As described by White (1986), at the outset, the film drawn across the hole will be several µms thick and will be in equilibrium with an annulus (Plateau-Gibbs border) formed as the lipid dispersion "wets" the septum. The film will thin spontaneously to form a bilayer. The primary instigation for thinning comes from Plateau-Gibbs border suction. As the film thins, London-van der Waals attraction between the aqueous phases on either side of the film contributes an additional driving force.

The material from which the cup is made can have a profound influence on the ease of formation and the stability of the planar bilayer. It is possible to manufacture cups from plastics such as Teflon^R, polycarbonate and polystyrene. Our experience has led us to use a styrene co-polymer. We have found that this material can be used to produce robust cups on which films thin readily to form stable bilayers. Similar results have been obtained by others using polycarbonate cups (Nelson *et al.* 1984; French *et al.* 1986), although in the past we have found it more difficult to induce film thinning on small holes in this material. Similarly, others have reported that film thinning is difficult to achieve with small holes (100 µm diameter) in Teflon^R (Alvarez, 1986).

Thinning of the film can be monitored in one of two ways:-

Observation under reflected light. It is possible to view the phospholipid film using a lens or a low power microscope. On formation, the film appears as a multicoloured structure reminiscent of the pattern seen when a thin layer of oil covers a puddle in the road. The bilayer starts to form at the base of the structure due to the buoyancy of the forming solution (White, 1986) and spreads steadily over the bulk of the hole. As thinning occurs the amount of light reflected from the film decreases until a bimolecular structure is formed (25-50 Å thick), which reflects essentially no light and appears black; hence the name black lipid membrane or BLM.

Monitoring membrane capacitance. Following ion channel incorporation, current flow through the bilayer is monitored using an operational amplifier as a current-voltage converter (Miller, 1982b). The *trans* chamber is clamped at virtual ground whilst the *cis* chamber can be clamped at a desired holding potential relative to ground. An indication of the capacitance of the membrane formed across the hole can be obtained using a low frequency (1 Hz), low amplitude (5-20 mV peak to peak) square wave. The application of this wave form to the system before the lipid film is applied to the hole results in large, saturating oscillations in current; the solutions in the *cis* and *trans* chambers are electrically coupled through the hole. On application of the lipid film to the septum, the hole is blocked by the forming solution and a small deflection from zero current is seen with each voltage clamp transition. As the film begins to thin the current deflection will increase in amplitude as the capacitance of the film increases. The final capacitance of a painted bilayer should be in the region of 0.4 $\mu F/cm^2$ (Alvarez, 1986). As hole size and geometry will vary from cup to cup, bilayer formation, as adjudged by capacitance, is often determined empirically; the operator learns the amplitude of the capacitance spike that will allow good channel incorporation and hence signifies stable bilayer formation. The height of the spike will be dependent upon the rise time of the square wave and the capacitance of the membrane.

What to do if the film will not thin

In practice we have all encountered times when application of the forming solution to the hole results in a blob of lipid which will not thin spontaneously to form a bilayer. In my experience this situation usually arises when the hole is surrounded with too large a quantity of forming solution. Under these conditions it is sometimes possible to induce the film to thin by applying a large voltage pulse to the system (e.g. ±100 mV) or by repeated painting of the hole with a clean painting stick. However, the best solution to this problem is to remove the cup and clean the septum before re-priming.

Chamber and solution preparation

As with all single-channel monitoring techniques, it is very important that the experimental chambers of the bilayer system be kept clean. The polystyrene chambers used in our laboratory are cleaned with a household dish washing detergent and are rinsed thoroughly under running water and dried before use. Similarly, chambers can be cleaned with methanol and again dried before use. Teflon[R] chambers can be cleaned in sodium dichromate-sulphuric acid as described by Alvarez (1986).

All solutions used in the bilayer system should be filtered before use, we routinely make experimental solutions with deionized water and pass them through 0.45 μm diameter pore Millipore filters before use.

5. Forming bilayers from monolayers (folded bilayers, bilayers on patch pipettes)

An alternative procedure for the formation of a planar phospholipid bilayer involves the apposition of two phospholipid monolayers formed at the interface of an aqueous solution and the air (White, 1986). This method has been adopted by some workers because the resultant bilayer contains somewhat less solvent than the equivalent bilayer formed from alkane dispersions of phospholipids by the painting method described above.

Monolayers can be formed by applying phospholipids in a volatile solvent such as pentane, chloroform or hexane to the surface of an aqueous solution; the solvent evaporates in minutes leaving a phospholipid monolayer at the air-solution interface (Montal and Mueller, 1972; Coronado and Latorre, 1983). Alternatively, monolayers can be formed by allowing phospholipid liposomes or mixtures of liposomes and native membrane vesicles to equilibrate with a monolayer at an air-water interface (Schindler, 1980; Schindler and Quast, 1980; Nelson et al. 1980). The latter method leads to the production of monolayers containing native membrane proteins including ion channels. These channels may then be incorporated directly into the membrane on bilayer formation (Schindler and Quast, 1980; Nelson et al. 1980; Suarez-Isla et al. 1983; Montal et al. 1986). Once a monolayer has been formed, there are two standard methods for producing a phospholipid bilayer.

(1) Monolayer folding

As is implied from its name, this method involves the folding together of two monolayers to form a planar bilayer. A brief outline of the method will be given here and interested readers are directed towards the following references for more detailed accounts (Schindler, 1980; Schindler and Quast, 1980; Nelson et al. 1980; Montal et al. 1986).

The apparatus used for bilayer formation via monolayer folding is in many respects similar to that used in the production of painted bilayers in that it involves two chambers separated by a septum containing a hole. A major difference exists in that the septum used for monolayer folding is not of rigid plastic but is composed of a very thin (10-25 μm) Teflon[R] membrane supported between two thicker Teflon[R] O-rings.

Holes can be formed in the septum by punching with a hypodermic needle. The tapered end of the needle should be removed and the level end sharpened either mechanically using fine abrasive paper or electrically by etching in 5 M HCl (Montal et al. 1986). Alternatively, holes can be produced with an electrical spark from a car ignition coil (Hartshorne et al. 1986). We have found this technique to be very useful; a small defect in the Teflon[R] membrane is created with a needle and the Teflon[R] can then be melted by the spark. The size of the hole produced in the membrane can be increased by discharging additional sparks across the membrane. With practice, it is possible to routinely produce holes with a smooth perimeter and diameters in the range 30-200 μm.

For bilayer construction, the Teflon[R] septum is clamped between two chambers (Fig. 2; see Montal et al. 1986 for details of chamber design). The hole on which the bilayer is to be constructed is then primed with 0.5% (v/v) hexadecane in hexane (Schindler and Quast, 1980; Montal et al. 1986). Once the priming solution has dried, the desired experimental solution is added to both chambers to a level below the hole in the septum, and a phospholipid monolayer is formed at the solution-air interface using one of the methods described above. The bilayer is formed by increasing the level of the solution, first in one and then in the other chamber so that each monolayer is raised to cover the hole (Fig. 3). Formation of the bilayer is monitored by measuring capacitance; the final membrane should have a capacitance of approximately 0.8 μF/cm^2 (White, 1986; Montal et al. 1986).

The production of planar phospholipid bilayers by folding monolayers offers the

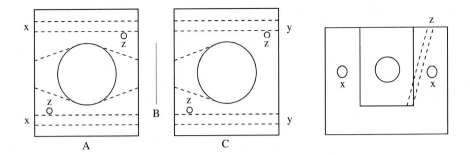

Fig. 2. Experimental chambers for the formation of folded planar phospholipid bilayers (based on a design provided by Montal *et al.* 1986.) Readers are advised to consult this reference for chamber dimensions). A hole is formed in a thin Teflon[R] film (B) as described in the text. This film is clamped between two Teflon[R] blocks A and C (viewed from the top in the left panel) with bolts that pass through the holes x-y. Each block contains a well which connects with the face of the block in contact with the film. The right panel shows a front view of the chamber. Solutions can be added and raised through ports in the blocks (z). Similar ports are used to connect the experimental chambers to the head amplifiers via salt bridges and silver-silver chloride electrodes.

possibility of making small membranes. I have found it difficult to routinely make painted bilayers on holes with a diameter of much less than 100 μm; with holes of this size it is very difficult to achieve reproducible film thinning. This should not be a problem with monolayer folding. Decreasing the surface area of the bilayer will increase the mechanical stability of the membrane and hence decrease the

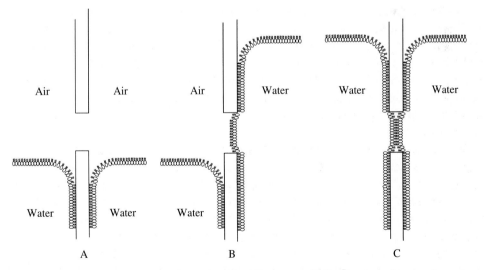

Fig. 3. Cartoon showing the formation of a bilayer on a Teflon[R] septum from preformed phospholipid monolayers. Monolayers are formed at the interface of an aqueous solution (water) and the air, using one of the methods described in the text. The bilayer is formed by the successive raising of the level of the solution in the chambers on either side of the septum (A-C). Channel proteins may be incorporated into the bilayer if they are present in either of the monolayers (see text). The septum, hole and phospholipids are not drawn to scale.

background noise of the system. The apposition of two monolayers to form a bilayer also allows the experimenter to construct asymmetric bilayers, for example one monolayer could be formed from an essentially uncharged phospholipid such as phosphatidyl ethanolamine whilst the other might contain a high proportion of a negatively charged phospholipid such as phosphatidyl serine. As stated at the beginning of this section, bilayers constructed from folded monolayers will contain somewhat less solvent than those cast from dispersions of phospholipids in a solvent. However, folded bilayers will not form in the absence of a hydrophobic environment for the formation of the annulus (White, 1986). An alternative method for the formation of bilayers from preformed phospholipid monolayers does permit the construction of truly solvent-free bilayers.

(2) Bilayers on patch-pipettes

The formation of planar bilayers on the end of patch pipettes was introduced by Wilmsen and colleagues (Wilmsen et al. 1983; Hanke et al. 1984) and has been used by a number of groups to investigate both native membrane channels and purified channel proteins (Suarez-Isla et al. 1983; Coronado and Latorre, 1983; Coronado, 1985; Ewald et al. 1985; Montal et al. 1986). The use of this method appears to have declined in recent years, however, I will discuss it here as it provides a method for the production of small solvent-free bilayers which may be required for particular reconstitution applications.

Bilayers are formed at the end of conventional patch-clamp pipettes (Sakmann and Neher, 1983; Corey and Stevens, 1983) with tip diameters in the range 0.5-5 µm either with or without fire polishing (Montal et al. 1986; Suarez-Isla et al. 1983; Coronado and Latorre, 1983). The tip of the pipette is immersed in the desired experimental solution in a compartment of a multi-well disposable tray (volume approximately 0.5 ml) and a phospholipid monolayer formed at the air-water interface using one of the methods described above (Fig. 4). A portion of the monolayer is transferred to the pipette tip by raising the pipette into the air. The polar head groups of the phospholipids orientate so that they interact with the aqueous pipette-filling solution and the glass wall of the pipette. The hydrocarbon chains of the molecules face the air. A bilayer is constructed by re-immersion of the pipette in the bath solution. As the tip of the pipette crosses the monolayer at the air-solution interface a second region of monolayer interacts with the monolayer in the pipette to form a bilayer. Bilayers can be formed from a range of purified phospholipids, however seal formation with phospholipids bearing a net negative charge may require the presence of divalent cations in the pipette and bath solutions (Coronado, 1985). If monolayers are created from suspensions of native membrane vesicles a certain proportion of bilayers formed using this method will contain channel proteins (Montal et al. 1986). As with all bilayer formation protocols, particular care must be taken to ensure that the pipette glass is clean and that all solutions are filtered before use. It may be desirable to use pipettes and experimental baths only once.

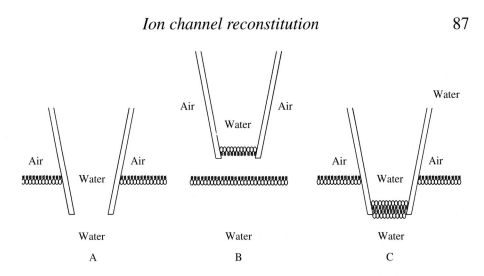

Fig. 4. Cartoon showing the formation of a bilayer from a phospholipid monolayer at the tip of a patch pipette. A bilayer is formed by the transfer of first one and then a second phospholipid monolayer to the pipette tip (A-C; see text for details). Channel proteins may be incorporated into the bilayer during formation if they are present in the original monolayer (see text). The pipette and phospholipids are not drawn to scale.

6. The incorporation of ion channels into planar phospholipid bilayers

The earlier sections of this chapter have provided an introduction to the methods currently available for the formation of planar phospholipid bilayers. I cannot emphasise too strongly the importance of the bilayer in the overall success or failure of an ion channel reconstitution experiment. If the bilayer is sub-standard, that is if it will not thin, or if the bilayer is leaky or unstable, there is absolutely no point in attempting to incorporate ion channels. No matter which of the above approaches is adopted to form the bilayer the resulting membrane must provide a stable, electrically quiet environment for the channel under investigation. Detailed studies of channel conduction or gating often take considerable periods of time, possibly up to an hour. If sufficient care and attention has been taken during bilayer formation, the stability of the bilayer should not be a limiting factor in experiments of this kind.

Some of the methods of bilayer formation described above offer the possibility of incorporating either native or purified channel proteins into the bilayer during formation. However in most cases, following the formation of a stable bilayer, channel proteins must be introduced into the membrane. Whilst there are recent reports describing the incorporation of purified ryanodine receptor-channel proteins into bilayers from detergent solutions (Imagawa *et al.* 1987; Lai *et al.* 1988; Smith *et al.* 1988; Anderson *et al.* 1989), the standard method for the incorporation of both native and purified channel proteins into pre-formed planar lipid bilayers involves the fusion of a channel-containing membrane vesicle with the bilayer; a procedure first described by Chris Miller in his studies of the sarcoplasmic reticulum K^+-selective channel (Miller and Racker, 1976; Miller, 1978).

Native membrane vesicles or proteo-liposomes containing purified channel proteins can be incorporated into planar phospholipid bilayers formed either by spreading phospholipid dispersions or from monolayers on Teflon[R] partitions or patch pipettes. The broad rules governing fusion are believed to be the same in all cases, although optimal conditions may vary slightly (see for example Cohen, 1986). Much of our understanding of the principles underlying vesicle-bilayer fusion has been derived from studies employing phospholipid vesicles in which fusion has been monitored either by following the transfer of vesicular contents across the planar bilayer (Zimmerberg *et al.* 1980; Woodbury and Hall, 1988a,b; Niles and Cohen, 1987) or by the incorporation of reconstituted VDAC or porin channels into the bilayer (Cohen *et al.* 1980; Woodbury and Hall, 1988a).

Fusion of membrane vesicles with a planar phospholipid bilayer is preceded by the development of a pre-fusion state in which the membrane vesicles become closely associated with, or bound to, the planar bilayer (Cohen, 1986). If either the planar bilayer or membrane vesicle contain a proportion of negatively charged phospholipids, the occurrence of the pre-fusion state can be encouraged by the inclusion of millimolar concentrations of divalent or trivalent cations in the experimental solutions (Cohen, 1986; Hanke, 1986).

Membrane vesicles in pre-fusion association with a planar bilayer will only fuse with the bilayer if they are induced to swell (Finkelstein *et al.* 1986). Vesicle swelling is most commonly induced by forming an osmotic gradient across the bilayer so that the osmotic pressure of the solution in the chamber to which the membrane vesicles are added (*cis*), is greater than that of the solution on the other side of the bilayer (*trans*). Under these conditions water will flow from the *trans* chamber to the *cis* chamber; some of this water will enter the membrane vesicles bound to the bilayer in the pre-fusion state. These vesicles will swell and some will burst leading to a coalescence of the vesicle with the bilayer (Fig. 5).

The osmotic strength of the *cis* solution should be raised using a solute that will readily cross the vesicle membrane and hence increase the osmotic pressure in the vesicle lumen. At the same time the substance should not be so permeant in the bilayer that the osmotic gradient is dissipated. In practice, glycerol and urea are efficient in stimulating vesicle fusion whilst more permeant substance such as ethylene glycol and formamide are less effective (Cohen, 1986). Osmotic gradients created with salt solutions will induce vesicle swelling if the vesicle contains channels permeable to one or both of the ions (Cohen, 1986). The hydrostatic pressure developed in the vesicle induces fusion (Niles *et al.* 1989), therefore the vesicle lumen should ideally also be hyperosmotic to the *cis* solution, this is often achieved by making or storing the membrane vesicles in concentrated sucrose solutions (0.3-1.0 M).

The efficiency of vesicle-bilayer fusion can also be influenced by other factors. The greater the surface area of the bilayer the more likely it is that membrane vesicles will come into pre-fusion contact and hence the greater the likelihood of fusion. The probability of vesicle fusion with a bilayer at the tip of a patch pipette may be so low as to make it impractical. The density of vesicles in the vicinity of the bilayer can be

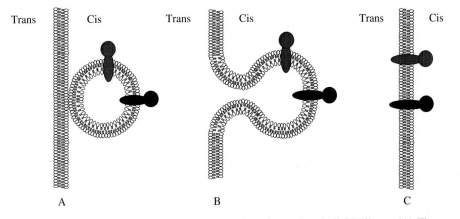

Fig. 5. The fusion of a membrane vesicle with a planar phospholipid bilayer. (A) The pre-fusion state. Membrane vesicles associate with, or bind to the planar phospholipid bilayer. (B) The establishment of an osmotic gradient across the bilayer induces the vesicle to swell, burst and incorporate with the bilayer. (C) Following fusion, the channel proteins (shaded and black areas) are aligned such that the face of the protein protruding from the vesicle faces the chamber to which the vesicles were added (*cis*), whilst the luminal face of the channel faces the other (*trans*) chamber. Not drawn to scale.

increased by adding them to the solution in the pipette rather than to the much greater volume of solution in the bath (Hanke *et al.* 1984). The fusion rate may also be influenced by the phospholipid composition of the bilayer, the presence of divalent cations, the size of the osmotic gradient, vesicle concentration and stirring (Labarca *et al.* 1980; Hanke, 1986; Cohen, 1986). It is sensible to employ a range of fusion conditions when attempting to investigate the channel content of a new vesicle population.

A general principle to emerge from the investigations outlined above, is that vesicles will fuse with planar bilayers in the presence of an osmotic gradient if the vesicle contains a permeability pathway for the solute; vesicles containing channels fuse more readily than channel-free vesicles (Woodbury and Hall, 1988b; Cohen *et al.* 1989). Woodbury and Miller (1990) have recently described a method for maximising vesicle fusion with planar bilayers in which nystatin is incorporated into membrane vesicles in the presence of ergosterol. Nystatin provides a weakly anion-selective permeability pathway in all vesicles so that they readily fuse in the presence of a salt gradient. Nystatin only forms functional conduction pathways in the presence of ergosterol, therefore if the bilayer into which the vesicles incorporate contains no ergosterol, fusion is marked by a transient increase in conductance, which decays as the ergosterol associated with the nystatin in the vesicle dissipates into the bulk of the bilayer. Using this method it is possible to assess the variety of channel species in a vesicle population and to determine the density of ion channels in a preparation of vesicles.

Having established the optimal conditions for vesicle fusion, the investigator is often faced with the problem of how to limit fusion. Studies of channel gating kinetics require the presence of a single channel in the bilayer and so it is important that the rate of vesicle fusion can be controlled and stopped at the appropriate point. In practice this can be done in a number of ways. If divalent cations are used to encourage fusion with negatively charged bilayers, chelation of the cation dramatically slows incorporation. Similarly, the dissipation of the osmotic gradient across the bilayer will eliminate fusion. This can be achieved either by increasing the osmotic strength of the *trans* chamber or by perfusing the *cis* chamber with a solution of lower osmotic strength. This method has the added advantage that unfused vesicles are removed from the solution.

As vesicles fuse, channel proteins will incorporate into the bilayer with an orientation that is dependent upon their orientation in the vesicle (Fig. 5). In other words, if the sample is made up of populations of vesicles with mixed orientation, or if individual vesicles contain proteins in both possible orientations, then both orientation of channel are likely, or at least possible, in the bilayer. A number of channel properties are side-specific, this is obviously so for agonist-activated channels such as the acetylcholine receptor, the plasmalemmal Ca^{2+}-activated K^+ channel or the Ca^{2+}-activated Ca^{2+}-release channel of muscle sarcoplasmic reticulum where the agonist binding site is located on a specific face of the channel protein. However, a number of channel conduction properties are also asymmetric, for example, it is common for blocking ions to have access to the conduction pathway from only one side of the channel (Tinker *et al.* 1992b,c). Therefore it is important to consider, and if possible to monitor, channel orientation following vesicle fusion with a planar bilayer. Fortunately, many of the channels that have been studied in bilayers, for example the voltage-dependent Na^+ channel (Moczydlowski *et al.* 1984), the skeletal muscle sarcolemmal Ca^{2+}-activated K^+ channel (Latorre, 1986) and the K^+ channel and Ca^{2+}-activated Ca^{2+}-release channel of sarcoplasmic reticulum (Miller and Rosenberg, 1979; Ashley and Williams, 1990), occur in isolated membrane vesicle populations which have a fixed orientation. As a result the channels incorporate into the bilayer with a defined alignment. Smooth muscle sarcolemmal vesicles appear to be randomly orientated and consequently Ca^{2+}-activated K^+ channels from this source incorporate into bilayers with random orientation (Latorre, 1986).

7. Incorporation of ion channels into liposomes suitable for patch-clamping

The final approach to be discussed in this chapter involves the incorporation of either native or purified channel proteins into small unilamellar liposomes and the transformation of these liposomes into structures suitable for conventional patch-clamp analysis.

Native membrane vesicles, isolated by differential or density gradient centrifugation following tissue homogenisation, are too small to patch-clamp (diameter 0.1-1.0 μm). Small unilamellar proteo-liposomes into which purified channel proteins are reconstituted by detergent removal are of a similar size. The size of these vesicles can be increased using either of the following methods.

Freeze-thaw

The formation of large liposomes by the successive freezing and thawing of small unilamellar vesicles was first demonstrated by Kasahara and Hinkle (1977). The use of this procedure to form channel-containing liposomes suitable for patch-clamping was introduced by Tank and Miller (1982, 1983). The method has been used to study chloride-selective channels from native membranes of *Torpedo* electroplax (Tank and Miller, 1982), the K^+-selective channel of native sarcoplasmic reticulum membranes (Tomlins and Williams, 1986), the purified voltage-dependent Na^+ channel (Agnew *et al.* 1986) and the purified acetylcholine receptor-channel from *Torpedo* (Tank *et al.* 1983).

Native membrane channel proteins are solubilised with a suitable detergent and separated from unsolubilised material by centrifugation. The solubilised membrane components or, where appropriate, the purified channel proteins are then incorporated into proteoliposomes. Solubilised proteins are mixed with excess phospholipid and the detergent removed. In the case of cholate, the detergent used in the studies quoted above, this is easily achieved by dialysis; other detergents may require different procedures (Jones *et al.* 1990). The channel-containing small unilamellar liposomes are then transformed into large liposomes by freeze-thaw. An aliquot of the sample is frozen and then allowed to thaw. Freezing can either be carried out using liquid nitrogen or alternatively the sample can be frozen at −80°C; the sample is allowed to thaw either on ice or at room temperature. Following this procedure, the initially clear suspension of small unilamellar proteoliposomes becomes turbid as the result of the production of larger multilamellar structures (Tank and Miller, 1983). The size of these structures may be increased still further by additional cycles of freeze-thaw and it is possible to produce structures with diameters in the range 10-30 μm. Freeze-thaw is believed to produce large membrane structures as the result of vesicle breakage and re-sealing induced by the creation and breakdown of ice crystals. The procedure will not work if the solution in which the vesicles are suspended contains any cryo-protectant; as little as 10 mM sucrose is sufficient to prevent the production of large patch-clampable structures (Tank and Miller, 1983).

Dehydration-rehydration

An alternative method for the production of large proteoliposomes suitable for patch-clamp investigation was described by Criado and Keller (1987) and has been used by them and others to monitor channel activity in a range of native and purified channel species (Keller *et al.* 1988; Riquelme *et al.* 1990a,b). As with freeze-thaw, the initial stage in this procedure as described by Criado and Keller is the solubilisation of the channel protein and its incorporation into small unilamellar vesicles. These authors employed Chaps (3-[(3-cholamidopropyl)-dimethylammonio]-1-propane sulphonate) to solubilise membrane preparations ranging from skeletal muscle sarcoplasmic reticulum to chloroplast envelopes. Solubilised channel proteins were then incorporated into small unilamellar vesicles by dialysis in the presence of exogenous lipids (Keller *et al.* 1988). Aliquots of the resulting proteoliposomes were

sedimented by centrifugation and the pellet resuspended in a small volume of 10%
Mops buffer (pH 7.4) containing 5% (w/v) ethylene glycol. A drop of this suspension
was applied to a glass slide and dehydrated at 4°C in a desiccator over $CaCl_2$. The
ethylene glycol in the suspension prevents complete dehydration. With a starting
volume of 20 µl a period of 3 hours dehydration was allowed before rehydration was
initiated by the addition of 20 µl of 100 mM KCl, or other experimental solution, to
the partially dehydrated suspension. Rehydration was allowed to continue overnight
at 4°C. At the end of this period large multilamellar liposomes ranging in diameter
from five to a few hundred µm were seen at the edges of the rehydrated film.

A variation of this technique has been used by Riquelme *et al.* (1990a). These authors
used membrane vesicles prepared from *Torpedo* electroplax, from which peripheral
proteins had been removed by alkaline extraction, to monitor acetylcholine receptor-
channel activity. These native membrane vesicles were added to phospholipid
liposomes and the mixture subjected to partial dehydration and rehydration as described
above. The same group have used dehydration-rehydration to monitor single-channel
events from purified glycine receptors (Riquelme *et al.* 1990b).

Patch-clamping large proteoliposomes

Irrespective of the method used to form proteoliposomes, single-channel properties
of channels incorporated into these structures can be investigated using conventional
patch-clamp procedures. Using fire polished pipettes (tip resistance 5-20 GΩ), seals
(10-200 GΩ) are readily obtained (Tank and Miller, 1983; Tomlins and Williams,
1986; Keller *et al.* 1988; Riquelme *et al.* 1990a). Channel activity is best monitored
following excision of the patch from the proteoliposome (Tank and Miller, 1983).

8. Which method should be used?

Clearly there are a number of methods available to a worker wishing to investigate
the properties of an ion channel in a reconstituted system. Before embarking on a
study using any of these approaches it is worth spending some time considering
which is the most appropriate for the task. Some factors worthy of consideration are:-

(1) *Ease of use and reliability.* It can probably be argued that the easiest and most
reliable method of ion channel reconstitution is the one with which any particular
investigator is most familiar. A brief survey of the literature would suggest that most
workers favour planar lipid bilayers formed on holes in partitions, and of these the
majority involve bilayers spread from a dispersion of phospholipids in n-decane
(painted bilayers). In my experience, painted bilayers are the easiest to make and will
provide a mechanically and electrically stable environment for either native or
purified channel proteins. However, this approach may not be suitable for all
applications. Bilayers produced in this way will contain some solvent which may, in
theory, affect channel performance. Although it should be noted that where
comparisons of channel activity have been carried out using solvent-containing and
solvent-free systems, bilayer solvent does not appear to have significant adverse
effects (Labarca *et al.* 1980; Latorre, 1986; Moczydlowski *et al.* 1984).

Both painted bilayers and folded bilayers formed on a partition have the important advantage that the investigator has ready access to, and can easily control, the constituents of the solutions on both sides of the bilayer.

(2) *Resolution.* The usefulness of a particular system of ion channel reconstitution may be limited by the relative size of the inherent background noise of the system and the signal under investigation. It is impractical to paint bilayers on holes with diameters much below 100 μm. The large surface area of such a bilayer in comparison with an equivalent bilayer formed at the end of a patch pipette with a diameter of approximately 1 μm means that the painted bilayer will be considerably more prone to electrical and mechanical noise. It may not be entirely coincidental that the majority of channels investigated in painted bilayers have high single-channel conductance.

Bilayer noise can be reduced by low-pass filtering but this will limit the resolution of channel open and closed lifetimes. Placing the experimental chamber in a metal box will screen the system from electrical interference and mechanical vibration can be reduced by siting the experimental chamber on some form of vibration isolation system (Alvarez, 1986).

(3) *Bilayer electronics.* The principles governing the measurement of single-channel current deflections are the same whether the channel is in its native cell membrane or reconstituted into an artificial bilayer, and the apparatus used for monitoring channels incorporated into bilayers on patch pipettes is identical to that used in conventional patch-clamp experiments. However some differences do arise when channels are incorporated into large planar bilayers. Under these conditions there is essentially no series resistance. The large surface area of the bilayer and hence the much larger capacitance of this system compared with a bilayer on a patch pipette means that capacity compensation is considerably more difficult to achieve.

The circuit layout used in our laboratory for the measurement of single-channel current fluctuations in large planar bilayers is shown in Fig. 6. The bilayer chambers are connected to the circuit via 2% (w/v) agar bridges in 3 M LiCl and silver-silver chloride electrodes (see Fig. 1). High ionic strength salt bridges are used to minimize liquid junction potentials for measurements under mixed ion conditions. The reed switch located between the input sockets connecting the silver-silver chloride electrodes to the circuit can be activated with a magnet. When activated, the reed switch shorts out the chambers and prevents the development of large voltages across the bilayer during chamber perfusion or additions to the chambers.

The requirements for the current to voltage converter A1 (OPA102BM from Burr-Brown, 1 Millfield House, Woodshots Meadow, Watford, Hertfordshire,WD1 8YX, UK), are: (a) low input bias current, (b) low noise, (c) high speed.

Alas these are to some extent mutually exclusive, resulting in some manufacturers using discrete components to fabricate head amplifiers. For simplicity's sake the use of a low noise electrometer operational amplifier together with a frequency compensation circuit yields a head amplifier with reproducible characteristics, simple construction and easy setting up.

R1 is a critical component, needing to be at least 1% tolerance, to ensure an

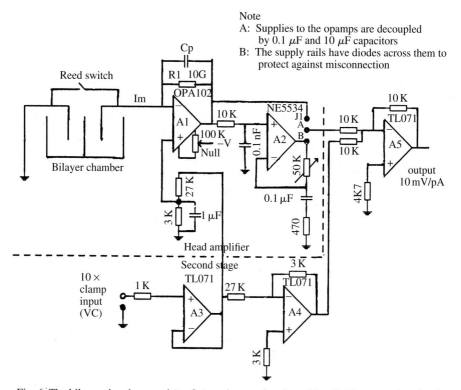

Note

A: Supplies to the opamps are decoupled
    by 0.1 μF and 10 μF capacitors

B: The supply rails have diodes across them to
    protect against misconnection

Fig. 6. The bilayer chamber consists of *cis* and *trans* chambers (Fig. 1). The *trans* chamber is held at virtual ground whilst the *cis* chamber may be clamped at a potential relative to ground. The bilayer current is converted to a voltage by amplifier A1 and resistor R1, to give an output of V=−(Im × R1) + (VC/10). Amplifier A2 applies frequency compensation. The output of this stage is passed to the second stage where it is summed with a voltage equivalent to - (VC/10), to give an output Im × R1.

accurate output, and needs to have a good 'lumped' parasitic capacitance. We have found resistors from KOBRA (123 Interstate Drive, W. Springfield, MA 01089, USA) to be satisfactory. The parasitic capacitance associated with R1 causes the output to be limited in rise time, as the capacitance has to be charged through R1. With R1=10 GΩ this capacitance gives a rise time of several ms! The frequency compensation circuit corrects this to approximately 200 μs. With smaller values of R1 the parasitic capacitance has less effect, allowing faster rise times, with concomitant lower output (1GΩ=1 mV/pA etc). The frequency compensation circuit is effectively a frequency sensitive amplifier, amplifying fast changing signals more than slow changing ones, and falling to unity gain at DC. This means that it will preferentially amplify any noise from the bilayer. R2 and C1 act as a low pass single pole filter to cut out high frequency noise before frequency compensation. The frequency compensated signal will still be more noisy than the non-frequency compensated (there is no such thing as a free lunch!). For noise reasons the command voltage to the head amplifier is supplied at 10× the required value and reduced to the required value at the head amplifier. This also reduces any noise impressed on the signal. The signal is also filtered by a capacitor across the 3 K resistor. This also limits the rise time of any command signal and therefore may be undesirable. As with the parasitic capacitance of R1 above, the membrane capacitance has to be charged through R1. This results in amplifier saturation on the application of a step change in VC. Due to the large capacitance associated with bilayers, around 47 pF for a 200 μm hole, as opposed to patch pipettes, capacity compensation for the membrane is difficult to implement in the usual fashion due to the limited voltage excursion of operational amplifiers.

The output of the head amplifier is a voltage equivalent to $-(\text{Im} \times \text{R1}) + (\text{VC}/10)$. To remove the contribution due to VC the second stage sums the head amplifier output with an inverted signal equivalent to VC/10.

The second stage supplies power to the head amplifier. This should be a low noise supply. A stirrer supply is also needed and can be run from the second stage. It is important to ensure that both leads to the stirrer are grounded when the stirrer is switched off. Star earthing should be used throughout to minimise earth related noise problems.

9. Conclusion

Ion channel reconstitution is an invaluable technique for the investigation of the properties of channels from intracellular membrane systems such as the endoplasmic or sarcoplasmic reticulum. It also provides an essential assay system for the characterisation of the properties of purified channel proteins. I hope that this chapter provides a useful introduction to the diversity and flexibility of the various approaches available to the potential ion channel "reconstituter".

Acknowledgements

I should like to thank Drs Chris Miller and Roberto Coronado for introducing me to the delights of ion channel reconstitution. I am also extremely grateful to Richard Montgomery who has designed and built much of the apparatus used in my laboratory. He made a number of useful comments on the manuscript and provided and described the circuit layout shown in Figure 6. The work from our laboratory cited in the chapter has been supported by grants from the British Heart Foundation, the Medical Research Council and the Wellcome Trust.

References

AGNEW, W. S., ROSENBERG, R. L. & TOMIKO, S. A. (1986). Reconstitution of the sodium channel from Electrophorus electricus. In *Ion Channel Reconstitution*, (ed. C. Miller), pp. 307-335. New York: Plenum.

ALVAREZ, O. (1986). How to set up a bilayer system. In *Ion Channel Reconstitution*, (ed. C. Miller), pp. 115-130. New York: Plenum.

ANDERSON, K., LAI, F. A., LIU, Q-Y., ROUSSEAU, E., ERICKSON, H. P. & MEISSNER, G. (1989). Structural and functional characterization of the purified cardiac ryanodine receptor-Ca^{2+} release channel complex. *J. Biol. Chem.* **264**, 1329-1335.

ASHLEY, R. H. & WILLIAMS, A.J. (1990). Divalent cation activation and inhibition of single calcium release channels from sheep cardiac sarcoplasmic reticulum. *J. Gen. Physiol.* **95**, 981-1005.

BELL, J. E. & MILLER, C. (1984). Effects of phospholipid surface charge on ion conduction in the K^+ channel of sarcoplasmic reticulum. *Biophys. J.* **45**, 279-287.

BEZPROZVANNY, I. B., WATRAS, J. & EHRLICH, B. E. (1991). Bell-shaped calcium-response curves of Ins(1,4,5)P_3- and calcium-gated channels from endoplasmic reticulum of cerebellum. *Nature* **351**, 751-754.

CATTERALL, W. A., SEAGAR, M. J., TAKAHASHI, M. & NUNOKI, K. (1989). Molecular properties of dihydropyridine-sensitive calcium channels. *Ann. New York Acad. Sci.* **560**, 1-14.

COHEN, F. S. (1986). Fusion of liposomes to planar bilayers. In *Ion Channel Reconstitution*, (ed. C. Miller), pp. 131-139. New York: Plenum.

COHEN, F. S., NILES, W. D. & AKABAS, M. H. (1989). Fusion of phospholipid vesicles with a planar membrane depends on the membrane permeability of the solute used to create the osmotic pressure. *J. Gen. Physiol.* **93**, 201-210.

COHEN, F. S., ZIMMERBERG, J. & FINKELSTEIN, A. (1980). Fusion of phospholipid vesicles with planar phospholipid bilayer membranes. II. Incorporation of a vesicular membrane marker into the planar membrane. *J. Gen. Physiol.* **75**, 251-270.

COREY, D. P. & STEVENS, C. F. (1983). Science and technology of patch-recording electrodes. In *Single-Channel Recording*, (eds. B. Sakmann & E. Neher), pp. 53-68. New York: Plenum.

CORONADO, R. (1985). Effect of divalent cations on the assembly of neutral and charged phospholipid bilayers in patch-recording pipettes. *Biophys. J.* **47**, 851-857.

CORONADO, R. (1986). Recent advances in planar phospholipid bilayer techniques for monitoring ion channels. *Ann. Rev. Biophysics Biophys. Chem.* **15**, 259-277.

CORONADO, R. & AFFOLTER, H. (1986). Insulation of the conduction pathway of muscle transverse tubule calcium channels from the surface charge of bilayer phospholipid. *J. Gen. Physiol.* **87**, 933-953.

CORONADO, R. & LATORRE, R. (1983). Phospholipid bilayers made from monolayers on patch-clamp pipettes. *Biophys. J.* **43**, 231-236.

CRIADO, M. & KELLER, B. U. (1987). A membrane fusion strategy for single-channel recordings of membranes usually non-accessible to patch-clamp pipette electrodes. *FEBS Lett.* **224**, 172-176.

EVANS, W. H. (1990). Organelles and membranes of animal cells. In *Biological Membranes: A Practical Approach* (eds. J. B. C. Findlay & W. H. Evans), pp. 1-35. Oxford: IRL Press.

EWALD, D. A., WILLIAMS, A. J. & LEVITAN, I. B. (1985). Modulation of single Ca^{2+}-dependent K^+-channel activity by protein phosphorylation. *Nature* **315**, 503-506.

FINKELSTEIN, A., ZIMMERBERG, J. & COHEN, F. S. (1986). Osmotic swelling of vesicles: its role in the fusion of vesicles in planar phospholipid bilayer membranes and its possible role in exocytosis. *A. Rev. Physiol.* **48**, 163-174.

FRENCH, R. J., WORLEY, J. F. III, BLAUSTEIN, M. B., ROMINE, W. O. JR., TAM, K. K. & KRUEGER, B. K. (1986). Gating of batrachotoxin-activated sodium channels in lipid bilayers. In *Ion Channel Reconstitution*, (ed. C. Miller), pp. 363-383. New York: Plenum.

HANKE, W. (1986). Incorporation of ion channels by fusion. In *Ion Channel Reconstitution*, (ed. C. Miller), pp. 141-153. New York: Plenum.

HANKE, W., METHFESSEL, C., WILMSEN, U. & BOHEIM. G. (1984). Ion channel reconstitution into lipid bilayer membranes on glass patch pipettes. *Bioelectrochem. Bioenergetics.* **12**, 329-339.

HARTSHORNE, R., TAMKUN, M. & MONTAL, M. (1986). The reconstituted sodium channel from brain. In *Ion Channel Reconstitution*, (ed. C. Miller), pp. 337-362. New York: Plenum.

IMAGAWA, T., SMITH, J. S., CORONADO, R. & CAMPBELL, K. P. (1987). Purified ryanodine receptor from skeletal muscle sarcoplasmic reticulum is the Ca^{2+}-permeable pore of the Ca release channel. *J. Biol. Chem.* **262**, 16636-16643.

JONES, O. T., EARNEST, J. P. & MCNAMEE, M. G. (1990). Solubilization and reconstitution of membrane proteins. In *Biological Membranes: A Practical Approach* (eds. J. B. C. Findlay and W. H. Evans), pp. 139-177. Oxford: IRL Press.

KASAHARA, M. & HINKLE, P. C. (1977). Reconstitution and purification of the D-glucose transporter from human erythrocytes. *J. Biol. Chem.* **252**, 7384-7390.

KELLER, B. U., HEDRICH, R., VAZ, W. L. C. & CRIADO, M. (1988). Single channel recordings of reconstituted ion channel proteins: an improved technique. *Pflugers Archiv. Eur. J. Physiol.* **411**, 94-100.

LABARCA, P., CORONADO, R. & MILLER, C. (1980). Thermodynamic and kinetic studies of the gating behaviour of a K⁺-selective channel from the sarcoplasmic reticulum membrane. *J. Gen. Physiol.* **76**, 397-424.

LAI, F. A., ERICKSON, H. P., ROUSSEAU, E., LIU, Q-Y. & MEISSNER, G. (1988). Purification and reconstitution of the Ca release channel from skeletal muscle. *Nature* **331**, 315-319.

LAI, F. A. & MEISSNER, G. (1989). The muscle ryanodine receptor and its intrinsic Ca²⁺ channel activity. *J. Bioenergetics and Biomembranes* **21**, 227-246.

LATORRE, R. (1986). The large calcium-activated potassium channel. In *Ion Channel Reconstitution*, (ed. C. Miller), pp. 431-467. New York: Plenum.

LEVITSKI, A. (1985). Reconstitution of membrane receptor systems. *Biochim. Biophys. Acta: Bio-Membranes* **822**, 127-153.

LINDSAY, A. R. G., MANNING, S. D. & WILLIAMS, A. J. (1991). Monovalent cation conductance in the ryanodine receptor-channel of sheep cardiac muscle sarcoplasmic reticulum. *J. Physiol.* **439**, 463-480.

MILLER, C. (1978). Voltage-gated cation conductance channel from fragmented sarcoplasmic reticulum: Steady-state electrical properties. *J. Memb. Biol.* **40**, 1-23.

MILLER, C. (1982a). Feeling around inside a channel in the dark. In *Transport in Biological Membranes* (ed. R. Antolini), pp. 99-108. New York: Raven Press.

MILLER, C. (1982b). Open-state substructure of single chloride channels from Torpedo electroplax. *Phil. Trans. Royal Soc. London B.* **299**, 401-411.

MILLER, C. (1983). Integral membrane channels: Studies in model membranes. *Physiol. Rev.* **63**, 1209-1242.

MILLER, C. & RACKER, E. (1976). Calcium-induced fusion of fragmented sarcoplasmic reticulum with artificial planar bilayers. *J. Memb. Biol.* **30**, 283-300.

MILLER, C. & ROSENBERG, R. L. (1979). A voltage-gated conductance channel from fragmented sarcoplasmic reticulum. Effects of transition metal ions. *Biochem.* **18**, 1138-1145.

MOCZYDLOWSKI, E., ALVAREZ, O., VERGARA, C. & LATORRE, R. (1985). Effect of phosopholipid surface charge on the conductance and gating of a Ca²⁺-activated K⁺ channel in planar lipid bilayers. *J. Membr. Biol.* **83**, 273-282.

MOCZYDLOWSKI, E., GARBER, S. H. & MILLER, C. (1984). Batrachotoxin-activated Na⁺ channels in planar lipid bilayers. Competition of tetrodotoxin block by Na⁺. *J. Gen. Physiol.* **84**, 665-686.

MONTAL, M. (1987). Reconstitution of channel proteins from excitable cells in planar lipid bilayer membranes. *J. Membr. Biol.* **98**, 101-115.

MONTAL, M., ANHOLT, R. & LABARCA, P. (1986). The reconstituted acetylcholine receptor. In *Ion Channel Reconstitution*, (ed. C. Miller), pp. 157-204. New York, London: Plenum.

MONTAL, M. & MUELLER, P. (1972). Formation of bimolecular membranes from lipid monolayers and a study of their electrical properties. *Proc. Nat. Acad. Sci. USA* **69**, 3561-3566.

MUELLER, P. & RUDIN, D. O. (1969). Bimolecular lipid membranes: Techniques of formation, study of electrical properties, and induction of ionic gating phenomena. In *Laboratory Techniques in Membrane Biophysics* (eds. H. Passow and R. Stampfli), pp. 141-156. Berlin: Springer-Verlag.

MUELLER, P., RUDIN, D. O., TIEN, H. T. & WESCOTT, W. C. (1962). Reconstitution of excitable cell membrane structure in vitro. *Circulation* **26**, 1167-1171.

NELSON, M. T., FRENCH, R. J. & KRUEGER, B. K. (1984). Voltage-dependent calcium channels from brain incorporated into planar lipid bilayers. *Nature* **308**, 77-80.

NELSON, N., ANHOLT, R., LINDSTROM, J. & MONTAL, M. (1980). Reconstitution of purified acetylcholine receptors with functional ion channels in planar lipid bilayers. *Proc. Nat. Acad. Sci. USA* **77**, 3057-3061.

NILES, W. D. & COHEN, F. S. (1987). Video fluorescence microscopy studies of phospholipid vesicle fusion with planar phospholipid bilayer membranes. Nature of membrane-membrane interactions and detection of release of contents. *J. Gen. Physiol.* **90**, 703-735.

NILES, W. D., COHEN, F. S. & FINKELSTEIN, A. (1989). Hydrostatic pressures developed by osmotic swelling vesicles bound to planar membranes. *J. Gen. Physiol.* **93**, 211-244.

RIQUELME, G., LOPEZ, E., GARCIA-SEGURA, L. M., FERRAGUT, J. A. & GONZALEZ-ROS, J. M. (1990a). Giant liposomes: A model system in which to obtain patch-clamp recordings of ionic channels. *Biochem.* **29**, 11215-11222.

RIQUELME, G., MORATO, E., LOPEZ, E., RUIZ-GOMEZ, A., FERRAGUT, J. A., GONZALEZ-ROS, J. M. & MAYOR, F. JR. (1990b). Agonist binding to purified glycine receptor reconstituted into giant liposomes elicits two types of chloride channel currents. *FEBS Lett.* **276**, 54-58.

ROUSSEAU, E., ROBERSON, M. & MEISSNER, G. (1988). Properties of single chloride selective channel from sarcoplasmic reticulum. *Eur. Biophys. J.* **16**, 143-151

SAKMANN, B. & NEHER, E. (1983). Geometric parameters of pipettes and membrane patches. In *Single-Channel Recording*, (eds. B. Sakmann & E. Neher), pp. 37-51. New York: Plenum.

SCHINDLER, H. (1980). Formation of planar bilayers from artificial and native membrane vesicles. *FEBS Lett.* **122**, 77-79.

SCHINDLER, H. & QUAST, U. (1980). Functional acetylcholine receptor from Torpedo marmorata in planar membranes. *Proc. Nat. Acad. Sci. USA* **77**, 3052-3056.

SILVIUS, J. R. (1992). Solubilization and functional reconstitution of biomembrane components. *A. Rev. Biophys. Biomol. Struct.* **21**, 323-348.

SMITH, J. S., IMAGAWA, T., MA, J. J., FILL, M., CAMPBELL, K. P. & CORONADO, R. (1988). Purified ryanodine receptor from rabbit skeletal muscle is the calcium-release channel of sarcoplasmic reticulum. *J. Gen. Physiol.* **92**, 1-26.

SUAREZ-ISLA, B. A., WAN, K., LINDSTROM, J. & MONTAL, M. (1983). Single-channel recordings from purified acetylcholine receptors reconstituted in bilayers formed at the tip of patch pipets. *Biochem.* **22**, 2319-2323.

TANK, D. W., HUGANIR, R. L., GREENGARD, P. & WEBB, W. W. (1983). Patch-recorded single-channel currents of the purified and reconstituted Torpedo acetylcholine receptor. *Proc. Nat. Acad. Sci. USA* **80**, 5129-5133.

TANK, D. W. & MILLER, C. (1982). Isolated-patch recording from liposomes containing functionally reconstituted chloride channels from Torpedo electroplax. *Proc. Nat. Acad. Sci. USA* **79**, 7749-7753.

TANK, D. W. & MILLER, C. (1983). Patch-Clamped Liposomes. Recording reconstituted ion channels. In *Single-Channel Recording* (eds. B. Sakmann & E. Neher), pp. 91-105. New York: Plenum.

TINKER, A., LINDSAY, A. R. G. & WILLIAMS, A. J. (1992a). A model for ionic conduction in the ryanodine receptor-channel of sheep cardiac muscle sarcoplasmic reticulum. *J. Gen. Physiol.* **100**, 459-517.

TINKER, A., LINDSAY, A. R. G. & WILLIAMS, A. J. (1922b). Block of the sheep cardiac sarcoplasmic reticulum Ca^{2+}-release channel by tetraalkyl ammonium cations. *J. Membr. Biol.* **127**, 149-159.

TINKER, A., LINDSAY, A. R. G. & WILLIAMS, A. J. (1992c). Large tetraalkyl ammonium cations produce a reduced conductance state in the sheep cardiac sarcoplasmic reticulum Ca^{2+}-release channel. *Biophys. J.* **61**, 1122-1132.

TINKER, A. & WILLIAMS, A. J. (1992). Divalent cation conduction in the ryanodine receptor-channel of sheep cardiac muscle sarcoplasmic reticulum. *J. Gen. Physiol.* **100**, 479-493.

TOMLINS, B. & WILLIAMS, A. J. (1986). Solubilisation and reconstitution of the rabbit skeletal muscle sarcoplasmic reticulum K^+ channel into liposomes suitable for patch clamp studies. *Pflugers Archiv.* **407**, 341-347.

TOMLINS, B., WILLIAMS, A. J. & MONTGOMERY, R. A. P. (1984). The characterization of a monovalent cation selective channel of mammalian cardiac muscle sarcoplasmic reticulum. *J. Membr. Biol.* **80**, 191-199.

WHITE, S. H. (1986). The physical nature of planar bilayer membranes. In *Ion Channel Reconstitution*, (ed. C. Miller), pp. 3-35. New York, London: Plenum.

WILLIAMS, A. J. (1992). Ion conduction and discrimination in the sarcoplasmic reticulum ryanodine receptor/calcium-release channel. *J. Muscle Res. Cell Motil.* **13**, 7-26.

WILMSEN, U., METHFESSEL, C., HANKE, W. & BOHEIM, G. (1983). Channel current fluctuation studies with solvent-free lipid bilayers using Neher-Sakmann pipettes. In *Physical Chemistry of Transmembrane Ion Motions* (ed. G. Spach), pp. 479-485. Amsterdam: Elsevier.

WOODBURY, D. J. & HALL, J. E. (1988a). Vesicle-membrane fusion. Observations of simultaneous membrane incorporation and content release. *Biophys. J.* **54**, 345-349.

WOODBURY, D. J. & HALL, J. E. (1988b). Role of channels in the fusion of vesicles with a planar bilayer. *Biophys. J.* **54**, 1053-1063.

WOODBURY, D. J. & MILLER, C. (1990). Nystatin-induced liposome fusion. A versatile approach to ion channel reconstitution into planar bilayers. *Biophys. J.* **58**, 833-839.

ZIMMERBERG, J., COHEN, F. S. & FINKELSTEIN, A. (1980). Fusion of phospholipid vesicles with planar phospholipid bilayer membranes. I. Discharge of vesicular contents across the planar membrane. *J. Gen. Physiol.* **75**, 241-250.

Chapter 6

Practical analysis of single channel records

DAVID COLQUHOUN

1. Introduction

The aim of this chapter is to discuss the sorts of things that can be measured in an experimental record of single channel currents, and how to make the measurements. What is measured will depend on the aims of the experiment. In some cases, interest may centre mainly on the *amplitudes* of the currents. For example, this might be the case (*a*) when channel conductance and subconductance levels are used as a criterion for a particular channel subtype, or (*b*) when measurements are made in solutions of different ionic composition and different membrane potentials, in order to investigate the mechanism of ion permeation through open channels. In other cases the *durations* of the open and shut times may be of primary interest, as when we want to know the nature of individual channel *activations* by a transmitter (the unitary event usually consists of more than one opening), or when the kinetic mechanism of channel operation is of primary interest.

The aims of a complete analysis are to measure (*a*) the amplitude(s) of the single channel currents, (*b*) the durations of shut periods, and the durations of sojourns at the various open channel current levels, and (*c*) the order in which the foregoing events occur. The amplitudes are, in the simplest cases at least, very nearly constant from one opening to the next. But, because we are looking at a single molecule, the durations of events and the order in which they occur are *random variables*; the information contained in them comes from measurements of their *distributions* (more strictly, their probability density functions - see Chapter 7 for more details).

Many measurements of individual durations have to be made in order to define these distributions properly. Single channel analysis must be one of the slowest known methods of generating an exponential curve from an experiment, because the averaging that normally results from having a large number of channels has not 'already been done for us'. Furthermore the analysis is particularly important; not much can be inferred by simply looking at the raw data, *because* of its randomness. These measurements can be rather time consuming; this leads to the temptation to use automatic or semiautomatic methods of analysis in which distributions are produced by a computer program with little intervention by the experimenter. It is not usual in

MRC Receptor Mechanisms Research Group, Department of Pharmacology, University College London, Gower Street, London WC1E 6BT, UK.

other fields for results to be produced without the data having been seen by the experimenter (though it is not entirely unknown). Personally I would be reluctant to accept for publication any paper in which the fit of each duration to the raw data had not been inspected by the experimenter, even if the experimenter had written the analysis program him or herself (and still more reluctant if the analyst was not even sure exactly what the program was doing to their data). Certainly it is important that reasonably complete details of analysis methods should be given in published papers so that the reader can make some sort of assessment of its reliability. There is a price for speed of analysis which is too high to be tolerable, and failing to look at your data exceeds that price, in my view. It would help if the writers of programs would not include options that allow this to be done.

2. Aims of analysis

It is usually supposed (with some reason) that the data can be well-represented as a series of rectangular transitions between discrete conductance states. The first aim of the analysis is to obtain an idealised version of the experimental record which resembles, as closely as possible, what would have been seen if the experiment had been free of noise and artefacts. This process is sometimes referred to as *restoration* of the observed record. The result will be a series of time intervals, each associated with the amplitude of the current during that interval, in the order in which the events actually occurred. Open and shut periods will not necessarily alternate; successive open intervals may occur if there are conductance sublevels, or if more than one channel opens simultaneously. Records in which two or more channels are open simultaneously are useful for checking the independence of channel openings (though tests of independence are insensitive - see Horn, 1991), but are *not* suitable for measuring lifetime distributions. A double opening may be omitted from the analysis of lifetimes by ignoring it (and simultaneously noting that the shut time between the preceding and following openings is to be ignored when forming the distribution of shut times); this procedure will bias both open and shut time distributions so it can be used only when double openings are rare.

The results of the analysis will not of course be entirely accurate. For example, if an opening is too short to be detected not only will its omission distort the open time distribution, but the shut periods on each side of it will appear to be one long shut period so the shut time distribution will be distorted too (and *vice versa* for missed shut times). This problem is discussed further below (section 10) and in Chapter 7.

Techniques are being developed that may allow direct fit of a mechanism to the original data without prior restoration (e.g. Chung *et al.* 1990; Fredkin & Rice, 1992), but these have not yet been developed to a point where they are useful in practice.

Computer programs for analysis

There are several commercial programs for single channel analysis. These are discussed briefly in Chapter 9. All of them can perform *some* of the types of analysis discussed below, but none of them can do *all* of these methods (or other types of analysis that are not discussed in this chapter). If you want to use the more advanced methods of analysis, or to develop new methods, then you have only two options: (1) write your own program (or modify an existing one if you can get hold of the source code), or (2) get a program from somebody who has already written it.

3. Filtering and digitization of the data

Whether data are recorded on magnetic tape during the experiment, or recorded on-line (see below), it will almost always be necessary to filter the data before any analysis is done.

Filtering the data

The purpose of filtering is essentially to reduce the amount of high frequency noise in the record, so a 'low-pass' filter should be used. This passes frequencies from 0 Hz (DC) up to an upper limit specified by cut-off frequency, f_c, set on the filter. Butterworth filters pass relatively little noise with frequencies above f_c - they have a 'steep roll-off', which is why they are used to prevent aliasing in noise analysis. However, they are quite unsuitable for filtering single channel data because the price paid for the steep roll-off is that they 'ring', i.e. produce a damped oscillation, in response to a rectangular input. Single channel currents are essentially rectangular, and will therefore be distorted by such a filter. The type of filter that is normally used is a Bessel filter (usually an 8-pole Bessel filter). The types labelled 'damped mode', or 'low Q', on some commercial filters are similar. This sort of filter produces little or no ringing in response to a step input, though it is less effective in removing frequencies above f_c. It should be noted that some commercial filters of the Bessel type are calibrated on the front panel with a number (the corner-frequency) that is twice the −3 dB frequency; it is preferable that the values of f_c stated in papers should always be the −3 dB frequency; see Chapter 16). The question of the optimum setting for f_c is discussed below.

   Filter risetime. Another convenient way to characterize the filter is by its rise time, t_r . This is given by

$$t_r = 0.3321/f_c , \tag{3.1}$$

so the higher the −3 dB frequency, the faster is the risetime. This expression, which is close to the 10-90% risetime, is actually derived for a type of filter known as a Gaussian filter, which behaves very like the 8-pole Bessel filter that is normally used in practice (see Colquhoun & Sigworth, 1983). Thus a 1 kHz filter has a risetime of 332 μs, and *pro rata* for other values of f_c.

   Combining filters. The relevant value of f_c in the discussions below is not that for the filter alone, but the effective value for the whole recording system. If, for example, the results were effectively filtered at $f_1 = 10$ kHz by the recording system, and were then filtered again at $f_2 = 5$ kHz, then the effective overall filtering would be given, approximately, by

$$\frac{1}{f_c^2} \approx \frac{1}{f_1^2} + \frac{1}{f_2^2},\qquad(3.2)$$

i.e. at $f_c \approx 4.47$ kHz in the present example. This expression is exact for a Gaussian filter and a good approximation for an 8-pole Bessel filter, but it may not give good results for steep roll-of filters such as Butterworth or Tchebychef type (usually used, for example, in tape recorders).

Digitization of the data

All practical forms of analysis have to be done on a computer, so the first stage is to convert the observed current into a series of numbers by means of an analogue-to-digital converter (ADC). This is usually done after filtering through an analogue filter; another possibility is to use a digital filter *after* sampling (though some pre-filtering may be needed to prevent saturation of the ADC by high frequency noise).

Agonist-activated channels will generally give rise to long records which can, most conveniently, be recorded on magnetic tape during the experiment, and then replayed later for digitization. The algorithm used for digitization of records should be capable of writing the numbers directly to magnetic disk as it goes, in order to avoid frequent breaks in the record. The sampling rate should be 10-20 times the −3 dB frequency of the filter in use (though a factor of 5 is sufficient if the resulting points are then supplemented by interpolation).

For example, for a record filtered with a Bessel-type filter with $f_c = 5$ kHz (−3 dB), a suitable sampling rate would be 50 kHz. Samples are normally stored as 16 bit (2 byte) integers, so this corresponds to 6 megabytes of data per minute. A computer with plenty of hard disk space is needed. Some routines (e.g. that supplied by Cambridge Electronic Design) can sample continuously to disk at up to 80 kHz, but others are limited to about 30 kHz. Further details can be found in Colquhoun & Sigworth (1983).

Sampling on-line

When the opening of channels is caused by a step change in membrane potential or agonist concentration it will usually be convenient to do the experiment on-line. The computer will supply the command signal, through a digital-to-analogue converter (DAC) output, to change the membrane potential or concentration, and then immediately sample the resulting current through an ADC input. This procedure avoids problems in defining the exact moment at which the command step was applied. The sample length will usually be much shorter than is needed for steady-state (e.g. agonist-activated) channel records. For example, a 300 ms depolarization might be applied every 10 s, or a 1 ms concentration jump might be applied and the resulting channel activity record for 1 s subsequently. In such cases the sampled values can easily be accommodated in the computer memory, and there will be no need to have a sampling routine that can write the numbers to a magnetic disc while sampling is in progress. In order to run DAC outputs and ADC inputs simultaneously, the computer routines should be interrupt-driven (see Chapter 9); the ability of different software/hardware combinations to do this varies greatly, but the latest version of the CED1401 interface (Cambridge Electronic Design) can do ADC sampling at 330 kHz simultaneously with DAC output at 105 kHz.

The length of the sample should be many times longer than the longest time constant that is to be investigated, so long samples will be needed for slow processes. If the sample is too short then the channel will appear sometimes to 'switch modes' between one sample and the next. Furthermore it is often forgotten in such experiments that it is not only the transient that follows the jump that is of interest, but also the equilibrium channel behaviour that is eventually attained. This is another reason for not making the sample too short. Preliminary analyses may be needed to determine how long the sample should be, and what recovery period is necessary between one pulse and the next.

If capacitative transients caused by the voltage step cannot be adequately compensated during the experiment, it may be desirable to do the compensation during the analysis. For example responses to pulses that happen to produce no channel openings can be averaged, and this average subtracted from each channel-containing record.

4. The measurement of amplitudes

As mentioned in the introduction, knowledge of channel amplitudes may be wanted as part of a complete analysis, or for particular purposes such as when measurements are made in solutions of different ionic composition and different membrane potentials, in order to investigate the mechanism of ion permeation through single open channels. The latter sort of study has been greatly facilitated by the ability to measure single channel currents. Before this was possible, such information had to be inferred indirectly from *instantaneous current-voltage relationships* in macroscopic experiments In the latter, currents are measured as soon as possible ('instantaneously') after a step change in membrane potential, so that changes in the macroscopic current that resulted from changes in channel permeation could be distinguished from the (generally slower) changes in macroscopic current that result from (potential-dependent) changes in the *number* of channels.

The amplitude of channel openings may appear (incorrectly) to be reduced by molecules that block the open channel, if the blockages are frequent and too short to be resolved (e.g., Ogden & Colquhoun, 1985).

The problems that arise in the measurement of single channel current amplitudes are as follows.

Attenuation of brief events

A brief event (opening or shutting) will, because of the filtering of the record, produce a response that does not reach full amplitude. This is illustrated in Figs 3 and 5-8. The fraction of the maximum amplitude that is attained (A_{max}/A_0) can be calculated as

$$A_{\mathrm{max}}/A_0 = \mathrm{erf}(2.668\, f_c w) = \mathrm{erf}(0.8860\, w/t_r) \qquad (4.1)$$

where erf( ) is the *error function*, f_c is the -3 dB frequency of the filter recording system, t_r is the risetime of the filter (see (3.1) above), and w is the duration of the event. The error function, which is a function that is closely related to the cumulative Gaussian distribution, can be obtained from Tables (e.g. Abramowitz & Stegun, 1965), or calculated by a computer subroutine (one is given by Colquhoun & Sigworth, 1983), and all mathematical subroutine libraries contain the error function (e.g. NAG library; Press *et al.* 1986). The function in (4.1) is plotted in Fig. 3; it is correct for a Gaussian filter, and is a good approximation for the eight pole Bessel-type filter that is most commonly used in practice (Colquhoun & Sigworth, 1983). It must be emphasised again that some commercial filters show a value of $2f_c$ on the front panel; the manual must be checked to find the correct -3 dB frequency (or the rise time determined empirically). Table 1 shows the duration of events (openings or shuttings) that are required to reach 25%, 50% and 90% of maximum amplitude (which correspond, respectively, to $0.25\, t_r$, $0.54\, t_r$ and $1.3\, t_r$).

Amplitudes of channel openings can be measured only for openings or shuttings that are long enough clearly to reach full amplitude (i.e. those with a length that is at least twice the rise time of the filter - see Fig. 3). There are two main approaches to

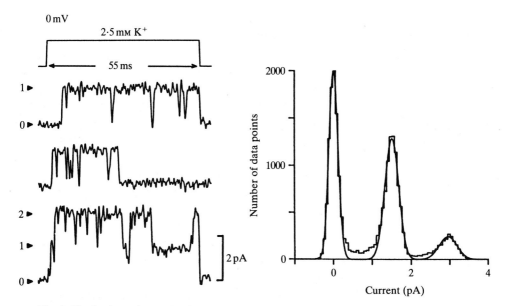

Fig. 1. Single channel records of potassium channels - delayed rectifier channels of skeletal muscle fibres - activated by a 55 msec depolarization from -100 mV to 0 mV. Two levels of opening are seen in the lowest record. On the right is shown a histogram of data points, showing peaks for the baseline and one and two levels of opening. Reproduced with permission from Standen *et al.* (1985).

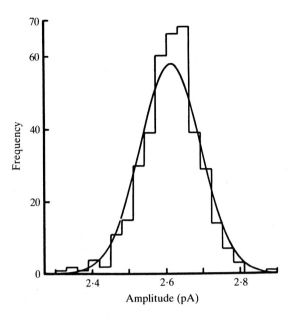

Fig. 2. An example of the distribution of the amplitude of single-channel currents. (Unpublished data of B. Sakmann & D. Colquhoun; R. *temporaria* end plate, $E_m = -91$ mV, suberyldicholine, 100 nM.) The mean of the 395 amplitudes was 2.61 pA. The continuous curve is a Gaussian distribution, which was fitted to the data by the method of maximum likelihood; it has a mean of 2.61 pA and a standard deviation of 0.08 pA. Note, however, that the observed distribution is rather more sharply peaked than the Gaussian curve. Reproduced with permission from Colquhoun & Sigworth (1983).

measurement of amplitudes, point amplitude histograms and amplitudes measured from each opening.

The point-amplitude histogram (distribution of data points)

The distribution of all the digitized current values can be plotted as a histogram. There will be a peak at the shut level and at each of the open levels. An example is shown in Fig. 1. The area under each peak is proportional to the *time spent* at that level.

This method has the advantage that the raw data are used without intervening processing, so the effects of the prejudices of the operator are minimized. It can also be calculated very rapidly. On the other hand, it has the disadvantages that (*a*) any drift in the baseline will distort the results (unless an appropriate correction is made), and (*b*) if there is more than one current amplitude there is no way to tell which *duration* (measured as below) is associated with which current amplitude.

Most real records contain some baseline irregularities, and often 'glitches' too, so in practice this method may be as slow to compute as the alternatives because of the necessity to correct properly for these imperfections. This method is also unsuitable if there are many openings or shuttings that are too short to reach full amplitude; in this case the histogram will be very smeared. It is useful, therefore, to confine the histogram to those points that correspond to periods when the channel is open, the

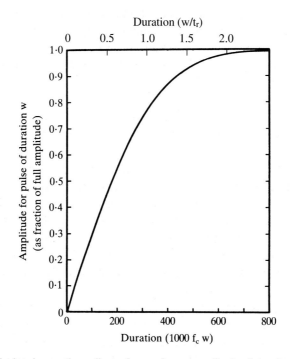

Fig. 3. Graph showing on the ordinate the maximum amplitude of the signal produced by a rectangular pulse of duration w. The amplitude is expressed as a fraction of the full amplitude (that produced by a long pulse). The abscissae are expressed in dimensionless units. The lower abscissa uses $1000 f_c w$ where f_c is the -3 dB frequency of the Gaussian filter (see text); These numbers represent the duration, w, in microseconds for $f_c = 1$ kHz, and *pro rata* for other f_c values. On the upper abscissa is plotted the duration of the pulse relative to the filter risetime, t_r. The fractional amplitude, up to a value of 0.5, is approximately equal to w/t_r (within 8%), e.g. for a fractional amplitude of 0.50, $w = 0.54 t_r = 0.179/f_c$. See also Fig. 5 and Table 1.

Table 1. *Duration (w, in μs) of rectangular pulses required to produce a response that reaches the specified fraction of the full amplitude, for Gaussian filter with various cutoff frequencies (f_c, -3 dB)*

| | | Fraction of full amplitude | | |
| | | 0.25 | 0.5 | 0.9 |
f_c (kHz)	t_r (μs)		w (μs)	
1	332	83	179	432
2	166	42	90	216
3	111	28	60	144
4	83	21	45	108
5	66	17	36	86

For fractional amplitudes of 0.25, 0.5 and 0.9, the values of w correspond to $w/t_r = 0.25$, 0.54 and 1.30 respectively. See also Fig. 3.

open point-amplitude histogram, so the smearing effects of the open-shut transitions are avoided; in order to be sure that the channel is fully open, it will be necessary to allow at least two risetimes to elapse before collecting points for the histogram. Similarly a *shut-point amplitude histogram* can be made by including only points from periods when the channel is shut.

Separate amplitude estimates for each opening

In this method a separate estimate of amplitude is made for each period for which the channel is open.

The amplitude may be estimated by averaging points at the shut level, and points at the open level (after allowing long enough after opening for the filter transient to be completed). However the mean of points at the open level is liable to be biased if the opening actually contains brief undetected closures. It may therefore be better (and will in any case be acceptable) to fit the open and shut level by eye, using cursors on the computer screen. Alternatively, for events that are sufficiently long, the amplitude can be estimated by a least squares fit, simultaneously with the duration, by the time-course method described in the next section.

The amplitude estimates thus found can be plotted as a histogram (as in Fig. 2). In this case the area under each peak represents the *number of sojourns* at that level (rather than the time spent at that level). The form of the expected distribution is not exactly Gaussian (see Colquhoun & Sigworth, 1983) but the deviation is usually sufficiently small that little harm is likely to come from fitting the results (by maximum likelihood - see below) with Gaussian curves, as illustrated in Fig. 2.

One great advantage of this method is that the *duration* of each individual opening is known, as well as the amplitude. This allows, for example, the examination of open time distributions that are restricted to openings that are in a specified amplitude range. This cannot be done if the only information about amplitudes comes from a point-amplitude histogram.

5. The measurement of durations

The information about channel mechanisms that is contained in a single channel record resides largely in the durations of the channel openings and shuttings. Measurement of open and shut times will, therefore, usually be required. There are two main problems to be solved; firstly transitions from one current level to another must be *detected*, then the duration of time between one transition and the next must be *measured*. For optimum results different criteria should be used for these two jobs.

Detection of transitions

The optimum methods for detection of transitions are described by Colquhoun & Sigworth (1983); they are complex, and involve, for example, knowledge of the spectral characteristics of the background noise. Such methods are virtually never used in practice at present; instead transitions are located as the points where the current crosses a preset threshold level. Fortunately, this simple method is not much worse than optimal methods.

In looking for transitions we aim to locate as many as possible of the genuine transitions, while rejecting, as far as is possible, any changes in current which, though they may look at first sight like transitions, are actually caused by random noise, or by door-slamming, tap-turning, seal breakdown or other such hazards of real life. Random noise can be coped with by setting the filter appropriately, and by imposition of a realistic resolution on the data, as described below. Disturbances resulting from things like door-slamming or refrigerators switching on and off, as well as baseline drift, can be dealt with *only by visual inspection of the data*. No foolproof method has yet been devised for automatically tracking a drifting baseline.

False event rate

The random noise in the record will, from time to time, result in fluctuations of the current sufficiently large to give the appearance of a transition (e.g. to cross a threshold level), even though no transition has actually occurred. Such *false events* can be kept to a minimum by filtering the data heavily, but if it is filtered too much important details may be obscured. The number of false events per second, λ_f, i.e. the number of times per second that the current departs from the baseline level by more than some specified amount, ϕ, is given, approximately (see Colquhoun & Sigworth, 1983), by

$$\lambda_f \approx f_c e^{-\phi^2/2\sigma_n^2}, \tag{5.1}$$

where f_c is the -3 dB frequency of the recording system, and σ_n is the background rms noise.

Note that the notation 'exp( )' is often used to denote 'e to the power', so that complicated expressions need not be written as superscripts; for example (5.1) would often be written as $\lambda_f \approx f_c \exp(-\phi^2/2\sigma_n^2)$.

In the context of threshold-crossing analysis ϕ would be the threshold current level (taking the shut level as zero). The false event rate thus depends on the filter setting, and on the ratio, ϕ/σ_n, of the threshold level to the rms baseline noise. Thus, when $f_c = 1$ kHz, a ratio, ϕ/σ_n, of 3 will result in about 11 false events per second, on average. Similarly $\phi/\sigma_n = 4$ corresponds to about 0.33 false events per second (one every 3 seconds), and $\phi/\sigma_n = 5$ corresponds to about one false event every 270 seconds. These rates change, *pro rata*, for other filter settings.

Measurement of transitions

Once transitions have been located, we then wish to estimate the time interval between adjacent transitions, i.e. to estimate the durations of openings and shuttings (and, possibly, to estimate the amplitudes of the events at the same time). Two methods are in common use (1) threshold-crossing and (2) time course analysis. The latter provides better resolution, but the former will usually be faster, though this depends on the amount of manual checking and error correction that is done.

Whichever method is used, it will be necessary, before starting the analysis, to measure the following two quantities.

(*a*) The amount of baseline noise should be measured by finding a stretch of baseline free from obvious events, and calculating the standard deviation of these points, i.e. the root mean square (rms) baseline noise, which will be denoted σ_n.

(*b*) A preliminary estimate of the channel amplitude (denoted A_0) should be made, by choosing some long openings that are easy to measure (this will be needed to position the threshold line and to fit the durations of events that are too short for their amplitude to be measured).

The threshold crossing method

Usually a threshold is set halfway between the fully open and the shut current levels; every time the observed current crosses this '50 percent threshold' a transition is deemed to have occurred, and the duration of an event is measured as the length of time for which the current stays above (or below) this threshold. It is important, therefore, to filter the data so that spurious transitions (false events - see (5.1) above) are rare.

Setting the filter for threshold-crossing analysis. A 50% threshold corresponds to $\phi/A_0 = 0.5$, where ϕ is the threshold level and A_0 is the channel amplitude (see above). The setting of the filter can be illustrated by an example, shown in Fig. 4, in which the channel amplitude was found to be $A_0 = 3.8$ pA. In this case, the 50% threshold will be set at $\phi = 1.9$ pA. *The filter setting, f_c, is now chosen so as to produce an acceptable false event rate.* In Fig. 4, the same channel opening (downward deflection, followed by two brief shuttings) is shown filtered at 1, 1.5, 2, 3 and 4 kHz (−3 dB). The standard deviation of the baseline noise (the r.m.s. noise) was, respectively, 0.10 pA, 0.14 pA, 0.19 pA, 0.27 pA and 0.33 pA. Thus, for the least filtered record (4 kHz), we have $\phi/\sigma_n = 1.9/0.33 = 5.8$. From (5.1), with $f_c = 4000$ Hz, we find $\lambda_f \approx 0.00025$ s^{-1}, i.e. roughly one false event in 66 minutes. This is a low rate, so filtering at 4 kHz would be suitable for threshold crossing analysis. It might be thought that this is an excessively low rate, and even less filtering would be safe. However, there are several reasons why it is better to be on the safe side. Firstly, the false event rate depends very steeply on ϕ/σ_n; in this case it was 5.8, but if ϕ/σ_n were to fall only to 5.0 (e.g. if the rms noise rose by only 15%, from 0.33 pA to 0.38 pA) the false event rate would go up from about one per hour to about one per minute. Secondly, this is worked out from the baseline (shut channel) noise, but the current is usually noisier when the channel is open so it is likely that there will be more false shuttings than predicted. And thirdly, most programs are not capable of holding the threshold accurately half-way between baseline and open level, which gives more scope for false events than calculated here. For time-course fitting the choice of filter is less critical; in a case such as that shown in Fig. 4, 2 or 3 kHz would be used.

Measuring the intervals between transitions. The time at which a transition occurred can be estimated by taking the data point on either side of the crossing of the threshold line, and interpolating between them to estimate the time at which the threshold is crossed. Some people simply count the number of data points between one threshold

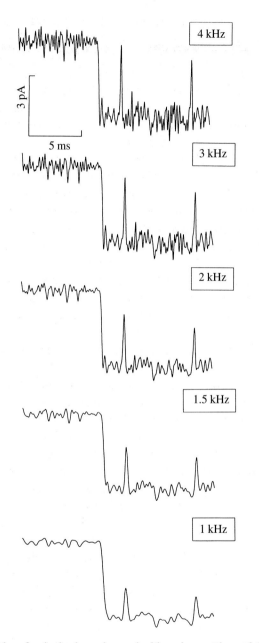

Fig. 4. Examples of a single channel record with various settings of the low pass filter. This channel comes from the record that was used for Fig. 10. The tape was replayed through an 8-pole Bessel filter set (top to bottom) at 4, 3, 2, 1.5 and 1 kHz (−3 dB). The channel amplitude is 3.8 pA, and the standard deviation of the baseline noise (the r.m.s. noise) is (top to bottom) 0.33 pA, 0.27 pA, 0.19 pA, 0.14 pA and 0.10 pA. The corresponding values of ϕ/σ_n are 5.8, 7.0, 10, 14 and 19 respectively. The data were originally recorded (after prefiltering at 20 kHz) on FM tape at 15 inches/second, giving a bandwidth up to 5 kHz. The output filter of the tape recorder was Tchebychef type, which is quite flat up to 5 kHz and then rolls off steeply. Nevertheless, the effective overall filtering, taking into account the patch clamp and tape recorder, will be somewhat more than is indicated (especially for the 4 kHz filter setting).

crossing and the next in order to estimate the duration, but, unless the data sampling rate is very fast, this method will be unnecessarily inaccurate for short events.

The *resolution* of the method, i.e. the shortest time interval that can be measured, is dictated by the signal-to-noise ratio of the data, and the filtering that must consequently be employed. Events that fail to reach 50 percent of full amplitude will, of course, be missed entirely by a 50 percent threshold detection method (see eq. 4.1, Fig. 3 and Table 1). Those that are just above 50 percent will be detected, but clearly the duration of intervals will be underestimated.

For example suppose that the filter is set at a −3 dB frequency of $f_c = 1$ kHz, so $t_r = 332$ μs (see (3.1) and Table 1). Events (openings or shuttings) of 179 μs or longer reach the 50 percent threshold (on average - the presence of random noise means that this will not happen every time). An event, of say, 190 μs duration would remain above the threshold for a short time only, and its duration would be seriously underestimated. It is shown by Colquhoun & Sigworth (1983) that the duration of events needs to be above roughly $1.3t_r$ (i.e. 430 μs in this example) before errors from this source become negligible. Although it is possible to correct for this effect, the correction is inexact in the presence of noise and is not usually used, so the resolution of the analysis is limited to 400-500 μs effectively, despite the fact that events much shorter than this can be *detected*.

Effect of duration on amplitude measurements. An opening must have a duration of at least $2t_r$ (of the order of 1 ms in the example above) before its amplitude can be measured reliably (see Fig. 3 - durations down to about t_r permit tolerable simultaneous fit of duration and amplitude, but an opening needs to last at least $2t_r$ before it is *clear* that the full amplitude has been reached; see Colquhoun & Sigworth, 1983).

For brief openings the response fails to reach full amplitude, as described above, and for any event much shorter than t_r, the shape of the response depends on the *area* of the pulse (i.e. the total charge passed during the opening), rather than on its amplitude. Any sort of brief current pulse will produce an observed response of the same shape; doubling the amplitude but halving the duration will result in indistinguishable responses (see Fig. 11-10 in Colquhoun & Sigworth, 1983). Brief events that do not reach full amplitude can therefore be fitted only if a value for the full amplitude is *assumed* (this applies equally to time course fitting). If there is only one sort of channel in the patch this is not a problem, but if the record contains large and small amplitude channel types then the duration of a brief event that falls short of the smaller amplitude cannot be estimated because there is no way to tell which sort of channel it originated from. In this case durations can be measured only for events that reach full amplitude so the resolution is drastically reduced (to about $2t_r$) for the whole analysis for the purposes of *fitting* of distributions to event durations, though the resolution for *detection* of events, and therefore the resolution to be imposed on the data (see below), may be much better than this.

The time course fitting method

The step-response function of the recording system. The time taken for the transition from a shut channel to a fully open channel is very short (less than 10 μs, Hamill *et al.*

1981), so the shape of the observed currents is almost entirely determined by the frequency response characteristics of the system. These characteristics may depend, for example, on the preparation itself, the patch clamp electrode, the patch clamp electronics, the tape recorder, and the filter that is used to limit high-frequency noise (see (3.2) above). The filter usually has the biggest effect. The response of a recording system to a step input can be measured as follows. A rectangular step input (intended to simulate the opening of a channel) can be induced by holding near to the headstage a wire connected to a high-quality triangular wave generator; the output is tape-recorded and filtered as in a real experiment. This output will be rounded, as illustrated in Figs 5 and 6. The result may depend on the tape speed (which controls the bandwidth of the recorder), and it may depend on whether the output filters on the

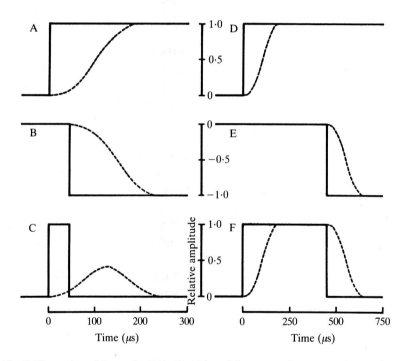

Fig. 5. Illustration of the method of calculation of the expected response to a step input. The left-hand column illustrates a short (45 μs) pulse, and the right-hand column a longer (450 μs) pulse. The dashed lines in A and D show (on different time scales) the experimentally measured response to a step input, shown schematically as a continuous line, for a system (patch clamp, tape recorder and filter) for which the final filter (eight pole Bessel) was set at 3 kHz (−3 dB). The rise time, t_r, of the filter is about 111 μs (see Table 1), so the pulse widths are 45 μs = 0.406t_r and 450 μs = 4.06t_r. (A) The response to a unit step at time zero is shown. B shows the same signal but shifted 45 μs to the right and inverted. The sum of the continuous lines in A and B gives the 45 μs unit pulse shown as a continuous line in C. The sum of the dashed lines in A and B is shown as a dashed line in C and is the predicted response of the apparatus to the 45 μs pulse. It reaches about 41% of the maximum amplitude, which is very close to the value of 39% expected for a Gaussian filter (see Table 1 and Fig. 3). D, E and F show, except for the time scale, the same as A, B and C but for a 450 μs pulse, which achieves full amplitude. Reproduced with permission from Colquhoun & Sigworth (1983).

tape recorder are set to Bessel or Tchebychef type (on recorders where they are switchable, such as the Racal recorders).

Fitting channel transitions. The measured step response can be used to calculate the response to any pattern of channel opening and shutting, by the method shown in Fig. 5. In the case of long openings (or long shuttings) the effect of the filtering is merely to round off the square corners of the transition, as illustrated by the first opening shown in Fig. 6A. But a variety of other patterns can be produced when short openings and shuttings occur; some of these are illustrated in Fig. 6B,C,D. Such calculated responses ('convolutions of step responses' in the usual jargon) can be superimposed on the observed current, and the time intervals and amplitudes adjusted until a good fit is obtained.

The fit may be judged by eye or by a least squares criterion. A least squares fit (even when amplitudes are estimated simultaneously) can be quite rapid on a modern fast PC-compatible computer. Examples of such fits are shown in Figs 7 and 8.

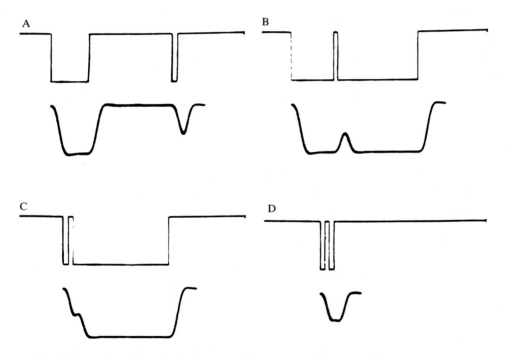

Fig. 6. Examples of the calculated output of the apparatus (lower trace) in response to two openings of an ion channel (upper traces). The curves are generated by a computer subroutine and were photographed on a monitor oscilloscope driven by the digital-to-analogue output of the computer. Openings are shown as downward deflection. (A) A fully resolved opening (435 µs) and gap (972 µs) followed by a partially resolved opening (67 µs). (B) Two long openings (485 and 937 µs) separated by a partially resolved gap (45.5 µs). (C) A brief opening (60.7 µs) and gap (53.1 µs) followed by a long opening (1113 µs); this gives the appearance of a single opening with an erratic opening transition. (D) Two short openings (both 58.2 µs) separated by a short gap (48.1 µs); this generates the appearance of a single opening that is only 55% of the real amplitude but which appears to have a more-or-less flat top, so it could easily be mistaken for a fully resolved subconductance level. Reproduced with permission from Colquhoun & Sigworth (1983).

Advantages and disadvantages of these two methods

Speed of analysis. As commonly practiced, the threshold crossing method is considerably faster than time course fitting, and when many channel openings have to be measured this is not a trivial consideration. However the speed difference depends very much on the amount of checking that is done. It obviously takes time to check visually the fit to every opening, and the position of the baseline before and after the opening. Time-course fitting forces you to make these checks, but threshold crossing

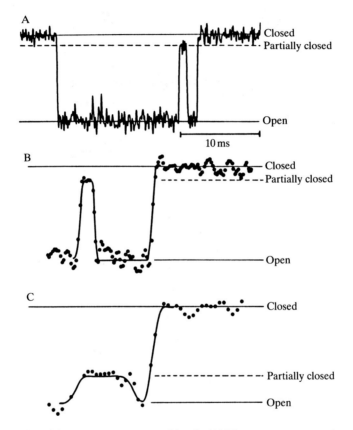

Fig. 7. Examples of single-channel current sublevels. (A) Elementary current activated by 100 nM ACh. −125 mV, 11°C. The two continuous horizontal lines, marked closed and open respectively, represent the patch current when a channel is either completely closed or completely open. The average amplitude of the current through the fully open channel is −3.71 pA. The dashed horizontal line represents the amplitude of a current sublevel. During the sublevel the channel is partially closed. The sublevel amplitude is −0.52 pA, i.e. 14% of the full amplitude. (B,C) Partial channel closures in another patch with 500 nM ACh at −178 mV and 10°C. The time course of the digitized current record is fitted (continuous line) by the same step response function as was used for full openings and closings. However, it was assumed that the amplitude of the current sublevels (marked partially closed) are 17% and 72%, respectively, of the full current amplitude. Sublevel amplitudes are indicated by the horizontal dashed line. The current through the fully open channel is −5.6 pA in (B) and −5.7 pA in C. The duration of the partial closures is 310 μs and 360 μs in B and C respectively. Filtered at 4 kHz (−3 dB). Reproduced with permission from Colquhoun & Sakmann (1985).

allows you to neglect them if you wish to do so. Furthermore, second generation time-course fitting programs will estimate both durations and amplitudes simultaneously, whereas threshold-crossing programs usually require separate estimation of amplitudes. Estimation of amplitudes requires an additional, separate, job, e.g. making a point amplitude histogram; this itself can take some time unless the

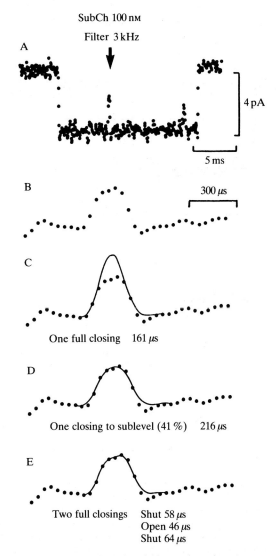

SubCh 100 nM

Filter 3 kHz

A

4 pA

5 ms

B   300 μs

C

One full closing   161 μs

D

One closing to sublevel (41 %)   216 μs

E

Two full closings   Shut 58 μs
Open 46 μs
Shut 64 μs

Fig. 8. Illustration of the problem of distinguishing subconductance states from multiple transitions. (A) Burst induced by SubCh, 100 nM, at −128 mV. Two putative brief gaps are visible. Low pass filter at 3 kHz (−3 dB). (B) The gap that is marked with an arrow in A shown on an expanded time scale. (C) Fit of data in B assuming a single complete closure of duration 161 μs. (D) Fit of data in B assuming a single closure to a subconductance state, of duration 216 μs. (E) Fit of data in B assuming that a full closure of 58 μs is followed by a full opening of 46 μs, and then another full closure of 64 μs. Reproduced with permission from Colquhoun & Sakmann (1985).

record is of such high quality that the baseline never drifts by more than a small fraction of the channel amplitude, and contains a negligible number of glitches. Even if this is done, information about the time spent at each amplitude level is lost.

Multiple conductance levels. Threshold-crossing methods are completely incapable of measuring channels that contain more than one conductance level. This is illustrated in Fig. 9, where a subconductance at about 50% of the full conductance level causes complete havoc. Lower subconductance levels would be missed entirely. The data in Fig. 9 are from a nicotinic channel; the problem is far more serious for GABA and glutamate channels which have frequent sublevel transitions.

Temporal resolution. The method of time course fitting clearly allows one to fit events that would be quite impossible with a threshold-crossing method, either (1) because they are too brief to reach the threshold at all (or too brief to have their duration measured accurately in this way, as discussed above) as in Fig. 6A,B, or (2) because brief events close to longer ones produce an apparent distortion as in Fig.

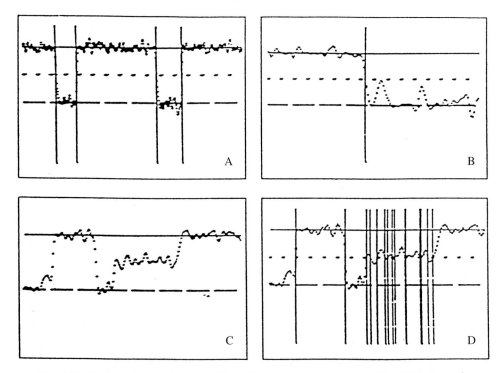

Fig. 9. Single channel currents recorded from *Xenopus* oocytes injected with mRNAs for α, β, γ and δ subunits of the BC3H1 nicotinic receptor. Cell-attached patch, acetylcholine 1 μM, filtered at 4 kHz (−3 dB) and sampled at 40 kHz. The solid horizontal line represents the baseline (shut level), the long-dashed line is the open level, and the short-dashed line is the threshold level. Vertical lines mark the positions of threshold crossings. In A and B, transitions are successfully located by the threshold-crossing method, but the channel shown in C and D happens to have a sublevel close to the threshold level, so, as shown in D, a large number of entirely spurious transitions are generated. If the program was operating automatically, these would all go into your results! (C. G. Marshall, A. J. Gibb & D. Colquhoun, unpublished data.)

6C,D. An automated threshold crossing method would, of course, produce *some* result when faced with such events, but the result would be meaningless; this is one reason for not using excessively automated methods. Time course fitting allows a resolution that is of the order of three-fold better than the threshold crossing method; briefer events can be fitted.

The most obvious danger of time course fitting is that of over-ambitiousness. The risk of false events arising from an attempt to fit events that are too short is just the same as described above for the threshold crossing method, but the temptation to fit them may be greater. Not only genuine channel transitions, but *any* disturbance (artefactual or random noise), if sufficiently brief, will produce a signal that can plausibly be fitted by the time- course method. For example Fig. 8 shows a case in which it is impossible to be sure whether a subconductance level should be fitted, or whether there are actually two full shuttings in quick succession. One way to deal with this part of the problem is to go through the data twice; the first time full closings are fitted whenever possible, the second time sublevels are fitted whenever possible. It may then be possible to decide which is the more plausible e.g. by seeing whether the sublevel fits produce a consistent amplitude estimate. Such ambiguities are, of course, not avoided by a threshold crossing analysis; they are merely brushed under the carpet.

Deciding the resolution

It is desirable, for several reasons, that the data should have a well-defined time resolution, i.e. that it be known that all events above the specified resolution, but none below it, have been fitted. For example, it is this time resolution that fixes the false event rate, it is the time resolution that fixes the minimum duration of an open or shut time that can be fitted (but see below). Furthermore, a well-defined time resolution is essential if any form of correction for missed events is to be applied (see below and Chapter 7).

It might be thought that the resolution is automatically fixed in the threshold-crossing analysis (as the pulse duration required to reach 50% of full amplitude), but (*a*) this is not constant (see Colquhoun & Sigworth, 1983), and (*b*) the safe resolution will be greater than this, as described above. In the case of time course fitting the shortest durations that are fitted are decided subjectively, and are unlikely to be constant throughout an experiment. It is therefore highly desirable that a fixed resolution should be imposed on the data *after* analysis, as described below. First, though, a decision must be made as to the appropriate value(s) for the resolution.

The false event rate (per second) that is acceptable obviously depends on the value of the true event rate. If we are looking for openings that occur at only one per 5 seconds on average, then the record contains a lot of shut baseline in which false events can occur. If a false event rate of about 2% of the true rate were thought tolerable then we might aim for a false event rate of about one per 270 seconds. The threshold would therefore be set to 5 times the baseline rms noise (see (5.1) above). In the example discussed above, with rms noise of 0.3 pA and a channel amplitude of A_0 = 3 pA, openings that have an amplitude less than $5 \times 0.3 = 1.5$ pA would therefore be ignored. Openings of this amplitude, which is 1.5/3.0 = 0.5 (50%) of the full amplitude, would (on average) have a duration of about $0.54t_r$, from (4.1) and Fig. 3. For example if the overall filtering of the data corresponded to $f_c = 1$ kHz (-3 dB), so $t_r = 332$ μs, then openings of $0.54 \times 332 \approx 180$ μs could be safely resolved (and *pro rata* for other values of f_c). Note, though, that although this resolution would be safe for *detection* of openings, it could be used for fitting them only if it was safe to *assume* an amplitude for such partially resolved openings; it requires a duration nearer to $2t_r$ for the amplitude to be resolved (see above).

For shut times, different values may apply. Suppose, in the example above, that openings occur in bursts, separated by short gaps, so there is relatively little open level in the record in which false gaps might be detected. For example Colquhoun & Sakmann (1985) observed brief gaps at a rate of about 500 per second of open time with acetylcholine. A false event rate of 2% of true rate, as before, now corresponds to about 11 false events per second, so the threshold could be set to only 3 times the baseline rms noise (see (5.1) above). In the example just discussed, shuttings that have an amplitude less than $3 \times 0.3 = 0.9$ pA would therefore be ignored. Shuttings of this amplitude, which is 0.9/3.0 = 0.3 (30%) of the full amplitude, would (on average) have a duration of about $0.31t_r$, from (4.1) and Fig. 3. For example if the overall filtering of the data corresponded to $f_c = 1$ kHz (-3 dB), so $t_r = 332$ μs, shuttings of $0.31 \times 332 \approx 100$ μs could be safely resolved (and *pro rata* for other values of f_c), though the safe resolution for open times is at least 180 μs, as just described.

These values change *pro rata* for other filter settings; e.g. a resolution of 25 μs for shut times might be obtainable at $f_c = 4$ kHz, as long as the signal to noise ratio of the data was good enough to give 0.3 pA rms noise at this filter setting. Such resolution can be obtained only by time course fitting. Some improvements might be obtained in threshold crossing analyses by using separate thresholds for openings and for shuttings, though this is not usually done, and would obviously cause problems if two or more successive events were brief.

Clearly the resolution that can be attained safely is not known until *after* the record has been analysed (so the optimum setting of thresholds for analysis is difficult). Our strategy with time course fitting evades this problem by fitting everything that could possibly be a real event while going through the data. At the end of the analysis the resolution must be specified, and this can be decided on the basis of several criteria: (*a*) the subjective feeling of the operator, during the analysis, concerning the shortest event that he or she can be sure is genuine, (*b*) calculations of false event rates of the sort illustrated above, based on the tentative analysis of the results, and (*c*) the appearance of the raw distributions of open and shut times, in particular the duration of interval below which there is 'obviously' a deficit of events. All of these criteria are, to some extent, subjective, but together they should allow realistic and safe values to be chosen for the best resolutions for open times and for shut times.

Imposition of the resolution. Once the resolution for open times has been decided then every open time shorter than the chosen value is treated, along with the shut time on each side of it, as one long shut period. An analogous procedure is followed for all shut times that are shorter than the chosen shut time resolution. The result is a record with a consistent resolution throughout; it contains no openings shorter than the chosen open time resolution and similarly for shut times. The open and shut times in this record are now ready to have distributions fitted, as described below.

This simple procedure is always used in our laboratory, though it does not seem to be widespread. Its neglect clearly leads, in principle, to inconsistency. Suppose, for example, that the open time resolution is decided (as is common) simply by looking at the distribution of raw open times and choosing the resolution as the duration below which there is an 'obvious' deficit of observations. This may well give a realistic estimate of the resolution for open times, but if we then go ahead and fit all open times in the raw data that are longer than the resolution so chosen we shall be fitting as distinct openings some pairs of openings that are separated only by a shut period that is shorter than the shut-time resolution, and which we therefore have no right to regard as well- defined separate openings. An analogous inconsistency obviously arises for the fitting of shut times. The problem is easily avoided if consistent open time and shut time resolutions are imposed on the data, as described above, before any fitting is attempted.

Measurement of P_{open}

The probability, P_{open}, that a channel is open (also known, for brevity, as the *open probability*) can be estimated as the fraction of time for which a channel is open, i.e. the total open time divided by the total length of the record.

A useful estimate of P_{open} can be obtained only when there is only one individual channel contributing to the record, and this is usually not the case. Sometimes, though, there are *sections* of the record that originate from one channel only. For example, at high agonist concentrations many channels show long silent periods during which all the channels in the patch are desensitized. Periodically one channel emerges from the desensitized state, and opens and shuts at a high rate (because the agonist concentration is high). The lack of double openings during such periods of high activity shows that they originate from one channel only. Therefore the silent desensitized periods can be cut out from the record (provided that they are so long as to be obviously desensitized), and P_{open} calculated from the periods of high activity only (e.g. Sakmann, Patlak & Neher, 1980; Colquhoun & Ogden, 1988). This has been used as a method for obtaining equilibrium concentration-response curves that are corrected for desensitization (the desensitized periods are cut out). This method also has the advantage that the response, P_{open}, is on an *absolute* scale (the maximum possible response is known, *a priori*, to be 1).

Once a list of all the open and shut times has been measured, it is simple to calculate P_{open} for any specified part of the record (see, for example, section on stability plots, below). However it is usually preferable not to measure P_{open} in this way, but rather to measure it by integrating the record (either numerically on the computer, or by playing the magnetic tape through an analogue integrator). Then P_{open} can be found as the area under the trace per unit time, divided by the single channel current amplitude. The great advantage of measuring P_{open} in this way is that it is insensitive to missed events (see section 10, below). The filtering of the signal attenuates and rounds the single channel signal (as shown in Figs 5-8), but *does not change the area* under the signal. Therefore the integration method is unaffected by the existence of undetected transitions.

6. The display of distributions

This section deals mainly with the display of measurements that have been made at equilibrium, so the average properties of the record are not changing with time (this can be checked by use of the stability plots described at the end of this section).

Following the analysis described above, we should have a list of our estimates of the durations of each (apparent) open period (together with the amplitude of the current) and of each (apparent) shut period in the order in which they occurred, each duration being greater than the chosen resolution. These durations are random variables so, in order to describe them quantitatively, a probability distribution must be fitted to them. We shall deal here mainly with the fitting of distributions that are described by the sum of one or more exponential (or geometric) components. This form of distribution is expected under the simplest (Markov) assumptions concerning the mechanism of channel opening; these assumptions are described in more detail in Chapter 7 and, for example, by Colquhoun & Hawkes (1983). Although this sort of distribution is what everybody uses it should be borne in mind that it will not be the correct form (a) when the resolution is such that many brief events are missed (though often this will not cause great deviations from exponential form) or (b) when the mechanism of channel opening does not obey the simple Markov assumptions (e.g. because membrane potential or ligand concentration are not held constant). Whatever the form of the distribution, it is described (for a continuous variable such as time) by a probability density function.

Probability density functions and histograms

The data consist of a list of times (e.g. open times or shut times or burst lengths); in order to display their distribution they must be displayed as a histogram. The histogram is usually described as showing the number of observations with durations that fall between the limits specified on the abscissa, i.e. its ordinate is dimensionless. The probability density function (p.d.f.), on the other hand, is a function such that the area under the curve (rather than the amplitude on the ordinate) represents probability, or number of observations. The p.d.f. therefore has dimensions of s^{-1}, and a total area of unity (see equation (1) below, and Chapter 7). Usually we will wish to superimpose a fitted p.d.f. on the histogram of the data, but the p.d.f. has dimensions that appear to be different from those of the histogram ordinate. The solution of this dilemma is that the histogram ordinate should be expressed not as frequency of events, but as a *frequency density*, for example, as 'frequency per 2 ms' (where 2 ms may be the bin width); thus it is expressed in reciprocal time units like the p.d.f. This is not only correct, but also makes it clear that *area* must represent frequency in the histogram if all bins are not of the same width. To achieve superimposition on the histogram, the area under the p.d.f. must be made to correspond to the *area* of the histogram boxes, $N\Delta t$, by multiplying the p.d.f. by $N\Delta t$ where Δt is the bin width and N is the total number of observations (including the estimated number that are below the chosen resolution and therefore not detected; see below). Further details are given by Colquhoun & Sigworth (1983).

A simple exponential p.d.f. for an interval of duration t is defined by

$$f(t) = \tau^{-1}e^{-t/\tau} \tag{6.1}$$

where τ is the mean of the distribution (which is the same thing as what would be called the 'time constant' of the curve if it were a decaying current rather than a p.d.f.). The initial constant, τ^{-1}, ensures that the p.d.f. has unit area. If there is more than one exponential component the distribution is referred to as a *mixture of exponential distributions* (or a 'sum of exponentials', but the former term is preferred since the total area must be 1). If a_i represents the area of the ith component, and τ_i is its 'mean' then

$$f(t) = a_1\tau_1^{-1}e^{-t/\tau_1} + a_2\tau_2^{-1}e^{-t/\tau_2} + \ldots$$
$$= \Sigma a_i\tau_i^{-1}e^{-t/\tau_i} \tag{6.2}$$

The areas add up to unity, i.e. $\Sigma a_i = 1$, and they are proportional, roughly speaking, to 'number of events' in each component. The overall mean duration is

$$\text{mean duration} = \Sigma a_i\tau_i . \tag{6.3}$$

The cumulative distributions. The *cumulative* form of this distribution, the probability that an interval is *longer than t*, is, for a single exponential,

$$P(\text{interval} > t) = \int_t^\infty f(t)dt = e^{-t/\tau}, \tag{6.4}$$

or, for more than one component, the sum of such integrals, *viz.*

$$P(\text{interval} > t) = \Sigma a_i e^{-t/\tau_i}. \tag{6.5}$$

Occasionally the data histogram is plotted in this cumulative form with the fitted function (6.5) superimposed on it. This presentation will always look smoother than the usual sort of histogram (the number of values in the early bins is large), but it should *never* be used, because the impression of precision that this display gives is *entirely spurious*. It results from the fact that each bin contains all the observations in all later bins, so adjacent bins contain nearly the same data. In other words successive points on the graph are not independent, but are strongly correlated, and this makes the results highly unsuitable for curve fitting.

To make matters worse, it may well not be obvious at first sight that cumulative distributions have been used, because the curve, (6.5), has exactly the same shape as the p.d.f. (6.2). There are no good reasons to use cumulative distributions to display data; they are highly misleading. In any case, it is much easier to compare results if everyone uses the same form of presentation.

Display of multi-component histograms

Figure 10A shows a histogram of shut times, with a time scale running from 0 to 1500 ms. This range includes virtually all the shut times that were observed.

The first bin actually starts at $t = 60$ μs rather than at $t = 0$, because a resolution of 60 μs was imposed on the data (see above) so there are no observations shorter than this. All that is visible on this plot is a single slowly-decaying component with a 'mean' of about 250 ms, though the first bin, the top of which is cut off on the display, shows that there are many short shut times too. The same data are shown in Fig. 10C, but only shut times up to 250 μs are shown here (so the 60 μs resolution is now obvious). There are many shut times longer than 250 μs of course, and these are

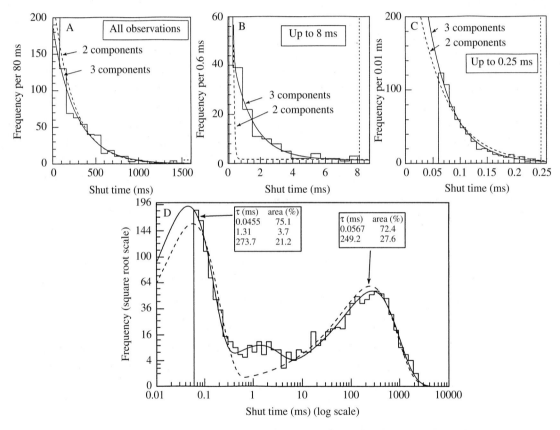

Fig. 10. Example of a distribution of shut times. In A, B and C the histogram of shut times is shown (on three different time scales), and in D the distribution of log(shut times), for the same data, is shown. The data are from nicotinic channels of frog endplate (suberyldicholine 100 nM, −130 mV). Resolutions of 80 μs for open times, and 60 μs for shut times, were imposed as described in the text; this resulted in 1348 shut times which were used to construct each of the histograms. The dashed bins (which are off scale in B, C) represent the number of observations above the upper limit. The data were fitted by the method of maximum likelihood with either two exponentials (dashed curve) or three exponentials (continuous curve). The same fit was superimposed on all of the histograms. The estimated parameters are shown in D. (D. Colquhoun and B. Sakmann, unpublished data.)

pooled in the dashed bin at the right hand end of the histogram (the top of which is cut off). Again the histogram looks close to a single exponential, but this time with a 'mean' of about 50 μs. Although it is not obvious from either of these displays, there is in fact a (small) third component in this shut time distribution. It is visible only in the display of the same data in Fig. 10B, in which all shut times up to 8 ms are shown, where an exponential with a mean of about 1 ms is visible. The data were not fitted separately for Fig. 10A,B,C, but one fit was done, to all the data (by maximum likelihood - see below) with either two exponential components (dashed line) or 3 exponential components (solid line). This same fit is shown in all four sections of Fig. 10. The deficiency of the 2 component fit is obvious only in the display up to 8 ms.

Clearly the conventional histogram display is inconvenient for intervals that cover such a wide range of values. The logarithmic display described next is preferable.

Logarithmic displays: the distribution of log(duration)

It was suggested by McManus, Blatz & Magleby (1987), and by Sine & Sigworth (1987), that it might be more convenient, when intervals cover a wide range (as in the preceding example), to look at the distribution of the logarithm of the time interval, rather than the distribution of the intervals themselves. Note that this is not simply a log transformation of the conventional display, because this would have bins of variable width on the log scale, whereas the distribution of log(t) is shown by bins of constant width on the log scale. In addition, Sine & Sigworth suggested using a square root transformation of the frequency density (to keep the errors approximately constant throughout the plot).

The distribution has the following form. If the length of an interval is denoted t, and we define

$$x = \log(t),$$

then we can find the p.d.f. of x, $f_X(x)$, as follows. First we note that if a t is less than some specified value t_1, then it will also be true that log(t) is less than log(t_1). Thus

$$\text{Prob}[t < t_1] = \text{Prob}[\log(t) < \log(t_1)] = P, \text{ say.} \qquad (6.6)$$

In other words the cumulative distributions for t and log(t) are the same. Now it is pointed out in chapter 7 (equations 3.7-3.9), that the p.d.f. can be found by differentiating the cumulative distribution. Thus, denoting the probability defined in (6.6) as P,

$$f_X(x) = \frac{dP}{dx} = \frac{dP}{d\log(t)} = \frac{dt}{d\log(t)} \cdot \frac{dP}{dt}$$

$$= t f(t)$$

$$= \Sigma a_i \tau_i^{-1} \exp(x - \tau_i^{-1} e^x) \qquad (6.7)$$

The second line here follows because dP/dt is simply the original distribution of time intervals, $f(t)$; it shows, oddly, that the distribution of $x = \log(t)$ can be expressed most simply not in terms of x, but in terms of t. When $f(t)$ is multi-exponential, as defined in (6.2), and we express $f_x(x)$ in terms of x by substituting $t = e^x$, we obtain the result in (6.7). This function is not exponential in shape, but is (for a single exponential component) a negatively skewed bell-shaped curve, the peak of which, very conveniently, occurs at $t = \tau$.

The same data, and the same fit, that was displayed in Fig. 10A,B,C, are shown in Fig. 10D as the distribution of log(shut times). The same fitted curves are also shown, and the three component fit shows three peaks which occur at the values of the three time constants. It is now clearly visible, from a single graph, that the two-exponential fit is inadequate. (The slow component of the 2-exponential fit also illustrates the shape of the distribution for a single exponential, because it is so much slower than the fast component that the two components hardly overlap.) This sort of display is now universally used for multi-component distributions. Its only disadvantage is that it is hard in the absence of a fitted line, to judge the extent to which the distribution is exponential in shape.

Bursts of channel openings

It is often observed that several channel openings (a *burst* of openings) occur in rapid succession, the individual openings being separated only by brief shut periods, and that then a much longer shut period is seen before the next burst. This may occur spontaneously, or as a result of brief channel blockages. This phenomenon is evident in the shut time distribution shown in Fig. 10, from which it is clear that about 75% of shut times are very short (around 50 µs on average), and almost all the rest are much longer. The former are the 'shut times within a burst', and the latter are the long shut times that separate one burst from the next ('shut times between bursts').

Definition of bursts in practice. If some critical time, t_c, is defined, such that shut times shorter than t_c are deemed to be 'within bursts', then the experimental record can be divided into bursts (the end of each burst being signalled by occurrence of a shut time longer than t_c. Such a division can never be totally unambiguous when dealing with a random process, but various criteria exist for choosing an optimum value for t_c (see, for example, Colquhoun & Sakmann, 1985). However, as long as the 'means' of the long and short components differ by a factor of at least 50, and preferably over 100, the dangers of misclassification are acceptable. In the example illustrated in Fig. 10, this criterion is met, even when the small component of shut times with a mean of around 1 ms is deemed to be 'within bursts'. In this case $t_c = 5$ ms result in less than 2% of shut times being misclassified.

Distributions based on bursts. Once bursts have been defined, it is possible to define many new sorts of distribution, which can be helpful in the interpretation of single channel records (see Chapter 7). One reason for their usefulness is that one can usually be sure that all the openings in a burst come from the same individual channel, so shut times within bursts can be interpreted in terms of channel mechanisms, even under conditions where there is a large and unknown number of channels in the patch (so consecutive bursts may not originate from the same channel, and the shut time separating them is therefore not interpretable).

For example, the distribution of the burst length, or of the total open time per burst, can be defined. For many sorts of channel it is the mean burst length, rather than the mean length of the individual opening, that constitutes the 'unitary event' for physiological purposes (it would be irrelevant for the function of a synapse that the burst actually contained some very short closures within it). Another advantage of measuring quantities such as these becomes clear when we consider the effect of failing to detect brief shuttings; the burst lengths will be far less sensitive to such failures than the individual open times (see below, and Chapter 7).

It is expected that multi-exponential distributions will also fit distributions such as those of burst length, open time per burst and so on. Some other forms of distributions are also encountered. For example the distributions of the sum of any *fixed* number of exponentially distributed intervals is described by a *gamma distribution* (used, for example, by Colquhoun & Sakmann, 1985), though the distribution of the sum of a *random* number of exponentially distributed intervals is itself exponentially distributed (which is why the burst length, in some cases, has an approximately exponential distribution).

But we can also define a different sort of distribution on the basis of division of the record into bursts. For example, the *number of openings per burst* is a discrete variable (it can take only integer values, 1, 2, 3, . . . etc), rather than a continuous variable like time. Under the simplest (Markov) assumptions it is expected to be described by a mixture of *geometric distributions* (see also Chapter 7). The geometric distribution is the discrete analogue of the exponential distribution, and is described next.

Geometric distributions

This sort of discrete distribution can be exemplified by the distribution of the number of openings per burst. The probability, $P(r)$, that a burst will contain r openings is given by

$$P(r) = \Sigma a_i(1 - \rho_i)\rho_i^{r-1}, \qquad r = 1, 2, 3, \ldots, \infty \qquad (6.8)$$

where a_i represents the area of each component, as before, and the ρ_i are constant coefficients (less than 1) that give the constant factor by which $P(r)$ is reduced each time r is increased by 1. The distribution therefore, apart from being discontinuous, has the same shape as an exponential distribution, as shown in Fig. 11.

The mean number of openings per burst for each component, μ_i say, is related to the ρ_i values thus

$$\mu_i = \frac{1}{1 - \rho_i} \qquad (6.9)$$

Fig. 11 shows the distribution of the number of openings per burst for similar data to that used for the shut time distribution in Fig. 10. A critical shut time of 3 ms was

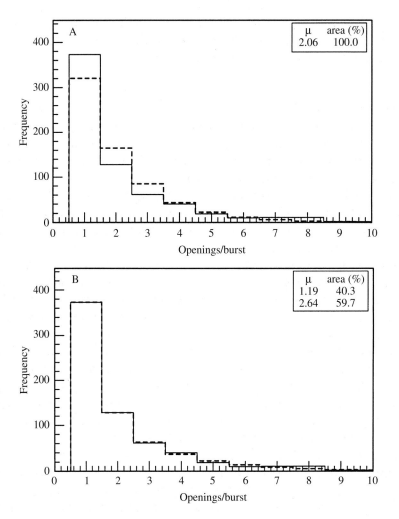

Fig. 11. Example of the distribution of the number of (apparent) openings per burst (frog muscle endplate, suberyldicholine 20 nM, −139 mV). Resolution was set as in Fig. 10; this resulted in 1355 resolved openings, and a critical shut time of 3 ms was used to divide the record into 659 bursts. The same data are shown in A and B. The data were fitted, by the method of maximum likelihood, with either a single geometric distribution (dashed line in A), or by a mixture of two geometric distributions (dashed line in B). The fitted parameter values are shown on the graphs. (D. Colquhoun and B. Sakmann, unpublished data.)

used to divide the record into bursts. As discussed above, many of the shut times are too short to be resolved, so this distribution should preferably be referred to as 'the number of *apparent openings* per burst'. If all shut times were detected there would be more shuttings, and more openings, than are detected here.

In Fig. 11A the distribution has been fitted with a single geometric component, and the fit is not good (there are too many values with one opening per burst, and too few with 2 or 3 openings per burst, for a good fit). In Fig. 11B the same data has been fitted with two geometric components, and the fit is good.

The areas and means can be predicted from a specified kinetic mechanism, as described in Chapter 7, so, conversely the fitted values of these parameters can be used to estimate the rate constants in the underlying mechanism.

Stability plots

This section has dealt mainly with the display of measurements that have been made at equilibrium, so the average properties of the record are not changing with time. In practice it is quite common for changes to occur with time. This can be checked by constructing a *stability plot* as suggested by Weiss & Magleby (1989). In the case, for example, of the measured open times, the approach is to construct a moving average of open times, and to plot this average against time, or, more commonly, against the interval number (e.g. the number of the interval at the centre of the averaged values). A common procedure is to average 50 consecutive open times, and then increment the starting point by 25 (i.e. average open times 1 to 50, 26 to 75, 51 to 100 etc). The overlap between samples smooths the graph (and so also blurs detail). An exactly similar procedure can be followed for shut times, and for open probabilities. In the case of open probabilities, a value for P_{open} is calculated for every each set of 50 (or whatever number is chosen) open and shut times, as total open time over total length.

Fig. 12A shows examples of stability plots for open times, shut times and P_{open} which was calculated from the same experimental record (from frog muscle nicotinic channels) as that used for the shut time distribution shown in Fig. 10. It can be seen at once that all three quantities are reasonably stable throughout the recording. In contrast Fig. 12B shows similar plots for a recording from the NMDA type of glutamate receptor channel (Gibb & Colquhoun, 1991). In this case, though the open times are stable throughout the recording, there are two periods when the shut times suddenly become short, and P_{open} correspondingly increases to nearly 1.

Plots of this sort can be used to mark (e.g. by superimposing cursors on the plot) to mark sections of the data that are to be omitted from the analysis. For example, this approach has been used to inspect, separately, the channel properties when the channel is in a 'high P_{open} period', and when it is behaving 'normally'.

It should be noted that when the average P_{open} value (the value for the whole record) is plotted on the stability plot, it can sometimes appear to be in the wrong position. This may happen when the record contains a very long shut period which reduces the overall P_{open}, but which affects only one point on the stability plot (which is normally constructed with 'interval number' on the abscissa, rather than time).

Amplitude stability plots. Exactly similar plots can be constructed for the channel amplitudes, in order to check whether they stay constant throughout the experiment.

7. The fitting of distributions

Fitting the results: empirical fits

At this stage we have a histogram that displays the experimentally-measured distribution of, for example, channel amplitudes, or open times, or number of

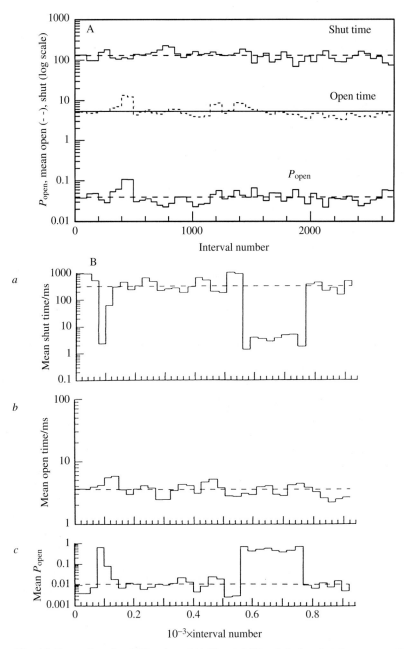

Fig. 12. Examples of stability plots. (A) The stability plots for shut times, open times and P_{open} are shown for the same data that were used for the shut time distribution in Fig. 10. All three plots are shown on the same graph by using a logarithmic ordinate. A running average of 50 values was calculated, the starting point being incremented by 25 values for each average. The overall average values of shut time, open time and P_{open} are plotted as horizontal lines. (B) Similar plots are shown (this time, as three separate graphs) for a recording from the NMDA-type glutamate receptor (outside-out patch from hippocampal CA1 cell, glutamate 20 nM + glycine 1 μM, −60 mV; Gibb & Colquhoun, 1991). In this case a running average of 20 values, with an increment of 10 values, was used.

openings per burst. The conventional approach is next to fit to these data a theoretical distribution (usually a mixture of Gaussian or exponential or geometric distributions, respectively). In the case of fitting exponentials, for example, the problem is how to find the values of the parameters (i.e. the 'means', τ_1, τ_2 . . ., and the relative areas, a_1, a_2. . .) that provide the best fit the experimental data. The question of what mechanism might account for the observations is then considered (if at all) retrospectively.

Fitting a mechanism

In cases where a specific kinetic mechanism is being postulated for the ion channel it is possible to do better that this; all the data can be fitted simultaneously, the parameters being the underlying rate constants (as defined by the law of mass action) in the mechanism, rather than a set of empirical values of τ and a. This more sophisticated approach can be used only if allowance is made, *during* the fitting process, for events that are too short to be detected, and methods for doing this are discussed in Chapter 7.

Fitting exponentials

For the moment we shall consider only the first case, and the problem will be exemplified initially by the case where exponentials are to be fitted.

If a simple exponential will suffice then the traditional way of estimating the time constant, from the slope of a semilogarithmic plot of frequency against time will, if not optimum, be quite satisfactory in practice. In practice, hardly any observations can be fitted by a single exponential, so the problem gets a bit more difficult; the traditional method of 'curve stripping' will rarely be satisfactory and is, in any case, little or no faster than doing the job properly.

All satisfactory methods involve the minimization (or maximization) of some function, so the first thing that has to be done is to get hold of a general purpose computer program that can minimize a specified function. The easiest methods to use are based on simple search procedures; for example Patternsearch (see Colquhoun, 1971) or the Simplex method (Nelder & Mead, 1965; O'Neill, 1971; Hill, 1978). Search methods will usually converge even when give rather poor initial guesses (they are 'robust'), and constraints can easily be incorporated into the fitting. However they are usually not as fast as more complex (but less robust) gradient methods. Most commercial subroutine libraries have a selection of minimization routines (e.g. the NAG PC library).

The next thing to do is to decide what is meant by '*best*' fit. Two methods are commonly used. The first of these, the minimum chi-squared method, used the histogram frequencies as the data. This method is described, for example, by Colquhoun & Sigworth (1983); although it is quite satisfactory in practice it will not be discussed further here because it is generally believed that the *method of maximum likelihood* is preferable to other criteria for best fit. This method produces the values for the parameters (τ_1, τ_2 . . ., and a_1, a_2 . . .) which make the observation of our particular set of data more probable than it would be with any other parameter values (see Edwards, 1972; Colquhoun & Sigworth, 1983). The principle is easy. Suppose that we have a set of measurements (e.g. open times) denoted t_1, t_2 . . t_n. If these observations are independent of each other then the probability (density) of making all n observations is proportional to the product of the separate probability (densities) for each observation, a quantity known as the likelihood (of a particular set of parameters), viz.

$$Lik = f(t_1)f(t_2) \ldots f(t_n) \tag{7.1}$$

where f is the p.d.f., defined in (6.1) or (6.2) as appropriate. For example, if a single exponential as in (6.1) is sufficient then $f(t_1) = \tau^{-1}e^{-t_1/\tau}$, and so on. Most commonly we work with the log(likelihood), L, which from (7.1) can be written as

$$L = \log(Lik) = \sum_{i=1}^{n} \log f(t_i) \tag{7.2}$$

The optimization program finds the values of the parameters that make L as large as possible (these same values will, of course, also make Lik as large as possible); if the program is designed for minimization (as most are) we simply minimize $-L$ in order to maximize L. Notice that for the purpose of this calculation the observations are regarded as constants (the particular values of t_1, t_2 . . . that we happen to have observed) while the parameter estimates are regarded as variables (the τ_i and a_i values are adjusted until L is maximized). The method is easily adapted to cope with the case that there are no observations below some specified minimum durations (e.g. the resolution), or above a specified maximum duration (see Colquhoun & Sigworth, 1983). Notice also that this method uses the original observed time intervals, *not* the histogram frequencies found from them. This has the advantages that (*a*) values of, say, 1.1 and 1.9 ms are not treated as though they were identical just because they happen to fall in the same histogram bin, and (*b*) the values found for the parameters will *not* depend on how we choose to divide up the observations to form a histogram. The histogram is still needed to display the fit once we have got it, but the fit is independent of how the bins are chosen. This adds considerably to the objectivity of the analysis.

How to superimpose a fitted curve on the histogram

The result of fitting two exponential components is shown in Fig. 13. In this case the data consisted of values of the *total open time per burst* for bursts of openings produced at the frog endplate by a very low concentration (4 nM) of suberyldicholine. Under these conditions there are many brief openings. In this experiment the resolution was set to 60 µs for openings (and hence for the data in Fig. 13), and 40 µs for shut times.

Maximum likelihood fitting gave the faster time constant as $\tau_f = 0.157$ ms, and the slower as $\tau_s = 22.8$ ms; the corresponding areas were $a_f = 0.686$ (i.e. 68.6 percent of area), and $a_s = 0.314$ (31.4 percent of area). The fitted p.d.f. was thus

$$f(t) = w_f e^{-t/0.157} + w_s e^{-t/22.8}, \tag{7.3}$$

where $w_f = a_f \tau_f^{-1} = 4369.4 \text{ s}^{-1}$ and $w_s = a_s \tau_s^{-1} = 13.8 \text{ s}^{-1}$ are the *amplitudes* (at $t = 0$)

of the components, and t is in milliseconds. All the data are shown in Fig. 13A, in a histogram with a bin width of $\Delta t = 4$ ms. The estimated total number of observations (i.e. the number of bursts in this case) was $N=279.7$. The area under the histogram is therefore $N\Delta t = 1.119$ s, and the continuous curve plotted in Fig. 13A is $g(t) = 1.119f(t)$, as explained above, i.e.

$$g(t) = 4888.5\ e^{-t/0.157} + 15.4\ e^{-t/22.8} \tag{7.4}$$

The continuous curve fits the slow component of the histogram quite adequately in Fig. 13A. However the fit to the fast component in Fig. 13A appears poor at first sight; the continuous curve does not pass centrally through the first bin, but is squashed up to the left hand side of it. In fact the fit is seen to be perfectly good when the *same* p.d.f., equation (7.3), is plotted on the expanded histogram in Fig. 13B; this shows data only up to 1 ms with 75 μs bins, i.e. it shows mainly the fast components of the distribution. This example shows that although only one fit is done (resulting in equation (7.3)), the results either have to be displayed as two separate histograms (on different time scales), or displayed in the logarithmic manner described in Fig. 10D, in order to be able to judge visually the goodness of fit. In fact the fit to the leftmost bin in Fig. 13A is perfectly good (though this is certainly not obvious); it is the *areas* rather than the

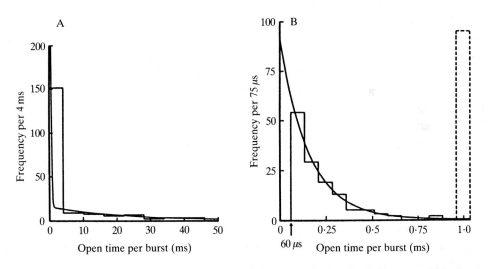

Fig. 13. Distribution of the total open time per burst (histogram) fitted, by the method of maximum likelihood, with a two-exponential probability density function (continuous line). The results are for frog endplate channels activated by suberyldicholine (4 nM, membrane potential, −188 mV). See text for details of the fitted parameters. The same data, and the same fitted curve, are shown in A and B on two different time scales. In A almost all of the data are shown (there were a few values longer than 50 ms). In B only values up to 1 ms are shown (the dashed bar represents all values greater than 1 ms). There are no observations below 60 μs, the open time resolution, because this resolution was imposed on the data before the histogram was formed (see text). In B the ordinate of the fitted curve at t = 0 is 91.9; in this graph the bin width is $\Delta t = 75$ μs so $N\Delta t = 0.0210$ s and $N\Delta t(a_f\tau_f^{-1}+a_s\tau_s^{-1}) = 91.9$; The area below 60 μs under the fitted curve suggests that there were 61.2 values that were too short to be seen. Modified from Colquhoun & Sakmann (1985) with permission.

ordinates that must match, and the amplitude of the continuous curve at $t = 0$ is, from (7.4), 4888.5 + 15.4 = 4903.9, i.e. it lies about 3 metres off the top of the page. The ordinate of the first histogram bin is 151 so its area is 151×4 ms = 604 ms. Now the open time resolution was 60 μs so the first bin extends from 0.06 ms to 4.06 ms (N.B. *not* from 0 to 4 ms); the area under the continuous curve over this range is

$$\int_{0.06}^{4.06} g(t)\mathrm{d}t = 524 + 56 = 580 \text{ ms} , \tag{7.5}$$

which is close to the observed bin area of 604 ms. Put another way, the continuous curve predicts a bin height of 580/4 = 145, close to the observed value of 151.

Errors in the parameters

The best way of assessing the error in, for example, a value of τ is to repeat the experiment several times and observe how consistent the values turn out to be. If this cannot be done, or an internal estimate of error from a single experiment is thought to be desirable for some other reason, then there are various methods of calculating errors. The two that have been used in practice are (*a*) calculation of approximate standard deviations and correlations and (*b*) calculation of 'support' or 'likelihood' intervals. The details of the calculations are given by Colquhoun & Sigworth (1983); the discussion here will be limited to some comments on the use of these methods.

Approximate standard deviations of parameter estimates. These will give realistic estimates of error only when there are rather a lot of observations, and the parameters are well-determined (i.e. when precise estimates of error are not usually needed). The same calculations also give an estimate of the *correlations* between parameter estimates, and these can be rather useful. A strong positive correlation between two parameters suggests, for example, that the quality of the fit will be little changed if the value of one parameter is increased as long as the value of the other is also increased. This may mean that the *ratio* of the two parameters is better determined than the values of either of them separately.

Likelihood intervals. When the maximum likelihood method is used the parameter estimates found are the most likely values (i.e. those that make our data most probable). A pair of values for a parameter can be found (one above the most likely value and one below it), that are less likely by some fixed amount, and this pair of values constitutes an interval within which we can reasonably expect the true value of the parameter to lie. For non-linear problems (such as fitting exponentials) we cannot attach an exact probability value to what we mean by 'reasonably'. However as a rough guide we can note that for a linear problem '0.5 unit interval' (values of the parameter for which L is 0.5 units below its maximum) would correspond to plus or minus one standard deviation, and a 2-unit interval would correspond to plus or minus two standard deviations.

The lower and upper likelihood limits will not generally be symmetrical about the best (maximum likelihood) value, and they are well-suited to expressing the range of possible values for rather ill-determined parameters. They have been found useful for this purpose in other curve-fitting problems also.

How many components are needed?

It is very often asked how one can decide whether a particular set of data requires, for example, three exponential components to fit it, or whether two components will suffice. A number of statistical methods (all approximate) have been proposed, to calculate 'whether the fit is significantly better' when an extra exponential is added (see, for example, Horn, 1987). In my view these methods are of very little value. The way to answer this question is to repeat the experiment several times, and each time fit the data with both two and three components, as illustrated in Fig. 10. In this example the third component, which had a mean of 1.3 ms, accounted for only 3.7% of the area under the distribution, so its reality might be doubted. However, many repetitions of this experiment revealed that a third component, *with approximately the same mean and area*, could be fitted in almost every case. It is this *consistency*, from one experiment to the next that provides convincing evidence that there are really three components. If in fact there were only two components, it would of course always be possible to fit the same data with three components, but the mean and area of the third component would be quite *inconsistent* from one experiment to the next, because we would not be fitting anything real, but just random noise in the data.

This approach shows, in another way, the undesirability of trying to decide the number of components by statistical tests on individual experiments. A small component such as that just illustrated, might well not reach 'statistical significance' in any one experiment, and so be missed altogether if one relied on such tests.

The example just discussed is interesting also because it shows that even a small extra component can have a surprisingly large effect on the fit of the major components. In Fig. 10 the mean for the fastest component was 56.7 µs when only two components were fitted, but addition of the small third component changed this estimate to 45.5 µs. Similarly, the mean for the component of long shut times was altered from 249 ms to 274 ms.

8. Correlations

The record of open and shut times that is used to fit distributions can also be used to calculate various correlation coefficients. For example one might be interested in the correlation coefficient between successive open times, successive burst lengths, or the lengths of successive openings within a burst. Such measurements can give information about the routes between various states of the system (see Chapter 7). When such correlations exist the distributions may *not* be the same for the first, second, etc. opening following a perturbation such as a voltage jump; care may therefore be needed in analysing such results. The display, and interpretation of correlations is considered in Chapter 7.

9. Transients: single channels after a voltage- or concentration-jump

So far the discussion has centred mostly around recordings made at equilibrium. Some new, and more complex, considerations arise when the record is not at equilibrium. For example, following a sudden change in concentration (a *concentration-jump*) or membrane potential (a *voltage-jump*), it will take some time for a new equilibrium to be established.

The reasons for wanting to measure the amplitude and duration of single channel currents after a jump include all the reasons already discussed, but such experiments also allow new sorts of question to be addressed.

A common motive for doing such jump experiments is simply to average a large number of responses to produce a more-or-less smooth curve. This average will be proportional to the probability that the channel is open as a function of time following the concentration or voltage step (though the absolute probability can be found only if it is known how many channels are active in the patch). It should, therefore, have the same shape as the macroscopic current relaxation found, for example, by the whole-cell clamp method. The big advantage of doing the experiment this way is that the individual single channel currents that underlie the macroscopic responses can be seen, and may be identifiable as a particular sort of channel. Thus the channels that carry the macroscopic current can be identified, with much greater certainty than could be done from the macroscopic current alone.

Another motive for measuring channel openings after a jump is to cast light on the channel mechanism. The sort of information about mechanisms that can be obtained from such experiments is different from, and complementary to, that which can be obtained from equilibrium measurements. Some of the principles involved are discussed in Chapter 7. For example, measurements of the *first latency* (the time from the moment of the jump to the first channel opening) can give valuable information about kinetic mechanisms that is not obtainable from equilibrium measurements. First latency measurements can also help to explain the time course of synaptic currents (e.g. Chapter 7, and Edmonds & Colquhoun, 1992).

In the case, particularly, of voltage-activated channels it has been common practice to investigate *only* voltage jumps. But, because the jumps are relatively brief, this precludes the measurement of any slow kinetic processes (which may consequently, and usually unnecessarily, be referred to as 'modes').

If correlations are present (section 8, above) the distributions of the 1st, 2nd, 3rd, . . . open times (shut times, or burst lengths etc) after the jump may not be the same (e.g. Colquhoun & Hawkes, 1987). Information from this source has yet to be exploited experimentally.

10. Effects of limited time resolution

It is very often true that some openings and shuttings are too short to be detected, and their omission will obviously cause errors. It is clear, for example, from Fig. 10 that many brief shuttings have not been detected. The three component fit in this case gave a mean of 45.5 μs for the fast component, but no shuttings shorter than 60 μs could be detected with confidence in this experiment. It follows from (6.4) that about 73% of the short openings were missed, and only 27% were detected and measured.

Corrections for missed events

When only *either* openings *or* shuttings (but not both) are short enough to be missed in substantial numbers, approximate corrections can be made, without having to know about the details of the channel mechanism. Take, for example, the case where most openings are long enough to be detected but substantial numbers of brief shut periods (gaps) are missed. In this case it is the *open* time distributions which will be in serious error because two openings separated by an unresolvable gap will be counted as a single opening. The shut time distributions will be accurate in the region where they can be measured, i.e. for those gaps that are long enough to be detected. The shut time distribution can therefore be extrapolated to zero time to get an estimate of the number of shut times that have escaped detection; this will of course work only if a sufficient number of the short shuttings (the ones that are longer than the resolution) are detected to allow accurate extrapolation. A corrected mean open time can then be found by taking the total observed 'open' time (minus the short time spent in undetected gaps) and dividing it by the total number of openings, i.e. by the corrected total number of gaps (those seen plus those undetected). Of course only the *mean* open time can be so corrected; the true *distribution* of open times cannot be found. If the open time distribution itself has two components, then we cannot say to what extent the inferred missed shut times were missed from 'short openings' or from 'long openings'. In order to say anything about this sort of question we must make some postulate concerning the underlying channel mechanism (see Chapter 7).

If the data contain openings and shut periods that are so short that substantial numbers of *both* are missed then there is no way of correcting the results without making some postulate about the detailed channel mechanism. When this can be done, methods for allowing for missed events have been developed, and these will be discussed in Chapter 7.

Details of the simpler corrections mentioned above, and the modifications that are needed when openings occur in bursts, are given, for example, by Ogden & Colquhoun (1985) and by Colquhoun & Sakmann (1985). When possible it is a good idea to work with values that are not sensitive to loss of brief events, such as the burst length (or, as in Fig. 13, the total open time per burst).

Minimizing the problem of missed events

In view of the problems surrounding corrections for missed events, it is desirable to circumvent the problem, whenever it is possible, by making measurements that are insensitive to the event omission. Some examples of such measurements follow.

(1) Measurements of burst length are clearly less sensitive than measurements of

open or shut times (e.g. failure to detect all the brief shuttings within a burst will have little effect on the measurement of its overall length).

(2) One motive for looking at the distribution of open times is that the number of components in this distribution is, in principle, equal to the number of different open states. However the same is true of the distribution of the total open time per burst. If there are many brief shuttings, the latter distribution can be measured more precisely, and so should be used in preference to the distribution of (apparent) open times.

(3) As was pointed out at the end of section 5, if P_{open} values are measured by integration of the experimental record, rather than by measurement of individual open and shut times, the errors resulting from missed events are largely eliminated.

References

ABRAMOVITZ, M. & STEGUN, I. A. (1965). *Handbook of Mathematical Functions.* Dover Publications, New York.

CHUNG, S. H., MOORE, J. B., XIA, L., PREMKUMAR, L. S. & GAGE, P. W. (1990). Characterization of single channel currents using digital signal processing techniques based on hidden Markov models. *Phil. Trans. Roy. Soc. London* B **329**, 265-285.

COLQUHOUN, D. (1971). *Lectures on Biostatistics.* Oxford: Clarendon Press.

COLQUHOUN, D. & HAWKES, A. G. (1983). The principles of the stochastic interpretation of ion channel mechanisms. In *Single Channel Recording* (ed. B. Sakmann & E. Neher). New York: Plenum Press.

COLQUHOUN, D. & HAWKES, A. G. (1987). A note on correlations in single ion channel records. *Proc. Roy. Soc. London* B **230**, 15-52.

COLQUHOUN, D. & OGDEN, D. C. (1988). Activation of ion channels in the frog end-plate by high concentrations of acetylcholine. *J. Physiol.* **395**, 131-159.

COLQUHOUN, D. & SAKMANN, B. (1983). Bursts of openings in transmitter-activated ion channels. In *Single Channel Recording* (ed. B. Sakmann & E. Neher). New York: Plenum Press.

COLQUHOUN, D. & SAKMANN, B. (1985). Fast events in single-channel currents activated by acetylcholine and its analogues at the frog muscle end-plate. *J. Physiol.* **369**, 501-557.

COLQUHOUN, D. & SIGWORTH, F. L. (1983). Fitting and statistical analysis of single channel records. In *Single Channel Recording* (ed. B. Sakmann & E. Neher). New York: Plenum Press.

EDMONDS, B. & COLQUHOUN, D. (1992). Rapid decay of averaged single-channel NMDA receptor activations recorded at low agonist concentration. *Proc. Roy. Soc. London B* **250**, 279-286.

EDWARDS, A. W. F. (1972). *Likelihood.* Cambridge University Press.

FREDKIN, D. R. AND RICE, J. A. (1992). Maximum likelihood estimation and identification directly from single-channel recordings. *Proc. Roy. Soc. London* B **249**, 125-132.

GIBB, A. J. AND COLQUHOUN, D. (1991). Glutamate activation of a single NMDA receptor-channel produces a cluster of channel openings. *Proc. Roy. Soc. London* B **243**, 39-45.

HAMILL, O. P., MARTY, A., NEHER, E., SAKMANN, B. & SIGWORTH, F. J. (1981). Improved patch- clamp techniques for high-resolution current recording from cells and cell-free membrane patches. *Pflügers Archiv.* **391**, 85-100.

HILL, I. D. (1978). A remark on algorithm AS47. Function minimization using a simplex procedure. *Appl. Statist.* **27**, 380-382.

HORN, R. (1987). Statistical methods for model discrimination. Applications to gating kinetics and permeation of the acetylcholine receptor channel. *Biophys. J.* **51**, 255-263.

HORN, R. (1991). Estimating the number of channels in patch recordings. *Biophys. J.* **60**, 433-439.

McMANUS, O. B., BLATZ, A. L. & MAGLEBY, K. L. (1987). Sampling, log-binning, fitting and plotting durations of open and shut intervals from single channels and the effects of noise. *Pflügers Arch.* **410**, 530-553.

NELDER, J. A. & MEAD, R. (1965). A simplex method for function minimization. *Computer J.* **7**, 308-313.

OGDEN, D. C. & COLQUHOUN, D. (1985). Ion channel block by acetylcholine, carbachol and suberyldicholine at the frog neuromuscular junction. *Proc. Roy. Soc. London* B **225**, 329-355.

O'NEILL, R. (1971). Algorithm AS47. Function minimization using a simplex procedure. *Appl. Statist.* **20**, 338-345.

PRESS, W. H., FLANNERY, B. P., TEUKOLSKY, S. A. & VETTERLING, W. T. (1986). *Numerical Recipes.* Cambridge University Press.

SAKMANN, B., PATLAK, J. & NEHER, E. (1980). Single acetylcholine-activated channels show burst-kinetics in presence of desensitizing concentrations of agonist. *Nature* **286**, 71-73.

SIGWORTH, F. J. & SINE, S. M. (1987). Data transformations for improved display and fitting of single-channel dwell time histograms. *Biophys. J.* **52**, 1047-1054.

STANDEN, N. B., STANFIELD, P. R. & WARD, T. A. (1985). Properties of single potassium channels in vesicles formed from the sarcolemma of frog skeletal muscle. *J. Physiol.* **364**, 339-358.

WEISS, D. S. & MAGLEBY, K. L. (1989). Gating scheme for single GABA-activated Cl-channels determined from stability plots, dwell-time distributions, and adjacent-interval durations. *J. Neurosci.* **9**, 1314-1324.

Chapter 7

The interpretation of single channel recordings

DAVID COLQUHOUN and ALAN G. HAWKES

1. Introduction

The information in a single channel record is contained in the amplitudes of the openings, the durations of the open and shut periods, and in the order in which the openings and shuttings occur. The record contains information about rates as well as about equilibria, even when the measurements are made when the system as a whole is at equilibrium. This is possible because, as with noise analysis, when the recording is made from a small number of channels the system will rarely be *exactly* at equilibrium at any given moment. If, for example, 50 percent of channels are open at equilibrium this means only that over a long record an individual channel will be open for 50 percent of the time; an individual channel will never be '50 percent open'.

In this chapter we shall consider how various aspects of single channel behaviour can be predicted, on the basis of a postulated kinetic mechanism for the channel. This will allow the mechanism to be tested, by comparing the predicted behaviour with that observed in experiments. We shall also discuss the effect that the imperfect resolution of experimental records will have on such inferences.

2. Macroscopic measurements and single channels

Measurements from large numbers of molecules will be referred to here as *macroscopic measurements*. For example the decay of a miniature endplate current (MEPC) at the neuromuscular junction provides a familiar example of a measurement of the rate of a macroscopic process. Several thousand channels are involved, a large enough number to produce a smooth curve in which the contribution of individual channels is not very noticeable. In this case the time course of the current is usually a simple exponential. Other forms of macroscopic measurements of rate include (*a*) observations of the reequilibration of the current following a sudden change in membrane potential (*voltage jump relaxations*), and (*b*) reequilibration of the current following a sudden change in ligand concentration (*concentration jump relaxations*).

D. COLQUHOUN, Department of Pharmacology, University College London, Gower Street, London WC1E 6BT, UK.
A. G. HAWKES, Statistics and OR Group, European Business Management School, University of Wales, Swansea SA2 8PP, Wales.

In these cases too, it is common to observe that the time course of the current can be fitted by an exponential curve, or by the sum of several exponential curves with different time constants. This is exactly the result expected on the basis of the law of mass action if (*a*) the system exists in several discrete states (see Colquhoun & Ogden, 1986, for a discussion), and (*b*) the transitions between these states occur at a constant rate. Before going further the terms used to define rates will be described.

Rate constants and transition rates

The *transition rate* between two states always has the dimensions of a rate or frequency, viz s^{-1}. For a unimolecular reaction that involves only one reactant (e.g. a conformation change) the transition rate is simply the *reaction rate constant* defined by the law of mass action. The law of mass action states that the *rate* of a chemical reaction (with dimensions s^{-1}) is directly proportional to the product of the reactant concentrations at any given moment, the *rate constant* being the constant of proportionality.

For a binding reaction, which is bimolecular (both receptor and free ligand are reactants), the reaction rate constant has dimensions $M^{-1}s^{-1}$; the actual transition rate in this case is the product of the rate constant ($M^{-1}s^{-1}$) and the free ligand concentration (M) and will have dimensions s^{-1} as required. The assumption that the transition rates are constant, i.e. they do not change with time, involves the assumption that the free concentration does not change with time; this may not always be true.

When the assumptions are satisfied the number of exponential components will, in principle, be one less than the number of states (e.g. Colquhoun & Hawkes, 1977). We must now ask how these relatively familiar measurements of rates are related to what is observed with a single ion channel.

Consider the simplest possible example, a channel that can exist in only two states, open and shut. The states will be numbered 1 and 2 respectively, so the rate constant for transition from shut to open is denoted k_{21}, and that for transition from open to shut is denoted k_{12}. The mechanism is conventionally written as

$$\text{shut} \underset{k_{12}}{\overset{k_{21}}{\rightleftharpoons}} \text{open} \tag{2.1}$$

state number   2   1

This is simple for two reasons. Firstly it is simple because there are only two states, and secondly it is simple because these two states are distinguishable by looking at the experimental record. We can see from the record which state the system is in at any moment. In the simple mechanism in (2.1), the opening rate is k_{21} times the fraction of shut channels (shut channels are the only reactant that participates in the opening reaction). For a simple binding reaction, $A+R \rightleftharpoons AR$ (A=ligand molecule, R=receptor), the association rate constant, k_{+1}, has dimensions $M^{-1}s^{-1}$ so when it is multiplied by the free concentration of A (to which the association rate is proportional) we get the actual transition rate for the forward reaction with dimensions s^{-1}.

Macroscopic behaviour

If the reaction in (2.1) is perturbed (e.g. by a sudden change in membrane potential or ligand concentration), the new equilibrium condition will be approached with a simple exponential time course (there is one exponential because there are two states). The rate of this process can be expressed as an *observed time constant*, τ, or as its reciprocal, the *observed rate constant* $\lambda = 1/\tau$. These values are given by

$$\tau = 1/(k_{12} + k_{21})$$

or $$\lambda = k_{12} + k_{21} \tag{2.2}$$

The equation that describes the time course contains an exponential term of the form $e^{-\lambda t}(=e^{-t/\tau})$. Fig. 1 shows an example, an endplate current for which $\tau=7.1$ ms. The observed current at time t will be directly proportional to the fraction of channels that are open (state 1) at time t, and this will be denoted $p_1(t)$. It is generally more convenient to work with the *fraction* of channels (denoted p) than with the *concentration* of channels (they differ only by a constant, namely the total concentration of channels in the system). The fraction of channels in a given state is also referred to as the *occupancy* of that state.

Observed rate constants

Notice that the observed or *macroscopic time constant* (τ), or its reciprocal the

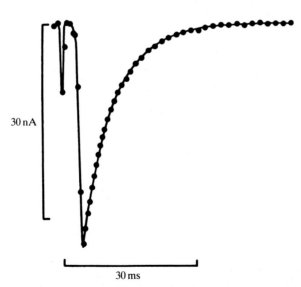

30 nA

30 ms

Fig. 1. Endplate current evoked by nerve stimulation (-130 mV, inward current shown downward). The observed current is shown by the filled circles; the continuous line is a fitted single exponential curve with a time constant of 7.1 ms. Reproduced with permission from Colquhoun & Sheridan (1981).

observed rate constant (λ) should be clearly distinguished from the transition rates (k_{12} and k_{21} in this case), and the reaction (mass action) rate constants such as k_{+1}, in the underlying reaction mechanism; this distinction is even more important in more complex cases. Not surprisingly the value of τ depends on *both* of the reaction transition rates; equilibration will be rapid (τ will be small) if *either* the opening rate (k_{21}) is fast or the shutting rate (k_{12}) is fast. Notice that this same expression holds for τ whether the re-equilibration process involves a net increase or a net decrease in the fraction of channels that are open.

The observed time constant can be written in terms of the individual rate constants if we note that the fraction of open channels at equilibrium (after long, effectively infinite time, and therefore denoted $p_1(\infty)$), is

$$p_1(\infty) = k_{21}/(k_{12} + k_{21}) , \qquad (2.3)$$

so the fraction of shut channels is $p_2(\infty)=1 - p_1(\infty)$. Hence we can write

$$\tau = p_1(\infty)k_{21}^{-1} = p_2(\infty)k_{12}^{-1} \qquad (2.4)$$

Therefore if only a small fraction of channels is open, so $p_2(\infty) \simeq 1$, then, to a good approximation, the observed value of τ gives us directly the value of one of the underlying transition rates, the shutting rate k_{12}. In general, however, the observed τ tells us only about some more or less complicated combination of the transition rates of the mechanism, not about any one of them alone.

So far the argument has shed little light on what is going on at the level of a single channel. The consideration of single channel behaviour will provide a much more concrete mental picture of what underlies the results given so far.

Randomness of single channel behaviour

There are several ways of interpreting the transition rates in terms of individual molecules. These are described in more detail by Colquhoun & Hawkes (1983) in a fairly informal way, and by Horn (1984), Colquhoun & Hawkes (1977, 1982, 1987) in a more formal way. The regular behaviour observed with large aggregates of molecules disappears when we look at a single molecule. The individual molecules behave in an entirely random way. Or, to be more precise, the length of time for which the molecule stays in one of its (supposedly) discrete states is quite random; the properties of the states themselves may be very constant (for example the amplitude of the current while the channel is open is usually very similar from one opening to another of the same channel, and from one channel to another).

The randomness of lifetimes is immediately obvious when looking at single channel records, and it should come as no surprise. Consider, for example, what would happen if the lifetime of individual openings was a fixed constant. During a MEPC a few thousand channels are opened almost simultaneously by a quantum of transmitter (which then quickly disappears); these channels would all stay open for their allotted time and then shut almost simultaneously so thc MEPC would be rectangular in shape. This is not what happens. The nature of the variability will be discussed below. First two simple interpretations of the transition rates in (2.1) will be given.

Rate constants as transition frequencies

The transition rates have the dimensions of frequency, and they can be interpreted as such. For example in scheme (2.1) k_{12} is the frequency of open $\rightarrow$ shut transitions. Such transitions can, of course, occur only when the channel is open so, to be more precise, k_{12} is the number of shutting transitions per unit open time. The channel is open for a fraction p_1 of the time (this is the single channel equivalent of the *occupancy* defined above), so the overall shutting frequency, the number of shutting transitions per unit time for a single channel, f_s say, is

$$f_s = p_1 k_{12} \tag{2.5}$$

Similarly the frequency of opening transitions for one channel, f_o say, is

$$f_o = p_2 k_{21} \tag{2.6}$$

At equilibrium f_o and f_s will be equal, thus implying the result given in (2.3). It is worth noticing that the rate constants do not tell us anything about how *long* the transition from 'shut' to 'open' takes once it has started, only about how *often* such transitions occur (see, for example, Colquhoun & Ogden, 1987). The transition itself is supposed to be instantaneous; in fact the open-shut transition is unresolvably fast (less than 10 μs, Hamill *et al.*, 1981).

Non-equilibrium conditions. When the macroscopic current is changing with time, as it will, for example after a voltage jump or concentration jump, then the opening and shutting frequencies, f_o and f_s, will change with time and will *not* be equal. The net excess of openings over shuttings per unit time, at time t, is

$$f_o - f_s = p_2(t)k_{21} - p_1(t)k_{12} = k_{21} - p_1(t)(k_{12} + k_{21}) \,.$$

This is exactly the expression that would be written down for the rate of change of the fraction of open channels, $dp_1(t)/dt$, by application of the conventional macroscopic law of mass action, but now it has been derived by considering the flipping rate of individual channel molecules. The proportionality constant that multiplies $p_1(t)$ in this expression, $(k_{12}+k_{21})$ is precisely the macroscopic observed rate constant, $1/\tau$, as in (2.2). Notice that the unequal frequencies result entirely from the fact that the *occupancies*, p_1 and p_2, are changing with time in (2.5) and (2.6); the fundamental transition rate constants (and hence the observed macroscopic rate constant that depends only on them) are not varying with time, i.e. the membrane potential (or agonist concentration) is held constant at all times following the application the step change (or so, at least, it is assumed in the conventional analyses).

Rate constants and mean lifetime. Another concrete way to think of the meaning of transition rates is to consider their relationship to how long the system stays in a particular state. This length of individual open time is variable, but the *mean* open time is a constant value and it can be shown (see below) that it is simply $1/k_{12}$ for scheme (2.1). Let us denote the mean lifetime by m; thus the mean open lifetime and the mean shut lifetime (with dimensions of seconds) in (2.1) are, respectively,

$$m_1 = 1/k_{12}$$

$$m_2 = 1/k_{21} \tag{2.7}$$

More generally the *mean lifetime in any individual state can be found by adding up all the values for transition rates that lead out of that state and then taking the reciprocal of this sum.*

By combining the results in (2.7) and (2.3) we find that the equilibrium fraction of open channels can be written in the form

$$p_1(\infty) = \frac{m_1}{m_1 + m_2} \tag{2.8}$$

i.e. it is simply the fraction of the time for which a channel is open, as asserted earlier. Also, from (2.4), the observed macroscopic time constant can be related to the mean lifetimes of the two states as follows:

$$\tau = p_1(\infty)m_2 = p_2(\infty)m_1 = 1/(m_1^{-1} + m_2^{-1}) \,. \tag{2.9}$$

Thus, if most channels are shut ($p_2 \approx 1$) then τ is approximately the mean open lifetime, and if most channels are open ($p_1 \approx 1$) τ is approximately the mean shut lifetime.

Rate constants and probabilities. Clearly the transition rate for shutting, k_{12}, tells us something about the probability that an open channel will shut. But k_{12} has dimensions of s^{-1} so it cannot itself be a probability (which must be dimensionless). In fact the required relationship (approximately)

$$\text{Prob[open channel will shut in a small interval } \Delta t] = k_{12}\,\Delta t \tag{2.10}$$

as long as Δt is small (see, for example, Colquhoun & Hawkes, 1983, for more details).

Numerical example. Suppose that in scheme (2.1) the opening transition rate is $k_{21}=250$ s^{-1} and the shutting transition rate is $k_{12}=1000$ s^{-1}. Thus, from (2.3), 20 percent of channels will be open and 80 percent shut at equilibrium (on average). The observed macroscopic time constant for re-equilibration (from equations (2.2), (2.4) or (2.9)) would be $\tau=1/1250=0.8$ ms. The fraction of open channels is not low enough for this to be very close to the mean open lifetime, $1/k_{12}=1$ ms. The mean shut time is $1/k_{21}=4$ ms. From (2.5) and (2.6), each channel would open (and shut) 200 times per second (on average) at equilibrium.

3. The distribution of lifetimes

At this stage we must consider more carefully the nature of the variability of lifetimes. Because lifetimes are random variables their behaviour must be described by probability distributions, and it is this necessity that makes rather unfamiliar the form in which the information is contained in a single channel record. First a distribution will be derived, and then used to illustrate a discussion of the nature of distributions and density functions.

The distribution of random lifetimes

The first thing to define is what we mean by 'random'. In the present context what we mean is that the probability that an open channel will shut during a short time interval (Δt) is constant, *viz.* $k_{12}\Delta t$ from equation (2.10) regardless of what has gone before (e.g. regardless of how long the channel has already been open). And events that occur in non-overlapping time intervals are independent of each other. In other words what happens in the future depends only on the present state of the system, and not on its past history. Statisticians call processes with this characteristic of 'lack of memory' *homogeneous Markov processes* (named after the Russian mathematician A. A. Markov, 1856-1922, who first studied them).

The conventional derivation of the required distribution starts with equation (2.10); it is given in full by Colquhoun & Hawkes (1983) so it will not be repeated here. Instead an alternative derivation will be given which, if less rigorous, provides greater physical insight (an even simpler version of the following argument is given by Colquhoun & Hawkes, 1983, p.142). This derivation can be regarded as an exploitation of the Markov characteristics of a series of tosses of a coin. The probability of getting 'heads' at each toss is the same, regardless of what has happened before (e.g. regardless of how many consecutive 'heads' have been observed in earlier tosses). The relevant analogy to tossing the coin is the random thermal movements of the channel protein molecule. It is the randomness of thermal motions that underlies the randomness of the lifetimes. The bonds of the protein will be vibrating, bending and stretching, and much of this motion will be very rapid - on a picosecond time scale. One can imagine that each time the open channel molecule 'stretches' there is a chance that all its atoms will get into a position where the molecule has a chance to surmount the energy barrier and flip into the shut conformation. Define p as the probability that the one channel will shut during each 'stretch', so $(1-p)$ is the probability that it will fail to shut. Suppose that this probability is the same at each attempt (each 'stretch') regardless of what has gone before. How many attempts (stretches) will be needed before the channel eventually shuts? The problem is just like asking how many tosses will be needed before the first 'heads' is encountered. The probability of success on the first attempt is p, and the probability of failure on the first attempt but success on the second is the product of the separate probabilities $(1-p)p$. The probabilities can simply be multiplied because the two attempts are supposed to be independent of each other. The probability $P(r)$, of success (i.e. the channel closing) at the rth attempt (i.e. $r-1$ failures followed by a success) is therefore

$$P(r) = (1 - p)^{r-1}p, \qquad r = 1, 2, \ldots \infty \tag{3.1}$$

This is called a *geometric distribution* and its mean, μ say, the mean number of attempts before the channel closes, is

$$\mu = \Sigma r P(r) = 1/p \tag{3.2}$$

Now the 'stretching' is on a picosecond time scale, whereas the channel stays open for milliseconds, so p is clearly small, and many 'attempts' will be needed before the channel shuts (*c.f.* coin tossing, for which $p=0.5$). Suppose further that a 'stretch' occurs every Δt picoseconds (this itself will be random but we shall treat it as though it were constant for the purpose of this argument - doing so does not give a wrong result). The time taken for r attempts is therefore $t=r\Delta t$, and the mean length of time before closure is $\mu\Delta t=\Delta t/p$. Furthermore, from the discussion above we know that the probability of closure during Δt is just $p=k_{12}\Delta t$, so the mean length of time before closure, $\Delta t/p$, is simply $1/k_{12}$, the mean open channel lifetime found before. When r is large we can put $t=r\Delta t$ and write (3.1) in the form

Prob[channel closes between t and $t + \Delta t$] = $P(r)$

$$\approx k_{12}\Delta t(1 - p)^r$$

$$= k_{12}\Delta t(1 - p)^{k_{12}t/p} \tag{3.3}$$

Now when Δt is small this geometric expression approaches an exponential form. The 'compound interest formula' states that $(1+1/n)^{nx}$ approaches e^x as n becomes very large (see, for example, Thompson, 1965), where e is the universal constant, the base of natural logarithms, 2.71828.... This formula can be rearranged to show that $(1-p)^{x/p}$ approaches e^{-x} as p becomes very small. Application of this form to (3.3) gives

Prob[lifetime between t and $t + \Delta t$] = $k_{12}\Delta t e^{-k_{12}t}$ \hfill (3.4)

as long as Δt is small.

In order to express this result as a probability density function (p.d.f.) we must divide by Δt to give the *exponential probability density function*

$$f(t) = k_{12}e^{-k_{12}t} \tag{3.5}$$

This curve is shown in Fig. 2C. Its mean is $1/k_{12}$ (see, for example, Colquhoun, 1971, appendix 1), which proves the result for the mean open lifetime already given in equation (2.7). The area under this curve between 0 and t gives the probability that a lifetime is *less than* the specified t. This is the cumulative exponential distribution, denoted $F(t)$ and is found by integration of (3.5) to be

$$F(t) = 1 - e^{-k_{12}t} \tag{3.6}$$

The last two steps, and the meaning of distributions and p.d.f.s in general will next be explained in a little more detail, before describing further the characteristics of the exponential distribution.

Distributions and density functions

Clearly it makes sense to talk about the probability that a lifetime is *less than* some specified length, t; this is just the long-run fraction of lifetimes that are less than t. It is referred to as the *cumulative distribution*, or *distribution function*, and is usually denoted $F(t)$. It starts at zero for $t=0$ and rises towards 1 for long enough times. Thus

$$\text{Prob[lifetime} < t] \equiv F(t) .\qquad (3.7)$$

For the exponential distribution this is $1-e^{-k_{12}t}$ as shown in (3.6). Clearly it also makes sense to talk about the probability that a lifetime lies in the interval Δt between two specified time values, t and $t+\Delta t$. Again this is the fraction of lifetimes that lie in this range in the long run, and it is given, in the above notation by

$$\text{Prob[lifetime between } t \text{ and } t + \Delta t] = F(t + \Delta t) - F(t) .\qquad (3.8)$$

Our experimental estimate of this is what goes into the histogram bin that stretches from t to $t+\Delta t$ (see Chapter 6).

What does *not* make sense is to ask what is the probability of the lifetime being t, e.g. the probability that it is *exactly* 5 ms, as opposed to 4.999 ms, or 5.001 ms say (we are dealing with a theoretical distribution so the question of how precisely values can be estimated experimentally does not arise). When Δt is made small then (3.4) and (3.8) approach zero; there is virtually no chance of seeing an *exactly* specified value. It is for this reason that we have to introduce the idea of a *probability density function* (p.d.f.). When Δt approaches zero, so t and $t+\Delta t$ converge on a single value, then (3.4) and (3.8) also approach zero. However the ratio of these two quantities does not. The p.d.f. is defined as this ratio, and is usually denoted $f(t)$. This ratio is

$$f(t) = \frac{F(t + \Delta t) - F(t)}{\Delta t} \longrightarrow \frac{dF(t)}{dt}\qquad (3.9)$$

For the exponential distribution this gives the result presented in equation (3.5) above. When Δt is very small this function becomes $dF(t)/dt$, the first derivative of the distribution function. *The p.d.f. is clearly not itself a probability (its dimensions are* s^{-1}*), but it is a function such that the area under the curve (which is dimensionless) between any specified time values is the probability of observing a lifetime between these values.*

The form of the variability is completely specified by either the cumulative distribution or the p.d.f., but the latter should normally be used (see comments in chapter on fitting of data). We now return to explore the properties of the exponential distribution in more detail.

Some properties of the exponential distribution

The exponential p.d.f. given in (3.5) has a rather unusual shape compared with other p.d.f.s that may be more familiar. For example Fig. 2A shows the symmetrical bell-shaped p.d.f. of the Gaussian ('normal') distribution.

Symmetry ensures that the *mode* (peak, or 'most frequent value'), the *median* (below which 50 percent of values lie) and the *mean* are all the same for the Gaussian. Fig. 2B shows a lognormal p.d.f. (for which the log of the variable is Gaussian); the

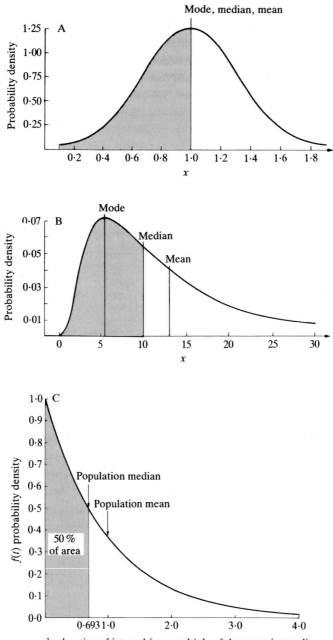

Fig. 2. (A) The p.d.f. for a Gaussian (or 'normal') distribution. This example has a mean of 1.0 and a standard deviation of 0.32. (B) The p.d.f. for a lognormal distribution (a distribution such that log x is Gaussian). (C) The p.d.f. for the exponential distribution (eqn (3.5)). The abscissa shows the value of the variable (e.g. open lifetime) expressed as a multiple of its mean.

positive skew of this p.d.f. means that mean > median > mode. Fig. 2C shows the exponential p.d.f. of lifetimes which is seen to be an extreme case of a positively skewed distribution. The modal lifetime is zero. The mean of the exponential distribution is seen to be what we would call the time constant if the exponential curve in Fig. 2C were describing a decaying current rather than a p.d.f. If we call the mean lifetime τ, then we can write the exponential p.d.f., (3.5), in an alternative form as

$$f(t) = \tau^{-1}e^{-t/\tau} \qquad (3.10)$$

The median of this distribution is 0.693τ (see below). Notice that in our simple example $\tau=1/k_{12}$ is *not* the same as the time constant, $1/(k_{12}+k_{21})$, from macroscopic measurements (they will be similar only if most channels are shut, as shown above). The single channel measurement is simpler in that it gives directly one of the underlying reaction transition rates (k_{12}) whereas the macroscopic measurement gives only a combination of them.

An exponential current decay and an exponential p.d.f. are quite different sorts of things, but it is not a coincidence that they both have the same shape. Fig. 3 shows a simulation of what happens during the decay of an MEPC. Many channels (five of which are shown in Fig. 3) are opened almost simultaneously by the transmitter, and each of them stays open for an exponentially-distributed length of time before closing.

Once closed they will not re-open because the transmitter concentration falls rapidly to zero - this means that opening rate is zero, so in this particular case, though not in general, the macroscopic time constant corresponds to the mean open lifetime. At any particular moment, the total current that flows is proportional to the total number of open channels, and this is shown in the lower part of Fig. 3. It shows an exponential decay of the current. This is expected because the fraction of channels that are open at t will be those that have a lifetime longer than t, i.e. $1-F(t)=e^{-t/\tau}$ (see equation (3.6)). The relationship between the smooth macroscopic behaviour and the underlying random behaviour of single channels should now be clear.

The median lifetime for an exponential p.d.f., 0.693τ, is defined such that 50 percent of lifetimes are shorter than the median and 50 percent are longer. This is the single-channel equivalent of the *half-time* for the decay of the MEPC. The proportion of lifetimes that a shorter than the mean is 63.2 percent, so 36.8 percent of intervals are longer-than-average lifetimes. However, although longer-than- average lifetimes are in a minority, their length is such that they actually occupy a greater proportion (75 percent) of the time than is occupied by the more numerous shorter-than-average lifetimes (see Colquhoun, 1971, for details). This fact means that if we stick a pin into the record at random we will, in the long run, tend to pick out longer-than average intervals. In fact those picked out will have, in the long run, twice the mean lifetime (a phenomenon known as length-biased sampling). It is this rather curious property that is responsible for some of the unexpected characteristics of random lifetimes.

The waiting time paradox. The property of length-biased sampling explains, for example, why we get the same macroscopic time constant whether the openings of channels are all lined up at $t=0$ (as in the MEPC illustrated in Fig. 3), or whether the channel is already at equilibrium with agonist before agonist is suddenly removed at $t=0$. In the latter case (but not the former) every channel that was open at $t=0$ (those whose shutting is subsequently followed) will already have been open for some time so one might think that half of their average lifetime had already been 'used up'. In a sense it has, but since the particular channels that were open at $t=0$ had a twice-normal lifetime because of length-biased sampling, the amount of time they have left to stay open is just on average, the overall mean open time, just as in the case of the MEPC.

The same sort of phenomenon explains why the distribution of the duration of the shut time that precedes the first opening after a voltage jump (the first latency) will be the same as the distribution of any other shut time. This is true, at least, for the simple mechanism discussed above, which has only one sort of shut state (it is not, however, always true - see discussion of correlations in section 6).

This waiting-time paradox, and other unexpected characteristics of random

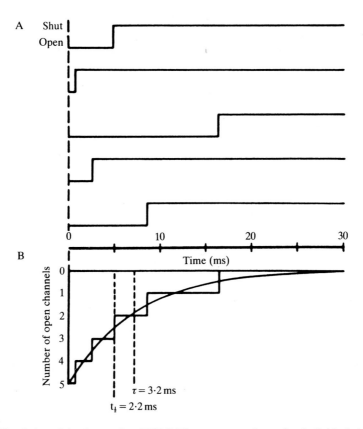

Fig. 3. Simulation of the decay of an MEPC. The upper part shows five individual channels. They have exponentially distributed lifetimes with a mean of 3.2 ms (and a median of 2.2 ms). The lower part shows the sum of these five channels with a smooth exponential curve superimposed on it. The curve has a time constant of 3.2 ms and a half-decay time of 2.2 ms. Reproduced with permission from Colquhoun & Hawkes (1983).

lifetimes are discussed in greater detail by Colquhoun (1971) and Colquhoun & Hawkes (1983).

The sum of several intervals - the idea of convolution

There are many occasions when it is necessary to consider not just a single interval, but the sum of several intervals. For example, when there are several shut states that intercommunicate directly (as in mechanism (4.1) below), a single observed shut period will generally consist of several oscillations between one shut state and another - the duration of the observed shut period will be the sum of the durations of all these (exponentially-distributed) sojourns. Similarly the duration of a *burst* of openings (e.g. section 5) will be the sum of each of the open and shut times that make up the burst. Another example occurs when the time course of synaptic currents is considered (see section 7 below); it is then important to consider the sum of the latency until the first channel activation plus the duration of the activation.

The distribution of the sum of several random intervals can be found by a method known as *convolution*. This method is central to much of single channel analysis, so the technique will be introduced here by the simplest case of its use.

An example of convolution: the distribution of the sum of two intervals. We shall derive the distribution of the sum of two exponentially distributed intervals. In the context of the simplest mechanism (2.1), we can derive the distribution of the sum of an open time and a shut time, i.e. the time from the start of one opening to the start of the next. Consider an individual open time; it is an exponentially- distributed variable, with mean $1/k_{12}$; its p.d.f. is

$$f_1(t) = k_{12}e^{-k_{12}t} \tag{3.11}$$

Similarly the mean shut time is $1/k_{21}$, and its p.d.f. is

$$f_2(t) = k_{12}e^{-k_{21}t} \tag{3.12}$$

Suppose that we now make a new variable by adding an open time and a shut time. These new values will, of course, be variable; their *mean* value will obviously be the sum of the separate means, $(1/k_{12})+(1/k_{21})$, but we wish to know what sort of distribution (p.d.f.) describes these new values. The p.d.f. of this variable, denoted $f(t)$ say, can be found by noting that any *specified* duration, t, of the sum, can be made in all sorts of different ways. If we denote the length of the open time as t_1, and the length of the shut time as t_2, then the total length is $t=t_1+t_2$. Now if the open time happens to have a length $t_1=u$, then, in order for the total length to be t, the length of the shut time must be $t-u$. And, for any fixed value, t, of the total length, u can have *any* value from 0 up to t (the former corresponds to the case where the total length consists entirely of t_2, the latter to the case where it consists entirely of t_1). The probability density that $t_1=u$ and $t_2=t-u$ can be found by multiplying the separate probability densities (because these two events are independent), i.e. from (3.11) and (3.12) it is $f_1(u)f_2(t-u)$. Since any value of u

(from 0 to t) will do, we must now add these probabilities for each possible value of u: since u is a continuous variable this addition takes the form of an integration, and the result is

$$f(t) = \int_{u=0}^{u=t} f_1(u)f_2(t-u) \, du \qquad (3.13)$$

and, in this case the integral can easily be evaluated explicitly to give (as long as k_{12} and k_{21} are different)

$$f(t) = \frac{k_{12}k_{21}}{k_{21}-k_{12}} (e^{-k_{12}t} - e^{-k_{21}t}) . \qquad (3.14)$$

This is plotted in Fig. 4.

Notice that, unlike the simple exponential distribution, this distribution goes through a maximum, at

$$t_{max} = \frac{\ln(k_{21}/k_{12})}{k_{21} - k_{12}} . \qquad (3.15)$$

In other words the most common (modal) values are no longer the shortest values, but values around t_{max}. This is not surprising because it is unlikely that t_1 and t_2 will *both* be very short. The distribution in (3.14) consists of two exponential terms with opposite signs; one exponential represents the rising phase of the distribution, and one the decay phase. Notice that (3.14) is completely unchanged if k_{12} and k_{21} are interchanged: whichever is the faster represents the rising phase, and the slower represents the decay phase. In other words the distribution of the sum of two intervals does not depend on whether the 'short one' or the 'long one' comes first.

Equations with the form of (3.13) are known as *convolution integrals*. Such equations occur whenever we consider the distribution of the sum of random variables, and are the basis for obtaining things like the distribution of burst lengths, though in cases more complex than this it is necessary to use Laplace transforms and matrix methods to obtain useful results (e.g. see Colquhoun & Hawkes, 1982). (Incidentally, convolution integrals also occur in single channel work in a rather different context: in linear systems such as electronic filters, the output of the system can be found by convolving the input with the impulse response function for the system, i.e. the input is represented as a series of impulse responses, the outputs for which superimpose linearly; see, for example, Chapter 6, and Colquhoun & Sigworth, 1983.)

Summary of results so far

The length of time for which the channel stays in an individual state of the system (the lifetime of this state) is a random variable. The variability is described by an

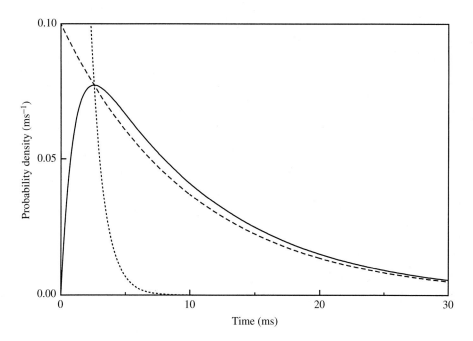

Fig. 4. Plot of two exponential distributions with rate constants $k_{12}=1$ ms^{-1} (mean lifetime=1 ms) (dotted curve), and $k_{21}=0.1$ ms^{-1} (mean lifetime=10 ms) (dashed curve). The former curve goes off-scale (intercept on the y axis is 1.0 at $t=0$). The solid line shows the convolution of these two distributions, from (3.14), i.e. the distribution of the sum of a '1 ms' amd a '10 ms' interval.

exponential p.d.f., $f(t)=\tau^{-1}e^{-t/\tau}$, the mean channel lifetime, τ, being the reciprocal of the sum of all transition rates for leaving the state in question. The exponential distribution of lifetimes is what underlies the exponential reequilibration seen in macroscopic measurements. The relationship between the observed time constants and the reaction transition rates in the underlying mechanism is indirect, but will usually (not always) be simpler for single channel measurements than for macroscopic measurements.

So far the only concrete example that has been considered, scheme (2.1), was simple because it had only two states, and because these two states could be distinguished on an experimental record. We must now consider what happens when there are more than two states, and, in particular, when not all of the states can be distinguished from one another by looking at the experimental record.

4. Some more realistic mechanisms: rapidly-equilibrating steps

We shall consider two simple mechanisms, each of which has two shut states and one open state.

An agonist activated channel

Consider, for example, a mechanism in which an agonist (A) binds to a receptor (R), following which an isomerization to the active state (i.e. the open channel, R*) may occur. This scheme, first proposed by Castillo & Katz (1957), can be written

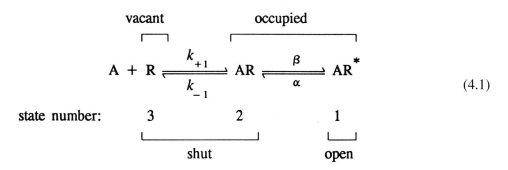

The rate constants are denoted by the symbols on the arrows, and $K_A = k_{-1}/k_{+1}$, is the equilibrium constant for the initial binding step. In fact, for many fast transmitters, more than one agonist molecule must be bound to open the channel efficiently, but the simple case in (4.1) will suffice to illustrate the principles. It has three states, and the lifetime in each of these states is expected to be exponentially distributed with, according to the rule given above, the mean lifetimes for states 1, 2 and 3 being

$$m_1 = 1/\alpha$$

$$m_2 = 1/(\beta + k_{-1})$$

$$m_3 = 1/k_{+1}x_A \tag{4.2}$$

respectively (where x_A is the free concentration of the agonist). The fraction of channels that are open at equilibrium, $p_1(\infty)$, is

$$p_1(\infty) = \frac{(x_A/K_A)(\beta/\alpha)}{1 + (x_A/K_A)(1 + \beta/\alpha)} \tag{4.3}$$

Now we can, in principle, measure the distribution of open lifetimes from the experimental record (but see discussion of bursts, and of missed events, below); the mean open time will provide an estimate of α directly. However a new problem arises when we consider shut times, because the two sorts of shut state (states 2 and 3) cannot be distinguished on the record (again see below). We therefore cannot say which of them a shut channel is in, so we cannot measure separately the mean lifetimes of states 2 and 3 in order to exploit directly the relationships in (4.2). This problem will be dealt with in §5.

A channel block mechanism

Another example with two shut states and one open state (but connected differently) is provided by the case where a channel, once open, can be subsequently plugged by an antagonist molecule in solution. This can be written

$$\text{shut} \underset{\alpha}{\overset{\beta'}{\rightleftharpoons}} \text{open} \underset{k_{-B}}{\overset{k_{+B}}{\rightleftharpoons}} \text{blocked.} \qquad (4.4)$$

state number: 3   1   2

By the same rule as before the mean lifetimes of sojourns in open, blocked and shut states will be, respectively,

$$m_1 = 1/(\alpha + k_{+B}x_B)$$

$$m_2 = 1/k_{-B}$$

$$m_3 = 1/\beta', \qquad (4.5)$$

where x_B is the free concentration of the antagonist. Once again, the two non-conducting states (shut and blocked) cannot be distinguished by inspection of the experimental record.

Before considering these two examples in a general way, it will be helpful to consider what happens when several states equilibrate with each other rapidly.

Pooling states that equilibrate rapidly

The analysis becomes simpler (but also less informative) if some states interconvert so rapidly that they behave kinetically like a single state.

Rapid binding. Consider, for example, what would happen if the binding and dissociation of agonist in scheme (4.1) were very rapid compared with the subsequent opening and shutting of the occupied channel. (This is probably not true for the nicotinic acetylcholine receptor, but it makes an interesting example.) The two shut states then appear to merge into one; this may be indicated by enclosing both in one box, and writing the mechanism thus

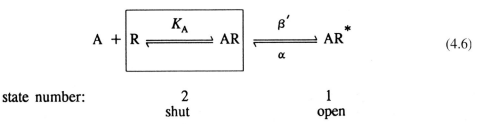

state number: 2   1
                               shut   open

This now looks formally just like the simple two-state case discussed first (equation (2.1)), and can be analysed as such. (If the channel being blocked in scheme (4.4) were an agonist activated channel then an approximation of this sort would be implicit in (4.4), which has been written with only one shut (as opposed to blocked) state.)

The channel shutting rate constant is α, just as in (4.1), but the opening rate constant needs some consideration. The real rate constant for opening is β in (4.1), the rate constant for AR→AR* transition. However, the compound 'shut' state in (4.6) is sometimes occupied (AR) and sometimes not (R), but the opening transition is possible *only* when it is occupied. Thus the *effective opening rate constant*, β', for transition from the compound shut state in (4.6) to the open state, must be defined as the true opening rate constant, β, multiplied by the fraction of time for which shut receptors are occupied (this fraction, and hence β', will increase with agonist concentration). Because the binding reaction has been supposed to be rapid this fraction will essentially not vary with time (it will always have its equilibrium value), so β' will not vary with time, and the reaction scheme is just like scheme (2.1). Hence, from results in (2.2) to (2.9), the open lifetimes and the shut lifetimes will both have simple exponential distributions, with means $1/\alpha$ and $1/\beta'$ respectively. The macroscopic time constant will be $\tau=1/(\alpha+\beta')$, which will approximate to the mean shut time if most channels are open, or to the mean open time if most channels are shut. The equilibrium fraction of open channels will be $p_1(\infty)=\beta'/(\alpha+\beta')$, which is another way of writing the result in (4.3). There will be $p_1(\infty)\alpha$ channel shuttings (and openings) per second at equilibrium.

    How fast is fast? The answer is that it is all relative. An interesting example is provided by a scheme like (4.4) in which the blocked state is very long- lived. An exactly similar example would be provided by a model in which the open state could lead to a very long-lived desensitized state (AD say). In such a case the shut ⇌ open equilibrium, while possibly slower than the binding reaction, might nevertheless be much faster than the desensitization reaction. In this case *all* of the states in (4.1) might be effectively pooled into a single 'active' (i.e. non-desensitized) state to give

$$R \underset{k_{-1}}{\overset{k_{+1}}{\rightleftharpoons}} AR \underset{\alpha}{\overset{\beta}{\rightleftharpoons}} AR^* \underset{k_{-D}}{\overset{k'_{+D}}{\rightleftharpoons}} AD \qquad (4.7)$$

state number: 1 ('active')   2 (desensitized)

Here k_{-D} is the rate constant for recovery from desensitization, and k'_{+D} is the *effective* rate constant for moving into the desensitized state (i.e. the true AR*→AD rate constant, multiplied by the equilibrium fraction of active channels that are open, and hence available to be desensitized). Again this looks just like the original two state scheme in (2.1), and the results derived for that scheme can be applied. For example the macroscopic rate constant for the desensitization will be $\tau=1/(k'_{+D}+k_{-D})$. The mean lifetime of the active state (each sojourn in which will consist of many openings) will be $1/k'_{+D}$, and the mean lifetime of the desensitized state will be $1/k_{-D}$. In fact long silent desensitized periods are observed in single channel records at high agonist concentrations (Sakmann, Patlak & Neher, 1980; Colquhoun, Ogden & Cachelin, 1986; see also Chapter 6). It often seems to be supposed that if, say, 10 second silent desensitized periods appear in the single channel record then we should see that macroscopic desensitization (at the same agonist concentration) develops with a similar time constant. However the argument just presented shows that, insofar as the desensitized periods are much longer than the non-desensitized active periods, then the macroscopic time constant should be closer to the mean length of the periods of activity than to the mean lengths of the silent desensitized periods.

5. Analysis of mechanisms with three states

Under the usual assumptions (see above) we would expect macroscopic observations on any mechanism with three states to reequilibrate along a time course described by the sum of two exponentials with different time constants (τ_s and τ_f say, for the slower and faster values respectively). In the cases just discussed, where some states equilibrate very rapidly with each other, one of the time constants would be so fast as to be undetectable, so we would appear to have only two states. We must now consider the case when this does *not* happen.

A channel block example

Fig. 5 shows an example where two exponentials are clearly visible: it is an endplate current recorded in the presence of a channel blocking agent, gallamine, for which τ_s =28.1 ms and τ_f=1.37 ms. Noise spectra also show two components and the results can be accounted for by a simple block mechanism in (4.4), as described by Colquhoun & Sheridan (1981).

As in the simplest case the macroscopic time constants (τ_s, τ_f) or their reciprocals, the macroscopic rate constants (λ_s, λ_f), are indirectly related to all of the transition

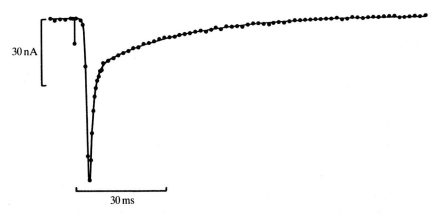

30 nA

30 ms

Fig. 5. An evoked endplate current as in Fig. 1 but recorded in the presence of a channel blocking agent (gallamine 5 μM). The current is shown by the closed circles the fitted double exponential curve has time constants τ_f=1.37 ms, τ_s=28.1 ms. Reproduced with permission from Colquhoun & Sheridan (1981).

rates in the underlying mechanism. The macroscopic rate constants in this case are found by solving a quadratic equation,

$$\lambda^2 + b\lambda + c = 0 .\tag{5.1}$$

The well-known solution of this quadratic is

$$\lambda_s, \lambda_f = 0.5\left(- b \pm \sqrt{b^2 - 4c}\right).\tag{5.2}$$

A less well-known alternative is

$$\lambda_s, \lambda_f = \frac{2c}{- b \mp \sqrt{b^2 - 4c}} ,\tag{5.3}$$

where

$$- b = \lambda_s + \lambda_f\tag{5.4}$$

and

$$c = \lambda_s\lambda_f .\tag{5.5}$$

At this point it will be useful to point out that when one of the rate constants is much bigger than the other (say $\lambda_f \gg \lambda_s$, where the subscripts denote 'fast' and 'slow'), i.e. when $b^2 \gg c$, then it follows from (5.2) and (5.4) that the faster rate constant, λ_f, is approximately

$$\lambda_f \approx - b\tag{5.6}$$

and, from (5.3) and (5.5), the slower rate constant, λ_s, is approximately

$$\lambda_s \approx - c/b\tag{5.7}$$

In this case the coefficients, b and c, are given by

$$- b = \alpha + \beta' + k_{+B}x_B + k_{-B} ,\tag{5.8}$$

$$c = \alpha k_{-B}\left[1 + \frac{\beta'}{\alpha}\left(1 + \frac{x_B}{K_B}\right)\right] ,\tag{5.9}$$

and $K_B = k_{-B}/k_{+B}$ is the equilibrium constant for binding of the blocking agent (concentration x_B) to the open channel. More generally when there are n states the $n-1$ macroscopic rate constants will be solutions of a polynomial of degree $n-1$, and their sum will be equal to the sum of all the underlying reaction transition rates, as in (5.4). A third parameter, the relative amplitudes of the fast and slow components, is needed to describe completely results such as those in Fig. 5. This too is related (in a still more complex way) to the reaction transition rates; it can be calculated by methods such as those described by Colquhoun & Hawkes (1977). It is a very important stage in the testing of a putative mechanism to ensure that not only the time constants, but also their amplitudes, are as predicted by the mechanism.

In order to test the fit of the channel block mechanism, (4.4), to experimental data we can, for example, see whether λ_s and λ_f (or, more conveniently, $\lambda_s + \lambda_f$ and $\lambda_s \lambda_f$), and the relative amplitude of the two components, vary with the concentration of blocker (x_B) and the concentration of agonist (which changes β' - see above) in the predicted way. If it does then values for the underlying transition rates, α, k_{-B} etc. can be extracted; in this process it is often useful to use approximations to the exact equations (such as (5.6) and (5.7)), and various procedures have been described (see, for example, Adams, 1976; Adams & Sakmann, 1978; Colquhoun, Dreyer & Sheridan, 1979; Colquhoun & Sheridan, 1981).

A simple agonist example

Similar arguments apply to the simple agonist mechanism in (4.1). Again two macroscopic rate constants are predicted, their values being found from the quadratic solution (5.2), but in this case the coefficients are given (e.g. Colquhoun & Hawkes, 1977) by

$$- b = \lambda_s + \lambda_f = \alpha + \beta + k_{+1}x_A + k_{-1} \tag{5.10}$$

$$c = \lambda_s \lambda_f = \alpha k_{-1} \left[1 + \frac{x_A}{K_A} \frac{(\alpha + \beta)}{\alpha} \right]$$

$$= \alpha k_{-1} + \alpha k_{+1}x_A + \beta k_{+1}x_A \tag{5.11}$$

where K_A is the microscopic equilibrium constant for the binding reaction, i.e. k_{-1}/k_{+1}. The problem with trying to use this result is that it is *not* generally possible to see experimentally the predicted two components in macroscopic responses to agonists. Usually, as in Fig. 1, a single exponential is a good fit. It is for this reason that it was postulated that agonist binding is very fast, as described in (4.6). However this is not the only possible explanation (and is probably not the correct explanation). In order to get further, the greater discriminating power of single channel measurements are needed.

Single channel results with more than two states

In practice it is usually observed, for virtually every type of channel, that the p.d.f. required to fit the distribution of shut times has more than one exponential component. For example Colquhoun & Sakmann (1985) found that three components were needed for the nicotinic channel even at low agonist concentrations (see Fig. 10 in Chapter 6). Under the usual assumptions, the number of components in this distribution provides a (minimum) estimate of the number of shut states (see Colquhoun & Hawkes, 1982). Likewise the number of components in the open time distribution indicates the (minimum) number of open states. These results are rather more informative than for macroscopic measurements for which the number of components (minus one) indicates the total number of states, but doesn't indicate how many of these are open states.

Thus both the examples above, the agonist mechanism in (4.1) and the channel block mechanism in (4.4), predict a double exponential shut time p.d.f. and a single exponential open time p.d.f. The open time p.d.f. will therefore be

$$f(t) = \tau^{-1}e^{-t/\tau} \tag{5.12}$$

where τ is the mean open time given in (4.2) and (4.5) for the two models (notice that this is decreased by increasing the blocker concentration for the channel block case). The shut time p.d.f. should have the double-exponential form (see Chapter 6)

$$f(t) = a_s\tau_s^{-1}e^{-t/\tau_s} + a_f\tau_f^{-1}e^{-t/\tau_f}, \tag{5.13}$$

where a_s and a_f are the relative areas under the p.d.f. accounted for by the slow and fast components respectively, and τ_s and τ_f are the observed 'time constants', or 'means', of the two components. We may also refer to their reciprocals, the observed rate constants for the shut time distributions, $\lambda_s=1/\tau_s$ and $\lambda_f=1/\tau_f$. These time constants are not, of course, the same as those found for macroscopic measurements. We must, therefore, now enquire how they are related to the underlying reaction transition rates.

Shut times for channel block. This case is particularly simple because no direct transition is possible between the two sorts of shut state in (4.4). Every shutting of the channel must, therefore, consist of *either* a single sojourn in the shut state (state 3), *or* a single sojourn in the blocked state (state 2). The two time constants in this case are therefore simply the mean lifetimes of these two states, $m_3=1/\beta'$ and $m_2=1/k_{-B}$ (the mean lifetime of a blockage). And the relative areas are determined simply by the relative frequency of sojourns in each of these states, i.e. on the relative values of α and $k_{+B}x_B$. These results are a great deal simpler than those for macroscopic measurements given in (5.8)-(5.9); the values of two transition rates can be found directly from τ_s and τ_f.

Shut times for the agonist mechanism. The scheme in (4.1) is a bit more complicated than that for channel block; the reason for this is that the two shut states intercommunicate directly. Thus an observed shut period *might* consist of a single sojourn in AR (state 2) followed by reopening of the channel, but it might also consist of a variable number of transitions from AR to R and back before another opening occurred (both states are shut so these transitions would not be visible on the experimental record). This complication means that we once again have to resort to solving a quadratic, as in (5.2), to predict the observed time constants for the shut time distribution. One way in which the distribution of shut times can be derived, for this particular mechanism, is given in appendix 1. By use of the matrix methods given by Colquhoun & Hawkes (1982), a general expression for the shut time distribution can be obtained that holds for *any* mechanism. In this case the observed rate constants are given by (e.g. appendix 1, and Colquhoun & Hawkes, 1981) by solving the quadratic, as in (5.1)-(5.7),

$$-b = \lambda_s + \lambda_f = \beta + k_{+1}x_A + k_{-1} \tag{5.14}$$

$$c = \lambda_s\lambda_f = \beta k_{+1}x_A . \tag{5.15}$$

This result, although more complex than for the channel block mechanism, is still a good deal simpler than the analogous result for macroscopic measurements given in (5.10) and (5.11). For example, the rate constant α does not appear at all here (only those transition rates that lead *away from* shut states appear), and the expressions are simpler. Nevertheless, as in the macroscopic case, the observed time constants are not *directly* related to individual reaction transition rates or to the mean lifetimes of states.

Although this statement is generally true, we can often find good approximations that lend a simple physical significance to the time constants. If, for example, we observe that one rate is very much faster than the other ($\lambda_f \gg \lambda_s$) then (5.14) will give a good approximation to λ_f; furthermore when the agonist concentration (x_A) is very low the term $k_{+1}x_A$ will be negligible, so under these conditions

$$\tau_f = 1/\lambda_f \approx \frac{1}{\beta + k_{-1}}, \tag{5.16}$$

and this is just the mean lifetime of a single sojourn in the AR state, as was pointed out in (4.2). Note, though, that while this is *exactly* the mean lifetime of AR, it is only *approximately* the fast time constant of the shut time distribution. The physical significance of this is that short shut periods (gaps) consisting of a single sojourn in AR are predicted to appear in between openings of the channel as a result of AR*→AR→AR* . . . transitions. If, while in AR, the agonist dissociates (AR→R), rather than the channel reopening (AR→AR*), then a much longer shut period will ensue, which will contribute to the slow component of the distribution of shut times. Thus the openings will appear to occur in rapid bursts.

Bursts of channel openings

Whenever the shut time distribution has one time constant that is much shorter than another the experimental record will contain short shut periods and some much longer ones (see, for example, the shut time distribution shown in Fig. 10 of Chapter 6). In this case several openings may occur, each opening separated from the next by a short shut period, before a long shut period occurs. The openings, will therefore appear to be grouped together in *bursts* of openings, each burst being separated from the next by a relatively long silent period. Examples of such bursts are shown, for example, by Neher & Steinbach (1978), Ogden, Siegelbaum & Colquhoun (1981), and Ogden & Colquhoun (1985) for channel blockers and by Colquhoun & Sakmann (1981, 1983, 1985) and Sine & Steinbach (1985) for agonists. Some bursts are illustrated in Fig. 6.

It may simplify the analysis considerably if the experimental record can be divided up, without too much ambiguity, into such bursts. A major advantage of doing this is that we can usually be sure that all the openings within one burst originate from the same individual channel, so the characteristics of shut periods *within* bursts can tell us something about the channel. On the other hand individual well-separated bursts may well originate from different channels so, in the absence

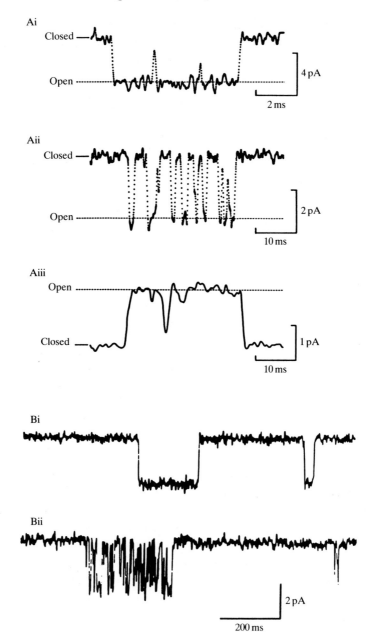

Fig. 6. (A) Burst-like appearance of single channel currents in different transmitter-activated ion channels. (Ai) Acetylcholine-activated single channel current recorded from the frog endplate. Downward deflection of the trace represents inward current. (Aii) Acetylcholine-activated single channel current recorded from a sinoatrial node cell of rabbit heart. Downward deflection represents inward K[+] current. (Aiii) Glycine-activated single channel current in soma membrane of a cultured spinal cord cell. Upward deflection of the trace corresponds to an outward current reflecting Cl[-] influx into the neuron. Reproduced with permission from Colquhoun & Sakmann (1983). (B) Bursts produced by rapid channel block. (Bi) Acetylcholine activated channels at frog end-plate. (Bii) The same in the presence of benzocaine (200 μM). Reproduced from Ogden, Siegelbaum & Colquhoun (1981).

of knowledge of how many channels there are in the membrane patch, it is impossible to interpret the lengths of gaps *between* bursts in any useful way. Another advantage of looking at the properties of bursts of openings is that the interpretation of the results may be simplified; long-lived shut states are, by definition, excluded from the bursts so we need consider a smaller subset of states than would otherwise be necessary.

Bursts with channel blockers

The appearance of bursts in the presence of a channel blocking agent will depend on how long the blockages last. If the blockages last for a long time (e.g. seconds, i.e. k_{-B} is small) the blocked periods will not be distinguishable from any other shut time, and the record will not look obviously bursty. At the other extreme, if blockages are very brief (i.e. k_{-B} is very large) so that they are mostly not resolvable, it will appear (incorrectly) that the single channel amplitude is depressed; this happens partly because such blockers will have a low affinity for the receptors, and so will be used in concentrations that make the openings very short (see (4.5)), as well as the blockages being short (see Ogden & Colquhoun, 1985, for an example).

Clear bursts will be seen in the presence of channel blockers when the blockages are of a duration that is comparable with the channel open times, so blockages appear as brief interruptions of a normal channel activation. Each individual opening will be shorter in the presence of the blocking agent, as shown in (4.5), and the predicted linear dependence of the reciprocal open time on blocker concentration provides a test of the model (and, if the test is passed, an estimate of the association rate constant for block of an open channel, k_{+B}). An example of the linear dependence of the reciprocal mean open time on blocker concentration is shown in Fig. 7 (filled circles).

The line is straight as predicted by (4.5) so the slope provides an estimate of the association rate constant for the blocking reaction ($k_{+B}=3.9 \times 10^7$ M^{-1} s^{-1} in Fig. 7). In this case the mean open times had been corrected for the fact that the brief interruptions, which the agonist (suberyldicholine) shows even in the absence of block, are only partially resolved. The intercept should therefore give an estimate of α. Another approach would be to ignore these spontaneous shuttings and look at the shortening of whole elementary event (the burst of openings produced by a single activation of the channel) by channel blocking; this should also give a slope of k_{+B} (Ogden & Colquhoun, 1985).

Fallacies in the interpretation of channel block

The shortening of the mean open time by a channel blocker might be regarded as resulting from openings being 'cut short' by the insertion of a blocking molecule into the open channel. This view seems consistent with the fact that the mechanism (4.4) predicts that the mean value of total open time in a burst is unaltered by the presence of the blocker (it is still $1/\alpha$ as it would be in the absence of the blocker). This prediction is also useful for testing the fit of the results to the postulated mechanism (e.g. Neher, 1983). The basis of this prediction is outlined by Colquhoun & Hawkes

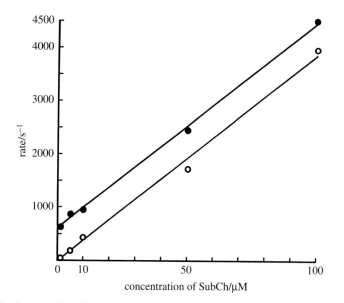

Fig. 7. Block of suberyldicholine-activated end-plate channels by the agonist itself at high concentrations. The blockages produce a characteristic component with a mean of about 5 ms in the distribution of shut times (so $k_{-B} \approx 1/5$ ms$=200$ s^{-1}). Blockages become more frequent (the relative area of the 5 ms component increases) with concentration. *Closed circles*: reciprocal of the corrected mean open time plotted against concentration (slope 3.9×10^7 M^{-1} s^{-1}, intercept 594 s^{-1}). *Open circles*: the 'blockage frequency plot' - the frequency of blockages per unit of open time is plotted against concentration (slope$=3.8 \times 10^7$ M^{-1} s^{-1}, intercept 4 s^{-1}). Reproduced with permission from Ogden & Colquhoun (1985).

(1983); it is as though the opening 'carried on as normal' after a blockage. However this way of thinking of the blocking effect is clearly incompatible with a memoryless process; the channel cannot 'know' how long it has been open during earlier openings in the burst. And it can lead to errors; for example it might be supposed that if we separated out from the record all of those channel activations in which no blockage happened to occur then such activations would not be 'cut short' so they would have a normal mean open time of $1/\alpha$ (the number of blockages per activation is random, there may be 0, 1, 2, . . . blockages; see Fig. 8). This is not true: in fact openings thus selected would have a shortened mean lifetime of $1/(\alpha+k_{+B}x_B)$ just like the openings that were terminated by a blockage (see Colquhoun & Hawkes, 1983 for a fuller discussion).

The blockage frequency plot

When activations of the channel are frequent (at high agonist concentrations) the beginning and end of a burst will not be clearly distinguishable, so the prediction that the mean total open time per burst is unaltered by the channel blocker cannot be tested. However a similar test can be achieved by doing a 'blockage frequency plot' (Ogden & Colquhoun, 1985), as long as a component that is attributable to blockages can be distinguished in the shut time distribution. By analogy with (2.5), the frequency of channel blockages should be $p_1 k_{+B} x_B$ where p_1 is the fraction of open channels. Thus, if the frequency of blockages per unit open time is plotted against the blocker concentration the slope of line should be k_{+B}. An experimental example of such a plot is shown in Fig. 7 (open circles); it is seen to behave as predicted by the simple channel block mechanism. The slope of this line gives $k_{+B}=3.8\times10^7\,\mathrm{M^{-1}\,s^{-1}}$, a very similar value to that estimated from the shortening of openings.

Relationship between bursts and macroscopic currents with channel block

The relationship between bursts at the single channel level, and double exponential relaxation at the macroscopic level is illustrated by the simulated MEPC in Fig. 8.

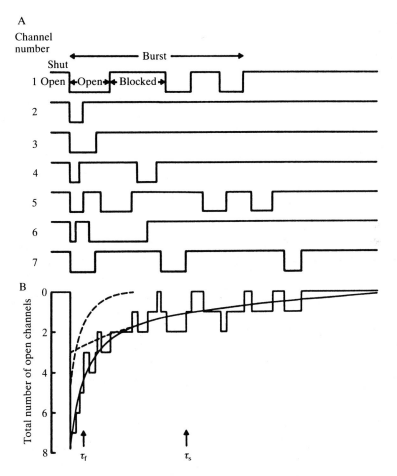

Fig. 8. Simulation of MEPC decay as in Fig. 3, but in this case a channel blocker is supposed to be present so some of the openings (all but channels 2 and 3) are interrupted by one or more blockages. The sum of all seven channels is shown in the lower part; a double exponential decay curve is superimposed on it (the two separate exponential components are shown as dashed lines).

As in Fig. 3, the seven channels that are illustrated open almost synchronously, but this time the opening may be interrupted by blockages before the agonist dissociates and the channel shuts. For example, channel 1 has two such blockages, channel 5 has three blockages, and channels 2 and 3 have no blockages. When the channels are summed the total current is seen to decay with a double exponential time course (as illustrated in the experiment in Fig. 5). There is an initial rapid decay as channels shut, or are blocked for the first time, and then a much slower decay the rate of which reflects (approximately) the length of the whole burst of openings. The latter follows from (5.7), and the definition of b and c in (5.8) and (5.9). These give, if the agonist concentration is small enough (i.e. β' is small enough), and the mean burst length is long compared with the length of a blockage,

$$\tau_s = 1/\lambda_s \approx -b/c$$

$$\approx \frac{1}{k_{-B}} + \frac{1 + x_B/K_B}{\alpha} \tag{5.17}$$

The second term on the right hand side is the mean burst length, as will now be shown (see equation 5.23). Before doing this, it is necessary to consider the number of blockages that occur in a burst.

The number of blockages per burst. First note that the number of openings in a burst must always be one greater than the number of blockages, so we shall actually work with the number of openings per burst. This will, like everything else, be random. The distribution of the number of openings per burst, and hence an expression for the mean number, can be derived as follows. Consider an open channel (state 1 in scheme (4.4)). Its next transition may be *either* blocking (going to state 2 with rate $k_{+B}x_B$), or shutting (going to state 3 with rate α). The relative probability of the former happening (regardless of how long it takes before it happens), which we shall denote π_{12}, is thus

$$\pi_{12} = \frac{k_{+B}x_B}{\alpha + k_{+B}x_B} \tag{5.18}$$

and the probability of the latter happening is therefore $\pi_{13}=1-\pi_{12}$. Notice also that a blocked channel (state 2) *must* unblock (to state 1) eventually, so $\pi_{21}=1$. The probability of blocking and then reopening is therefore $\pi_{12}\pi_{21}=\pi_{12}$. A burst will contain r openings (and $r-1$ blockages) if an open channel blocks and unblocks $r-1$ times (probability $\pi_{12}{}^{r-1}$), and then returns after the last opening to the long- lived shut state, state 3 (with probability π_{13}). The probability of seeing r openings (i.e. $r-1$ blockages) in a burst is thus

$$P(r) = \pi_{12}{}^{r-1} \pi_{13} . \tag{5.19}$$

This form of distribution is known as a *geometric distribution*. It is the analogue, for a discontinuous variable, of the exponential distribution, and has the same shape as an exponential distribution except that it decays in steps, rather than continuously.

One or other of the possible outcomes (r=1, 2, 3, . . .,∞) must occur, so these probabilities must add to unity, i.e., from (5.19),

$$\sum_{r=1}^{\infty} P(r) = 1 .$$ (5.20)

The mean number of openings per burst can be found from the general expression for the mean, μ, of a discontinuous distribution, *viz.*

$$\mu = \sum_{r=1}^{\infty} rP(r) .$$ (5.21)

In the present example, the mean number of openings per burst is therefore

$$\mu = \frac{1}{1 - \pi_{12}} = 1 + \frac{k_{+B}x_B}{\alpha} .$$ (5.22)

As expected this depends simply on the relative probabilities of leaving the open state for the blocked, or the shut, states. The mean number of blockages per burst, $k_{+B}x_B/\alpha$, is directly proportional to the blocker concentration.

We can now work out the mean length of a burst. The average burst consists of μ openings, each of mean length m_1, and μ−1 blockages, each of mean length m_2 (as defined in (4.5)). The mean burst length is therefore

$$\mu m_1 + (\mu - 1)m_2 = \frac{1 + x_B/K_B}{\alpha} .$$ (5.23)

This should increase linearly (from $1/\alpha$) with the blocker concentration.

Bursts with agonists alone

One reason for expecting openings to occur in bursts has been discussed above, *viz.* multiple openings during a single occupancy. Whatever the reasons the phenomenon is certainly common, as illustrated in Fig. 6. A typical histogram of all shut times for an agonist activated channel is given, and discussed, in Chapter 6 (Fig. 10).

If we assume that a mechanism like (4.1) is what underlies these observations, then we can extract information from the results by two sorts of measurements. It has already been mentioned (4.2) that the mean length of the short gaps within a burst will be approximately $1/(\beta+k_{-1})$. This alone does not allow us to estimate the separate values of β and of k_{-1} . However it is also true (see below) that the mean number of openings per burst should be $1+\beta/k_{-1}$. If this is measured too, then the channel opening rate constant β, and the agonist dissociation rate constant k_{-1} can both be separately estimated (e.g. Colquhoun & Sakmann, 1985).

The number of openings per burst for agonist alone. The distribution of the number of openings per burst can be derived in much the same way as in the case of bursts caused by channel blockages, in (5.18) - (5.22). Consider a channel in the intermediate state AR (state 2 in scheme (4.1)). Its next transition may be *either* reopening (rate β) or agonist dissociation (rate k_{-1}). The relative probability of the former happening (regardless of how long it takes before it happens), which we shall denote π_{21}, is thus

$$\pi_{21} = \frac{\beta}{\beta + k_{-1}} \tag{5.24}$$

and the probability of the latter happening is therefore $\pi_{23}=1-\pi_{21}$. A burst must contain at least one opening; for the channel to *re*-open $r-1$ times before dissociation occurs the former event must occur $r-1$ times before the latter happens. The probability of r openings in a burst ($r=1, 2, 3, ..., \infty$) is thus

$$P(r) = \pi_{21}^{r-1} \pi_{23} \tag{5.25}$$

This is the geometric distribution of the number of openings per burst. The mean number of openings per burst is thus, from (5.21)

$$\mu = 1/\pi_{23} = 1 + \beta/k_{-1}, \tag{5.26}$$

as asserted above.

The actual molecular transitions that underlie the bursting behaviour in scheme (4.1) are illustrated explicitly in Fig. 9. Notice the 'invisible' oscillations between the two shut times (mentioned above) that result from agonist occupancies that fail to produce opening of the channel. A more elaborate version of this diagram is given by Colquhoun, Ogden & Cachelin (1986) who use it to illustrate what happens when

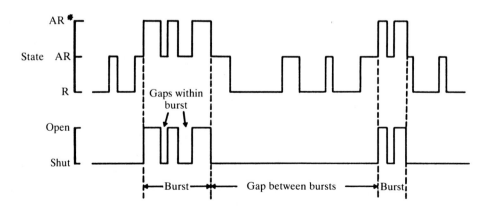

Fig. 9. Schematic illustrations of transitions between various states (top) and the corresponding observed single channel currents (bottom) to illustrate the origin of bursting behaviour for the simple agonist mechanism in (4.1). Notice that many transitions between shut states are invisible on the experimental record.

either the binding step is very fast (see above) or, alternatively, when the open-shut conformation change is very fast.

6. Correlations in single channel records

It was assumed above that the future evolution of a system depends only on its present state and not on its past history. For the simple open-shut model in (2.1) this implies, for example, that the length of an opening must be quite independent of (and therefore uncorrelated with) the length of the preceding shut time and of the lengths of preceding open times. The same is true of the 3-state mechanisms in (4.1) and (4.4); furthermore there will be no correlation between the length of one burst of openings and the next for either of these mechanisms. However more complex mechanisms may show such correlations; they are still 'memoryless', so sojourns in *individual* states will be independent, but correlations can arise because it is not possible to distinguish one sort of shut state from another on the record (they all have zero conductance), and similarly open states of equal conductance are not distinguishable. In particular, there will be correlations if there are (*a*) at least two open states (*b*) at least two shut states and (*c*) at least 'two routes' between the shut states and the open states (Fredkin, Montal & Rice, 1985; Colquhoun & Hawkes, 1987; Ball & Sansom, 1988).

Correlations between open times and between shut times

The last condition for the appearance of correlations, that there should be at least 'two routes' between the shut states and the open states, must now be put rather more precisely. Correlations will be found if there is no single state, deletion of which totally separates the open states from the shut states. The number of states which must be deleted to achieve such a separation is the *connectivity* of open and shut states, so correlations will be seen if the connectivity is greater than one. The mechanisms in (6.1) show three mechanisms each with

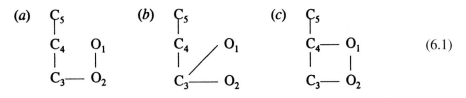

two open states (denoted O) and three shut states (denoted C). In schemes (*a*) and (*b*) there will be no correlations; deletion of state C_3 (or of state O_2) in (*a*) separates the open and shut states, as does deletion of C_3 in (*b*). In (*c*), on the other hand, the connectivity is 2 (e.g. deletion of C_3 *and* C_4 will separate open and shut states) so correlations between open times may be seen. Even in this case correlations between successive open times will be seen only if the two open states, O_1 and O_2, have different mean lifetimes. The correlations result simply from the occurrence of several $C_4{\rightleftharpoons}O_1$ oscillations followed by a $C_4{\rightarrow}C_3$ transition and then several $C_3{\rightleftharpoons}O_2$ oscillations, so runs of O_1 and runs of O_2 openings occur. The effect will clearly be most pronounced if the $C_4{\rightleftharpoons}C_3$ reaction is relatively slow.

For example, most of the properties of the nicotinic receptor are predicted well by (*c*): in this case O_2 has a long mean lifetime compared with O_1 (but it has the same conductance), whereas C_3 has a very short lifetime. Thus long open times tend to occur in runs (so there is a positive correlation between the length of one opening and the next), but long openings tend to occur adjacent to short shuttings, giving a negative correlation between open time and subsequent shut time Colquhoun & Sakmann (1985).

In the earlier work in this field, it was usual to measure correlation coefficients from the experimental record. However it is visually more attractive, and in some respects more informative, to present the results as graphs, as suggested by McManus, Blatz & Magleby (1985), Blatz & Magleby (1989), and Magleby & Weiss (1990). An example of such a plot is shown in Fig. 10.

This graph illustrates correlations found for the NMDA-type glutamate receptor (Gibb & Colquhoun, 1992). To construct this graph, five contiguous shut time ranges were defined (each range being centred around the time constant of a component of the shut time distribution). Then, for each range, the average of the open times was calculated for all openings that were adjacent to shut times in this range, and this average open time was plotted against the mean of the shut times in the range. The graph in Fig. 10 shows a continuous decline, so it is clear that long open times tend to be adjacent to short shut times, and *vice versa*.

In principle the connectivity between open and shut states can be measured experimentally because the correlation between an open time and the *n*th subsequent

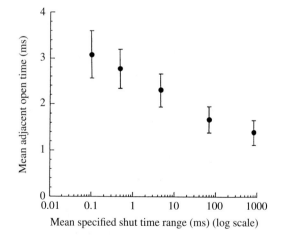

Fig. 10. Relationship between the mean durations of adjacent open and shut intervals. The graph shows the mean ± standard error of the open times in sixteen different patches, plotted against the average of the mean adjacent shut time ranges used in each patch. Reproduced from Gibb & Colquhoun (1992).

open time (for a single channel) will decay with increasing lag (n) towards zero as the sum of m geometric terms, where m is the connectivity minus one (Fredkin *et al.*, 1985; Colquhoun & Hawkes, 1987). However such a quantitative interpretation of correlations has not yet been achieved in practice.

These results can be extended to correlations between the lengths of bursts of openings, and between the lengths of openings within a burst (Colquhoun & Hawkes, 1987). There will be correlations between bursts when the connectivity (as defined above) between open states and *long-lived* shut states is greater than one. There will be correlations between openings within a burst when the *direct* connectivity between open states and *short-lived* shut states is greater than one (the term *direct connectivity* refers only to routes that connect open and short-lived shut states directly, not including routes that connect them indirectly *via* a long lived shut state, entry into which would signal the end of a burst). Thus, for the examples in (6.1), taking C_5 to be the long lived shut state, neither (*a*) nor (*b*) would show any such correlations, whereas (*c*) would show correlations within bursts but no correlations between bursts (as observed experimentally by Colquhoun & Sakmann, 1985). The following scheme (in which C_5 and C_6 both represent long lived shut states), on the other hand, would show all three types of correlation

$$
\begin{array}{cc}
C_5 & C_6 \\
| & | \\
C_4 & O_1 \\
| & | \\
C_3 & O_2
\end{array}
\qquad (6.2)
$$

7. Single channels after a voltage- or concentration-jump

The lifetime of a sojourn in a single state will always (under our assumptions) follow an exponential distribution. However some sorts of measurements may be dependent on whether the recording is made at equilibrium or not. So far we have supposed that recordings are made at equilibrium. Strictly speaking the assumption is a bit less strong - we assume that there is a steady state; however, for the sort of mechanism we are discussing this amounts to the same thing (Colquhoun & Hawkes, 1982, 1983).

Some new, and more complex, considerations arise when the record is not at equilibrium. For example, following a sudden change in concentration (a *concentration-jump*) or membrane potential (a *voltage-jump*), it will take some time for a new equilibrium to be established. It is usual to look at the channel openings (or shuttings, or bursts) that follow such a perturbation as though the 1st, 2nd, . . . opening (or shutting, or burst) were all directly comparable. This will be the case if there is only one sort of shut state and one sort of open state as in the simplest scheme (2.1) that was discussed earlier (Colquhoun & Hawkes, 1987).

Consider, for example, the case where the channels are shut initially but may open following a depolarizing voltage step applied at $t=0$. If there is more than one sort of shut state the latency until the first opening may depend on how the system was distributed among the various shut states at $t=0$, so measurements of the first latency can give information about this, and about the lifetimes of these shut states. Thereafter, however, every open time and shut time will be exactly comparable, each having exactly the same distribution as it would have in an equilibrium record. This simple result will happen in precisely those cases in which open and shut times are not correlated (as described in the preceding section).

When open and shut times are correlated then the lengths of the 1st, 2nd, . . . etc. openings (or shuttings, or bursts) will not all have the same distribution, though they will approach the equilibrium distribution eventually. Their distributions should, however, all have the same observed time constants; these time constants are functions (though possibly quite complicated functions) *only* of the underlying transition rates and, according to our assumptions, these remain constant throughout the post-jump period, because the membrane potential and/or ligand concentrations are supposed to remain constant. What is happening is that the occupancies of each state, and hence, from (2.5) and (2.6), the actual transition frequencies, are varying with time. This means that the relative *areas* of the components associated with each time constant may be different for the 1st, 2nd, . . . *etc.* open time following the jump. This phenomenon is discussed and exemplified by Colquhoun & Hawkes (1987), but it has not yet been exploited experimentally.

It is worth noting that voltage jumps and concentration jumps are usually quite brief (often in the range 1-300 ms), so they are not a good way of investigating slow kinetic processes (such as the so-called mode changes), which are necessarily obscured in short recordings.

The time course of synaptic currents

In many (though possibly not all) cases it seems that the time course of the post-synaptic current elicited by nerve stimulation is quite long compared with the length of time for which the transmitter is present in the synaptic cleft (e.g. Anderson & Stevens, 1973). In other words the presynaptic ending provides a very brief concentration jump of transmitter; this opens some channels which subsequently shut in the absence of transmitter. In the simulations in Figs 3 and 8 it was supposed that channels open essentially synchronously, i.e. that the latency to the first opening (*first latency*, for short) following exposure to the agonist is very short. This is probably close to the truth for fast channels such as the muscle-type nicotinic acetylcholine receptor (e.g. Franke *et al.*, 1991; Liu & Dilger, 1991), and may well be true for at least some AMPA-type glutamate receptors in the CNS (Colquhoun, Jonas & Sakmann, 1992).

However it is clearly *not* true that channels open synchronously for the NMDA-type glutamate receptor (Edmonds & Colquhoun, 1992); in this case the first opening may occur hundreds of milliseconds after a brief pulse of glutamate is applied. The consequences of this for the time course of the synaptic curent can be illustrated by the following oversimplified example.

Consider a hypothetical channel which, after brief agonist application, produces an activation consisting of a single opening, after the first latency has elapsed (for the NMDA receptor, the activation is actually a great deal more complicated than a single opening). In Figure 11A, nine examples are shown of simulated channels with a mean first latency of 1 ms, and a mean open time of 10 ms (the variability of both being described by simple exponential distributions).

The average current (shown at the top), is seen, not surprisingly, to have a rising phase that can be fitted with an exponential with a time constant of about 1 ms, and the decay phase has a time constant of about 10 ms. Apart from being about 10 times too slow, this example is similar to what happens at a neuromuscular junction.

More surprising, perhaps, are the results shown in Figure 11B, in which the numbers are reversed, and simulated channels have a mean first latency of 10 ms, and a mean open time of 1 ms. The averaged current shown at the top is seen to have essentially the same shape as in Figure 11A (though it is ten times smaller, and

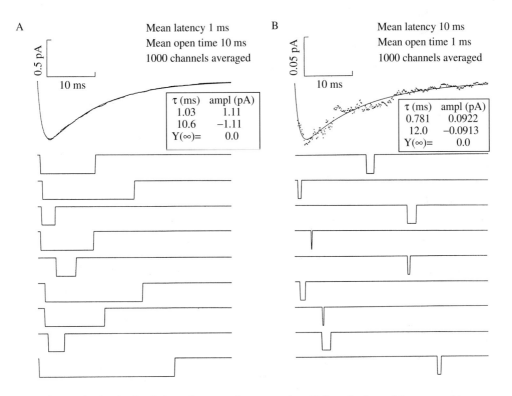

Fig. 11. A simple simulation of a synaptic current, in which each channel is supposed to produce a single opening only, after an exponentially-distributed latency. (A) Mean latency 1 ms, mean open time 10 ms. The lower part shows nine examples of simulated channels. The top trace is the average of 1000 such channels; the double-exponential curve fitted to the average has $\tau=1.03$ ms (amplitude 1.11 pA), and $\tau=10.6$ ms (amplitude=−1.11 pA). (B) Similar but with a mean latency of 10 ms and a mean open time of 1 ms. The double-exponential curve fitted to the average has $\tau=0.76$ ms (amplitude 0.092 pA), and $\tau=12.0$ ms (amplitude=−0.091 pA).

considerably noisier relative to its amplitude). Thus, in this case, *the rate of decay reflects the duration of the first latency, whereas the rate of rise represents the mean channel open time* (this case resembles observations on sodium channels: Aldrich, Corey & Stevens, 1983). The reason for this result, which seems paradoxical at first sight, can be seen from the simulations (e.g. the exponential distribution of first latencies means that short latencies are more common than long ones), and from the relevant theory which was outlined above. The distribution of the time from the stimulus until the channel shuts finally is simply the distribution of the sum of (*a*) the first latency (mean length $1/k_{21}$ say), and (*b*) the length of the channel opening (mean length $1/k_{12}$ say). Since both have been taken to be simple exponentials, as defined in (3.12) and (3.11) respectively, this distribution is given by the convolution in (3.13). The result, $f(t)$, has already been given in (3.14), and illustrated in Fig. 4. It has the form of the difference between two exponentials, which is what has been fitted to the averages in Fig. 11.

In this particular simple case, though not in general, there is a very simple relationship between the distribution, $f(t)$, of the total event length, and the shape of the averaged current. The time course of the current is given, apart from a scale factor, by the probability that a channel is open at time t. Now a channel will be open at time t if (*a*) the first latency is of length u, *and* (*b*) the channel stays open for a time *equal to or greater than $t-u$*. The probability that a channel stays open for a time $t-u$ *or longer*, is, from (3.11), the cumulative distribution

$$F_1(t-u) \equiv e^{-k_{12}(t-u)} \tag{7.1}$$

(see Chapter 6, equations 6.4-6.5), so, by an argument exactly like that used to arrive at (3.13), the probability that a channel is open at time t is

$$P_{\text{open}}(t) = \int_{u=0}^{u=t} f_2(u)F_1(t-u)\,\mathrm{d}u \tag{7.2}$$

This differs from (3.14) only by a factor of $1/k_{12}$=mean open lifetime, so

$$P_{\text{open}}(t) = f(t)/k_{12} \tag{7.3}$$

which is, apart from its amplitude, unchanged when k_{12} and k_{21} are interchanged The amplitudes of the two exponential components are equal and opposite, being, from (3.14),

$$a = k_{21}/(k_{21} - k_{12}) , \tag{7.4}$$

with a maximum at t_{max} defined in (3.15). The simulated average currents in Fig. 11 are indeed well-fitted by these values.

It is clear from this discussion that observations on the average (macroscopic) current alone cannot tell us whether the slow decay of an observed average results from a long first latency, or from a long channel opening (see, for example, Edmonds & Colquhoun, 1992, for experimental results). In order to distinguish between these possibilities it is necessary to measure either the first latency distribution, or the length of channel activations (or preferably both) independently. In practice this may not be as easy as it sounds (*a*) because the channel activations are usually more complicated than the single openings assumed here (in the case of the NMDA type of glutamate receptor they are *much* more complicated), and (*b*) because the measurements must be made on a patch of membrane that contains only one ion channel molecule (or at least a known number of molecules), and this is not easy to achieve in practice.

8. The problems of missed events and inference of mechanisms

Inferring reaction mechanisms

The discussion so far has concentrated on looking at the behaviour of specified reaction schemes. However the most important question of all comes at an earlier stage, namely, what is the qualitative nature of the reaction mechanism? The first stage is usually to inspect the distributions of open times and shut times. If these distributions can be fitted well with a sum of exponential components then the number of components required can be taken as a (minimum) estimate of the number of open states and the number of shut states, respectively. If the record can clearly be divided into bursts of openings then other distributions can be investigated, for example the distribution of the burst length, the distribution of the number of openings per burst, and the distribution of the total open time per burst. The last two should both have a number of components equal to the number of open states, and the distribution of the total open time per burst is of particular interest because, unlike the distribution of open time, it is relatively little affected by inability to detect brief shuttings of the channel.

The next question concerns how the open and shut states are connected to each other. The measurement of correlations between openings, bursts etc. may cast light on this problem, as may investigation of the distributions at various times after a voltage jump or other perturbation. In addition the presence or absence of a component of isolated openings in the distribution of the number of openings per burst may be informative (Colquhoun & Hawkes, 1987). However the single channel record does not, in principle, contain sufficient information to allow all of the connections to be worked out unambiguously. Single channel analysis, just like all other experimental work (whether biochemical or physiological) procedes by erecting hypotheses and then trying to demolish them with experimental results that are incompatible with the predictions of the hypotheses.

Missed-event corrections based on a postulated mechanism

The problems that arise from the inability to detect, in practice, the briefest openings and shuttings has already been discussed, from the practical point of view, in Chapter 6 (§10). It was pointed out there that if substantial numbers of both openings and shuttings were missed, then it was possible to correct for this imperfection only in cases where a kinetic mechanism could be postulated for the channel. Even then there may be no unique solution of the problem (Colquhoun & Sigworth, 1983). Corrections based on a postulated mechanism will be discussed next.

When openings and shuttings are missed in substantial numbers, the distributions of the *observed* (inaccurate) open and shut times are no longer expected to be described by a mixture of exponentials, as has been assumed throughout, and the theory gets a good deal more complicated. Several approximate methods for dealing with the problem have been suggested, e.g. Roux & Sauvé (1985), Wilson & Brown (1985), Blatz & Magleby (1986), Ball & Sansom (1988), Milne *et al.* (1989) and Crouzy & Sigworth (1990). However an exact solution to the problem (as usually formulated) was found by Hawkes, Jalali & Colquhoun (1990) (their calculations suggest that the best of the approximations is that of Crouzy & Sigworth). Calculation of the exact result for short times (where it is simple), combined with use of an asymptotic approximation (Hawkes, Jalali & Colquhoun, 1992) at longer times, allows accurate prediction of the *observed* distributions for any specified mechanism and time resolution. Such calculated distributions will be referred to as HJC distributions.

Fitting a mechanism directly to data

The ability to calculate the distribution of the quantities that are actually observed provides a way to correct for missed events. But in addition, it also opens the way to fitting a specified mechanism directly to the data. Previously all one could do (as in the examples in Chapter 6) was to fit empirical mixtures of exponentials separately to open times, shut times, burst lengths etc., which yielded values for *observed* time constants and areas; the corrections for missed events, the interpretation of the results in terms of a mechanism, and the estimation of the underlying 'mass action' rate constants (see §2) for the mechanism, all had to be done retrospectively. Now that missed events can be allowed for, it is possible to fit the HJC distributions directly to the *measured* open and shut times, with the adjustable parameters in the fit *not* being empirical time constants and areas, but the underlying rate constants for the model. Since the time resolution must be specified in order to do this, it becomes particularly important to ensure that this is known and consistent, as discussed in Chapter 6 (§5).

Simultaneous maximum likelihood fitting

In fact it may be possible to do much better than this. It is a problem with the conventional approach to inference of mechanisms that there are many different sources of information to be collated. For example one may have distributions of open times, shut times, burst lengths, numbers of openings per burst, correlations, and many others. Furthermore, the information from some sorts of fit overlaps heavily (e.g. the distribution of total open time per burst may be very similar to the distribution of burst length) so it can be a problem to decide how many such things to calculate.

However, once one can calculate the probability (density) for the *observed* open and shut time, by means of the HJC distributions, or approximations to them, then it becomes feasible to calculate the *likelihood* (see Chapter 6) of an entire single channel record (see Horn & Lange, 1983; Horn & Vandenberg, 1984; Fredkin & Rice, 1992). This means that, in order to do the fitting, we do not have to fit separately all of the sorts of distribution mentioned above; instead we simply adjust the parameters (mass action rate constants) so as to maximise the likelihood for the entire record. This takes into account *simultaneously* the information from the open times, from the shut times and from *the order in which they occur* (i.e. from the burst structure and from correlations between events). We have found this to be quite feasible on a fast PC, at least for a few thousands of observations. Further than this, it also becomes feasible to fit several different sorts of measurement simultaneously, for example recordings made with different concentrations of agonist (Hawkes *et al.,* in preparation).

This approach can be used to judge the relative merits of alternative postulated mechanisms. If each of the proposed mechanisms is fitted to the same data, the relative plausibility of each mechanism can be assessed from how large its likelihood is. This has been done, for example, by Horn & Vandenberg (1984) (without missed event correction).

An example of fitting with missed events

An example of maximum likelihood fitting by the HJC method (to simulated single channel data) is shown in Fig. 12.

The data in the histograms in Fig. 12 were simulated using the 5-state model for the nicotinic acetylcholine receptor, with the rate constants and concentration used by Colquhoun & Hawkes (1982) as shown in (8.1). The numbers are transition rates in s^{-1} (see discussion following 2.1).

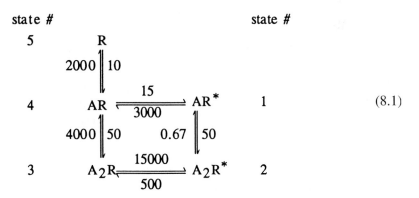

$$(8.1)$$

A resolution of 100 µs (for both open and shut times) was then imposed on the simulated sequence of open and shut times (see Chapter 6), to produce a sequence of *apparent* open and shut times from which the histograms were constructed. These histograms were not themselves fitted, but the likelihood of the entire sequence of *apparent* open and shut times was calculated, as outlined above, and the free parameters adjusted (by the simplex method) to maximize this likelihood. The estimates of the transition rates, from this *single* fitting procedure, were close to the values shown in (8.1), as hoped. The true values from (8.1) were then used to calculate the HJC distributions for the *apparent* open times and shut times. These are shown as continuous lines superimposed on the histograms in Fig. 12. It can be seen that they fit the 'observations' quite well. The same values were then used to calculate

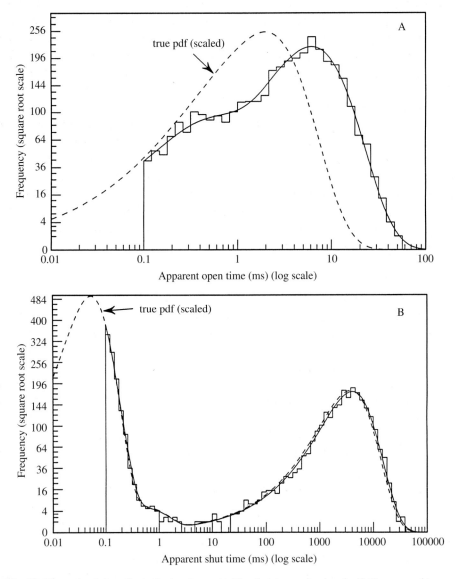

Fig. 12. Example of the effect of missed events. The 5-state mechanism in (8.1) was used to simulate 10235 channel transitions. A resolution of 100 μs was imposed on the simulated channels resulting in 3419 resolved intervals, from which the histograms of apparent open tines (A), and apparent shut times (B) were constructed. The continuous lines are the HJC distributions calculated for a resolution of 100 μs with rate constants as in (8.1); the exact solution is used for intervals up to 300 μs, and the asymptotic solution thereafter. The dashed lines show the true open and shut time distributions.

the predicted open and shut time distributions by the standard methods (e.g. Colquhoun & Hawkes, 1982), with no allowance for missed events. These distributions, which are what would be expected if we had perfect time resolution, are shown as dashed lines in Fig. 12A,B.

The open time distribution in Fig. 12A shows that the openings appear to be considerably longer than their true values, because many brief *shuttings* are below the 100 μs resolution, and therefore not detected. The true fast shut time component has $\tau=52.6$ μs, so 85% of this component is missed. With perfect resolution the open time distribution would, with the rate constants in (8.1), have two exponential components with $\tau=2.0$ ms (93% of area) and $\tau=0.33$ ms (7% of area). The asymptotic HJC distribution of apparent open times, on the other hand, has components with $\tau=6.1$ ms (81% of area) and 0.33 ms (19% of area).

The shut time distribution in Fig. 12B shows, in contrast, that the distribution of apparent shut times is close to the true distribution, except of course that there are no shut times below 100 μs: this happens because, in this particular example, relatively few *openings* are below 100 μs. This shows, incidentally, that for results of this sort, the number of missed shuttings can be estimated quite accurately by extrapolating the shut time distribution to zero length. Thus the crude form of retrospective missed event correction used by Colquhoun & Sakmann (1985) (see also Chapter 6) should have been reasonably accurate.

Problems: the number of channels

Perhaps the biggest single problem that hinders the interpretation of single channel records stems from the fact that one rarely knows how many functional channels were present in the membrane patch from which the record was made. Shut time distributions, including first latency distributions, must obviously depend on the number of channels that are present.

If it is observed, at any time during the recording, that more than one channel is open, then the patch must contain more than one channel, but it is notoriously difficult to estimate how many there are, at least when P_{open} is low (see Horn, 1991, and Chapter 6, §5). A method based on the length of runs of single openings has been proposed (Colquhoun & Hawkes, 1990), which may be useful if the burst structure of the data is not too complex.

Perhaps the most common way of managing this problem is to confine attention only to subsections of the record that *can*, with confidence, be attributed to the activity of only one channel. Such subsections may consist of individual bursts (e.g. at low agonist concentrations), or may consist of much longer clusters of openings when P_{open} is high (see Chapter 6). This means, of course, that information from longer shut periods is lost. This approach can be incorporated into maximum likelihood fitting with HJC distributions (Hawkes, Jalali, & Colquhoun, in preparation).

Problems: indeterminacy of parameters

It is unlikely that any individual experimental record will contain enough information to allow estimation of *all* the rate constants in the postulated mechanism. This problem can be minimized if several sorts of experiment (e.g. equilibrium records with different ligand concentrations, jump experiments, P_{open} curves etc.) can be analysed simultaneously, as described above. But in practice it will often be necessary to estimate some rate constants form one sort of experiment, and then to treat these values as fixed while estimating others from a different sort of experiment (see, for example, Colquhoun & Sakmann, 1985; Sine, Claudio & Sigworth, 1990).

Problems: indeterminacy of mechanisms

It is well-known that kinetic mechanisms are not unique. Whatever mechanism is proposed, it will virtually always be possible to find another mechanism that fits the experimental results just as well. Some classes of mechanism predict results that are too similar for them to be distinguished in practice. Worse still, some classes of mechanism are indistinguishable in principle; this has been discussed very elegantly by Kienker (1989). Nevertheless, combination of different sorts of biophysical experiment (as in the last paragraph), together with biochemical and structural information, can do much to reduce the ambiguity.

These problems lead some people to be pessimistic about the possibility of interpreting experiments in terms of physical mechanisms. One extreme reaction is to point out that proteins have an essentially infinite number of conformations (which is, no doubt, true), and must therefore be analysed by fractal methods, with consequent denial that anything can be gleaned about the physical mechanisms involved (see. for example, Korn & Horn, 1988). We think that it is perverse to suggest that experimental results have not been able to cast light on topics such as mechanisms of ion channel block, or the fine structure of bursts; the proposed mechanisms are doubtless approximations, but they are not baseless.

Appendix 1. Derivation of the shut time distribution for the Castillo-Katz mechanism

The Castillo-Katz mechanism defined in (4.1) will be written again here, for easy reference.

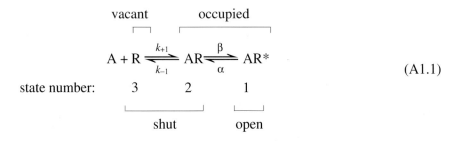

$$A + R \underset{k_{-1}}{\overset{k_{+1}}{\rightleftharpoons}} AR \underset{\alpha}{\overset{\beta}{\rightleftharpoons}} AR^* \tag{A1.1}$$

The derivation is fairly lengthy, even for this mechanism with only three states.

This complexity results, to a great extent, from the fact that the two shut states can interconvert directly (unlike the simple channel block mechanism), so that there may be any number of oscillations between them during any individual shut period. It can easily be imagined that derivations such as that to follow are not generally feasible for mechanisms with more than three states, and even with three states the complexity is great enough to provide a considerable incentive to program the quite general matrix results (Colquhoun & Hawkes, 1982) which will give numerical results for *any* mechanism.

First, notice that when the channel is at equilibrium, *every* shut period follows an opening, and so must start with a sojourn in the intermediate complex, AR (state 2). After leaving AR the channel may reopen immediately, or it may go through any number of AR⇌R oscillations before reopening. Furthermore every shut period must end in state 2 (AR) also, because the shut period ends when an opening occurs, and opening is possible only from state 2.

We start by defining the probability, denoted $P_{22}(t)$, that the channel starts in state 2, stays shut (state 2 or 3) *throughout* the time period from 0 to t, and is in state 2 at t. The shut lifetime will be t if, having stayed shut throughout the time from 0 to t, the channel then opens. The probability that a channel opens during a small time interval, Δt, is, by analogy with (2.10), $\beta \Delta t$. The probability density function for all shut times, $f_s(t)$ say, can be found by multiplying these two probabilities (the probabilities are independent, because the probability of a transition from state 2 to state 1 in the interval between t and $t+\Delta t$ is independent of the previous history of the process): it is, therefore, the limiting value, for very small Δt, of the following expression.

$$f_s(t) = \text{Prob[channel shut from 0 to } t \textit{ and} \text{ opens during } t, \, t+\Delta t]/\Delta t$$

$$= P_{22}(t)\beta \tag{A1.2}$$

We can now get an expression for $P_{22}(t)$, in the following way. The channel will be in state 2 (AR) at time $t+\Delta t$ if *either* (1) it was in state 2 at time t, *and* fails to leave state 2 during Δt, *or* (2) it was in state 3 (R) at time t, *and* moves from 2 to 3 (i.e. binds a ligand molecule) during Δt. To evaluate the first of these contingencies, we note that the probability of leaving state 2 (for either state 3 *or* state 1) during Δt is $(\beta + k_{-1})\Delta t$, so the probability of *not* leaving is 1 minus this quantity. To evaluate the second term we must first define $P_{23}(t)$ as the probability that a channel that is in state 2 at $t=0$ then remains shut throughout the time from 0 to t and is in state 3 at t; the probability of a transition from 3 to 2 in Δt is then $k_{+1}x_A\Delta t$. We can now assemble all the possibilities to write

$$P_{22}(t + \Delta t) = P_{22}(t)[1 - (\beta + k_{-1})\Delta t] + P_{23}(t)k_{+1}x_A\Delta t . \tag{A1.3}$$

Thus

$$\frac{P_{22}(t+\Delta t) - P_{22}(t)}{\Delta t} = - P_{22}(t)(\beta + k_{-1}) + P_{23}(t)k_{+1}x_A \tag{A1.4}$$

and so, letting $\Delta t \to 0$, we obtain

$$\frac{dP_{22}(t)}{dt} = - P_{22}(t)(\beta + k_{-1}) + P_{23}(t)k_{+1}x_A \tag{A1.5}$$

This equation contains two unknowns, $P_{22}(t)$ and $P_{23}(t)$. To solve this problem we must go through an exactly analogous argument to obtain a second equation, viz.

$$\frac{dP_{23}(t)}{dt} = P_{22}(t) k_{-1} - P_{23}(t)k_{+1}x_A . \tag{A1.6}$$

These two equations allow the two unknowns to be evaluated. To achieve this we eliminate $P_{23}(t)$ from the equations by using (A1.5) to obtain an expression for $P_{23}(t)$, which is then substituted into (A1.6). The result is a second order equation that involves only $P_{22}(t)$, thus

$$\frac{d^2P_{22}(t)}{dt^2} + (\beta + k_{-1} + k_{+1}x_A) \frac{dP_{22}(t)}{dt} + k_{+1}x_A\beta P_{22}(t) = 0 \tag{A1.7}$$

If we define the (constant) coefficients, b and c, as

$$b = \lambda_1 + \lambda_2 = \beta + k_{-1} + k_{+1}x_A \tag{A1.8}$$

$$c = \lambda_1\lambda_2 = k_{+1}x_A\beta \tag{A1.9}$$

then (A1.7) can be written in the form

$$\frac{d^2P_{22}(t)}{dt^2} + b \frac{dP_{22}(t)}{dt} + cP_{22}(t) = 0 \tag{A1.10}$$

We are now in a position to obtain expressions for the observed time constants. A solution of (A1.10) is needed, and we proceed, in the irritating manner of mathematicians, to propose, for no apparent reason, that there is a particular solution of (A1.10) that has the form $P_{22}(t)=e^{-\lambda t}$. If this is the case then $dP_{22}(t)/dt=-\lambda e^{-\lambda t}$, and $d^2P_{22}(t)/dt^2=\lambda^2 e^{-\lambda t}$. Substitution of these into (A1.10) gives:

$$\lambda^2 e^{-\lambda t} - b\lambda e^{-\lambda t} + ce^{-\lambda t} = 0 \tag{A1.11}$$

This result must be true at all times, including at $t=0$. Putting $t=0$ in (A1.11) gives

$$\lambda^2 - b\lambda + c = 0 \tag{A1.12}$$

This is a quadratic equation and therefore there will generally be *two* values of the *observed* rate constants, λ_1 and λ_2 say, that satisfy it, and these can be found as the two solutions of (A1.12), namely

$$\lambda_1, \lambda_2 = 0.5\left(b \pm \sqrt{b^2 - 4c}\right) \tag{A1.13}$$

or

$$\lambda_1, \lambda_2 = \frac{2c}{b \mp \sqrt{b^2 - 4c}} , \tag{A1.14}$$

where the coefficients, b and c, were defined in (A1.8) and (A1.9). Notice that these coefficients are rather simpler than those given in (5.4) and (5.5) for the macroscopic relaxation, because we are now dealing only with the shut states. They no longer involve *all* of the fundamental rate constants (α does not appear).

To complete the distribution of shut times, we must next find the *areas*, a_1 and a_2, of the two components. The p.d.f. has the form (see Chapter 6)

$$f_s(t) = \beta P_{22}(t) = a_1\lambda_1 e^{-\lambda_1 t} + a_2\lambda_2 e^{-\lambda_2 t} \tag{A1.15}$$

where $a_2 = 1 - a_1$, because the total are must be 1. Thus, from (A1.2) and (A1.15),

$$\left.\frac{df_s(t)}{dt}\right|_{t=0} = \left.\frac{\beta dP_{22}(t)}{dt}\right|_{t=0} = -a_1\lambda_1^2 - a_2\lambda_2^2$$

$$= a_1(\lambda_2^2 - \lambda_1^2) - \lambda_2^2$$

$$= -\beta(\beta + k_{-1}) \tag{A1.16}$$

The last line follows from (A1.5) because $P_{22}(0)=1$ (state 2 cannot be left in zero time), and $P_{23}(0)=0$ (cannot get from state 2 to 3 in zero time). From this, together with the fact that $\lambda_1 + \lambda_2 = \beta + k_{-1} + k_{+1}x_A$ and $\lambda_1\lambda_2 = k_{+1}x_A\beta$ (see A1.8 and A1.9), we find, after some manipulation, that the areas of the two components are given by

$$a_1 = \frac{\beta(k_{+1}x_A - \lambda_1)}{(\lambda_2 - \lambda_1)\lambda_1}, \quad \text{and} \quad a_2 = 1 - a_1 . \tag{A1.17}$$

This completes the derivation of the distribution of all shut times. The same result was derived, in a quite different way, by Colquhoun & Hawkes (1981), but if all that is needed is numerical values for the time constants and areas of the components it is obviously much easier to use the entirely general result (Colquhoun & Hawkes, 1982), rather than go through a derivation like that above for every mechanism of interest.

Approximations for the time constants. If $b^2 \gg 4c$ then it follows from (A1.13) that one of the observed rate constants will be much larger than the other, say $\lambda_2 \gg \lambda_1$. This will be the case, for example, when the ligand concentration, x_A, is very low, so from (A1.8), we find that the faster time constant is approximately

$$\tau_2 = 1/\lambda_2 \approx \frac{1}{\beta + k_{-1}} \tag{A1.18}$$

The right hand side of this is simply the mean lifetime of a single sojourn in the intermediate complex AR (state 2). This is, therefore, an example of a case where, contrary to the general rule, a physical interpretation *can* be placed on an observed time constant, as an approximation. In this case the sojourns in AR that occur as the channel oscillates between AR⇌AR*, can be seen (approximately) as the fast component of the distribution of all shut times.

References

ADAMS, P. R. (1976). Drug blockade of open end-plate channels. *J. Physiol.* **260**, 531-552.

ADAMS, P. R. & SAKMANN, B. (1978). Decamethonium both opens and blocks endplate channels. *Proc. Natl. Acad. Sci. U.S.A.* **75**, 2994-2998.

ALDRICH, R. W., COREY, D. P. & STEVENS, C. F. (1983). A reinterpretation of mammalian sodium channel gating based on single channel recording. *Nature* **306**, 436-441.

ANDERSON, C. R. & STEVENS, C. F. (1973). Voltage clamp analysis of acetylcholine produced end-plate current fluctuations at frog neuromuscular junction. *J. Physiol.* **235**, 655-691.

BALL, F. G. & SANSOM, M. S. P. (1988). Aggregated Markov processes incorporating time interval omission. *Adv. Appl. Prob.* **20**, 546-572.

BALL, F. G. & SANSOM, M. S. P. (1988). Single channel autocorrelation functions. The effects of time interval omission. *Biophys. J.* **53**, 819-832.

BLATZ, A. L. & MAGLEBY, K. L. (1986). Correcting single channel data for missed events. *Biophys. J.* **49**, 967-980.

BLATZ, A. L. & MAGLEBY, K. L. (1989). Adjacent interval analysis distinguishes among gating mechanisms for the fast chloride channel from rat skeletal muscle. *J. Physiol.* **410**, 561-585.

CASTILLO, J. del & KATZ, B. (1957). Interaction at end-plate receptors between different choline derivatives. *Proc. Roy. Soc.* B **146**, 369-381.

COLQUHOUN, D. (1971). *Lectures on Biostatistics.* Oxford: Clarendon Press.

COLQUHOUN, D., DREYER, F. & SHERIDAN, R. E. (1979). The actions of tubocurarine at the frog neuromuscular junction. *J. Physiol.* **293**, 247-284.

COLQUHOUN, D. & HAWKES, A. G. (1977). Relaxation and fluctuations of membrane currents that flow through drug-operated ion channels. *Proc. Roy. Soc.* B **199**, 231-262.

COLQUHOUN, D. & HAWKES, A. G. (1981). On the stochastic properties of single ion channels. *Proc. Roy. Soc.* B **211**, 205-235.

COLQUHOUN, D. & HAWKES, A. G. (1982). On the stochastic properties of bursts of single ion channel openings and of clusters of bursts. *Phil. Trans. Roy. Soc.* B **300**, 1-59.

COLQUHOUN, D. & HAWKES, A. G. (1983). The principles of the stochastic interpretation of ion channel mechanisms. In *Single Channel Recording* (ed. B. Sakmann & E. Neher). New York: Plenum Press.

COLQUHOUN, D. & HAWKES A. G. (1987). A note on correlations in single channel records. *Proc. Roy. Soc.* B **230**, 15-52.

COLQUHOUN, D. & HAWKES, A. G. (1990). Stochastic properties of ion channel openings and bursts in a membrane patch that contains two channels: evidence concerning the number of channels present when a record containing only single openings is observed. *Proc. Roy. Soc. London* B **240**, 453-477.

COLQUHOUN, D., JONAS, P. & SAKMANN, B. (1992). Action of brief pulses of glutamate on AMPA/kainate receptors in patches from different neurones of rat hippocampal slices. *J. Physiol.* **458**, 261-287.

COLQUHOUN, D. & OGDEN, D. C. (1986). States of the acetylcholine receptor: enumeration characteristics and structure. In *Nicotinic Acetylcholine Receptor: structure and function* (ed. A. Maelicke). Berlin: Springer-Verlag. pp. 197-232.

COLQUHOUN, D., OGDEN. D. C. & CACHELIN, A. B. (1986). Mode of action of agonists on nicotinic receptors. In *Ion Channels in Neural Membranes* (ed. J. M. Ritchie, R. D. Keynes & L.Bolis), pp. 255-273. New York: A. R. Liss.

COLQUHOUN, D. & SAKMANN, B. (1981). Fluctuations in the microsecond time range of the current through single acetylcholine receptor ion channels. *Nature* **294**, 464-466.

COLQUHOUN, D. & SAKMANN, B. (1983). Bursts of openings in transmitter-activated ion channels. In *Single Channel Recording* (ed. B. Sakmann & E. Neher). New York: Plenum Press.

COLQUHOUN, D. & SAKMANN, B. (1985). Fast events in single-channel currents activated by acetylcholine and its analogues at the frog muscle end-plate. *J. Physiol.* **369**, 501-557.

COLQUHOUN, D . & SHERIDAN R. E. (1981) . The modes of action of gallamine. *Proc. Roy. Soc. London* B **211**, 181-203.

CROUZY, S. C. & SIGWORTH, F. J. (1990). Yet another approach to the dwell-time omission problem of single-channel analysis. *Biophys. J.* **58**, 731-743.

EDMONDS, B. & COLQUHOUN, D. (1992). Rapid decay of averaged single-channel NMDA receptor activations recorded at low agonis concentration. *Proc. Roy. Soc. London* B, **250**, 279-286.

FRANKE, CH., HATT, H., PARNAS, H. & DUDEL, J. (1991). Kinetic constants of the acetylcholine (ACh) receptor reaction deduced from the rise in open probability after steps in ACh concentration. *Biophys. J.* **60**, 1008-1016.

FREDKIN, D. R., MONTAL, M. & RICE, J. A. (1985). Identification of aggregated Markovian models: application to the nicotinic acetylcholine receptor. In: Proceedings of the Berkeley Conference in honor of Jerzy Neyman and Jack Kiefer, vol. I (ed. L. M. Le Carn & R. A. Olshen). pp. 269-289. Wadsworth Press.

FREDKIN, D. R. & RICE, J. A. (1992). Maximum likelihood estimation and identification directly from single-channel recordings. *Proc. Roy. Soc. London* B **249**, 125-132.

GIBB, A. J. & COLQUHOUN, D. (1992). Activation of NMDA receptors by L-glutamate in cells dissociated from adult rat hippocampus. *J. Physiol.* **456**, 143-179.

HAMILL, O. P., MARTY, A., NEHER, E., SAKMANN, B. & SIGWORTH, F. J. (1981). Improved patch-clamp techniques for high-resolution current recording from cells and cell-free membrane patches. *Pflugers Archiv.* **391**, 85-100.

HAWKES, A. G., JALALI, A. & COLQUHOUN, D. (1990). The distributions of the apparent open times and shut times in a single channel record when brief events can not be detected. *Phil. Trans. Roy. Soc. London* A **332**, 511-538.

HAWKES, A.G., JALALI, A. & COLQUHOUN, D. (1992). Asymptotic distributions of apparent open times and shut times in a single channel record allowing for the omission of brief events. *Phil. Trans. Roy. Soc. London* B **337**, 383-404.

HORN, R. (1991). Estimating the number of channels in patch recordings. *Biophys. J.* **60**, 433-439.

HORN, R. (1984). Gating of channels in nerve and musclc: a stochastic approach. In *Ion Channels: Molecular and Physiological Aspects* (ed. W. D. Stein). pp. 53-97. New York: Academic Press.

HORN, R. & LANGE, K. (1983). Estimating kinetic constants from single channel data. *Biophys. J.* **43**, 207-223.

HORN, R. & VANDENBURG, C. A. (1984). Statistical properties of single sodium channels. *J. Gen. Physiol.* **84**, 505-534.

JACKSON, M. B., WONG, B. S., MORRIS, C. E., LECAR, H. & CHRISTIAN, C. N. (1983). Successive openings of the same acetylcholine receptor channels are correlated in open time. *Biophys. J.* **42**, 109-114.

KIENKER, P. (1989). Equivalence of aggregated Markov models of ion-channel gating. *Proc. Roy. Soc. London* B **236**, 269-309.

KORN, J. S. & HORN, R. (1988). Statistical discrimination of fractal and Markov models of single-channel gating. *Biophys. J.* **54**, 871-877.

LABARCA, P., RICE, J. A., FREDKIN, D. R. & MONTAL, M. (1985). Kinetic analysis of channel gating: application to the cholinergic receptor channel and the chloride channel from Torpedo californica. *Biophys. J.* **47**, 469-478.

LIU, Y. & DILGER, J. P. (1991). Opening rate of acetylcholine receptor channels. *Biophys. J.* **60**, 424-432.

MAGLEBY, K. L. & WEISS, D. S. (1990). Identifying kinetic gating mechanisms for ion channels by using two-dimensional distributions of simulated dwell times. *Proc. Roy. Soc. London* B **241**, 220-228.

McMANUS, O. B., BLATZ, A. L. & MAGLEBY, K. L. (1985). Inverse relationship of the durations of open and shut intervals for Cl and K channels. *Nature* **317**, 625-627.

MILNE, R. K., YEO, G. F., EDESON, R. O. & MADSEN, B. W. (1989). Estimation of single channel kinetic parameters from data subject to limited time resolution. *Biophys. J.* **55**, 673-676.

NEHER, E. (1983). The charge carried by single-channel currents of rat cultured muscle cells in the presence of local anaesthetics. *J. Physiol.* **339**, 663-678.

NEHER, E. & STEINBACH, J. H. (1978). Local anaesthetics transiently block currents through single acetylcholine-receptor channels. *J. Physiol.* **277**, 153-176.

OGDEN, D. C. & COLQUHOUN, D. (1985). Ion channel block by acetylcholine, carbachol and suberyldicholine at the frog neuromuscular junction. *Proc. Roy. Soc. London* B **225**, 329- 355.

OGDEN, D. C., SIEGELBAUM, S. A. & COLQUHOUN, D. (1981). Block of acetylcholine- activated ion channels by an uncharged local anaesthetic. *Nature* **289**, 596-599.

ROUX, B. & SAUVE, R. (1985). A general solution to the time interval omission problem applied to single channel analysis. *Biophys. J.* **48**, 149-158.

SAKMANN, B., PATLAK, J. & NEHER, E. (1980). Single acetylcholine-activated channels show burst-kinetics in presence of desensitizing concentrations of agonist. *Nature* **286**, 72-73.

SAKMANN, B. & TRUBE, G. (1984). Voltage-dependent inactivation of inward-rectifying single channel currents in the guinea-pig heart cell membrane. *J. Physiol.* **347**, 659-683.

SINE, S. M. & STEINBACH, J. H. (1986). Activation of acetylcholine receptors on clonal BC3H-1 cells by low concentrations of agonist. *J. Physiol.* **373**, 129-162.

SINE, S. M., CLAUDIO, T. & SIGWORTH, F. J. (1990). Activation of *Torpedo* acetylcholine receptors expressed in mouse fibroblasts: single channel current kinetics reveal distinct agonist binding affinities. *J. Gen. Physiol.* **96**, 395-437.

THOMPSON, S.P. (1965). *Calculus Made Easy.* London: Macmillan.

WILSON, D. L. & BROWN, A. M. (1985). Effect of limited interval resolution on single channel measurements with application to Ca channels. IEEE Trans. *Biomed. Eng.* **32**, 780-797.

Chapter 8

Analysis of whole cell currents to estimate the kinetics and amplitude of underlying unitary events: relaxation and 'noise' analysis

PETER T. A. GRAY

1. Introduction

The membrane currents evoked in voltage-clamped cells by a voltage step or by a pulse of neurotransmitter, released from nerve terminals or applied in vitro, are made up of the sum of many small unitary currents that flow through ion channels, aqueous pores in the cell membrane. Such summed, macroscopic, currents contain a component of 'noise' that results from the addition of many independent, randomly occurring unitary events. The time course of the evoked currents reflects the kinetics of the underlying channels and the channel kinetics determine the rate at which the relaxation to a new equilibrium occurs after a perturbation, such as a step of membrane potential.

The first direct evidence that these whole cell currents resulted from the summation of many smaller unit currents, flowing through ion channels, came from studies of the membrane voltage noise evoked by acetylcholine (ACh) at the frog neuromuscular junction, by Katz and Miledi (1970). Since that time more refined analysis of the noise components of current signals has allowed the amplitude and mean lifetime of the ACh activated channels at the frog neuromuscular junction (NMJ) to be measured (Anderson & Stevens, 1973); in turn allowing estimation of an upper limit of the lifetime of the pulse of ACh in the synaptic cleft (Magelby & Stevens, 1972). These techniques and relaxation analysis following a voltage step have also allowed investigation of the mechanisms of blockade of ion channels (Adams, 1977; Colquhoun et al. 1979).

More recently patch clamp techniques have allowed the kinetic analysis of both voltage-gated and transmitter activated conductances to be carried much further, as is described elsewhere in this book (Chapters 4 to 7). Though patch clamp studies can in principle provide much greater detail about the mechanisms of a conductance change voltage clamp techniques remain a valuable tool under many circumstances. Firstly, by their nature, studies of macroscopic currents involve the investigation of the averaged behaviour of all active ion channels in the cell membrane. By comparing the results of noise analysis with predictions obtained from single channel analysis it is

52, Sudbourne Rd., London SW2 5AH, UK.

possible to check that the observed single channel behaviour can indeed explain the behaviour of the whole cell. This may not happen if, for example, more than one population of channels is present but unevenly distributed so that a single type is preferentially present in patches used for analysis, or as a second example, if a channel population has too small an amplitude to be resolved in single channel recordings but contributes appreciable current. Estimates of single channel amplitude from noise analysis that are much smaller than those obtained from single channel recording could be indicative of such problems, though they could also arise, for example, if a high frequency component of channel 'flicker' were resolved in single channel records, but was too fast to be resolved by whole cell noise recording. In addition, there are situations in which analysis of macroscopic currents may reveal details that single channel analysis cannot, for example, when the single channel amplitude is too small to be resolved (e.g. Gray & Attwell, 1985).

In general these forms of analysis of macroscopic signals provide information about two aspects of the underlying unitary events, their amplitude and kinetics. Two forms of analysis will be considered here in detail, (1) 'Noise', or fluctuation, analysis allows determination of information about both amplitude and kinetics from analysis of the fluctuations of a signal around the mean level. All electrical systems give rise to noise, for example 'shot' noise due to flow of charges in electronic components. The analysis of noisy signals was originally developed in relation to noise sources in electrical circuits and applied to biological systems by Katz & Miledi (1970). One of the principal difficulties to be faced when performing 'noise analysis' is the isolation of the biological noise from unwanted noise signals, such as shot noise in the recording circuitry and periodic interference such as 50 Hz mains and radio. (2) Whereas noise analysis gives information about the amplitude and time course of the underlying unitary events relaxation analysis of the kinetics of re-equilibration after a voltage or concentration step produces information solely about the kinetics of the events. In either case the data is analysed by comparing the observed data with the predictions of models of the unitary event, e.g. unitary current amplitude and channel kinetics.

While the signals analysed most frequently are membrane currents that result from the combination of single channel currents flowing through a population of ion channels, noise analysis can be applied to other forms of signal. For example the initial electrophysiological application of noise analysis by Katz & Miledi (1970) was a determination of the pulse of membrane potential that resulted from the opening of a single ACh activated channel at the frog NMJ. In general, however, meaningful kinetic information is much harder to obtain from voltage recordings, as the kinetics of the signals are determined by the effects of cell capacitance if the cell is not voltage clamped.

2. Using noise analysis to estimate unitary amplitudes

Stationary noise analysis

Noise analysis is based upon the fact that a steady state signal that is made up from a population of randomly occurring identical unitary events, such as single channel currents, exhibits fluctuations, or noise, about its mean level. Analysis of these fluctuations allows the properties of the underlying events to be investigated. Fig. 1 illustrates how the combination of a large number of unitary current events produces a signal that has both a DC component and fluctuates about that level.

Determination of the amplitude of the unitary events from the noise signal requires calculation of the mean and variance of the signal. The variance of the signal is given by:

$$\text{var } (x) = \frac{1}{N} \sum_{i=1}^{N} (x_i - \bar{x})^2 \tag{1}$$

and the mean ($\bar{x}$) by:

$$\text{mean } (\bar{x}) = \frac{1}{N} \sum_{i=1}^{N} x_i \tag{2}$$

where N is the number of points sampled.

In practice the variance of a recorded signal consists of both the biological noise, the required signal, and background noise, for example instrument noise. However, the variance of two summed signals is given by:

$$\text{var } (a+b) = \text{var } a + \text{var } b + \text{covariance } (a,b).$$

Therefore, provided that the biological noise is independent from other noise sources (so that covariance $(a,b) = 0$) then the test variance is obtained simply by subtracting

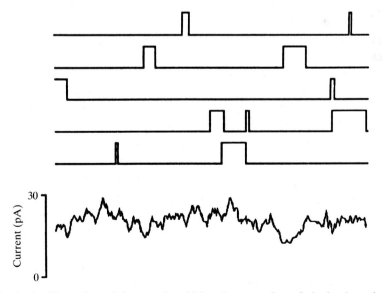

Fig. 1. An illustration of the way in which a large number of single channels opening randomly produce a noisy signal in addition to a DC current component. The upper five traces show computer generated current traces, modelled assuming a single channel current of 1 pA and a two-state model with $\alpha=200$ s^{-1} and $\beta=50$ s^{-1}. The bottom trace is the sum of 100 such stretches of modelled single channel activity.

the background variance, which may be obtained from a control recording, from the total variance.

The relationship between the variance and mean and the amplitude of the unitary events depends upon the nature of those events. When studying the voltage noise generated by ion channels at the frog NMJ Katz and Miledi (1970, 1972) applied Cambell's theorem, which describes the noise generated by the summation of 'shot' events, each described by a function $f(t)$. In this case the mean and variance are given by:

$$\text{mean} = n \int_{-\infty}^{+\infty} f(t)\, dt \qquad (3)$$

$$\text{var} = n \int_{-\infty}^{+\infty} f^2(t)\, dt \qquad (4)$$

where n = frequency of events.

In the particular case of membrane voltage noise, if the unitary events are assumed to have the same amplitude then $f(t) = ae^{-t/\tau}$; where a is the amplitude and τ is the membrane time constant (this describes an elementary voltage 'blip' similar in waveform to the m.e.p.p., but of smaller amplitude). In this case the mean and variance are given by:

$$\text{mean } (V) = na\tau \qquad (5)$$

$$\text{var } (V) = \frac{na^2\tau}{2} = \frac{a.\text{mean } (V)}{2} \qquad (6)$$

and the unitary amplitude is given by:

$$a = 2\frac{\text{var } (V)}{\text{mean } (V)}. \qquad (7)$$

Alternative assumptions discussed by Katz and Miledi (1972) are when the unitary events had random amplitude or that they had different waveforms, $f(t)$.

Where the signal is current noise from a cell at constant voltage, i.e. data from voltage clamp recordings, then, provided that a single type of channel contributes to the noise, the mean current is given by:

$$\text{mean } (I) = NP_o i \qquad (8)$$

where N is the number of channels, i the unit conductance and P_o the steady state open probability (which will be a function of, for example, neurotransmitter concentration and membrane potential). If the channel may be only in an open or a closed state then, from binomial theory, the current variance is:

$$\text{var } (I) = Ni^2 P_o(1-P_o) \qquad (9)$$

where $(1-P_o)$ is the probability of the channel being closed. Thus:

$$\frac{\text{var}\,(I)}{\text{mean}\,(I)} = i(1 - P_\text{o}). \tag{10}$$

The plot of variance against mean gives a parabolic relationship as P_o varies from 0 to 1; however, for low values of P_o (i.e. $P_\text{o} \ll 1$) the relationship simplifies to:

$$i = \frac{\text{var}\,(I)}{\text{mean}\,(I)}. \tag{11}$$

Data preparation for noise analysis. The unit amplitude of an underlying unitary signal is most easily estimated by obtaining the mean and variance from a stretch of signal recorded under steady state conditions, i.e. in which the mean amplitude is constant, or at least changing slowly, over the duration of the recording. Once such a record has been obtained it can be analysed using a hardwired variance meter, in which case the mean and variance can be measured by hand from a chart record of the signal and its variance, and then plotted. Alternatively the signal may be sampled by a computer and analysed using software to calculate the variance, display the plot and fit a slope or parabola as needed (see Chapter 9). In practice the latter method is to be preferred, and with laboratory computer equipment becoming ever cheaper there are no longer any advantages to using a variance meter. In either case the raw data should be recorded on a DC tape recorder such as an FM or DAT recorder to allow gain and filter settings to be selected after completion of the experiment.

On play-back for analysis two signals must be prepared from the original recording (Fig. 2A). Firstly, a low pass filtered, low gain signal is needed for the measurement of mean current. Low pass filtering is essential to remove high frequency noise components that may obscure the biological noise. Noise generated by channel activity drops off at high frequencies, while, in contrast, electrical noise sources in the recording set-up increase with frequency, so substantial benefits in signal to noise ratio are obtainable by careful low pass filtering. Secondly, a high gain signal that has been both low and high pass filtered is needed for the calculation of variance. The high pass filtering of the second trace AC couples it, so that the signal contains only deviations from the mean level. This allows the signal from which the variance will be calculated to be amplified to a high gain, ensuring full use of the input resolution of the laboratory interface (see Chapter 9). The amplified signal should be monitored on an oscilloscope as it is sampled to ensure that it is sufficiently amplified to be well resolved, but not so large that it overshoots the input range of the ADC.

The choice of filter and sample frequencies must be made to ensure that the filter pass band used for the AC coupled trace encompasses all significant frequencies in the signal. This will depend upon the kinetics of the channels. If, for example, the low pass filter is set too low, then a significant amount of high frequency signal noise may be lost; this will result in an underestimate of the variance, and hence of i. The low pass filter setting must also be selected to ensure that 'aliasing' of the data does not occur. A set of data points sampled at frequency f points/sec cannot unambiguously represent a signal that contains periodic frequencies greater than $f/2$. It is impossible to distinguish, in the sampled record, between a signal with a frequency below $f/2$ and a signal with a frequency the same amount above $f/2$. This is known as signal aliasing, and the frequency $f/2$ is known as the Nyquist frequency (see Chapter 16). It is essential that the signal is filtered at $\frac{1}{2}$ the sample rate, or below, in order to prevent distortion of both estimates of i and of any kinetic information that is obtained. For this reason Butterworth filters which have a very sharp roll off, but which may distort transient signals (see Chapter 16), are generally used for noise analysis.

In general, where the kinetics of the underlying events are not already well

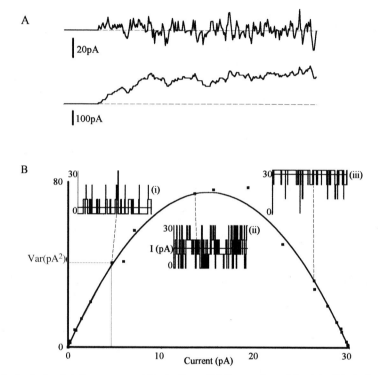

Fig. 2. Analysis of stationary current fluctuations. (A) Preparation of the data. The bottom trace shows data generated by a model that mimics the application of agonist to a cell possessing 60 channels, each of which have one closed state and one open (agonist bound) state in which an outward current of 10 pA passes under the experimental conditions. As the agonist washes on channels open and a noisy signal, fluctuating about a steady state, is evoked. The upper trace shows the same data AC coupled by high pass filtering at a higher gain. In both traces the zero current level is shown by the horizontal dashed line. (B) Current variance plotted against mean current for data generated by a model in which 3 identical channels are recorded from under a range of conditions. The channels have two states, a closed state and an open state, which carries a current of 10 pA under the experimental conditions. Data were generated for steady state open probability values ranging from 0 to 1, by varying the values of the opening rate (α) and the closing rate (β), used to generate the current records. The continuous line gives the predicted relationship between variance and mean current, calculated by the equation:

$$i\bar{I} - \bar{I}^2/N,$$

where $i = 10$ pA and $N = 3$.

The three inset traces show samples of the generated data, linked to their corresponding points on the variance/mean current plot by dashed lines. The mean current level is shown in each case by a continuous horizontal line.

described, it is better to perform both a kinetic and an amplitude analysis of the data, even if only the amplitude is of interest. This is so for the reason given above, i.e. that sample rates and filter frequencies must be chosen so that all significant frequency components in the signal are included in the variance calculation. This is hard to ensure unless the frequency characteristics of the signal are analysed. If a kinetic analysis is also carried out then an additional check on the estimated single channel current amplitude may be made from the parameters of the Lorentzian function fitted to the power spectrum (see below).

An additional advantage obtained by using a computer for the analysis is that the data may be readily edited. In general the long stretches of steady state data that are needed for noise analysis will contain a number of artefacts, such as spikes caused by neighbouring equipment turning on or off, or lower frequency artefacts caused by vibration. If analysed unedited such artefacts will lead to an over-estimation of variance and hence i. Once data has been sampled by the computer it may be displayed and parts selected for omission from analysis.

One source of unwanted noise that is hard to eliminate is 50 Hz mains 'hum'. In general this frequency falls within the frequency range of interest, so cannot be filtered out. Low levels of mains noise that are not visible by inspection of the raw data may still be sufficient to generate a peak at 50 Hz (and sometimes at the harmonics 100 and 150 Hz) in a noise spectrum. For this reason particular care in eliminating mains pickup is required when recording data for noise analysis.

Data analysis. Once the variance has been calculated the unit current amplitude, i can be calculated. The initial slope of the relationship between variance and mean current described by equation 10 is i. Thus for currents that are small relative to the maximum, i.e. under conditions where the steady state open probability (P_o) of any channel is low, then the relationship between variance and mean current is a straight line with an intercept of 0. Under such circumstances i is given by equation 11, i.e.:

$$i = \frac{\text{var}(I)}{\text{mean}(I)}$$

and can be calculated directly from the data. Inset (i) in Fig. 2B illustrates such a case.

However, under conditions where the assumption that P_o is small does not hold then it is apparent that equation 11 will yield erroneous results. In such cases, or where it is not known if the assumption holds, it is necessary to obtain recordings at different mean currents and to produce a plot of variance against mean current to which a parabola can be fitted (Fig. 2B), or a straight line if it proves that the assumption of low P_o is accurate. Such data may be obtained by applying different agonist concentrations, as is modelled in Fig. 2B, or by analysing the rising or falling phases of a response (e.g. Gray & Attwell, 1985). In the latter case it is important that the rate of change in mean current should be sufficiently slow that the change in mean current during the period over which each variance sample is calculated is small.

Non-stationary noise analysis

The analysis described above requires recordings to be made of noise signals whose characteristics remain steady over a period of time, or at least where the changes are slow relative to the length of the periods over which variance and mean current are calculated (for example analysis of transmitter activated noise where the transmitter is applied at low concentration to the preparation by bath perfusion). However, some events have properties that change too rapidly for such methods to be applied. Most notably, many voltage activated conductances activate and inactivate rapidly after a voltage step. This makes it impossible to use stationary noise analysis to estimate the single channel current amplitude. An alternative approach is possible and was described by Sigworth (1980). The method relies on recording a large number of responses to an identical stimulus, in this case a voltage step, though the method has also been applied to synaptic currents by Traynelis *et al.* (1993). The variance is calculated at identical time points after the step by summing many consecutive responses. Thus, the mean and variance are:

$$\bar{I}(t) = \frac{1}{n} \sum_{k=1}^{n} y_k(t) \tag{12}$$

$$\sigma^2(t) = \frac{1}{n-1} \sum_{k=1}^{n} (y_k(t) - \bar{I}(t))^2 \tag{13}$$

where $\bar{I}(t)$ is the mean current at time t for all n records, $\sigma(t)$ the variance at time t and $y_k(t)$ is the signal amplitude of the kth record at time t after the stimulus or start of the step. This process is illustrated in Figs 3 and 4. Fig. 3A shows three traces generated by computer simulation to illustrate the inactivation phase of a voltage activated conductance which has an outward single channel current of 10 pA. For simplicity it is assumed that the activation by the voltage step at time 0 is instantaneous, the open probability $P_o=0.5$ immediately after the step and that open channels only close to an inactivated state from which they do not reopen. The model for each channel can be represented as:

$$t \geqslant 0: \text{Open} \rightarrow \text{Inactivated}$$

with a rate constant of 40 s^{-1}. The records are from a patch of membrane containing 3 such channels, though in practice such an analysis would generally be performed on preparations containing many more channels. Fig. 3B shows the mean current obtained by averaging 250 traces, and Fig. 3C illustrates the calculation of the deviation, i.e.

$$y_k(t) - \bar{I}_k(t)$$

by subtracting the mean current averaged from all traces from the kth raw data trace. In practice, to allow for slow changes with time during the course of a long experiment, such records would be averaged in groups so that mean current and variance were calculated repeatedly for several sets of data. This would reduce the likelihood of over-estimation of the variance due to slow drift in experimental parameters.

Once the records have been averaged, and the deviations calculated then the variance for each time point can be calculated. Fig. 4A shows the variance plotted against time for the 250 simulated traces. By plotting variance for each time point against the mean current for the same time point (Fig. 3B) a plot of variance against mean current is obtained (Fig. 4B). In the case modelled, the initial value of P_o was 0.5 so i cannot be estimated from the initial slope of the parabola, but must be obtained by fitting equation 10 to the data. In practice, when fitting such data it is often convenient to rearrange equation 10 by substituting

$$P_o = \frac{\bar{I}}{N.i}$$

giving:

$$\text{var } I = i\bar{I} - (\bar{I}^2/N)$$

(where N is the number of channels) to obtain estimates of parameters i and N. The

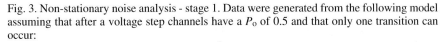

Fig. 3. Non-stationary noise analysis - stage 1. Data were generated from the following model assuming that after a voltage step channels have a P_o of 0.5 and that only one transition can occur:

$$40 \text{ s}^{-1}$$
$$\text{open} \rightarrow \text{inactivated}$$

i.e. the channels can only close to an inactivated state, from which they do not reopen. The channel lifetimes are randomly distributed with a mean of 25 ms. 250 sets of data were generated from this model, assuming a patch containing 3 such channels, and that the experimental conditions gave an outward current of 10 pA through the open channel. Three of these traces are shown in A(i)-A(iii). B shows the mean current in response to the step, obtained by averaging all 250 sweeps. Trace C shows a sample trace in which the mean current has been subtracted from one of the sweeps, the data shown is that from trace A(iii).

open probability at each current level can be obtained by subtitution ($I=NiP_o$). In Fig. 4B this parabola has been superimposed on the data, with $i=10\times10^{-12}$ A and $N=3$.

For the same reasons as are described above in relation to steady state analysis it is important that the raw data can be edited prior to analysis to eliminate artefacts, and that allowance is made for variance that originates from sources other than those of interest. If the channels being studied can be blocked fully by an applied drug then a control can be obtained by repeating the experiment in the presence of that drug, for example when studying sodium channels they may be blocked with TTX.

3. Kinetic analysis of macroscopic currents

Relaxation analysis

When a population of channels is perturbed by an event such as a pulse of agonist or a step change in membrane potential, the current through those channels will change (relax) to a new equilibrium level at a rate which reflects the underlying kinetics of the channels. The relationship between the rate constants for the transitions of the channels between states and the current through a large number of channels can be derived from the law of mass action. In the simple case of a voltage gated channel

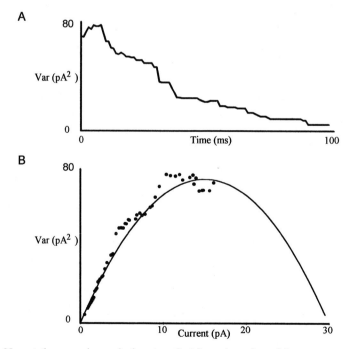

Fig. 4. Non-stationary noise analysis - stage 2. After subtraction of the mean current from the set of traces the variance at each time point is calculated, allowing a plot of variance against time to be made (A). Finally, the values for variance are replotted against mean current (B). A parabolic curve can be fitted to this data. In this case the values used for the parameters of the curve are unit current (i) = 10 pA, number of channels (N)=3.

having only two states, closed (C) and open (O), the rate constants for the opening (β) and closing (α) being voltage dependent, the analysis is as follows:

$$C \underset{\alpha}{\overset{\beta}{\rightleftharpoons}} O. \qquad \text{(Scheme A)}$$

Following a step change in potential, which modifies the rate constants, the change in probability of the channel being in the open state (P_o) with time after the step is given by the sum of the fluxes into and out of the open state, i.e.:

$$\frac{dP_o}{dt} = \beta P_c - \alpha P_o \qquad (14)$$

where α and β are the rate constants at the new potential. Since the probability of being in the closed state $P_c = 1 - P_o$ this can be rearranged as the differential equation:

$$\frac{dP_o}{dt} = \beta - (\alpha + \beta)P_o \qquad (15)$$

which integrates to give:

$$P_o(t) = P_o(\infty) + [P_o(0) - P_o(\infty)]\, e^{-t/\tau} \qquad (16)$$

where $\tau = 1/(\alpha+\beta)$ and $P_o(\infty) = \beta/(\alpha+\beta)$, the steady state probability of being in the open state.

Thus relaxation to a new steady state after a perturbation follows an exponential time course having a time constant of $1/(\alpha+\beta)$. Where P_o is low (i.e. $\alpha \gg \beta$) the rate constant of the relaxation gives an estimate of the mean channel lifetime ($1/\alpha$).

In the case of transmitter activated channels the simplest possible scheme is slightly more complex, as both binding and opening must be modelled, i.e.:

$$A + R \underset{k_2}{\overset{k_1}{\rightleftharpoons}} AR \underset{\alpha}{\overset{\beta}{\rightleftharpoons}} AR^*. \qquad \text{(Scheme B)}$$
$$\qquad\qquad\quad \text{closed} \quad\ \text{open}$$

However, if binding is assumed to be much more rapid than opening, so reequilibration is determined by the opening reaction alone, then this can be simplified to:

$$AR \underset{\alpha}{\overset{\beta'}{\rightleftharpoons}} AR^* \qquad \text{(Scheme C)}$$

where $\beta' = \beta.A/(A + K)$, with $K = k_2/k_1$ and A= concentration of agonist. It can be seen that with this assumption $\beta' = \beta$ times the steady state probability of the $A+R \rightleftharpoons R$ equilibrium being in the AR state, so the opening rate depends on the transmitter concentration. In this case equation 16 still holds, with β' replacing β. Similarly, for a transmitter activated channel, where the transmitter concentration is low ($\alpha \gg \beta'$) the relaxation will still have a time constant that approximates to the mean channel lifetime ($1/\alpha$).

This result is made use of in voltage-jump relaxation experiments and has been applied to a number of preparations in which the channel opening and closing rate constants are voltage dependent. It has been particularly useful in cases where the opening and closing rate constants of a transmitter activated channel are voltage dependent (such as the nicotinic ACh channel of the endplate, e.g. Adams, 1975; Neher & Sakmann, 1975).

Though concentration jumps would in principle perturb the equilibrium proportion of open transmitter activated channels, allowing kinetic information to be obtained in a similar way, voltage-jump relaxation experiments have in general proved more useful as voltage clamp circuits allow good control of membrane potential. In contrast, rapid concentration jumps are difficult to produce on even a single cell because of the time taken to diffuse across 'unstirred' regions adjacent to the surface (10-20 ms on single

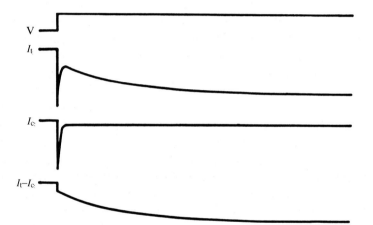

Fig. 5. Illustrates the protocol of a voltage-jump experiment. The upper trace (V) represents the voltage step and I_t, (the total current) a hypothetical recording of the current evoked in response from a preparation such as the frog neuromuscular junction in presence of ACh. Four components to the response can be seen. The first two of these components are indistinguishable in this trace. They are both step changes in the current level, due to the increased driving force of the voltage step causing a larger current to flow both through those channels that were open at the instant of the voltage step and through the leakage conductance. Thirdly, there is a fast transient current superimposed on the step change, this is the current that flows into the cell capacitance as a result of the change in potential gradient. Lastly, there is a slow exponential relaxation to a new steady current level. This is due to the proportion of open channels changing to a new level with a time constant of $1/(\alpha+\beta')$, where the values of α and β' are those corresponding to the new potential. The effects of the leakage conductance and cell capacitance are often corrected for as shown in the lower two traces. In the case of transmitter activated channels the jump protocol can be repeated in the absence of agonist. In this case the recorded current (I_c) is in principle the sum of the same leakage and capacitative components that are present when the agonist is. By subtracting the currents in the presence and absence of agonist the current that is carried by the transmitter activated channels in response to the voltage step is found (I_t–I_c). Similar corrections are possible when studying voltage activated channels by comparing the response to voltage steps in opposite directions. If the channels studied are activated by voltage steps of one polarity but not the other then again the leakage and capacitative components can be subtracted by adding currents evoked by jumps in both directions.

neurones). Furthermore, the binding of the drug to the receptors being studied may further slow diffusion in thick preparations (see Colquhoun & Ritchie, 1972). Rapid concentration changes have proved possible using outside out patches, where the unstirred layer is small, and diffusion times can be reduced to about 1 ms. In combination with fast perfusion systems this approach has allowed study of the response of transmitter activated channels to step changes in concentration (Franke *et al.* 1987; Lui & Dilger, 1991; Maconochie & Knight, 1989). A special case in which a concentration jump can be said to occur is in the initiation of the endplate current, when rapid release and hydrolysis of the pulse of ACh means that it is short lived relative to the channel activated lifetime and so channels do not in general reopen once closed (Magelby & Stevens, 1972). Thus the timecourse of the e.p.c. is determined by the duration of individual activations of the receptor (see Chapter 7).

Fig. 5 shows the protocol of a voltage-jump experiment on nicotinic ACh receptors. At the holding potential (V_H) a steady current flows in the presence of a low and constant agonist concentration. As shown in the upper trace, the voltage is stepped to a test potential (V_t) after a delay. This evokes a step change in the recorded current (I_t), with a superimposed rapid capacity transient. Following these changes there is a slow increase in the inward current to a new steady state level. The steady state change consists of two components, the change in current flowing through the leak conductance and the change in current flowing through that proportion of channels open at equilibrium at V_H. The slow change in the current signal represents the relaxation of the population of channels towards a new equilibrium open probability, different from that at V_H as both the opening and closing rate constants are voltage dependent. The time constant of the relaxation is $1/(\alpha + \beta')$, at V_t. In practice the capacity transient and channel related relaxation are often not so clearly separated so it is generally necessary to record the response to the voltage step in both the presence and absence of the agonist. The difference of the two signals then, in principle, represents only the current flowing through drug activated channels (lower traces). In practice many such responses will be averaged to minimise the effects of noise sources.

Where data cannot be fitted by a simple model that has only two resolvable rate constants, e.g. if there is inactivation or desensitisation, the relaxation may need to be fitted with the sum of two or more exponentials.

Noise analysis

Theoretical basis. Noise analysis can be used to investigate the kinetic properties of unitary events. It is intuitively clear that the time or frequency characteristics of the noise must reflect, in some way, those of the underlying events. This relationship is most easily understood by considering the autocovariance of the signal. The autocovariance measures the correlation between values of a noisy signal measured at time intervals of Δt apart, and for a signal with a mean value of zero it is given by

$$C(\Delta t) = \Sigma \frac{x(t).x(t + \Delta t)}{n - 1} \qquad (17)$$

where $x(t)$ is the signal value at any time t, $x(t + \Delta t)$ is the value a given time increment (Δt) later and n is the number of values of x sampled. For a current signal the autocovariance would be measured as shown in Fig. 6. The current amplitudes are measured relative to the mean current level. Thus if Δt is very large relative to the cycle length of the major frequency components of the noise signal then the pairs of current values will be uncorrelated and the value of C(Δt) will on average be very small. In contrast, if Δt is very small the pairs of current values will be well correlated, i.e. similar in value, as little change in the current signal would be expected if the time increment is short (compared to the cycle lengths of the predominate frequency components in the signal), and the value of C(Δt) will be high. Thus C(Δt) will be high for small values of Δt and approach zero as Δt approaches infinity. Clearly the way in which C(Δt) falls with increasing time increments (Δt) will be determined by the frequency characteristics of the noise.

What is the relationship between C(Δt) and the single channel kinetics? Firstly, we need only consider the behaviour of a single channel, if the N channels present have the same properties and behave independently of each other the autocovariance for all N channels is simply N× the autocovariance of a single channel. Thus

$$C(\Delta t) = N \Sigma \frac{x(t).x(t + \Delta t)}{n - 1} \tag{18}$$

$$= N \times \text{mean of } [i(t) . i(t + \Delta t)]$$

where $x(t)=i(t)$ the current through a single channel at time t and $i(t+\Delta t)$ is the current in the same channel at $t+\Delta t$. As the single channel current is either i or zero, $i(t).i(t + \Delta t)$ has a value of either i^2 or zero. The value is i^2 if the channel is open at both t and at

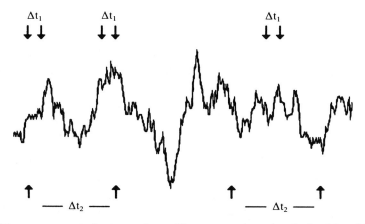

Fig. 6. The measurement of autocovariance. The autocovariance is calculated by taking the product of the current value at time t and the value at time $t + \Delta t$. The sum of these products for each value of t is then found. It is apparent that when the time interval is small relative to the major frequency components of the noise (Δt_1) then the autocovariance will be high (the values within each pair will be highly correlated). However, if the interval is long (Δt_2) then the autocorrelation will be low.

$t + \Delta t$. For a single channel the mean value of $i(t).i(t + \Delta t)$ is given by the probability of the channel being open at *both* times t and $t+\Delta t$ multiplied by i^2. Thus

$$C(\Delta t) = Ni^2P[\text{open at } t \text{ and open at } t + \Delta t]. \qquad (19)$$

$P[\text{open at } t \text{ and open at } t + \Delta t]$ can be restated by the rule that the probability of both of two events is the product of the probability of each of them. $P[\text{open at } t \text{ and open at } t + \Delta t]$ is the product of $P[\text{open at } t]$ and $P[\text{open at } t + \Delta t \mid \text{open at } t]$. The second term is the conditional probability that the channel is open at $t + \Delta t$ given that it was open at t. Thus:

$$C(\Delta t) = Ni^2P[\text{open at } t].P[\text{open at } t + \Delta t \mid \text{open at } t] \qquad (20)$$

$$= Ni^2P_o. P[\text{open at } t + \Delta t \mid \text{open at } t]$$

as $P[\text{open at } t]$ is simply the open probability, P_o. $P[\text{open at } t + \Delta t \mid \text{open at } t]$ is the same as the probability of the lifetime being $\geqslant \Delta t$ provided that there is a negligible probability of the channel closing and reopening during Δt. The statistical distribution of random lifetimes Δt with a mean value τ is:

$$P[\text{lifetime} \geqslant \Delta t] = e^{-\Delta t/\tau} \qquad (21)$$

(see Chapter 7). Thus τ is the mean open lifetime ($=1/\alpha$ in Scheme C) and equation 20 can be rewritten as:

$$C(\Delta t) = Ni^2P_o\,e^{-\Delta t/\tau} \qquad (22)$$

and since $\bar{I} = NiP_o$:

$$C(\Delta t) = \bar{I}ie^{-\Delta t/\tau}.$$

If the channel has an Ohmic I/V relation so $i = \gamma(V-V_{eq})$ where γ is the single channel conductance and $(V-V_{eq})$ the electrochemical driving potential then

$$C(\Delta t) = \bar{I}\gamma(V-V_{eq})e^{-\Delta t/\tau}. \qquad (23)$$

Therefore the autocovariance of a current noise signal with gating described by Scheme C, when plotted against Δt, decays exponentially with a time constant of $1/\alpha$ (Fig. 7A).

Most commonly such data is presented not in the form of an autocovariance plot but as a power spectrum (Fig. 7B) in which the power carried by the signal at each frequency is plotted against frequency, f. The 'power' is effectively the (current)^2 flowing in a $1\ \Omega$ resistor and the ordinate of the power spectrum has dimensions $A^2Hz^{-1}=A^2s$. This is the Fourier transform of the autocovariance plot, which gives us an expression in terms of frequency instead of time. The reason for doing this is that the handling of the data is simplified as the Fourier transform of the data can be easily obtained by the use of one of the fast Fourier transform computer routines available (see Chapter 9). The Fourier transform of $e^{-\Delta t/\tau}$ is

$$\frac{2}{1 + (2\pi f\tau)^2}$$

thus

$$G(f) = 4\bar{I}\gamma(V-V_{eq})\frac{1}{1+(2\pi f\tau)^2} \qquad (24)$$

where $G(f)$ is the spectral density function; this has been multiplied by two to obtain the 'single sided' spectral density function (Bendat & Piersol, 1986). We can define $\tau=1/2\pi f_c$, for convenience, thus

$$G(f) = 4\bar{I}\gamma(V-V_{eq})\frac{1}{1+(f/f_c)^2} \qquad (25)$$

This is a Lorentzian function and when plotted on log/log coordinates as a power spectrum has a characteristic shape (Fig. 7B). The terminology applied to power spectra is important (see Chapter 16); f_c is the 'cut-off' or 'half-power' frequency, which is the frequency at which the power carried has dropped to $\frac{1}{2}$ of the zero frequency value [$G(0)$]. $G(0)$ is usually known as the 'zero frequency asymptote' as the zero frequency point is not measured (measurement of the zero frequency point would require infinite sample lengths). Beyond f_c the curve falls away with a slope of -2 which gives rise to the term '$1/f^2$ noise' which is sometimes applied to signals that fit the high frequency slope of the Lorentzian function.

A simple Lorentzian spectrum is predicted by models of channel opening with single open and closed states. More complex models predict the sum of 2 or more Lorentzians, in general the number of Lorentzians = the total number of states (open and closed) -1. However, one or more of these states may have very small amplitude and be unresolvable. For a full treatment of the theoretical basis of fitting data with the sum of multiple Lorentzians see Colquhoun and Hawkes (1977).

As discussed in the previous section, noise analysis can also provide information about the amplitude of the single channel currents that underlie the macroscopic current, the single channel current being obtained by dividing the variance of the signal by the mean current. The variance of a noise signal that follows a Lorentzian relationship is given by the area under the spectral density function. Thus the single channel conductance can be estimated from the integral of the fitted Lorentzian between zero and infinity, which can be shown to be

$$\text{var} = \frac{\pi G(0)f_c}{2} \qquad (26)$$

where $G(0) = 4\bar{I}\gamma(V-V_{eq})$, the zero frequency asymptote. Thus

$$\gamma = \frac{\pi G(0)f_c}{2\bar{I}(V-V_{eq})} \qquad (27)$$

as $\gamma = i/(V-V_{eq})$ from Ohm's law, where i is the unit current and $i=\text{var}/\bar{I}$ (equation 11), provided that P_o is low. If γ is to be estimated from the spectra it is essential that the mean current is constant during the period over which all data used for calculation of spectra were collected.

In situations where the noise spectrum is fitted by the sum of more than one Lorentzian the conductance can be estimated from the sum of a series of terms relating to each of the Lorentzian terms, provided the assumption is made that all the

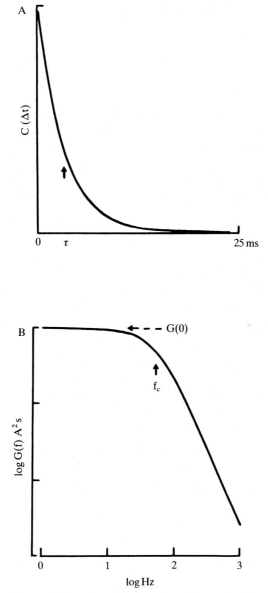

Fig. 7. The autocovariance of the noise signal produced by a 2 state model falls off exponentially as the time interval is increased (A). Generally the results of noise analysis are plotted in the form of a power spectrum on log/log coordinates (B). The form of the curve is a Lorentzian, the half-power frequency $(f_c) = 1/2\pi\tau$, where τ is the time constant of the autocovariance plot (in this case 3.18 ms, so $f_c = 50$ Hz).

open states (if more than one) have the same conductance, i.e. when the noise signal can be fitted by the sum of two Lorentzian components then γ can be estimated from

$$\gamma = \frac{\pi G(0)_1 f_{c1}}{2\bar{I}(V - V_{eq})} + \frac{\pi G(0)_2 f_{c2}}{2\bar{I}(V - V_{eq})}. \tag{28}$$

This approach to the estimation of γ may be useful at times, particularly where a significant fraction of the noise is outside the bandwidth of the recording equipment, provided there is good reason to be confident about the fitted model. An alternative approach to the problem of restricted bandwidth is to calculate the predicted proportion of unresolved variance, after fitting a spectrum, and to adjust the value of the directly measured current variance accordingly (Colquhoun *et al.* 1979).

For more complete treatments of the theoretical basis of noise analysis see Bendat & Piersol (1986), Colquhoun & Hawkes (1977), Conti & Wanke (1975), Neher & Stevens (1977) or DeFelice (1981).

Data preparation

Preparation of data for kinetic analysis using noise analysis is essentially the same as that described above when using noise analysis for investigating unit amplitudes. However, two additionals factors must be considered. Firstly, the sampled noise signal will contain, in addition to the biological noise of interest, noise from a number of other sources. All electrical recording equipment generates noise, significant elements of which will depend upon the characteristics of the cell or other preparation being recorded from, in particular the cell capacitance will determine the amplitude of significant elements of noise. Furthermore, other channels or carriers in the cell membrane may generate signals, which themselves contain fluctuations. For these reasons, it is essential that allowance is made for those components of noise that are not generated by the events of interest. This is generally done by determining the spectrum of a control signal and subtracting this from the total spectrum prior to fitting any model. Where the events being studied can be readily turned on by application of some external factor, such as an agonist, then the net variance is readily calculated by subtracting the variance in the absence of the agonist from that in the presence of the agonist.

Secondly, the data must be split into records whose length depends upon the lower and upper frequencies required in the power spectrum. The lowest frequency point obtained is given by the inverse of the sample length, thus a 2 second long sample is needed if the minimum frequency point is to be at 0.5 Hz. This also determines the frequency resolution of the power spectrum, power values will be obtained for frequencies incrementing by the same value, every 0.5 Hz in the example. The high pass filter setting used to prepare the AC coupled record should be chosen to allow data at the lowest frequency point in the spectrum to pass unattenuated, but to attenuate all frequencies below this. The high frequency limit of the spectrum is the same as the Nyquist frequency, i.e. half the sample rate. In general, steady state stretches of both experimental and control data would be obtained. These would be edited to remove artefacts, split into records and analysed. A number of spectra, usually 10–20, would be obtained for both sets of conditions and all spectra in each set averaged, to improve the signal to noise ratio of the final spectra. Finally the mean control spectrum is subtracted from the mean experimental spectrum before any curve fitting takes place.

Estimates of the parameters of a power spectrum are obtained by fitting Lorentzian curves to the data. In general this will be done using a least squares algorithm. Procedures for curve fitting and the estimation of errors are discussed in Chapter 9.

References

ADAMS, P. R. (1975). An analysis of the dose-response curve at voltage clamped frog endplates. *Pflugers Arch.* **360**, 145-153.

ADAMS, P. R. (1977). Voltage jump analysis of procaine action at frog endplate. *J. Physiol., Lond.* **268**, 291-318.

ANDERSON, C. R. & STEVENS, C. F. (1973). Voltage clamp analysis of acetylcholine produced end-plate current fluctuations at frog neuromuscular junction. *J. Physiol., Lond.* **235**, 655-691.

BENDAT, J. S. & PIERSOL, A. G. (1986). *Random Data: Analysis and Measurement Procedures.* New York: Wiley Interscience.

COLQUHOUN, D., DREYER, F. & SHERIDAN, R. E. (1979). The actions of tubocurarine at the frog neuromuscular junction. *J. Physiol., Lond.* **293**, 247-284.

COLQUHOUN, D. & HAWKES, A. G. (1977). Relaxation and fluctuations of membrane currents that flow through drug-operated channels. *Proc. R. Soc. Lond.* B **199**, 231-262.

COLQUHOUN, D. & RITCHIE, J. M. (1972). The kinetics of the interaction between tetrodotoxin and mammalian nonmyelinated nerve fibres. *Molec. Pharmacol.* **8**, 285-292.

CONTI, F. & WANKE, E. (1975). Channel noise in nerve membranes and lipid bilayers. *Q. Rev. Biophys.* **8**, 451-506.

DEFELICE, L. J. (1981). *Introduction to Membrane Noise.* New York: Plenum Press.

FRANKE, C., HATT, H. & DUDEL, J. (1987). Liquid filament switch for ultra-fast exchanges of solution at excised patches. *Neurosci. Lett.* **77**, 199-204.

GRAY, P. T. A. & ATTWELL, D. (1985). Kinetics of light-sensitive channels in vertebrate photoreceptors. *Proc. R. Soc. Lond.* B **223**, 379-388.

KATZ, B. & MILEDI, R. (1970). Membrane noise produced by acetylcholine. *Nature, Lond.* **226**, 962-963.

KATZ, B. & MILEDI, R. (1972). The statistical nature of the acetylcholine potential and its molecular components. *J. Physiol.; Lond.* **224**, 665-699.

LUI, Y. & DILGER, J. P. (1991). The opening rate of ACh receptor channels. *Biophys. J.* **60**, 424-432.

MACONOCHIE, D. & KNIGHT, D. E. (1989) A method for making solution changes in the submillisecond range at the tip of a patch pipette. *Pfulgers Arch.* **484**, 589-596.

MAGELBY, K. L. & STEVENS, C. F. (1972). A quantitative description of endplate currents. *J. Physiol., Lond.* **223**, 173-197.

NEHER, E. & SAKMANN, B. (1975). Voltage-dependence of drug-induced conductance in frog neuromuscular junction. *Proc. Natl. Acad. Sci. USA* **72**, 2140-2144.

NEHER, E. & STEVENS, C. F. (1977). Conductance fluctuations and ionic pores in membranes. *Ann. Rev. Biophys. Bioeng.* **6**, 345-381.

SIGWORTH, F. J. (1980). The variance of sodium current fluctuations at the node of Ranvier. *J. Physiol. Lond.* **307**, 97-129.

TRAYNELIS, S. F., SILVER, R. A. & CULL-CANDY, S. G. (1993). Estimated conductance of glutamate receptor channels activated during EPSCs at the cerebellar mossy fibre-granule cell synapse. *Neuron* **11**, 279-289.

Chapter 9
Computers

ARMAND B. CACHELIN, JOHN DEMPSTER and
PETER T. A. GRAY

> *Hardware is the parts of a computer you can kick.*
> *Computers don't save time, they redistribute it.*

This chapter is a short introduction to computers and their use in the laboratory. Not surprisingly, this chapter has become far more outdated than others in the six years since the publication of the first edition of Microelectrode Techniques - The Plymouth Workshop Handbook. The potential role of desktop computers was just beginning to emerge in 1987. Six years on, two computer generations later, the personal computer reigns undisputed in the office and in the lab. Thus the aim of this chapter is not so much to familiarise readers with PCs than to draw their attention to important points when choosing a computer or commercially available software for their lab. The chapter starts with a brief outline of some of the most important issues to consider when selecting a system. This is followed by sections which provide a more detailed consideration, firstly, of the basic hardware issues relating to the acquisition and storage of signals on a computer system and, secondly, of the functional requirements for software used for the analysis of electrophysiological data.

1. Selecting a system

Selection of a computer system when starting from scratch requires balancing a large number of factors. In general a guide-line widely accepted when specifying commercial systems, which applies as well here, is that the first thing to choose is the software. Choose software that will do what you want, then select hardware that will run it. In many cases, of course, the hardware choice may already be constrained by the existence, for example, of a PC of a particular type that must be used for cost reasons. However even in these cases the rule above should be applied to selection of other parts of the hardware, such as laboratory interfaces.

When selecting software it is essential to bear in mind likely future requirements as

A. B. CACHELIN, Pharmacology Department, University of Berne, 3010 Berne,
Switzerland
J. DEMPSTER, Department of Physiology and Pharmacology, University of Strathclyde,
Glasgow G1 1XW, UK
P. T. A. GRAY, 52 Sudbourne Road, London SW2 5AH, UK

well as current requirements. This can be very hard to do in any environment, as the introduction of a computer system always changes working patterns in ways that are hard to predict in advance. In a scientific laboratory, where the nature of future experiments is hard to predict anyway, this is doubly true. This is one reason for trying to select an adaptable system if at all possible. Maximum flexibility is obtained by writing your own software, but the cost in terms of time is high, and cannot generally be justified where off-the-shelf packages are available. However, if you are developing novel techniques it may be essential to have total control over your software. For most users applying established techniques, the choice is best limited to off-the-shelf systems, but the flexibility of the system should be a primary factor in the decision.

Flexibility in an off-the-shelf system can be assessed in many ways. For example, at the level of data acquisition are maximum sample rates, sample sizes, available complexity of output control pulses (e.g. for controlling voltage jump protocols) likely to match future requirements. At the next stage, factors such as the forms of data analysis available must be considered. How flexible are they? In many systems inflexibility here can often be partially overcome by exporting data to an external analysis package, provided that the software selected allows for data export in a suitable form. At the highest level of flexibility, can the system be used to control, sample and analyse data from completely new types of experiment? In most cases the answer, unsurprisingly, will be no. In this case, one must simply wait for off-the-shelf software to become available unless one writes the program oneself.

PART ONE: HARDWARE

2. Of bytes and men

Before going further it is worth introducing some basic concepts and terms which are widely used in computer-related discussions. Decimal numbers are a natural form of representation for ten-fingered humans. According to this system, numbers are composed of digits which take values between 0 and 9. Digital computers make extensive use of a different, so called binary number system. The elementary piece of information used in the computer is the bit which can take only two values (0 or 1). This binary system fits the ON-OFF nature of the computer's digital logic circuits. A number expressed in binary form consists of a string of '1's and '0's. For instance, the decimal number 23 ($2\times10^1+3\times10^0$) has the binary form 10111 which is equal to $1\times2^4+0\times2^3+1\times2^2+1\times2^1+1\times2^0$.

Computers use a 'language' in which all words have the same length. Thus both operands (data) and operations (instructions) used by computer systems are encoded as fixed length binary numbers. Common word lengths are 16, 32 and more recently 64 bits. An 8 bit word is known as a byte and the term is often used as a unit of data storage capacity. One kilobyte (KB) is equivalent to 1024 bytes (2^{10} bytes)[1], a MB to 1048576 (2^{20}) bytes. The address bus width sets the limits to the address space. The

---

[1]By convention, we shall use "K" as abbreviation for 1024 and "k" for 1000.

wordsize sets the width of data that a processor can manage conveniently. For example, 2^{16} (64 KB) is the largest number of memory locations that can be addressed directly by a 16 bit wide address bus. Additional bits must be used to store and fetch data from larger RAM. Of course 32 bit processors are much less limited being able to address up to 4 GB of RAM.

A further consequence of the use of a fixed length binary number system is that, at the most fundamental level, a digital computer can only store and manipulate integer numbers (0, 1, 2,...), so the highest integer number that could be stored within 16 bits is 2^{16}. The computer reserves one bit to represent the sign of a integer. Thus whole numbers in the range $\pm2^{15}$ can be stored within 2 bytes. Although almost any kind of data can be stored and processed by a computer system, some forms of data are easier to handle than others. Large numbers ($>2^{15}$) and real numbers (i.e. with fractional parts, e.g. 1.23) cannot be directly stored in memory locations. How this and other problems are solved is the subject of the next section.

3. Computer systems

The digital computer forms the core of the data acquisition system, providing the following three basic functions: data input, processing and output. In the context of a laboratory these functions could be for example:

Input: acquisition of signal(s) (analogue to digital conversion)
                          reading back data files
                          input of commands by keyboard or mouse

Processing: qualitative signal analysis (preparation of 'idealiscd' data)
                          quantitative signal analysis (statistics, curve fitting, estimation of
                             parameters of models)

Output: command signals (digital to analogue conversion)
                          visual display of stored data
                          storage to data files on disk or tape
                          production of hard copies on paper
                          connection to other computer systems via a network

The advances in computer technology are such that most readily available personal computers within the IBM personal computer ('PC'[2]) and Macintosh families are more than adequate for the above tasks, in terms of computing speed and graphical display quality. Given the pace of change in computer technology, it is futile to provide particular specifications as they must become rapidly obsolete. In the following, we present certain general observations which should be borne in mind when choosing a computer.

Central processor unit

As can be seen from Fig.1, the central processing unit (CPU) forms the core of the

---

[2]From now on and throughout this chapter we will use the term 'PC' to designate a computer that is compatible with the IBM PC family of computers regardless of the model (PC/XT/286, 386...) or manufacturer (IBM, Compaq, Hewlett-Packard etc.).

computer system. It is, essentially, a machine for processing binary numerical data, according to the instructions contained in a binary program. The operation of the CPU is governed by a master clock which times the execution of instructions, the higher the clock rate the faster the CPU[3]. The performance of a CPU also depends upon the size of the data that a single instruction can handle, calculations being more efficient if they can be performed by a single instruction rather than a group of instructions. Typical personal computer CPUs (c. 1993) can handle 32 bit numbers in a single instruction and operate at clock rates of 25-66 MHz. The two major personal computer families (PCs and Apple Macintosh) are based upon different CPU designs, PCs use the 80×86 family of Intel processors (or clones by AMD or Cyrix) while the Macintosh uses the Motorola 680×0 family. Over the years, each CPU family has evolved with the introduction of newer and faster CPUs, while retaining essential compatibility with earlier models. Currently used members of the Intel CPU family include the 80386, 80486 and the recently introduced Pentium ('80586'). Macintosh computers are built around the Motorola 68030 and 68040 processors. The first computers built around the PowerPC 601/603 chips (see BYTE article) will be available by the time this book is printed. The clear trend towards graphical user interfaces makes it advisable to chose a computer system which uses at least an 80486 processor. Modern CPUs have more than enough power for most PC applications. However, certain tasks, such as program development, lengthy numerical analysis, such as curve fitting, and graphics intensive applications, do benefit from the fastest possible processor.

Floating point unit

The arithmetic operations within the instruction set of most CPUs are designed to handle integer numbers of a fixed size. Scientific calculations, however, require a more flexible number system capable of representing a much wider range of numbers including the complete set of real numbers as well as handling trigonometric, transcendental and other mathematical functions. For this purpose, computers use an internal notation convention called 'floating point' in which real numbers are approximated as product of a power of 2 times a normalised fraction. Thus:

$$R = 2^k \times f,$$

where k is the exponent and f is the mantissa. The most common laboratory computers, PCs and Macintoshes, both represent these numbers internally following the IEEE standard 754. In both cases, ignoring details of ordering of bytes in memory, these are stored internally as follows:

←most significant bit   least significant bit→

S	exponent	significand

The sign bit S is set if the number is negative. The exponent may be either positive or negative (see Table 1), however, it is internally stored as a number in the range

---

[3]This is of course only true when comparing identical CPUs. Different CPUs running identical clock rates need not perform the same operations at the same speed.

Table 1. *Characteristics of floating point numbers (IEEE 754)*

Type	Size bits	Exponent bits	Exponent range	Significand bits	Representational range*
Single	32	8	−126 to 127	23	$1.2{\times}10^{-38}$ to $3.4{\times}10^{38}$
Double	64	11	−1022 to 1023	52	$2.3{\times}10^{-308}$ to $1.7{\times}10^{308}$
Extended	80	15	−16382 to 16383	64	$1.7{\times}10^{-4932}$ to $1.1{\times}10^{4932}$

*Positive or negative numbers within these ranges can be represented.

between 0 and some positive number (whose value depends on the floating point format used). The IEEE 754 standard establishes 3 types of floating point number, which differ in the amount of memory they require for storage and the range and precision of the numbers that they can represent. These characteristics are enumerated in Table 1.

This increased complexity makes floating point calculations much slower than integer calculations, particularly if they have to be constructed from the basic CPU integer arithmetic operations. However, the floating point performance of a computer can be greatly improved by using a floating point unit (FPU; also called a maths coprocessor). This is a device dedicated to the performance of floating point operations which works in close co-operation with the CPU, in effect adding a set of floating point arithmetic operations to the basic CPU instruction set. Floating point calculations performed using an FPU can be an order of magnitude faster than the same operations implemented in software using CPU instructions.

Computers in both the PC and Macintosh families can be fitted with an optional maths coprocessor. Intel 80x86 CPUs in PCs can be supplemented by an 80x87, a maths coprocessor, though computers based on the 80486DX and Pentium CPUs have these built into the CPU. Macintosh computers, based on the Motorola 68×00 family of CPUs can also be fitted with maths coprocessor; the 68881 is used with 68020 CPUs, and the 68882 with 68030 and some 68040 CPUs.

RAM memory

Random access memory (RAM) is the high speed integrated circuit memory used to store a computer program and its data while it is being executed by the CPU. The amount of RAM determines the size of program that can be executed and/or the amount of storage space available for data. At present, (c. 1993) a standard personal computer contains 4 MB of RAM. However, the requirements for RAM continue to rise, as increasingly sophisticated and, unfortunately also, larger programs are developed, so we recommend 8-16 MB RAM, which is likely to become the norm in the next few years.

Disc storage

Modern applications software, such as word processors, spreadsheets, etc., take up large amounts of disc space, some consuming as much as 15 MB of disc space. The storage of laboratory data also requires large amounts of magnetic disc space. A single two channel digitised record, containing 1024 samples per channel, requires 4 KB of disc space (2 channels, 2 bytes per sample). A day's experiment might produce

1000 of such records, occupying up 4 MB of disc space. The same amount of disc space will be taken up by less than 2 minutes of continuous sampling at 20 kHz on one channel. It is important, therefore, to ensure that there is sufficient disc storage space to comfortably accommodate the quantity of experimental work envisaged, particularly when the computer is being used as the only storage device for experimental signals. Overall, 200-300 MB of disc capacity should be available and for the more demanding laboratory applications, 1000 MB might be preferable.

The hard disk system should also allow rapid access. Software disk caches, which use some of the system's RAM are effective in speeding system response, but it is better, when buying new hardware, to ensure that the disk performance is as good as

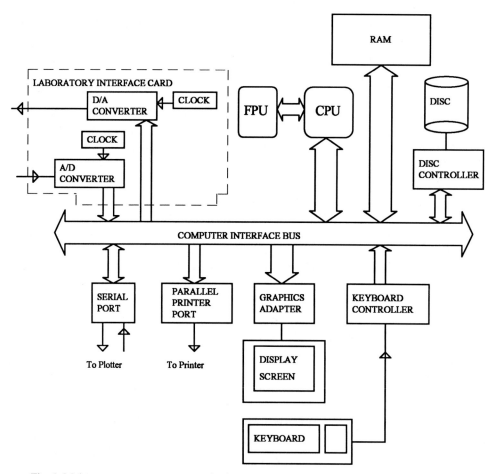

Fig. 1. Main computer components and sub-systems. Computer operation is governed by the central processing unit (CPU) which executes programs stored in the random access memory (RAM). Important sub-systems (graphics display, disc, keyboard) have their own semi-independent controllers. Data is transferred between the CPU, RAM, and these sub-systems, via the computer interface bus. The laboratory interface card, installed in an expansion slot, is connected to this bus and thus can exchange data between itself and RAM, and respond to instructions from the CPU.

can be afforded. A fast disk speeds up many applications more than a faster processor. On PCs a range of disk drives is available. The choice of disk interface may play a significant role in determining disk drive performance. The most common form is the IDE interface which is standard on many systems. SCSI drive interfaces generally offer better performance than other types, in particular for large drives. Performance may also be enhanced by using interfaces which have on-board memory, which is used to cache recently accessed data. Better performance still can be achieved on modern PCs with a local bus version of these cached interfaces, which pass data to the processor by a direct high speed link, instead of using the much slower general purpose interface bus that older models do.

Archive storage

Although it is convenient to store experimental data on the computer's internal hard disc, even the largest discs cannot accommodate more than a small number of experiments. Some form of removable data storage medium is therefore required for archival storage of the data. The secure storage of experimental data is an important issue. Experimental recordings are extremely valuable, if one takes into account the cost of person-hours required to design and perform the experiments. At least one 'back up' copy of the recorded data should be made on a different medium, in case the hard disc fails. It may also be necessary to preserve this archive data for many years, for scientific audit purposes or to allow future re-evaluation of results.

A variety of removable computer tape and disc media are available for archive storage. This is one area where rapid improvements in technology (and the consequent obsolescence of the old) does not help since it becomes difficult to determine which media are going to remain in widespread use. However, at present there are a number of choices. A number of tape forms exist, using cartridges of varieties of sizes and capacities (DC300, 45 MB; DC600, 65 MB; DAT (Digital Audio Tape) 1200 MB). Removable magnetic discs exist (e.g. the Bernoulli disc) but these are relatively expensive and have small capacities (40-80 MB) compared to alternatives.

In general, high capacity removable disc media use one of a number of optical data storage technologies. WORM (Write Once Read Many) discs, for instance, use a laser beam to vaporise tiny pits in the reflective surface of a metal-coated disc, which can be read by detecting the reflections of a lower-power laser light trained on the surface. As the name suggests, data can be written to a WORM disc only once, but read as many times as necessary. Rewritable optical discs are also available, which use the laser beam to effect a reversible phase change in the disc surface rather than a permanent pit. Two sizes of optical disc are available, 5.25 inch with a capacity of 640 MB (e.g. Panasonic LF-7010), and 3.5 inch supporting 128 MB (e.g. Sony S350 or Panasonic LF3000). The 3.5 inch discs are rewritable while the 5.25 inch drives support either WORM or rewritable discs.

A strong argument can also be made for storing a complete recording of the **original** experimental signals on some form of instrumentation tape recorder, designed for storing analogue signals. FM (Frequency Modulation) tape recorders, such as the Racal Store 4, with a DC-5 kHz recording bandwidth (at 7.5 inches per second tape speed) have been used for many years. Analogue signals can also be stored on standard video cassette recorders, using a PCM (pulse code modulation) encoder to convert the signals into pseudo-video format. The Instrutech VR10B, for instance, can be used to store two DC-22.5 kHz analogue channels (or 1, and 4 digital marker channels) on a standard VHS video tape. These devices may gradually be replaced by modified DAT (Digital Audio Tape) recorders, such as the Biologic DTR1200 or DTR1600 which provide a similar capacity and performance to the PCM/video systems. DAT tape recorders are simple to operate compared to most computer systems and have a high effective storage capacity (a single 60 min DAT tape is equivalent to over 600 MB of digital storage)[4]. They provide additional security in case of loss of the digitised data, and allow the data to be stored in an economical and accessible form (compared to the non-standard data storage formats of most electrophysiology software). They thus provide the opportunity for the signals to be replayed and processed by different computer systems. An additional advantage is that a voice track can often be added during the experiment, offering a valuable form of notation. It MUST be borne in mind that ALL forms of magnetic tape (i.e. also DAT recorders) have a limited bandwidth though the reason for this limited bandwidth differs from one type of tape recorder to the other.

Networks

Laboratory computer systems are often connected to local area networks which allow data to be transferred between computers systems at high speed, permitting shared access to high capacity storage discs and printers. Often, the network can also be used to exchange electronic mail, both locally and, if connected to a campus network, internationally. Networks can be categorised both by the nature of the hardware connecting computers (e.g. Ethernet) and the software protocols (Novell Netware, Windows, NetBios) used to control the flow of data[5]. Within the PC family, the most commonly used network hardware is Ethernet followed by Token Ring. Ethernet permits data to be transferred at a maximum rate of 10 Mbps (bit per second). An Ethernet adapter card is required to connect a PC to an Ethernet network. Low end Macintosh computers are fitted with a LocalTalk network connector as a standard feature. However, LocalTalk, with a transfer rate of 230 kbps, is noticeably slower than Ethernet. On the other hand, high end Apple computers (Quadra family) and printers (Laser Writer IIg, Laser Writer Pro 630) are factory-equipped with serial, LocalTalk and Ethernet ports.

A wide variety of network software is available from a number of competing suppliers, the most commonly used being Novell Netware, PC-NFS (a system which allows PCs to use discs and printers on UNIX workstations), Windows for Workgroups on PCs or Asanté (and other programs) on Apple networks. Netware and PC-NFS are server-based systems where a dedicated file server is required to provide

---

[4]Two types of DAT recorders are currently available: "straight" data recorders and file-oriented back up devices. The data format and hence the access mode and time to the data is different.

[5]This is a little oversimplified since for example Ethernet is also a data transfer protocol (IEEE 802.3).

shared discs and printers. Windows for Workgroups, can act as a peer-to-peer network where any computer on the network can allow its discs and printers to be shared by any other. Macintosh computers also use a peer-to-peer system. In general, peer-to-peer systems are easier to set up, and are well-suited to the needs of a small group of 4-10 users within a single laboratory. Server-based systems, on the other hand, are better suited to the needs of large groups (10-100) on a departmental or institutional scale. They are complex to set up and administer, but have many features needed to handle electronic mail, archiving procedures, security, and remote printers which the simple peer-to-peer systems lack. Such networks require professional management, and large academic and research institutions may provide central support, particularly for Netware and PC-NFS. Differences between client/server and peer-to-peer are becoming blurred as time passes. Windows for Workgroups, for example, can access data on Novell file servers and can coexist with PC-NFS.

One final issue concerning networks is that there is often a potential conflict between the needs of the network adapter card and the laboratory interface unit. Particular care must be taken to ensure that the boards do not attempt to use the same DMA channel or interrupt request (IRQ) line. A more difficult problem, with running network software on a PC, is the 50-100 KB of RAM memory taken up by the network access software. The program has to be resident in the memory all the time leaving only 400-500 KB free memory (DOS limits the amount of memory for running programs to the lower 640 KB) which may not be sufficient to load some of the more memory-demanding data acquisition programs currently in use. DOS 5.xx (and above) by allowing loading of devices drivers and short programs in the so-called 'high' memory (i.e. the address range between 640 KB and 1 MB) frees room in the 'low' memory for running other programs. These memory limitations problems are not known to Macintosh computer users because directly addressable memory is not limited by the operating system. Furthermore (slow) local network support for up to 32 users is provided from System 7.xx on. On PCs memory limitations are likely to become a thing of the past for good with the introduction of 'Windows 4' (currently code-named Chicago, which could be the end of DOS) or are already abolished by Windows NT.

However, while it may be advantageous to have a laboratory computer linked to a network (for example, to allow for exchange of data, text and figures between PCs), in general, timing constraints mean that it is not possible for a PC to act as a server during data acquisition or experimental control.

4. Data acquisition

Acquisition is the most important operation performed on the experimental data. It transforms the original analogue experimental signal into a data form (binary digits) that can be handled by a computer. Some of the relevant steps are discussed below. A general introduction to data acquisition and analysis can be found in Bendat and Piersol (1986).

Analogue to digital conversion
The outputs from most laboratory recording apparatus are analogue electrical signals,

i.e. continuous time-varying voltages. Before this analogue signal can be processed by a computer system it must be brought into a computer-readable form. A digitised approximation must be made, by using an analogue to digital converter (ADC) which takes samples (i.e. measurements) of the continuous voltage signal at fixed time intervals and quantifies them, i.e. produces a binary number proportional to the voltage level, as shown in Fig. 2. The frequency resolution of the digitised recording depends upon the rate at which samples are taken. The signal to noise ratio and amplitude resolution depend on the number of quantization levels available (machine-dependent) and actually used (signal preparation!). The choice of the optimal sampling frequency and the importance of the resolution are discussed in Chapter 6 (see also Colquhoun and Sigworth, 1983; Bendat and Piersol, 1986).

The choice of commercially available ADC performances is very broad with resolutions available between 8-20 bits and sampling rates as high as 1 MHz (1 Hz = 1 sec^{-1}). However, for general use within the laboratory, an ADC with a sampling rate of 100 kHz and 12 bit resolution, allowing the measurement of 2^{12}=4096 discrete voltage levels, with ±5 volts input range[6] is quite sufficient (though ±10 V might better match the output range of certain voltage-clamp amplifiers, see Table A1).

ADC facilities can be added to most personal computers by installing a specialised electronic device. This is either an internal circuit card or a specialised external laboratory interface unit. Either form supplies the ADC, digital-analogue converter (DAC) and timing hardware necessary to allow the computer to interact with laboratory data and equipment. The circuit card is usually installed in one of the slots of the computer expansion bus (see Fig. 1). A data 'bus' is actually more a data highway with multiple lanes used to transfer data between CPU and computer peripherals. An external laboratory interface unit is connected to the computer via a specialised interface card which sits in the expansion bus. Software running on the personal computer can thus both control the operation of the laboratory interface and data transfers to and from computer memory.

Analogue signal conditioning

Matching the data to the ADC

It is important to ensure that the voltage levels of the analogue signal are suitably matched to the input voltage range of the ADC. A number of ADCs are designed to handle voltage ranges of ±5 V. With 12 bits of resolution, the minimum voltage difference that can be measured is 2.44 mV (10000 mV/4096). However, many laboratory recording devices supply signals of quite small amplitude. For instance, a microelectrode amplifier (e.g. World Precision Instruments WPI 705) recording miniature synaptic potentials from skeletal muscle might provide signals no more than 1-2 mV in amplitude at its output. If such signals are to be measured accurately, additional amplification (e.g. ×1000) is required to scale up the signals so that they

---

[6]Signal input can is either be single-ended (se) or differential (di). The signal of a single-ended input is measured relative to ground. Alternatively using-differential inputs the difference between two inputs is measured.

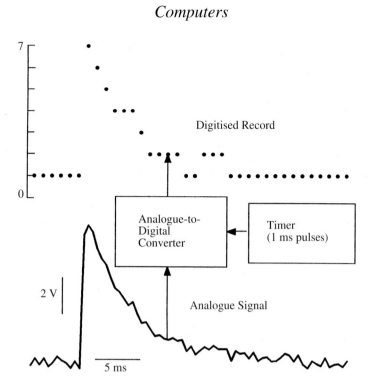

Fig. 2. The digitised record (upper trace) has been prepared from the continuous analogue signal (lower trace) by taking samples of the signal at 1ms intervals using a 3 bit A/D converter. The resulting record consists of 36 integer numbers with values in the range 0-7. Under normal circumstances, a much higher sampling rate and a 12 bit A/D converter would be used, allowing the values of individual data points in the digitised record to range from 0-4095.

span a significant fraction of the ADC voltage range. If the signals are superimposed upon a relatively large DC level, such as the −90 mV resting potential in the case of the mepps, it is also be necessary to subtract this level off before amplification, in order to keep the signal level within the working range of the amplifier. A DC-offset facility is therefore required.

Filters

It may sometimes be necessary to filter the analogue signals to remove some of the high or low frequency signal components prior to digitisation and analysis. For example when performing noise analysis, all signal components with frequencies higher than half of the sampling rate MUST be removed from a signal in order to avoid aliasing of the power spectrum[7]. Low-pass filtering may, additionally, be required to improve the signal-noise ratio or to diminish the false-event detection rate (see

---

[7]At least two points per cycle must be sampled to define a frequency component in the original data, so the highest frequency that can be resolved is half the sampling frequency (the so-called Nyquist or folding frequency). Frequencies above the Nyquist frequency in the sampled data will be folded back (aliased) into frequencies between 0 and the Nyquist frequency. In practice 5-10 points per cycle are better.

Chapter 6 on single channel analysis). The low-pass filter used for this purpose should be of a design which provides a frequency response with as sharp a roll-off as possible above the cut-off frequency[8], but without producing peaking or ringing in the signal. High-pass filters are less frequently needed, but are used when performing noise analysis to eliminate frequencies below those of interest. The roll-off characteristics of filters vary e.g. *Bessel,* used for single channel data and *Butterworth,* used for noise analysis. *Tchebychev* filters are not used for analysis but are found in FM tape recorders (e.g. RACAL) to remove carrier frequencies (see 'Filters' in Chapter 16).

Event detection

For some types of experiments, particularly those including random events, there may also be need for an event discriminator. This is a device which detects the occurrence of large amplitude waveforms within the incoming analogue signal. Typical uses include the detection of the signals associated with the spontaneous release of quantal transmitter packets (miniature synaptic currents or miniature synaptic potentials) at central and peripheral synapses or the opening of ion channels during agonist application. The detection of events by the software is discussed below. A hardware discriminator monitors the incoming analogue signal level and generates a digital TTL (transistor-transistor logic) pulse each time the signal crosses an adjustable threshold. TTL signals represent binary ON / OFF states by two voltage levels, 0 V (LOW) and about 5 V (HIGH). The TTL output of a discriminator is often held at 5 V, with a short pulse step to 0 V signifying that an event has been detected.

The TTL signal can be used to initiate an A/D conversion sweep by the laboratory interface to capture the event. In practice, it is not as simple as this since there is often a requirement for obtaining 'pre-trigger' data samples so that the leading edge of the waveform is not lost. Discriminators are also useful for matching trigger pulses from a variety of devices (e.g. Grass stimulators, Racal FM tape recorders) which produce signals incompatible with the TTL trigger input requirements of most laboratory interfaces. Suitable event discriminators include the World Precision Instruments Model 121, Axon Instruments AI2020A, Neurolog NL201 module or CED 1401-18 add-on card for the 1401.

All-in-one signal conditioner

In summary, a flexible signal conditioning system should provide one or more amplifiers with gains variable over the range 1-1000, facilities to offset DC levels over a ±10 V range, a low/high pass filter with cut-off frequencies variable over the range 1 Hz-20 kHz and an event discriminator. Such systems can be constructed in-house with operational amplifier circuits (Discussed in Chapter 16). However, it is often more convenient to purchase one of the modular signal conditioning systems, available from a number of suppliers. Digitimer's (Welwyn, UK) Neurolog system, for instance, consists of a 19 inch rack mountable mains-powered housing, into which an number of differential

---

[8]The frequency at which the ratio of the output to the input voltage signal amplitudes is 0.7 (-3dB). 1 dB (decibel) $= 20 \times \log$ (output amplitude / input amplitude), see Chapter 16.

amplifiers, filters and other modules can be inserted. Similar systems can be obtained from Fylde Electronic Laboratories (Preston, UK). The Frequency Devices LPF902 low-pass Bessel filter unit can also be used as a simple signal conditioner, providing a differential amplifier with gains of ×1, ×3.16, ×10, and an 8-pole Bessel filter with a cut-off variable over the range 1 Hz-100 kHz (noise analysis requires a Butterworth filter). The characteristics of certain commercial filters can be switched between Bessel and Butterworth and from low to high pass (e.g. EF5-01/02, Fern Developments).

Whatever signal conditioning system is used it is important that the exact amplifier gain and other factors are available to the computer software in order that the digitised signals can be related to the actual units of the signal being recorded (e.g. pA or µA for membrane currents and mV for membrane potentials). In the past, this has had to be done by the user manually entering calibration factors into the program after reading switch settings from the signal conditioners - a rather inconvenient and potentially error-prone exercise. Recently however, a number of suppliers have introduced computer-controlled signal conditioners, such as the Axon Instruments CyberAmp series and the Cambridge Electronic Design 1902, which can be directly programmed from the computer (usually via an RS232 serial port). In these devices, the amplifier gain, DC offset levels, and filter cut-off frequencies can all be directly programmed and read from the software running on the computer system. Such systems have the potential to greatly simplify the calibration of experimental recordings by allowing the computer to automatically monitor any changes to the amplifier gain and DC offset levels. Similarly, some modern voltage and patch clamps now have facilities to 'telegraph' their amplifier gain settings to computer systems, either using RS232 communications lines (Axon Axopatch 200, GeneClamp 500) or as fixed voltage levels (Axoclamp 2A). A particularly advanced example of this approach can be seen in the EPC-9 patch clamp (Heka Elektronik) where the controlling computer and patch clamp are closely integrated, allowing patch clamp parameter settings (including capacity and series resistance compensation), to be automatically controlled by the software.

Data and control outputs

In addition to sampling data it is also often desirable to generate output signals that may be used, for example, to control the experiment, to present a display of the sampled data on an XY monitor scope or to trigger the sweep of a conventional YT oscilloscope so that the display is synchronised with stimuli sent to the experimental preparation. Such outputs fall into two main classes, digital outputs and analogue outputs. Digital outputs are single bit digital signals, that can be either Off or On. These generally take the form of TTL signals. Signals like this can be used to provide trigger signals for oscilloscopes or to control two state devices such as taps. Analogue outputs are converted from an internal digital representation into an analogue signal by a digital to analogue converter (DAC). A DAC essentially performs the reverse transformation to that provided by an ADC, thus a number (between 0 and 4095 with a 12 bit DAC) held in the computer is converted to an output voltage that represents an appropriate fraction of the total output range, usually +/−5 or +/−10 V. Analogue

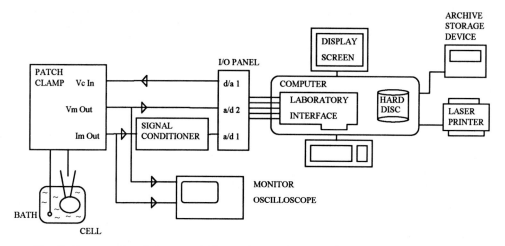

Fig. 3. Computer-based data acquisition and control system for an electrophysiological experiment. Cell membrane current (Im) and potential (Vm) signals from a patch clamp amplifier are fed into A/D converter channels 1 and 2 of the laboratory interface, digitised and stored on the computer's hard disc. The current signal is amplified and low-pass filtered using a signal conditioning unit. Voltage command pulse waveforms can be generated by the computer and output to the patch clamp (Vc In) using a D/A converter. Hard copies of waveforms are produced on a laser printer and digitised signal records are archived on a removable disc or tape storage device.

signals can be used, for example, to control the holding potential of a voltage clamp amplifier, or to plot a hard copy of the incoming data to a chart recorder.

The laboratory interface

A laboratory interface used for electrophysiological data acquisition should have at least the following specification:

- 8 analogue input channels
- 12 bit resolution
- 100 kHz sampling rate (with hardware provision to enable sampling to disk)
- direct memory access (DMA) data transfer
- 2 analogue output channels (3 if a scope is to be driven by DACs)
- independent timing of analogue to digital conversion and digital to analogue waveform generation
- 4 digital inputs and 4 digital outputs[9]
- interrupt-driven operation[10]

A wide range of laboratory interfaces, meeting most of these specifications, are

---

[9]Though often one finds 16 digital input / outputs due to word length.

[10]Certain computer actions (sampling) should not be held up by any other computer operation. Precisely timed actions are obtained using so-called interrupt routines. Devices (eg a clock) issue an interrupt request - a little like the telephone ringing. The CPU acknowledges the interrupt request, saves the current program location and starts the interrupt routine (just like answering the phone) at a predetermined priority level. Upon completion of the interrupt routine, program location is restored and normal activity resumed.

available for both PC and Apple Macintosh families of computers. In the electrophysiology laboratory, the devices most commonly used in combination with PCs are the Scientific Solutions Labmaster DMA card, the Cambridge Electronics Design 1401*plus* interface (also for Mac) and - to a lesser extent - the Data Translation DT2801A card (Table A1 contains information on a more recent DT card). On the Apple Macintosh, the Instrutech ITC-16 is quite widely used.

The Labmaster card was one of the first laboratory interfaces to become available for PCs and has been in use for more than 10 years. Its popularity stems from its use with the pCLAMP software supplied by Axon Instruments. However, its old design limits its future even though it was updated in 1988 to support DMA - direct memory access, a hardware technique for transferring digitised data into the computer memory, essential for high speed data acquisition operations (Table A1 contains documentation on the follow up card, the Labmaster AD). Axon Instruments has now replaced it with an interface of their own design, the Digidata 1200.

The CED 1401 interface has also been in use for a number of years, particularly in Europe. It is a quite different design from the Labmaster card, being more a special purpose data acquisition computer in its own right, with an on-board CPU and RAM memory capable of executing programs and storing data independently of the host computer. The 1401 has recently been superseded by the much higher performance CED 1401*plus* which uses a 32 bit CPU rather than the original 8 bit 6502. An unusual feature of the CED 1401 is that it can be connected to a wide range of different types of computer system, including not only PC and Macintosh families, but also minicomputers such as the VAX and Sun workstations, via an IEEE 488 standard connection (a specialised parallel port).

Quite often the choice of laboratory interface is determined more by the choice of electrophysiological analysis software, rather than basic specifications. However, these specifications can be critical when writing one's own software. Unlike many other computer system components (e.g. disc drives, video displays), there are currently no hardware standards for the design and operation of laboratory interfaces. Each device has its own unique set of I/O ports and operating commands. Since most commercially available software packages work with only one particular interface card (usually manufactured by the supplier), it is not normally possible to substitute one type for another (an exception here is the Strathclyde Electrophysiology Software which supports several of the commonly available devices). The availability of suitable software has been one of the reasons for the widespread use of the CED 1401 and the Labmaster in electrophysiology labs.

5. Examining and exporting data

Once the data has been sampled, it must be examined and analysed. A video display unit is the minimum peripheral required for visual inspection. Permanent documentation requires some sort of hard copy device.

Video graphics display

Due to the lack of cheap high-resolution graphical display facilities in earlier computer systems (e.g. minicomputers in the 1980s), it was necessary to display digitised signals on an oscilloscope screen driven through DACs. Most modern personal computers are now capable of displaying data on the computer screen itself with enough resolution. Images are created on these screens by generating a bit map array in special video RAM in which each element determines whether a point on the screen is illuminated, in which colour (Red / Green / Blue) and at which intensity. The quality of the image is determined by the display resolution, measured as the number of independent pixels (picture elements) that can be illuminated. A typical screen display (c.1993) has a screen resolution of 640 (horizontal) × 480 (vertical) pixels each pixel capable of being illuminated in 16 colours from a palette of 64. A resolution of 800×600 is getting common and is more convenient for running Windows. Higher resolutions (up to 1024×768 pixels, or more, and up to 16 million colours) are available but not usable by all software. In general a trade off must be made between higher resolution or more colours on one hand and both the speed and cost of the graphic display on the other.

With the increasing use of graphical user interfaces, the performance of the graphics display hardware has become at least as important as that of the main processor. In order to maintain adequate display rates, an increasing degree of intelligence has been built into video graphics cards, allowing basic operations such as the filling of areas of the screen with blocks of colour and drawing of lines, to be done by the display hardware rather than the CPU (see BYTE reference). The updating of a high resolution display also requires large amounts of data to be transferred from the main RAM into video RAM, making great demands on the interface over which data passes. In particular, the standard ISA (Industry Standard Architecture) interface bus found in most PCs, with its 8 MHz 16 bit data path, has become inadequate to support modern video displays. This problem has been overcome using special 'local bus' connectors, such as the VL or PCI buses, allowing data to be transferred to the video card over 32 bit paths and at much higher speeds. Many PCs now come equipped with both ISA interface slots and one or more local bus slots. The PCI bus is currently gaining acceptance for wider uses with its endorsement by IBM and Apple for use in their Power-PC based computers.

Static data display

Overall, the ability to combine text and graphics in colour, and its convenience, make the use of the bit mapped computer display preferable to using a separate oscilloscope screen, but important limitations should be noted. The 480 pixel vertical resolution of most screens is not sufficient to fully represent the 4096 levels of a 12 bit ADC sample. Similarly, the 640 pixel horizontal resolution limits the number of samples that can displayed at one time, e.g. with a 20 kHz sampling rate, the screen is a mere 32 ms 'wide'. The displayed image is thus of much lower resolution than the digitised signal and does not contain (by far) as much information as is contained in the data itself. Brief/transient signals may be missed on a slow time scale. However, the lack of resolution can be partially overcome by using software with the ability to selectively magnify sections of the displayed signal.

Dynamic data display

The transient nature of the oscilloscope image, which persists only for a fraction of a second without refreshment, is well suited to the scrolling of signals across the screen, since only the new trace needs to be written. Images written to bit map, on the other hand, remain on the screen once written, supported by the graphics adapter display sub-system. This in turns means that to create a moving display the old trace must be erased (usually pixel by pixel because the fast screen erase erases ALL information from the screen) before a new trace can be drawn. Thus, implementing a fast scrolling display on a graphics display requires considerable handling of data and proves to be a difficult programming task. At present, few electrophysiological analysis programs have succeeded in implementing a scrolling display of the same dynamic quality as can be achieved using an oscilloscope. The problem is not insoluble, however, and it is likely that with the current rate of improvement in graphic display technology good scrolling signal displays will soon be realised on a computer screen. It is possible to generate scrolling displays using some existing hardware if the interface card is used to generate the X, Y and intensity ('Z') signals needed to generate such a display on an XY oscilloscope. For example, the support library supplied with the CED 1401 interface provides a command which allows programmers to generate such a display.

Hard copy graphics

The production of hard copies of stored signals, of sufficient quality for reproduction in a journal, is at least as important as the dynamic display of signals on screen. Hard copy devices fall into two main categories: printers which generate an image using a bit-map, like display screen graphics, and digital plotters which create the image by drawing lines (vectors) with a pen.

Until recently, the digital plotter was the most commonly used device for producing high quality plots of digitised signals. These devices, typical examples being the Hewlett Packard 7470A or 7475, are fitted with a number of coloured pens and can draw lines and simple alphanumeric characters on an A4 sheet of paper. Data is sent to the plotter in the form of a simple graph plotting language HPGL (Hewlett Packard Graphics Language) which contains instructions for selecting, raising and lowering pens, drawing lines etc. Although digital plotters can address over a thousand points per inch (1016 plotter units per inch) they position their pens to within a much larger fraction of a millimetre due to mechanical and other limitations (pens and penwidth). They can fully reproduce a 12 bit digitised record on an A4 page (10900×7650 addressable points). Plotters, however, are slow devices taking several minutes to produce a single trace with 1-2000 points. Also, the text produced by plotters was not quite of publication quality. Using HPGL has however the advantage that such files can be imported by a number of graphics programs (as will be discussed later).

Several laser printers can also emulate HPGL plotters[11], in particular the HP Laserjet III and IV series. Others can also do so provided they have been equipped with the

---

[11]Availability of line thickness adjustment is an important prerequisite for publication-quality figures.

appropriate hardware or using emulation software running on the PC. This is a particularly useful feature since some (older) software packages only provide graphical output in HPGL form. To a large extent, plotters are being replaced by printers that can produce hard copy graphics of a similar quality to the plotter, and also very high quality text. The laser printer is ideally suited for general laboratory work, being fast, capable of producing high quality text in a wide variety of typefaces. Laser printers are also reliable and easy to maintain in operation.. The latest printers, with 600 dpi resolution, are capable of fully reproducing a 12 bit digitised record without loss of precision (7002×5100 dots for an A4 page). Ink jet printers (e.g. Hewlett Packard Deskjet, Canon Stylewriter/Bubblejet, Epson Stylus) provide a similar quality of text and graphics (300 DPI) but a lower cost. The most recent members of the HP Deskjet family make colour accessible at a low (Deskjet 550C incl. PostScript for just less than US$ 1000.0) to moderate price (Deskjet 1200C incl. PostScript). They are still very slow (up to several minutes per page) and their use is recommended for final copies only.

As in many other areas of the computer industry, competitive pressures have lead to the creation of de facto standards. The most important emulations are PCL5 (HP Laserjet III), PCL5e (HP Laserjet 4) and Postscript Level I or II (see BYTE November 1993 p. 276 ff). Other frequent emulations include ESC-P (EPSON LQ/FX), IBM Proprinter and HPGL. The various emulations are differentiated by the codes used to transmit formatting and graphics instructions to the printer. Postscript is a specialised Page Description Language (PDL) licenced from Adobe Inc. designed for the precise description of the shape and location of text characters and graphics on the printed page. One of its prime advantages is its portability. A Postscript description of a page of text should be printable on any printing device supporting PostScript PDL (not just a laser printer). It is in widespread use within the publishing and printing industry. The main practical difference between the Postscript and other graphic images is that Postscript's images are defined using vectors which are be sent directly to the printing device while PCL graphics must be first created as bit-maps within host computer memory (a more complicated process) then sent to the printer.

In practice, there is little to choose between either type of laser printer. Postscript printers, perhaps, have the edge in terms of variety of type styles, while PCL printers are usually faster at printing bit mapped images. However, both of these families are widely supported by applications software such as word processors, spreadsheets and graph plotting programs. In addition, many printers now have both PCL and Postscript capabilities (e.g. HP Laserjet 4MP), while Postscript facilities can be easily added to older HP Laserjet printers using an expansion cartridge (e.g. PacificPage Postscript Emulator).

PART TWO: SOFTWARE

6. Operating systems

The operating system is a collection of interlinked programs which perform the basic operations necessary to make a computer system usable; reading data from the keyboard, creating, erasing, copying disc files, displaying data on the screen. The operating system acts as a environment from which to invoke applications programs. It defines the characteristics of the computer system at least as much as the hardware does.

The operating system appears to the user as the set of commands and menus used to operate the computer. It acts as an interface between hardware and the user, hence the term 'user interface'. From the point of view of the programmer, the operating system is the set of subroutines and conventions used to call upon operating systems services - the applications programmer interface (API). The debate about the relative merits of different computer families (e.g. Macintosh vs. PC) is very much a discussion about the nature of the operating system and what features it should provide to the user, rather than about the hardware, which is essentially similar in performance.

Although this chapter is primarily concerned with the applications of the computer to data acquisition in the context of electrophysiology, a brief consideration of the development of operating systems for personal computers is worthwhile since they provide the basic environment within which our software has to function.

Character-based OS

Most electrophysiology software runs on PCs, and uses MS-DOS (Microsoft Disc Operating System). MS-DOS is a relatively basic, single task system; primarily designed for running only one program at a time. This simplicity of design, compared to more complex systems like the Macintosh OS or UNIX, has made it relatively easy to develop efficient data acquisition software for PCs. However, the simple nature of MS-DOS, which was well matched to the performance of computers of the 1980's, is increasingly becoming a major limitation in making effective use of the modern processors which are 100 times more powerful[12].

MS-DOS, like its ancestor CP/M, uses a command line interface (CLI), i.e. the user operates the system by typing series of somewhat terse commands (e.g. DIR C: to list the directory of files on disc drive C:). This approach, though simple to implement, requires the user to memorise large numbers of command names and use them in very syntactically precise ways. A number of utility programs (e.g. PCTools, Norton Commander, Xtree) simplify often used file operations (e.g. copy, move, delete) and launching of programs, as well as providing a simple, mouse-driven graphical user interface (GUI). A very primitive GUI called DOSSHELL is also present in DOS (since version 4). It enables starting several programs which can be run alternatively (but not simultaneously)[13]. Modern operating systems are expected to be able to run several programs at the same time and provide a GUI. Windows 4 ('Chicago') will eventually replace DOS (but emulate it).

---

[12]Most PC users have confronted, at least once, the 640KB limit on memory dictated by DOS and the architecture of the first PCs. Most PCs, with 386 and above processors, overcome this problem by using 32 bit memory addressing to support up to 4GB of memory. This feature (32 bit memory addressing) is supported by recent versions of Windows. Though this is also supported by so-called DOS extenders, we do not recommend this practice is it is likely to cause compatibility problems.

[13]It is of limited practical use. Windows (3.1 and above) is a much better alternative.

Graphical user interfaces

The major *external* difference between character-based and graphical user interfaces is the way the user interacts with the computer. Within a GUI, the user uses a hand-held pointing device (the mouse) linked to a cursor on the screen in order to point to pictorial objects (icons) on the screen, which represent programs or data files, or to choose from various menu options. The first widespread implementation of a GUI was on the Apple Macintosh family, and contributed markedly to the success of that machine. The concept has now been widely accepted and, with the introduction of the Microsoft Windows and IBM OS/2 for PCs, most personal computers are now supplied with a GUI.

Windows, OS/2 and Macintosh System 7 provide very similar facilities, being mouse-driven, with pull-down menus, 'simultaneous' execution of multiple programs linked to movable resizable windows. The main difference between these three systems is that on the Macintosh and in OS/2, the GUI is closely integrated with the rest of the operating system whereas the Windows GUI is separate from, but makes use of, MS-DOS. Apple, IBM and Microsoft are in close competition with each other to develop newer and more powerful operating systems for the 1990's. Microsoft, has recently released Windows NT, a new operating system which integrates the Windows GUI with a new underlying operating system, better able to support multi-tasking and without many of MS-DOS's limitations.

These newer operating systems also all provide some support for 32 bit applications, that is programs that can directly access the full 4GB of memory that the CPU can access. The degree of integration of this support varies, OS/2 and Windows NT providing full support, Windows 3.xx only partial support. The additional overheads involved in supporting the graphical user interface do, however, mean that there is a cost both in terms of performance and memory requirement in using these operating systems. The entry level requirements for hardware to run these operating systems (in terms of memory size and processor power) are greater than those for a basic DOS system: e.g. a DOS based system running on a 80286 processor and with 640 KB of memory would provide adequate performance for most tasks. In contrast the minimum possible system to take full advantage of the current version of Windows (3.1) is an 80386 processor and 2MB of memory. In fact even this configuration would quickly be found to be severely limiting if any use was made of the multitasking capabilities of Windows.

7. Electrophysiology software

Electrophysiology covers a wide range of experiments and produces many different types of data. The electrical signal produced by biological preparations in such experiments has mainly two components: time and amplitude. Thus, analysis of electrophysiological records is primarily concerned with monitoring the amplitude of a signal as a function of time, or the time spent at a given amplitude. Most electrophysiology programs are like a 'Swiss army knife': although they can deal with a lot of experimental situations, they usually fall short when the proper instrument is needed. The 'perfect program', which incorporates only the best subroutines to deal with each type of data obviously does not exist. This is usually the reason why some expertise in programming is important. Writing one's own software may sometimes be essential to allow those experiments to be done that should be done and not those dictated by the software. It is the aim of the present section to mention which options should be present in a good electrophysiology program (see also Table A3-A5).

Most widely used electrophysiology packages for PCs (e.g. pCLAMP, CED EP Suite, Strathclyde Software) currently run under DOS. GUI based multitasking operating systems for PCs, such as Windows and OS/2 are gradually replacing DOS, though it may some time before character based OSs disappear altogether. This shift is reflected in a migration towards GUI versions of data acquisition programs: e.g. Axon Instruments' Axodata or Instrutech's PULSE for the Macintosh OS or CED's Spike2 for Macintosh, Windows 3.1 and Windows NT. Many standard applications (e.g. word processors, spreadsheets, graphic packages) are now available in a GUI-based version (e.g. for Windows, OS/2; Macintosh). How far electrophysiology packages follow the trend towards being GUI based depends to some extent upon the continued availability of character-based shells within future GUI OSs, such as the 'DOS box' within current versions of Windows and OS/2. There are undoubtedly great advantages that come along with GUIs, e.g. consistency of the user interface, ability to exchange data with other programs, multi-tasking[14]. Unfortunately, there is also a price to pay for the increased functionality of GUIs. It proves to be more difficult to develop programs for a graphical, multi-tasking environment than for a character-based OS. Fortunately, software publishers (e.g. CED) already provide direct-link libraries (DLLs) which greatly simplify communication with data acquisition hardware (e.g. 1401) or access to data files using proprietary file formats under Windows.

Different programs for different experiments

In the real world, most electrophysiology programs originate from teams which have designed their software to perform optimally with one type of experiment. Electrophysiology software has evolved in mainly two directions. Some programs were written to analyse short stretches of data often sampled in a repetitive way. Such programs were originally written to investigate the properties of voltage activated channels (e.g. sodium, potassium or calcium channels) in response to polarisations, such as voltage steps, applied by the experimenter. For example the AxoBASIC set of subroutines was uniquely tailored to perform repetitive tasks on a small number of data points.

Other programs have been designed to deal with random data at equilibrium; for example programs specifically conceived to sample and analyse long stretches of data obtained from agonist-activated ion channels. As already mentioned, the type of experiments that are likely to be done may often determine the choice of the program

---

[14]True multi-tasking, allowing several different programs to be active at once, by time sharing the CPU, is available in Windows enhanced mode. However, real time laboratory applications, such as sampling and experimental control, should normally be run exclusively, or there may be problems such as missed data points or timing errors.

and as a consequence the hardware e.g. requirements for interrupt driven data acquisition and large capacity data storage. It is important that experimental requirements determine software, and not vice versa; in early voltage-jump experiments hardware limitations in sample rates and durations had a profound effect on the information extracted from the data produced.

Fig. 3 shows the major components of a typical computer-based data acquisition system, as it would be applied to the recording of electrophysiological signals from a voltage or patch clamp experiment. The membrane current and potential signals from the voltage clamp are amplified and/or filtered via a signal conditioning sub-system and fed into the analogue inputs of the laboratory interface unit. Computer-synthesised voltage pulses and other waveforms can be output via DACs and applied to the command voltage input of the voltage clamp in order to stimulate the cell. On completion of the experiment, the digitised signals stored on disc can be displayed on the computer screen and analysed using signal analysis software. Hard copies of the signal waveforms and analysis results can be produced on a laser printer or digital plotter. As digitised signals accumulate, filling up the hard disc, they are removed from the system and stored on a archival storage device such as digital magnetic tape or optical disc.

Short events

The currents produced by voltage-operated channels during membrane depolarisation are typical short events lasting a few hundreds of milliseconds, though steady state may be reached only after much longer periods of time. It is important that the duration of each sample should be long enough for the relaxation of the properties of the channels to a new equilibrium to be complete during the sampling period. Essential program options are simple menus to adjust sample frequency and duration. Advanced program options include the ability to sample during several intervals within a repetitive pattern. The optimal sample frequency is determined by several parameters such as the required time resolution, and the filter-cut-off frequency is adjusted accordingly e.g. Chapter 6, Colquhoun and Sigworth (1983). Typically such experiments require sampling on two channels, e.g. voltage and current. Sometimes one may wish to monitor an additional signal such as the output of a photomultiplier tube, which allows measurement of the intracellular concentration of various mono- and divalent ions when used with appropriate indicators. It should be borne in mind that the maximum sample frequency (f_{max}) quoted for a laboratory interface applies to sampling on a single channel. The maximum frequency per channel when sampling on multiple channels is usually equal to f_{max} divided by the number of channels.

Watching the data on-line

One should **always** monitor the original, unmodified data on a scope. During repetitive sampling, this requires the software to keep triggering the scope even during episodes in which no data is being sampled (or manually switching the scope to a self-triggered mode). Also, it is often useful and sometimes essential to be able to watch the conditioned, sampled data on-line during acquisition.

Watching large amplitude signals such as whole-cell response to an agonist on-line should not be problem provided the data has been correctly amplified prior to the acquisition (see Signal Conditioning above). The situation is entirely different if the relative amplitude of the data is small compared to the underlying signal. This is the case for single channel openings of voltage-activated channels. Rapid changes of the membrane potential are accompanied by capacity transients which can be much larger than the actual single channel amplitude. The amplitude of capacity transients can be at least partially reduced by adjusting appropriate compensatory circuits in the voltage or patch clamp amplifier.

The channel openings are also superimposed on a holding current whose size may be much larger than the amplitude of a single channel opening. Before details can be observed, it is necessary to subtract such offsets, usually due to (unspecific) leak currents and capacity transients, from the original data. If this cannot performed by external hardware prior to or during acquisition, it is necessary to prepare a leak trace without openings to be subtracted from the sampled data. A good electrophysiology program should include options to do this on-line leak and capacity transient subtraction.

Leak current and capacity transient subtraction

A method often used for leak subtraction is the following. Let us assume that the channel is activated by stepping from a holding potential of −100 mV to a potential of −10 mV, a protocol often used to activate sodium channels. The individual channel openings will be much smaller than holding current and capacity transients. A correction signal is obtained by averaging N traces measured while inverting the command pulse thus hyperpolarising the membrane potential from −100 mV to −190 mV, a procedure which does not activate ion channels but allows measurement of the passive membrane properties, the silent assumption of this protocol being that these passive electrical membrane properties are purely ohmic, i.e. symmetrical and linear over a wide range of potentials. When the step potential is large, for example to −240 mV in order to compensate a step from −100 to +40 mV, a fraction of the voltage step is used instead (e.g. from −100 mV to −120 mV) and a linear extrapolation made.

Assuming that the leak current is linear, the sampled response is then scaled (in our example by a factor of 7) and may be used for correction of multiple voltage-step amplitudes ('scalable' leak in Table A3). This method was originally introduced by Armstrong and Bezanilla (1974) and called 'P/4' because the pulse amplitude was divided by 4. More generally the method is called P/N to indicate that an arbitrary fraction of the pulse is used. However, the assumption of linear leak is not always correct and the leak trace many need arbitrary adjustment ('adjustable' leak in Table A3). Most procedures for leak and capacity transient subtraction are not perfect so it is good practice to store both the original, unaltered data, as well as the leak subtraction and capacity compensation signal, so that a later, better or more detailed analysis can be performed as required. Another method of signal correction which, however, does not work on-line, consists in marking and averaging traces which do not contain single channel openings ('nulls') to prepare the leak and capacity transient subtraction signal. This method has been used to analyse data obtained from voltage-dependent calcium channels.

It is important that the method used for preparation of leak and capacity transient compensation signals used by a software package should be clearly explained in the program documentation. Ideally, it should be possible for the user to modify the method used (for example by having access to the code).

Long samples

Some types of data require sampling for long periods of time. These are for example the slow responses of a cell following application of an agonist, or random data, such as the single channel openings or membrane current noise during agonist application or spontaneous minature synaptic events.

Triggered events

Predictable events, such as evoked synaptic currents, are easy to deal with provided they have undergone adequate conditioning (e.g. amplification) prior to sampling. A number of programs exist which turn the computer into a smart chart recorder. These programs can record at moderate sampling frequencies on up to 16 and more channels. They include options for entering single character markers. These programs can usually play back the data but their analysis options are normally limited when they exist at all. Examples of such programs are CED's CHART, AXON's AXOTAPE or Labtech's Notebook programs. Their limited analysis capabilities make it essential that programs can export data in various formats for analysis by other software (see below).

Random events

Random events are more difficult to deal with. An important question is to decide whether all of the data should be recorded on-line as well as on an analogue recorder. The considerable amount of disk storage space required to store all of an experiment makes it worth considering detection hardware and algorithms in order to sample only the data which is actually relevant. Obviously, the event detector must be capable of keeping track of time between meaningful events (such as minature synaptic events or single channel openings), as time intervals are one of the variables in the data. The event detector may be implemented in hardware, as described briefly above, or in software. Software event detectors sample data into an 'endless', circular, buffer. The size of the buffer determines how much data prior to the triggering event will be available.

A problem with the use of event detectors is that the experimental record between events is not seen. Signal properties, such as the noise levels, or the size of the event that fires the trigger, may change during an experiment. In the latter case, the event detector may fail to trigger and data will be lost. It is therefore essential if using this technique to record all original data to tape, so that re-analysis with, for example, different trigger parameters may be carried out if needed. In addition, the voice track on a tape recorder can contain far more useful information than a few comments typed quickly in a file.

Driving external devices

Several types of devices can be activated by a computer-generated signal. The simplest devices (taps, and other electromechanical equipment) require a simple TTL (on-off) signal which can be generated either via a DAC or parallel digital port. Triggering the monitor scope usually only requires a TTL pulse. One usually wishes to trigger the oscilloscope at a given time before say depolarising a cell so the response can be monitored during the entire duration of the depolarisation. A good electrophysiology program should allow triggering of external devices (oscilloscope, magnetic tap, amplifier, signal conditioner, etc.) at predetermined times.

The next level of complexity requires variable signals to be sent to a voltage clamp amplifier. Although the membrane potential or current level can be dialed manually on the voltage-clamp amplifier the command pulses need to coincide with data sampling so it is better to rely on the computer to generate voltage steps, though complex voltage-step protocols can also be generated by a programmable stimulator provided a trigger signal is provided to initiate sampling. In general, use of the computer to generate command pulses makes it easier to create a wide range of complex command signals such as voltage ramps, multistep protocols, and playback of pre-recorded samples, such as action potentials. Command signals are generated by a DAC. The amplifier is treated in these protocols as an entirely passive device, though the latest generation of patch clamp amplifiers, such as the Axon Axopatch 2000 and the List EPC9, are capable of two way communication with the computer.

At the highest level of sophistication smart devices such as digital oscilloscopes and digital plotters can interpret simple, human-readable, commands of alphanumeric data sent in ASCII form via a serial or parallel port. Examples are Hewlett Packard, and other HPGL compatible, plotters, some Nicolet and Tektronix oscilloscopes and the CED 1401, all of which use a proprietry command language, just as printers do.

Synchronous command pulses

Investigating the properties of voltage-activated ion channels often means using complex pulse protocols. For example, the inactivation of sodium, potassium and other voltage-activated conductances may be studied by use of prepulses applied before a test depolarisation to determine the proportion of channels that can be activated (see Chapter 2). The conductance of ligand gated channels, that show little time dependence can be tested with slowly increasing ramp commands to plot I/V curves directly. Alternatively, the conductance of voltage-gated channels may be investigated by applying a series of incrementing, or decrementing, pulses from a holding potential. We can see from these two examples that there should be menu options to prepare pulse protocols in advance.

For voltage-ramps, it is important that the program should maximize the number of voltage steps so as to minimize their size. The amplitude of the smallest V_{CMD} step is a function of both resolution and voltage range of the DACs: with 12 bit resolution and a voltage range of ±5 V, one computer unit is converted to a V_{CMD} step of 2.44 mV which translates to a step of 49 μV at the pipette tip after attenuation by the amplifier (50-fold for Axon amplifiers). The size of the smallest V_{CMD} step is correspondingly larger (244 μV) with amplifiers which use a 10-fold attenuation (e.g. List). The latter size is such that each update of the amplitude (by one computer unit) will cause a small capacity transient and thus blur the current response. This problem can be minimised by low-pass filtering the DAC output before passing it to the voltage clamp amplifier V_{CMD} input.

Data processing

Processing electrophysiological data includes three operations: data preparation, statistical analysis and model fitting.

Data preparation

The aim of data preparation is to isolate those parts of the data that will be further analysed and to remove unwanted data stretches e.g. those containing artefacts or spurious signals. Editing a trace is accelerated if graphical tools (pointers, cursors), mouse support or a graphical user interface with pull-down menus, etc. are available. Although few of the currently available electrophysiology programs (e.g. Spike2 and PULSE for the Mac) provide mouse or true GUI support, it is likely that future releases of these programs will run under some form of GUI (Windows, OS/2 or Mac OS).

Some of the data preparation, such as baseline subtraction, capacity transient suppression and event capture may have already happened during acquisition. For slow responses the size and position of peaks or troughs of current as a function of membrane potential or other experimental parameters may be of interest. Alternatively, it may be desirable to isolate regions of a signal for analysis. Examples are fitting a sum of exponential curves to membrane currents obtained during prolonged agonist application, in order to quantify the time course of desensitisation, or to minature synaptic currents.

When dealing with fast events, in particular with single channel data, event detection normally is very time consuming because a large number of shut-open and open-shut transitions can occur within a short period of time. Thus 'manual' analysis of a 30 seconds stretch of single channel data may take several hours. The simplest method for detecting transitions in the conductance state relies on the crossing of an (arbitrary) threshold for event detection. A commonly chosen threshold level lies halfway between baseline and single channel amplitude. However, the limited bandwidth of any recording system reduces the apparent amplitude of fast events[15] and leads to problems with their detection, in particular with threshold-crossing detection methods. These problems in single channel analysis are discussed in Chapter 6.

Automated event detection

Most detection methods can be automated (see Sachs, 1983) and the temptation to use them is large because serious single-channel analysis is so time consuming. However, even simple algorithms such as automatic baseline and threshold determination are much less trivial than at first anticipated. The threshold is chosen with respect to a

---

[15]It may be worth remembering at that stage that tape recorders have a limited bandwith and may be placed in the recording pathway.

reference value (baseline). This baseline is in theory easily determined by the program (for example as mean current in the absence of agonist-elicited single channel openings or whole-cell current). Unfortunately, real baselines often drift because deterioration of the cell-pipette seal causes a slow increase of the holding current. Real baselines may also contain fast artefacts caused by switching on/off of electrical equipment. Thus automatic computer-assisted baseline and threshold settings are much more difficult in practice than in theory. However, the main objection to automated analysis is the fact that the data are not inspected by the experimenter.

Overall, simple detection methods, such as the half amplitude threshold crossing, are useful in circumstances where the data shows good signal-noise ratios, no sub-conductance levels are present and the majority of the single channel openings are long compared to the limits imposed by the bandwidth of the recording system. In contrast, time course fitting (see below), while yielding data of consistently high quality, cannot easily be automated (though a number of operations can be speeded up by careful programming). Whatever method is used the result of this first step in the analysis of single channel data is an idealised trace which consists of a succession of paired amplitude and duration parameters (with or without additional information such as validation parameters etc.) in place of the original data.

Statistical analysis

Statistical analysis of electrophysiological data is concerned with estimating parameters that characterise the idealised data, such as mean and standard deviation of the amplitude and kinetic properties of the event being studied.

Peak or steady state amplitude. The methods used to estimate amplitude are varied and dependent upon the types of data being analysed. Moveable cursors superimposed on a computer-generated display can be used for many forms of data, such as single channel data, with clear defined open and closed levels, or slow events, such as a calcium peak evoked by an agonist.

Other data may require more computationally intensive methods. For example, accurate estimation of the amplitude of small signals, with a poor signal-to-noise ratio, may require averaging of several responses prior to amplitude measurement. When working with single channel data, amplitudes can be estimated, not only by direct measurement, but also by methods which rely on analysis of the distribution of an entire set of data points. For example, calculation of point amplitude histograms (see Chapter 6) saves considerable labour and provides some information about the proportions of time spent at each current level. When dealing with data containing many open levels such analysis may be less useful. In this case more precision may be obtained by calculating the current variance and mean current amplitude for a set of windows of n data points (Patlak, 1988). This window is moved through the data one point at a time, giving values for variance and mean current for n points around each point. The variance is then plotted against mean current at each point. Groups containing a transition between levels have high variance and clusters of points of low variance correspond to current levels at which the trace is stable for some time and thus correspond to discrete current levels.

Time course. As with amplitude measurements, much data can be dealt with by taking measurements with superimposed cursors on a computer-generated display, e.g. the latency of responses to applied agonist. However, in many cases, this method will either be too slow in dealing with large amounts of data, will be unable to produce the information required or will fail to achieve the degree of precision required. In these cases, methods that make greater use of the power of the computer can be used. In the case of synaptic currents, the kinetic information can generally only be obtained by fitting exponential curves to the data (see Chapter 8). In the case of single channel data greater temporal precision can be obtained by fitting the time course of each transition, whereas greater speed but less precision may be achieved with the threshold-crossing method. The relative advantages and disadvantages of these methods are discussed in Chapter 6.

Noise analysis. The unitary events, such as channel gating or receptor potentials, underlying a response occur as random events of characteristic amplitude, lifetime and frequency. The statistical noise they introduce can be analysed to provide information about the amplitude and kinetics of underlying events in signals in which the individual events themselves cannot be resolved (see Chapter 8). Analysis of noise, beyond the most basic studies is impossible without the use of a computer. The required resolution determines the sample rate and length required. The lowest frequency point that can be analysed is given by inverse of the sample length, while the highest possible frequency is half the sample rate to avoid aliasing (Chapter 16). Kinetic analysis of noise requires calculation of the Fourier transform of the signal. This is always done using a Fast Fourier Transform (FFT) algorithm (see Press et al., 1986). Such algorithms impose a limitation upon the way the data is transformed in that they will only operate on arrays whose size is a power of 2, so data can only be processed in blocks of 2^n points. In the past the size of available memory could impose tight limits upon the size of the array that could be transformed. Thus, if each number is represented by a 4 byte floating point number, a sample of 2 seconds at 4069Hz, allowing analysis from 0.5Hz to 2KHz, would take up 32 KB. Using compilers that allow access to the full addressable memory of an 80386 or 80486 processor effectively removes this limitation as arrays much larger than this can be handled.

A second issue concerns the use of integer or floating point numbers to represent the individual data points. Integer FFTs were much faster than floating point FFTs, as CPUs can perform integer arithmetic more rapidly. However, the drawback is that the power spectrum produced, the plot of the signal power vs frequency, is distorted at the bottom end by the limited numerical resolution, the points being clustered at discrete levels. In practice, in a recent computer fitted with a FPU the difference in performance of integer over floating point FFTs is minimal, and not worth the loss of resolution.

In general, analysis of noise requires sampling on two channels, one channel records a low gain, low pass filtered record, for calculation of mean current levels. The second channel records the noise at high gain, after removal of DC components by high pass filtering, and removal of all frequencies above the folding (Nyquist) frequency by low pass filtering. Calculation of spectrum and variance of the noise record allow information about the amplitude and kinetics of the underlying events to be obtained, as described in Chapter 8.

Curve fitting. The term curve fitting means the process of adjusting the parameters of some equation (e.g. the slope and intercept of a straight line) so as to make the calculated line the best possible fit to the experimental data points. The first thing to decide is what equation is to be fitted and which of its parameters are to be estimated. This, of course, depends entirely on the scientific problem in hand. The second thing to decide is what 'best fit' means; in practice, it is usually taken to mean the 'method of least squares'. Thus, we define a weighted sum of squared deviations:

$$S = \sum_{i=1}^{N} w_i(y_i - Y_i)^2$$

where y_i is the ith observed value (that at x_i), Y_i is the corresponding value calculated from the fitted equation (using the current values of its parameters), and w_i is the **weight** for the ith observation. Thus S is a measure of the 'badness of fit' of the calculated values. The procedure is to adjust the parameters (by means of a minimization algorithm) and so alter the Y_i values so as to make S as small as possible.

The weights are a measure of the precision of the data point and should - if possible - be defined as:

$$w_i = 1/\text{var}(y_i)$$

where var (y_i) is the square of the standard deviation of y_i. For the purpose of parameter estimation, it is only the **relative** values of the weights that matter. A constant coefficient of variation can, for example, be simulated by taking $w_i = 1/Y_i^2$. If, however, internal estimates of error of the fit are to be calculated, the real standard deviations must be used. For digitized data, it is not usually possible to attach useful weighting to the data points; in this case the weights are ignored (i.e. they are implicitly taken to be the same for each point). In any case, it is generally not possible to calculate the errors of the fit for digitized data, because adjacent points are not independent, so the problem cannot be described by the usual curve fitting theory.

The fitting of distributions to single channel data is a different sort of problem (there is no conventional graph of y against x). It is best dealt with by the method of maximum likelihood (see Chapters 6 and 7).

Minimisation algorithms. Whatever method is used it will always involve changing the values of the parameters so as to minimise some quantity such as S. There are many algorithms available for doing this job, and it is one part of the program that you should not normally need to write. There are two types of minimisation algorithms: the **search** methods (e.g. simplex or patternsearch) and **gradient** types (Gauss-Newton or Levenberg-Marquardt). The search methods tend to be more robust and it is much easier to build constraints (i.e. forcing a parameter to be positive) into the fitting with them, but they are usually slow (especially when they are approaching convergence). Gradient methods tend to require more accurate initial guesses, are difficult to use with constraints, but are often faster. Note that the importance of speed depends on the speed itself and on how often you are going to run the program: a 10-fold speed difference may be trivial if it means the difference between taking 1 and 10 seconds to do a fit. It may be rather important if it means the difference between 1000 seconds and 10000 seconds. Most algorithms will find a minimum most of the time: it depends on both the data and the function being fitted. Some data may exhibit local minima when fitted with some functions, in this case no algorithm can be guaranteed to find the smallest of them. However, fitting repeatedly with different initial values may help to obtain a clear estimate of the parameters that give the true minimum.

Understanding the model. The law of computer use 'garbage in, garbage out' applies nowhere better than in curve fitting. Each set of data can be fitted by several equations. Thus it is essential to know what the implications of a given equation are. As an example the Hill-Langmuir equation may be used to fit concentration-binding curves and concentration-response curves. The proportion of ligand bound to a population of receptors is

$$p = \frac{x^n}{(x^n + EC_{50}^n)}$$

where x is concentration, EC_{50} concentration for $p = 0.5$ and n estimates the number of molecules of ligand bound per site. In the case of simple binding where the receptor does not change conformation (e.g. open an ion channel) the EC_{50} is an estimate of the dissociation constant (affinity) of binding and n is the number bound per site. However, in the case of concentration-response curves the data may fit perfectly well but the EC_{50} measures an apparent affinity of binding plus conformational equilibrium and n underestimates the number of ligands bound for receptor activation. Thus, there is mechanistic information in the former, but not in the latter case.

We can, for example, use an equation which links the response to agonist concentration to the occupancy of the open state in a four state model in which consecutive binding of two agonist molecules is followed by an additional conformation change to open the channel (see Colquhoun and Hawkes, 1981):

$$A + T \underset{k_{-1}}{\overset{2k_{+1}}{\rightleftharpoons}} AT \underset{2k_{-2}}{\overset{k_{+2}}{\rightleftharpoons}} A_2T \underset{\alpha}{\overset{\beta}{\rightleftharpoons}} A_2R,$$

where T represents the unoccupied receptor, A the agonist molecule, AT the singly, A_2T the doubly bound receptor and A_2R the active, conducting, receptor configuration. The transitions between the states are directed by the rate constants k_{+1}, k_{+2}, k_{-1}, k_{-2}, α and β which are the parameters for which we search estimates. In a case such as this it is unlikely that any single experiment will generate enough information to determine all the parameters. In this case one can either (*a*) determine some parameters from one sort of experiment and then treat these as known values while getting estimates of the others from different experiments, or (*b*) fit several experiments simultaneously. These parameters are now linked to a mechanism and may have an actual physical correlate. The equation describing the occupancy of the open state as function of the agonist concentration for the model above is:

$$p_{A2R}(\infty) = \frac{c_1 c_2 \beta/\alpha}{[1+2c_1+c_1 c_2(1 + \beta/\alpha)]}$$

where

$$c_1 = \frac{x_A}{K_1}, \; c_2 = \frac{x_A}{K_1}, \; K_1 = \frac{k_{-1}}{k_{+1}} \text{ and } K_2 = \frac{k_{-2}}{k_{+2}}$$

where x_A is the agonist concentration. The issues involved in composing models to fit data, selecting the most appropriate model and estimating errors are discussed by Colquhoun and Ogden (1988).

Finally, sometimes, one may deliberately fit an equation whose parameters do not have biological correlates in order to get an estimate of a biologically meaningful parameter. For example current-voltage curves may be fitted with a nth degree polynomial in order to get an estimate of the reversal potential V_R for I=0.

A large number of different equations may be fitted to both primary (original) as well as secondary data (obtained after analysis). A few examples are listed in Table 2 (for fitting of distributions see Chapter 6).

Many electrophysiology programs contain a curve fitting routine which can handle only a limited number of pre-defined equations. Usually, they cannot handle arbitrary, user defined equations. As a consequence, a good electrophysiology program should include options for exporting all primary and secondary data it produces (see below). A number of fitting programs capable of handling curve fitting with up to 9 and more parameters are now commercially available. Some of them will cope with limited constraints on parameters. Examples include GraphPad, Instat, FigP or Sigmaplot for DOS, Kaleidagraph for the Macintosh. However, many cannot cope with special constraints and in some cases, given time, it may be a good idea to write one's own curve fitting program using published minimising routines. Such subroutines (FORTRAN) have been published by Bevington (1969), Press *et al.* (1986 and references therein) or are commercially available in a compiled form (e.g. NAG library).

Output: exporting the data

Sampled and processed data can be sent to three destinations: screen, printer and data

Table 2. *Some equations used in curve fitting procedures*

Data	Fitted curve
current decay	sum of exponentials
IV curves	n^{th} degree polynomial, Goldman-Hodgkin-Katz
inactivation curve	Boltzman curve
power spectrum	sum of (1/f), Lorentzians
conc-response	Langmuir curve, Hill curve
conc-p_{open}	p_{A2R} (see above)

files. Hardware aspects of graphic displays and printers have been discussed in the first part of this chapter. The necessity to export data to data files arises from at least three reasons: the need for detailed statistics or specialised analysis (e.g. noise analysis), the possibility to fit arbitrary, user-defined curves to the data and finally the necessity to produce publication-quality figures. In the following, we shall discuss various ways to export data and present specific advantages / disadvantages of a few data file formats (see also Table A5).

Text (ASCII) files

The simplest format used to export data consists in transforming all information into user-readable text using the ASCII code. This format is universally readable even across various computer types (DOS, OS/2, Macintosh OS). Most if not all commercially available programs (spreadsheets, statistics, graphics) can read ASCII files. ASCII export / import routines are easily implemented in user-written programs. Finally, ASCII files can easily be read or modified using simple text editors (DOS edit, Windows Notepad, etc.). ASCII files are unfortunately quite uneconomic in terms of storage space: Numbers are no longer encoded in a fixed number of bytes but require a byte per digit. It is worth checking how the various programs interpret ASCII files since some programs (e.g. Lotus 1-2-3) cannot directly handle ASCII files containing both text and numbers. Additional data parsing may be required to restore the original appearance of the data. Other programs (e.g. Kaleidagraph) cannot contain text and numbers within the same column. If the exported data consists of several columns of numbers or text, they must be separated by a character that is recognised as such by the reading program. Commonly used separators are space, comma and tab characters.

Lotus (*.WK%) file format

This format is a custom binary file format used by Lotus Development to encode data within its Lotus 1-2-3 spreadsheet program. Lotus 1-2-3 is a widespread program which runs on multiple platforms (DOS, Windows, OS/2, Macintosh). Furthermore it can be read by several other spreadsheet programs such as MS-Excel or Borland Quattro Pro. Several graphics programs (Lotus Freelance, Harvard Graphics) also can read Lotus 1-2-3 (*.WK%) files. These files can contain text, numbers and graphics information for plots. The disadvantage is that it is not as universally recognised as ASCII files. Determined programmers can incorporate code to read and write Lotus 1-2-3 files into their programs (see Lotus reference).

Proprietary file formats

A recent and promising development is the use of standardised file formats for digitised signals by commercially available software. Both CED and AXON use structured binary data files so the user may extract whatever information he or she requires. CED (and others) have gone a step further by writing and distributing a set of subroutines written in Pascal or C which can read and write standard CED Filing System ('CFS') files (and other files using proprietary file formats). An immediate advantage of standardisation is that data can be easily exchanged between programs and users. Thus data sampled with CED's CHART program can be analysed with PATCH or Spike2. Similarly, a number of researcher-written programs for single channel analysis, such as IPROC and LPROC, can read the Axon pCLAMP data file structure (ABF: Axon Binary File). Unfortunately, the ABF and CFS file structures are quite different. Programs designed to handle CFS files cannot directly read ABF files, (a file conversion program ABF to CFS is shipped with CED's EPC suite) and vice versa. So far, no single standardised data file structure has yet been widely accepted.

Graphics file formats

The two most important types of graphics files are bit-mapped (raster) and vector-graphics (object-oriented) files. The main difference between those two file types is that vector-based files can be enlarged without loss of precision.

Bit-mapped. Common bit-mapped file formats are TIFF (Tag Image File Format, codeveloped by Aldus) and PAINT files (common on Macintoshes). Raster files are a little like photocopies or photographs. They can be modified but usually not at the level of individual elements of the picture.

Vector-graphics. Vector-based graphics files can more easily be scaled and modified. PostScript files are text (ASCII) files which describe a vector image. Encapsulated PostScript Files (EPS Files) include a PostScript text file and an optional low resolution preview of the undelying image. The data file can be written as a text (ASCII) or binary file. Interestingly, EPS files can also be used to save raster images though they do not perform a raster to vector transformation. The HPGL graphic file format is also a vector-based graphic file format which enjoys a certain amount of universality (see hardware section) and is used by certain electrophysiology programs to export graphs (see Table A2).

Metafiles. This graphic files format contains both raster and vector graphics. Such files are PICT files on Macs and Windows Metafile Format (WMF) on PCs. They are the Clipboard standard files. Another metafile format is the CGM (Computer Graphics Metafile) format, a format defined both by the International Standards Organisation (ISO) and the American National Standard Institute (ANSI).

There is a certain amount of compatibility and exchangeability in that some graphic file formats are shared by several programs. This is true of Metafiles within DOS (CGM) and Windows (WMF). HPGL (and Lotus PIC) files are recognised by a number of DOS and Windows programs too. If export across platforms (DOS /Windows to Mac and vice-versa) is required the best bet may be to use TIFF and EPS graphic files. No rule can be made though about which file types can be exported or imported by any given program (see Sosinsky, 1994).

8. Writing your own program

Thus far, the discussion has concentrated upon the nature of a good program and the extent to which those currently available meet the criteria. Although these programs cover most of the common forms of electrophysiological experimentation, there can be no doubt that occasions still exist where it is necessary for the experimenter to write his/her own program.

One language does not fit all

Computer programming can be done at three levels: Low level language (Assembler) programming is for the experts. Assembler reaches parts of the computer other languages cannot reach. It is hardest to write, but fastest in execution time. At a moderate level of programming expertise, 'high level' programming languages (BASIC, FORTRAN, Pascal, C, Modula etc.) enable all (arithmetic) data manipulation usually required in a lab. Solving given mathematical problems requires for example, understanding of the mathematical / statistical background behind the data manipulation. The reader should be aware that knowing that solving a given problem requires inverting a matrix is only the first part of the solution of this problem. It is usually just as important to have access to specialised literature on so-called 'numerical recipes' (e.g. Press *et al.*, 1986) to perform given mathematical tasks (e.g. matrix inversion). Specialised applications such as writing software to convert binary file formats require additional knowledge on proprietary file formats which is usually provided by the software companies (as manuals) or published (e.g. Lotus File Formats). Most commercially available programs were written in one these high level languages. At the highest (language) level, some commercial programs enable the user to simplify repetitive tasks in data manipulation so they can run in a predetermined, 'programmed' way, using so-called macros (e.g. Lotus 1-2-3). Recently, a number of programs have appeared which can perform analytical mathematics (as opposed to numerical). Such programs include Wolfram Research's *Mathematica*, MathSoft's *Mathcad* or Waterloo Maple Software's *Maple*.

It is striking how many of the existing commercial programs originated in research laboratories (e.g. Axon's pCLAMP and Axograph, Instrutech PULSE software or CED PATCH & VCLAMP modules). A case can therefore be made for the need to retain programming skills within at least some electrophysiological laboratories. New forms of signals such as those obtained from fluorescent dyes often require the recording of signals from new classes of devices (e.g. photomultipliers or video cameras) and different analysis algorithms.

A number of programming languages are available for both the DOS and Macintosh platforms. They include numerous brands of BASIC, Pascal, FORTRAN, Modula 2, Assembler and C, to name but the most important. Programs in a high level language can be pepped up with low-level subroutines: this so called mixed language programming may be an excellent approach for solving certain problems. A program written in a 'high level' language makes use of compiled subroutines written in 'lower level' languages (Assembler, C) which run very fast. This removes the task of writing and checking specialised subroutines from the occasional programmer while giving him/her access to high powered software. The reader is referred to specialised literature for this sort of programming (e.g. Microsoft mixed language programming guide).

One critical application of this approach to anyone writing data acquisition or experimental control software is the provision of libraries of low level procedures that handle communication with the interface. Programming these aspects generally involves lengthy Assembler programming. Most laboratory interfaces are supplied with libraries of user callable procedures (e.g. Axon's set of subroutines for the Digidata 1200 or the LabPac set of subroutines for the Labmaster DMA card), or other support mechanisms, that handle those low level aspects. When selecting an interface with the intention of writing one's own software, the quality and functionality of the programmer support is a major factor to consider. Other examples of user callable subroutine packages (libraries) are: the CFS set of subroutines to access CFS files, the scientific NAG (National Algorithms Group) library (state of the art statistics and mathematics package), or the MetaWindows set of graphic subroutines.

The widespread adoption of GUI based systems, such as Microsoft Windows and the Macintosh OS calls for programming languages adapted to these new environments. Several programming languages are available for the Windows platform and the Macintosh OS. Microsoft's Visual Basic, Borland's Turbo Pascal for Windows (and other similar tools) are fairly easily learnt languages which enable writing of true Windows applications (if somewhat slower than applications written in C). These programming environments, together with the latest versions of Microsoft and Borland C++, all support a visual programming metaphor for the construction of the Windows user interface. This removes from the programmer a major part of the load of developing a Windows application, allowing the programmer to concentrate on coding the underlying functionality in a traditional programming language. Mixed language programming with all its advantages is possible (also from Visual Basic) by calling so-called direct link libraries (DLLs) which perform specialised tasks (e.g. data acquisition or display). Some of the laboratory interfaces presented in Table A1 are supported by DLLs. However, as discussed earlier, the development of such programs is more difficult than writing straight DOS programs. This may become unavoidable when multi-tasking systems such as Windows NT become standard. Using the Macintosh toolbox is already now a difficult task for non-professional programmers. Another approach may be to use DOS compilers which take advantage of the protected (true 32 bit) processor mode. With the exception of the latest generation of C++ programming environments, these may have the disadvantage of not being mainstream.

A further solution to the problem of developing customised laboratory software within the new GUI environments is to make use of a specialised software development system which provides the basic software components necessary for the acquisition, display and analysis of analogue signals. One of the most advanced examples of this approach is the National Instruments LabView package which allows the development of data acquisition software for both Windows and the Macintosh. With this package, 'virtual instruments' can be created by graphically 'wiring up' acquisition, analysis and display icons on the screen. National Instruments also produce the Lab-Windows package which allows the development of similar systems (for PCs only), but with a more conventional programming approach, using enhanced versions of the C and QuickBasic programming languages.

In each case, the merits of developing a new program in-house should be carefully weighed, with consideration given to whether there might be better uses for one's time. Software development, even with the assistance of a good development environment, is a skilled and time consuming activity. There is little value in producing a poor and unreliable imitation of an existing program. On the other hand, the existing software that is widely available is far from perfect, and there is always a place for faster, more versatile, easier to use, software.

9. Concluding remarks

The preceding sections are intended to provide sufficient basic information and guidance to allow a suitable laboratory computer system to be selected. However, it is without doubt a good idea to take additional advice, where possible, from colleagues in your own field who have experience of performing experiments and analysis similar to those that you intend to undertake, with one or more of the packages under consideration. In addition, it is crucial that before making a selection, a specification of requirements should be drawn up which details what functionality is needed such as the number of signals to be recorded in parallel, maximum sample rates (which will be determined by signal bandwidth), signal length, short and long term storage requirements (which will be related to the sample rate and length), provision of control signals and data or trigger outputs to monitor scopes, forms of analysis required, forms of hard copy needed. Other factors such as cost limitations or constraints imposed by existing hardware must also be considered.

The field of personal computers is moving fast. It is quite obvious that by the time this chapter is published or soon after, new, better and faster processors will be available. This may be important to satisfy the increasing demands of number crunching applications and of graphical user interfaces. The ease of use and affordability of personal computers has created a trend for people to buy ready to use 'turn-key' systems. This is a clear change from the time of mini-computers such as the PDP-11 when all software had to be custom made. There is a danger that electrophysiologists do only those experiments that can be done with commercial software instead of doing those experiments necessary to answer a (biological) question. The clear trend towards GUIs may aggravate this situation by further decreasing the number of people both willing and able[16] to write their own software.

This means that electrophysiologists have to rely on programs written by people who are not necessarily familiar with the biology behind the numbers. Electrophysiologists using such programs should be aware of this fact and critical about what computers tell them. To those who have not got access to the program source, the hardware and software should be treated as a black-box whose behaviour should be carefully assessed and critically analysed to be sure its limitations are appreciated. Numbers need not be correct just because they appear on a terminal or a computer print out. Users should never forget that while computers are infinitely

---

[16] The limits often being set by time, budget and publishing considerations.

diligent they are also infinitely dumb. By taking away the tedious task of number 'crunching' computers have if anything increased the responsibility of users to know and understand the theories and models underlying their calculations.

Acknowledgements

Several discussions with David Colquhoun, David Ogden, Jörg Stucki, as well as with Peter Rice and Greg Smith, using various means of communication (phone, fax and electronic mail) have greatly enhanced the accuracy of this chapter. People often identify strongly with their computers: the opinions expressed here are those of the authors or a hard fought compromise thereof.

References

Original articles and books

ARMSTRONG, C. M. & BEZANILLA, F. (1974). Charge movement associated with the opening and closing of the activation gates of the Na channels. *J. Gen. Physiol.* **63**, 533-552.

BENDAT, J. S. & PIERSOL, A. G. (1986). Random data: analysis and measurement procedures. New York: Wiley Interscience.

BEVINGTON, P. R. (1969). Data reduction and error analysis for the physical sciences. New York: McGraw Hill.

COLQUHOUN, D. & HAWKES, A. (1981). On the stochastic properties of single ion channels. *Proc. R. Soc. Lond. B.* **211**, 205-235.

COLQUHOUN, D. & SIGWORTH, F. J. (1983). Fitting and statistical analysis of single-channel records (Ch.11) In Single-Channel Recording, (ed. B. Sakmann and E. Neher). New York, London: Plenum Press.

COLQUHOUN, D. & OGDEN, D. (1988). Activation of ion channels in the frog end-plate by high concentrations of acetylcholine. *J. Physiol. (Lond)*, **395**, 131-159.

DEMPSTER, J. (1993). Computer analysis of electrophysiological signals. London: Academic Press.

LOTUS FILE FORMATS FOR 1-2-3 SYMPHONY & JAZZ. File structure descriptions for Developers. Lotus Books. Reading, MA, USA: Addison-Wesley Publishing Company.

PATLAK, J. B. (1988). Sodium channel subconductance levels measured with a new variance-mean analysis. *J. Gen. Physiol.* **92**, 413-430.

PRESS, W. H., FLANNERY, B. P., TEUKOLSKY, S. A., & VETTERLING, W. T. (1986). *Numerical Recipes, The Art of Scientific Computing.* Cambridge University Press.

SACHS, F. (1983). Automated analysis of single-channel records. (Ch. 12) In *Single-Channel Recording.* (ed. B. Sakmann and E. Neher). New York: Plenum Press.

SOSINSKY, B. (1994). Graphically speaking. *MacUser* Jan. 1994. pp.145-149.

Useful articles in BYTE

Although BYTE is not the only source of information on computers it was chosen because it has become an authority in the field and is likely to be available in libraries.

FPU Face-off. *BYTE* Vol **15** (12), 194-20. (Nov. 90)

Managing gigabytes. *BYTE* Vol **16** (5), 150-215 (May 91).

Operating systems trends. *BYTE* Vol **17** (10), 158-197 (Oct. 92).

PowerPC performs for less. *BYTE* Vol **18** (9), 56-78 (Aug. 93).

Printers face off. *BYTE* Vol **18** (12), 276-307. (Nov. 93).

State of the art mass storage: magnetic, optical, holographic and more. *BYTE* Vol **15** (12), 272-341 (Nov. 90).
Windows accelerators. *BYTE* Vol **18** (1), 202-206 (Jan. 93).

Suppliers

Axon Instruments Inc.,1101 Chess Drive, Foster City, CA 94404, USA
T: +1 415 571 9400, F: +1 415 571 9500
Cambridge Electronic Design Ltd, The Science Park, Milton Road, Cambridge, CB4 4FE, UK.
T: +44 223 420186, F: +44 223 420 488
Data Translation,100 Locke Drive, Marlboro, MA 01752. USA
T: +1 508 481 3700, +1 508 481 8620
Dagan Corporation, 2855 Park Avenue, Minneapolis, MN 55407, USA
T: +1 612 827 6535
Digitimer Ltd, 37 Hydeway, Welwyn Garden City, Herts., AL7 3BE, UK
T: +44 707 328347.
Frequency Devices Inc., 25 Locust Street, Haverhill, MA 01830, USA
T: +1 508 374-0761, F: +1 508 521 1839
Fylde Electronic Laboratories Ltd, 49/51 Fylde Road, Preston, PR1 2XQ, UK
T: +44 772 57560.
Instrutech Corp., 475 Northern Blvd, Suite 31, Great Neck, NY 11021, USA
T: +1 516 829 5942, F:+1 516 829 0934
NAG Ltd, Wilkinson House, Jordan Hill Road, Oxford, OX2 8DR, UK
T: +44 865 511 245, F: +44 865 310 139
National Instruments, 6504 Bridge Point Parkway, Austin, TX 78730-5039, USA
T: +1 512 794 0100, F: +1 512 794 8411
Scientific Solutions, 6225 Cochran Road, Solon, OHIO 44139, USA
T: +1 216 349 4030, F: +1 216 349 0851

Appendix

The tables in the following appendix summarise typical commercially available hardware and software solutions available for electrophysiological (and other lab) data acquisition and analysis, as of January 1994. The data presented were either obtained directly from the manufacturers, or taken from their literature. As with all issues relating to computer hardware and software, these systems and packages are subject to rapid change. Therefore, this appendix is intended to provide a guide to the type of solutions that may be available, and the data provided should be checked with the manufacturers before any purchasing decisions are made.

Manufacturer's addresses are listed at end of chapter. All boards are supported by various software classified as bundled (B) with the product, additional (A), i.e. available at additional cost and third party (T).

Table A1. *Comparison of commonly used laboratory interfaces*

Interface	CED 1401*plus*	Digidata 1200	ITC-16	DT 2839	Labmaster AD	AT/NB[1]-MIO-16
Supported by electrophysiology software	yes	yes	yes	no[2]	no[3]	no
Computer type	PC, PS/2, MacIntosh	PC	PC, MacIntosh, Atari	PC	PC	PC, PS/2, MacIntosh
Analog input(s)	16 SE[4] 32 option	32 pseudo DI	8 DI[5]	16 expandable to 224	16 SE / 8 DI expandable to 64 se or 32 di	16 SE / 8 DI
ADC range(s)	±5 V standard ±10 V option	±10.24 V	±10.24 V	0-1.25, 0-2.5, 0-5, 0-10 V, ±1.25, ±2.5, ±5, ±10 V	±10 V	0-10 V, ±5 V, ±10 V
AD converter	12 bit 333 kHz	12 bit 333 kHz	16 bit 100 kHz / 12 bit 200 kHz	12 bit 224 kHz (1 MHz 1 ch)	12 bit 330 kHz	12 bit 200 kHz[6]
Programmable input gain	option[7]	1, 2, 4, 8	no	1, 2, 4, 8	PGH: 1,2,4,8 **or** PGL: 1, 10, 100,1000	0.5, 1, 2, 5, 10, 20, 50,100
DA converters	4× 12 bit 444 kHz	2× 12 bit 333 kHz	4 × 16 bit 200 kHz	2 × 12 bit 130 kHz; 260 kHz aggregate	2× 12 bit 130 kHz full scale max 500 kHz	2× 12 bit 250 kHz[8]
DAC range(s)	±5 V standard ±10 V option	±10.24 V	±10.24 V	0-5 V, 0-10 V, ±2.5 V, ±5 V, ±10 V	±10 V (5 mA)	0-10 V, ±10 V
Asynchronous digital input(s)	16× TTL	8× TTL	16× TTL	16× TTL	8× TTL	8× TTL
Asynchronous digital output(s)	16× 24 mA	8× 24 mA (or 4 sync.)	16× 24 mA	16× 24 mA	16× 24 mA	8× 24 mA
Clock(s)	4× 32 bit 250 ns 1× 48 bit 1 μs	48 bit 250 ns	250 ns	2× 16 bit 250 ns programmable	16 bit 250 ns	16 bit 200 ns programmable
Counter/timer	2 channels	9513A 5 channels	no	9513A 5 channels	9513A 5 channels	9513A 5 channels

[1]AT indicates PC cards, NB (NuBus) Mac cards. [2]Older board (DT2801A) is supported by Strathclyde software and other packages. [3]Older board (Labmaster DMA) is supported by Strathclyde software and other packages. [4]SE: single-ended, DI: differential. See section 4 for definition. [5]Optically isolated. [6]NB-MIO: 100 kHz. [7]Gains: 1, 2, 5, 10, 20, 50, 100. [8]Option 64 F5: 1 MHz.

CED 1401*plus* (CED Ltd, UK)

Hardware

Connection to computer via 1 short ISA or MCA card for PCs, 1 NuBus card for Macs. Stand alone operation possible thanks to 32 bit 10 MIPS built-in microprocessor. Up to 17 MB on-board memory. All input/outputs (ADC, DAC, Clock, Digital I/O) are brought to BNC and D type connectors on front panel of a 19 inch box for connection to lab equipment.

Software

B: utility programs with source in C, QBASIC, PASCAL, commands for internal processor

A: Spike2, Chart, Patch, Signal Averager, EEG, ECG, FFTs, Windows 3.1 and Windows NT DLLs, device drivers and support. C support and device driver support for Macintosh.

Digidata 1200 (Axon Instruments, USA)

Hardware

Consists of 1 ISA bus card. Connection to lab equipment via optional: BNC interface, screw panel terminal and Cyberamp adapter(s).

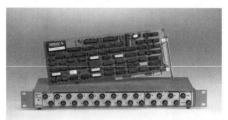

Software

B: utility programs, support library (+ source in C and BASIC), Windows DLL in development

A: pCLAMP (patch clamping / microelectrode recording), AxoTape (strip chart recorder with real time display), AxoBasic (development environment for creating acquisition and analysis programs).

ITC-16 (Instrutech, USA)

Hardware

Connection to computer via 1 NuBus (MacIntosh) or 1 ISA card (PCs). Partial stand-alone operation via local microprocessor. Connection to lab equipment via rack mount case with BNC connectors.

Software

B: IGOR Xops, QBasic drivers, C Drivers
A: PULSE (Heka), AXODATA (Axon), SYNAPSE, Pulse Control.

DT2839 (Data Translation, USA)

Hardware

Consists of 1 ISA/EISA bus card. Connection to lab equipment via simple screw terminal or expander board.

Software

B: Gallery, example programs, Windows 3.1 DLL, C software developer's kit for Windows 3.1, Tookit subroutine libraries
A: Visual Basic Programming tools, custom controls for data acquisition and high speed plotting Labtech Notebook, Control
T: P-DAAS (Pemtech), QEDIesing 1000 (Momentum), Experimenter's Workbench/Discovery(Brainwaves).

Labmaster AD (Scientific Solutions, Inc, USA)

Hardware

Consists of 1 16 bit ISA (EISA compatible) card. Connection to lab equipment via 'LabRack' optional BNC termination panel.

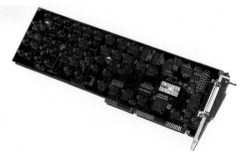

Software

B: utility and test software, software library for most MS DOS languages, data streaming oscilloscope software
A: paperless strip chart recording, data logging, storage scope emulation
T: Snapshot, Notebook/Control, DADisp, EP/EEG Scan, DriverLINX, RT-DAS, Data Sponge, ASYST. Supported by Software Toolkit (not by LabPac).

AT/NB MIO-16

Hardware

Consists of 1 ISA/MCA (PCs) or 1 NuBus (MacIntosh) card. Connection to lab equipment via optional screw terminal (BNC) or conditioning modules.

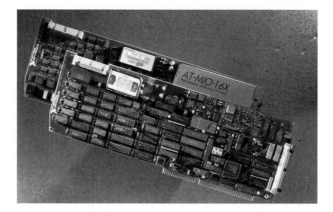

Software

B: DAQWare (integrated package), NI-DAQ for DOS (most MS DOS languages), NI-DAQ for Windows (DLLs). Library of functions & utilities for Macintosh OS
A: Labwindows for DOS, LabVIEW for Windows/MacIntosh/Sun, LabTech Notebook, HT Basic. Measure (Add on for Lotus 1-2-3).

Table A2. *Hardware requirements of some commercially available electrophysiology packages*

Author/manufacturer	Axon Instruments	CED Ltd	HEKA Elektronik	John Dempster
Name/Version	pCLAMP 6.0 June 1993	PATCH+VCLAMP 6.03 June 1993	PULSE + PULSEFIT 7.0 October 1992	STRATHCLYDE EP SUITE
Source code	MS Basic PDS 7.1, Borland Pascal, 7.0 some routines in MS C / Assembler	Borland Pascal 7.0 + some Assembler	Modula-2	Microsoft FORTRAN & Assembler
Source code available	Source code for analysis provided free of charge with package, source code for acquisition not available	enquire	yes, non disclosure agreement required	yes
Approx price	US$ 2000.	UK£ 1000.	US$ 1500.-	Free to academics (shareware)
Site licence available (multi-user)	by department	yes	no	no
1. Computer equipment				
Operating system / Version	DOS-3.1 , DOS 5.0 or higher recommended	DOS 2.0 or higher	Apple OS 6.03 or higher	DOS 3.3 or higher
Computer type / (Processor)	AT 33 MHz 80386 DX or higher recommended	PC/AT 80×86	Macintosh II, Quadra	PC/AT 80×86
Maths co-processor	required	recommended	required	recommended
RAM minimum / recommended	640 kB (+ 1 MB disk cache recommended)	640 kB / 1 MB	8 MB	512 kbyte/640 kbyte
Diskspace minimum	7 MB for pCLAMP files	1.5 MB	3 MB	2 MB
Max screen resolution	VGA (640×480)	VGA 640×480 Wyse 1280×800	unlimited	640×480
ADC interface(s) supported	Axon Instruments Digidata 1200, Scientific Soltns Labmaster DMA or MC/DAS (PS/2)	CED1401 / 1401*plus*	Instrutech ITC16	CED 1401/1401*plus*, Data Translation DT2801A, DT2821, DT2812, Scientific Solutions Labmaster, National Instruments LAB-PC, ATMIO-16X.
2. Printers/plotters				
Plotters	HPGL (≤ 6 pens)	HPGL	yes[1]	
Dot matrix printer	many types	Epson or -compatible	yes[1]	any Epson compatible
Inkjet printer	DeskJet	DeskJet	yes[1]	DeskJet
Laser printer	many types incl. HP LaserJet	HP-LaserJet or compatible	yes[1]	HP LaserJet or compatibles
PostScript laser printer	yes	no	yes[1]	yes
3. Electrophysiology equipment				
Recommended amplifier(s)	any	any	EPC-9	any
4. Computer controlled lab equipment				
Amplifier / signal conditioner	CyberAmp[2]	no[3]	no	

[1] any Macintosh OS compatible.
[2] CyberAmp: Axon Instruments' signal conditioner. See text.
[3] CED1902 signal conditioner and Axon CyberAmp supported by programs SIGAV, CHART and SPIKE2.

Table A3. *Comparison of 4 commercially available electrophysiology programs.*
Data Acquisition characteristics

	pCLAMP	PATCH + VCLAMP	PULSE + PULSEFIT	STRATHCLYDE SUITE[1]
1. General				
Min interval	3-10 μsec[2]	3 μsec	5 μsec	10 μsec[2]
Max interval	1 sec	1 sec	65 msec	15 msec-10 sec[2]
Split clock sampling	yes	no	no	no
Input channels	8	rep.samp.: 4 cont.samp.: 1	out:2, in:2	6
Amplifier monitoring	yes	yes	yes	no
Filtering during acquisition	no	no	yes	no
2. Repetitive sampling				
(a) acquisition				
Exact episode timing	yes	yes	yes	
Linking of stimulus families	yes	yes	yes	
Seal test	automatic	standard pulse	standard pulse	standard pulse
Max number of points/pulse	16 k samples	10 k samples	16 k samples	2 k per channel
Adjustable trigger pulse	yes	yes	yes	yes
On-line leak subtraction	yes	yes	yes	no[3]
Leak subtraction method	P/N[4]	P/N	P/N	P/N
On-line averaging	yes	yes	yes	yes (1 channel)
(b) Command pulses				
Command types	steps/ramps $\leq$ 8 transitions	steps/ramps $\leq$ 2 transitions	steps/ramps/ sine waves 16 k transitions	steps/ramps $\leq$ 3 transitions
Free definable command	up to 16 k points memory/file	$\leq$100 dig pulses on 8 ch / sweep	file	
Auto increment/decrement	amplitude +duration	amplitude + duration	amplitude + duration	amplitude + duration
Arbitrary order of V steps	user-defined list	on line user control	interleaved alternate	interleaved alternate
Conditioning pulses	yes	yes	yes	yes
3. Continuous sampling				
Max rate to disk	80 kHz[5]	80 kHz[2]	200 kHz[2]	30 kHz (PAT)
Max data file size	4 GB	4 GB	2 GB	32 MB
Watch - write mode switch	yes	yes	yes	yes
Event-triggered acquisition	yes	yes	yes	yes
Adjustable pre-trigger	yes	no[6]	no	yes
Markers during sampling	no	yes	no	no

[1]Consists of following programs: WCP: whole cell analysis, PAT: single channel analysis, SCAN: synaptic current analysis, VCAN: voltage clamp analysis and SPAN: power spectrum/variance analysis.
[2]Depends on interface used.
[3]Implemented off-line.
[4]P/N: Pulse amplitude P is divided by arbitrary N, applied N times and scaled by N to obtain leak signal.
[5]Up to 333 kHz using appropriate hardware.
[6]Achievable by adjusting data block size.

Table A4. *Comparison of 4 commercially available electrophysiology programs.*
Data Analysis characteristics

	pCLAMP	PATCH + VCLAMP	PULSE + PULSEFIT	STRATHCLYDE SUITE
1. General				
On-line analysis	yes[1]	yes[1]	yes[2]	no
Mathematical operations between episodes	yes	no	yes	no
2. Amplitude domain				
Point histogram (max bins)	max 256 bins	max 1024 bins	yes	max 512 bins
Probability density-amplitude histogram	yes	yes	yes	yes
Amplitude vs time histogram	yes	yes	yes[3]	no
Automatic peak - trough follow	yes	yes	yes[3]	?
3. Time domain				
Open time histogram	yes	yes	yes[3]	yes
Shut time histogram	yes	yes	yes[3]	yes
Sigworth Sine histogram	max 512 bins	no	yes	see below
Rise time to peak in whole cell current	10-90% rise time	yes	yes	10-90% rise time
First latency of single channel data	yes	yes	yes	no
4. Curve fitting				
Search / optimizing algorithm	various[4]	Simplex	Simplex	Marquardt-Lev.
Gauss to point histogram	yes up to 6	yes up to 6	yes	yes
Exp. to dwell times	yes up to 6	yes up to 2	yes	yes (PAT)
Exp. to current response	yes up to 6	yes	yes	yes (WCP)
Other curves	yes[5]	yes[6]	yes[7]	yes[8]
5. Single channel analysis/detection method/burst analysis				
Threshold crossing	yes	yes	yes	yes[9]
Automated analysis	yes	yes	yes	yes
Idealized trace	yes	yes	yes	yes
T-crit adjustable		yes	yes	yes
Burst length	yes	yes	yes	yes
Closings per burst	yes	yes (openings)	yes	yes
Burst count	yes	yes	yes	yes
6. Power spectrum/noise analysis				
Available	no	no[10]	yes	yes

[1]Partial; [2]All aspects except curve-fitting possible; [3]Program Mac TAC reads pCLAMP (ABF) and PULSE files; [4]Marquardt-Levenberg, LSQ, Simplex (SSQ/max likelihood), Chebychev; [5]Boltzman, power, linear regression, exponential with sloping baseline; [6]Geometrics for burst analysis; [7]Polynomial up to 9th degree, double exp., Hodgkin & Huxley, Boltzman, Linear (IV)GHK, Activation, high voltage block, low-voltage block; [8]Hodgkin & Huxley; [9]Uses this method only to compute an amplitude histogram; [10]Spectral analysis functions in Spike2 and Waterfall (window definable).

Table A5. *Comparison of 4 commercially available electrophysiology programs.*
Data handling characteristics

	pCLAMP	PATCH + VCLAMP	PULSE + PULSEFIT	STRATHCLYDE SUITE
DATA FILE INFO				
Session logging/ protocol file	user actions recorded in log file with time stamping	short table in data file	yes, incl. text editor	time stamped session log program WCP
Automatic filename implementation	yes	yes	no	no
Notepad in data file	yes	yes	yes	yes
DATA IMPORT / EXPORT				
1. Data file format				
Proprietary file format	yes[1]	yes[2]	yes	yes
Documentation	yes	CFS User guide	Pulse User guide	yes
Dedicated user software	yes[3]	yes[4]	yes[5]	no[6]
Export formats	ABF, ASCII	ASCII, CFS	ASCII, IGOR text/binary, spreadsheet, PICT	ABF, ASCII, CFS
Import formats from	ABF, ASCII	ABF, CFS	Atari	ABF, ASCII, CFS
2. Handling of results of analysis				
Idealised trace editable	yes	yes	yes	yes
Histogram data	yes	yes	yes	yes
IV data	yes	yes	yes	yes
Other data		yes[8]	yes[7]	
3. Available XY Plots				
Choice of X	yes[9]	yes[10]	yes[11]	yes[12]
Choice of Y	yes[9]	yes[10]	yes[13]	yes[12]
4. Graphics				
Export file format	HPGL	HPGL	PICT File max32kB	HPGL
Screen dump	to printer	to printer	no[14]	no

[1]ABF (Axon Binary File), see text
[2]CFS (CED Filing system), see text
[3]Source in QBasic, C and PASCAL included
[4]Libraries for MS-DOS languages as well as WINDOWS DLL
[5]Into IGOR, ASCII spreadsheet, sample code for user programmable exports
[6]Program source code available
[7]Any parameter (experimental or computed) can be plotted vs most others and fitted
[8]Raw ADC, curve fit param.
[9]Results of analyses, ADC channels, data files
[10]I and V differences, index, time, peak/trough value, user defined
[11]Voltage, duration, real time, index, user defined
[12]Peak, average, rise time, rate of rise, variance, inter-record interval
[13]Extreme, min, max, mean, var, charge, Cslow, Gseries, user defined
[14]Requires utility program

Chapter 10

Patch clamp recording from cells in sliced tissues

ALASDAIR J. GIBB and FRANCES A. EDWARDS

1. Introduction

Before the introduction of gigohm seal patch clamp techniques (Hamill *et al.* 1981) the best resolution that could be achieved when measuring ionic currents in cell membranes was of the order of 100 pA. The combination of the gigohm seal and the placement of the current-voltage converter clamping amplifier in the headstage gave a dramatic improvement in signal resolution allowing currents of around a pA or less to be resolved (Neher, 1992; Sakmann, 1992). This great improvement in technique and instrumentation was followed by a tremendous expansion in the variety of cell types accessible to electrophysiological investigation. This was particularly true in relation to experiments on small cells (e.g. adrenal chromaffin cells: Fenwick *et al.* 1983; and central neurones in tissue culture: Nowak *et al.* 1984; Cull-Candy & Ogden, 1985; Bormann *et al.* 1987) and should also have been true for neurones in brain slices. However, the problem of obtaining a clean access for the patch electrode to the surface of neurones in brain slices seemed insoluble and so the potential advantages of applying patch clamping to brain slices were not immediately achieved.

Signal resolution in patch clamp recordings

The signal resolution achieved in any particular experiment depends on several related factors but the basic point in this relates to the noise inherent in any resistor (e.g. Neher, 1992; see also Chapters 4 and 16). The rms thermal noise in a simple resistor is given by $\sigma_n = (4kT\Delta f/R)^{0.5}$ where σ_n is the rms of the current through the resistor, k is Boltzmann's constant, T the absolute temperature, Δf is the bandwidth and R is the resistance. Basically, what this equation says is that the current noise in the recording is inversely proportional to the source resistance. In order to measure currents of the order of a pA at a bandwidth of 1 kHz, the resistance of the source needs to be of the order of 2 GΩ (2×10^9 Ω). The 'source resistance' in this case is mainly the

A. J. GIBB, Department of Pharmacology, University College London, Gower St, London WC1E 6BT, UK
F. A. EDWARDS, Department of Pharmacology, University of Sydney, NSW 2006, Australia

combination of the feedback resistor in the amplifier, the seal resistance, and the resistance of the preparation itself (cell or isolated patch). In addition, noise associated with the electrode and cell or patch capacitance add to the total (see Rae & Levis, 1992 for a discussion of noise sources). The noise contributions from each source add together as the square root of the sum of the squared rms noise for each component.

Some background to patching cells in slices

In principle the small cell size and small membrane currents of cells in the central nervous system make them ideal for patch clamp recording. However, in the years after patch clamping was first described, cells in intact central nervous system tissue were thought to be inaccessible for patch clamping because there was no apparent way to maintain the intact structure of brain tissue and yet achieve the clean cell membrane that is essential for forming a gigohm seal between glass pipette and cell (Hamill *et al.* 1981). In the meantime neurones and glia were studied in primary culture. Unfortunately, primary cultures have the disadvantage that changes in gene expression and synaptic connections, and even the identity of cells all become unknown factors. An alternative was to study neurones and glia after acute dissociation using enzymes such as papain or trypsin (Numann & Wong, 1984; Gray & Johnston, 1986; Kay & Wong, 1986; Barres *et al.* 1990). Although these preparations allowed the study of receptors and ion channels on adult central neurones (e.g. Kay & Wong, 1987; Huguenard *et al.* 1988; Sah *et al.* 1988; Gibb & Colquhoun, 1992), the possibility remained that the receptors or ion channels of interest could be altered by the enzymes used and synaptic transmission, of course, could not be studied.

Meanwhile, during the 1980s the techniques for recording from brain slices with intracellular electrodes were greatly improving, aided by the development of the discontinuous single electrode voltage clamp (Finkel & Redman, 1985). Several studies were published where simultaneous recordings from pairs of synaptically connected neurones in hippocampal slices were achieved (e.g. Knowles & Schwartzkroin, 1981; Miles & Wong, 1984; Scharfman *et al.* 1990; Sayer *et al.* 1990; Mason *et al.* 1991). However, these recordings were still limited by the lower resolution of intracellular microelectrode recording.

The problems of cell isolation and signal resolution were overcome by the introduction of techniques to allow patch clamp recordings to be made from brain slices (Edwards *et al.* 1989, adapted by Blanton *et al.* 1989) and even from the *in vivo* brain (Xing Pei *et al.* 1991; Ferster & Jagadeesh, 1992). The tremendous significance of this advance is seen by the fact that these techniques have rapidly been applied to many fundamental questions in neurobiology such as the mechanism of synaptic transmission at inhibitory (Edwards *et al.* 1990; Takahashi, 1992) and excitatory synapses (Hestrin *et al.* 1990; Silver *et al.* 1992; Stern *et al.* 1992, Hestrin, 1992a), as well as allowing single channel recording of the properties of receptors and ion channels in identified neurones from the brain (e.g. NMDA receptors, Gibb & Colquhoun, 1991; GABA$_A$ receptors, Edwards *et al.* 1990; glycine receptors, Takahashi & Momiyama, 1991).

2. Preparation of brain slices for patch clamp recording

Brain slices have been widely used for both biochemical and electrophysiological studies (for review see Alger *et al.* 1984), and the methods used to prepare brain slices are widely documented (e.g. Langmoen & Andersen, 1981; Cuello & Carson, 1983; Alger *et al.* 1984; Madison, 1991). Although different labs have their own individual variants, below is a brief description of the procedures that we use, which seem to give good results with a variety of brain areas (see also Edwards *et al.* 1989; Konnerth, 1990; Edwards & Konnerth, 1992).

(i) The animal is decapitated and the brain removed and placed in ice-cold physiological solution within 60 seconds of decapitation (the solution should be so cold that it contains a few ice crystals and to maintain this temperature the container should be sitting on ice).

(ii) Pause for 3-5 minutes while the tissue cools down.

(iii) Trim or block off the tissue using clean cuts with a sharp scalpel blade in preparation for gluing to the stage of the tissue slicer. Avoid squeezing or otherwise deforming the tissue at this stage.

(iv) Apply a <u>thin</u> layer of cyanoacrylate glue (Super-glue[R]) to the stage of the tissue slicer and then gently place the tissue at the correct orientation onto the glue (the most common orientation is to have the region of interest near to the blade, or at least try to avoid cutting through white matter before reaching the area to be sliced). Immediately, pour ice-cold solution over the tissue until it is submerged.

(v) Cut slices (100-300 μm thick) with vibrating slicer. Fine dissecting scissors or two hypodermic needles can be used to dissect out an area of brain from each whole brain slice.

(vi) Using a Pasteur pipette cut and fire-polished to an opening of 3-5 mm across, transfer each slice as it is produced to the holding chamber which should be in a water bath at 32-35°C with a good steady flow of O_2/CO_2 bubbling through the solution.

(vii) Incubate the slices at 32-35°C for at least 30 minutes before beginning recordings.

Equipment check-list

About 250 ml of ice-cold 'slicing Krebs' sitting on crushed ice

Large scissors for decapitation

Small scissors to cut open the skull

Curved, blunt forceps to remove top of skull

No. 11 scalpel to hemisect the brain

Small spatula to remove brain halves from skull

Large weighing boat or similar shallow container of ice-cold Krebs sitting on crushed ice to cool the brain halves

Cyanoacrylate glue

Large spatula to lift blocked-off piece of brain onto tissue block

Two fine hypodermic needles or fine dissecting scissors for dissecting small regions from the brain slice

Broken and fire-polished pasteur pipette (opening 3-5 mm)

3. Notes on making slices

The time factor

It is critical that the time from decapitation till immersion of the brain in cold solution is kept short (<1 min). Partly because of this, it seems easier to make healthy slices from younger animals (e.g. less than 3 weeks) where the skull is soft and can be removed more rapidly. In addition, the smaller brain of younger animals will cool more rapidly than a larger adult brain and may be more resistant to anoxia. To improve cooling some people remove the skull with the whole head submerged in ice-cold Krebs. Bubbling the Krebs during the cooling period may also improve cooling.

The whole process of making slices should preferably not take more that about 30 minutes. However, we have observed that tissue kept ice-cold for half an hour (e.g. the second half of the brain when making hippocampal slices) can still be glued to the slicer and good slices prepared from it. This can be useful if two people wish to slice different parts of the brain or as a backup if something goes wrong during slicing such as the tissue block coming off the slicer stage during slicing (this may happen occasionally although less often with practice: perhaps the block was moist before the glue was applied, or too thick a layer of glue was used, or the glue itself was too thick in consistency, or the surface of the tissue block is not flat).

Tissue slicers

The tissue slicer should be able to vibrate at sufficient frequency (around 10 Hz) and with a long enough stroke (1-2 mm) to cut cleanly through the tissue. Most commercially available slicers will do this when set at their maximum settings. Care must be taken that there is an absolute minimum of play or vibration in the mechanism driving the cutting blade. Use of a rotating blade for cutting slices, rather than an oscillating blade, has recently been described and a rotating blade slicer is now marketed by Dosaka (Model DTY 8700). However, we have no personal experience of rotating blade slicers as compared to oscillating blade slicers.

Depending on the brain area being sliced, it may be useful to view the slicing using a low-magnification dissecting microscope or large magnifying glass (some slicers come fitted with a magnifying glass). It is always useful to have a good light source available to illuminate the tissue block (e.g. using fibre optic light guides).

The simplest slicers have a manual movement of the tissue block towards the oscillating blade (e.g. Cambden Vibroslice, UK) and in our experience these work very well for a variety of different brain areas. Some slicers have a Peltier-cooled stage to maintain the tissue close to 0°C during slicing. However, it is perfectly adequate to have frozen Krebs in the bottom of the slicing chamber (or make Krebs ice-cubes) to ensure the tissue stays cool during slicing.

Some slicers have a motor drive to advance the blade or tissue during cutting (e.g. Dosaka 1500E, Japan; Camden Vibroslice, UK; FTB Vibracut, Germany; Technical Products Inc. Vibratome 1000, USA). An annoying feature is that some slicers automatically reverse at the end of the cut, when what is often needed is to stop the blade in that position until a piece of tissue of interest is dissected free from the whole brain slice. The Vibracut has a useful innovation in that the tissue bath mounts on a magnet allowing it to be rotated to any angle, which avoids the difficulty of placing the tissue on the glue at exactly the right angle.

There are quite a variety of tissue slicers available with the tissue block inside a tissue bath and so suitable for cutting living slices. These slicers vary in sophistication and price. However we find that a simple slicer such as the Camden Vibroslice works very well. If a more sophisticated slicer is preferred, we recommend the vibrating Dosaka slicer or the Vibracut.

Slicer blades

The blades used for slicing should be as sharp as possible. High-carbon steel blades are preferable to stainless-steel (a high-carbon steel is magnetic and brittle and will break with a sharp snap). Stainless-steel razor blades are probably not as sharp.

Slicing different brain areas

Different brain areas are more or less difficult to slice and in the original description of the technique a large variety of different brain regions were successfully recorded from (Edwards *et al.* 1989). As well as taking the age of the animal into account, a high degree of myelination and vascularization of a particular area tends to make slicing more difficult (areas like this seem to require a particularly slow forward speed during cutting). The spinal cord and brain stem are regarded as difficult to slice, particularly in older animals, but in the last few years successful patch clamp experiments have been made with both young (e.g. Takahashi, 1992) and adult spinal cord (Yoshimura & Nishi, 1993) and with several parts of the brain stem (e.g. Forsythe & Barnes-Davies, 1993; Kobayashi & Takahashi, 1993).

Getting the right angle

Many neurones have their dendritic tree angled in a particular orientation or plane. A big improvement in cell survival is obtained if care is taken to cut the slices at an angle that will preserve the dendrites (e.g. transverse slices of hippocampus to maintain the pyramidal cell apical dendrites). Of course, the angle of slicing could also be important in maintaining synaptic connections. Obviously, if a particular input is to be stimulated then the angle of slicing must be arranged to avoid cutting the incoming axons. Alternatively, it may be necessary to stimulate locally (e.g. with a patch electrode pushed into the slice) if the desire is to stimulate a local interneurone.

In general it is harder to obtain healthy, large neurones in slices compared to obtaining healthy small neurones, probably because of the problem of neuronal death if some of the dendrites are cut, but perhaps also due to differences in resistance to anoxia between different cell types. Thus, in hippocampal slices, even when CA1 and CA3 cells look poor, it is often possible to see healthy granule cells. Likewise in cerebellar slices it is more difficult to obtain healthy Purkinje cells than healthy granule cells.

Slicing solutions

Slices are made with a standard extracellular Krebs solution. For example we use (in mM) NaCl 125, KCl 2.5, $CaCl_2$ 2, $NaHCO_3$ 26, NaH_2PO_4 1.25, $MgCl_2$ 1, Glucose 25, of pH 7.4 when bubbled with 95% O_2 and 5% CO_2. Although the exact composition of the Krebs solution varies between laboratories, particularly in the concentrations of Ca^{2+}, $NaHCO_3$ and glucose, it is generally considered important that the K^+ concentration is less than 3 mM (to avoid epileptiform activity in the slice) and that a high glucose concentration is used.

Efforts to improve cell survival during slicing include the substitution of sucrose for 50% of the NaCl in the Krebs (Aghajanian & Rasmussen, 1989), inclusion of Hepes buffer as well as HCO_3^- buffer in the Krebs, raised extracellular Mg^{2+}, use of NMDA channel blockers and excitatory amino-acid antagonists.

Slice incubation chamber

A good incubation chamber must be able to provide a good circulation of freshly oxygenated solution since the slices must be kept in good condition in the chamber for the whole of the experimental day (10-12 hours). It must be stable not only to prevent mechanical disturbance of the slices but also so that slices can be placed in or removed from the chamber without disturbing any of the other slices. Preferably it should be simple to make and clean. Different laboratories use different types of incubation chamber. Here we describe a simple construction illustrated in Fig. 1 which we find easy to make and use.

This incubation chamber uses a standard 100 ml beaker. It contains a piece of light cotton clamped across two rings made for example using the base and lid of a 35 mm Petri dish which have had the top and bottom broken out (Falcon dishes seem to work best). This makes a tight net of cotton on which the slices will rest. The cotton clamp is then wedged halfway down the beaker using a piece of stiff plastic tube about 3-4 cm long. This plastic tube should reach from almost the bottom of the beaker to about 5 mm <u>below</u> the surface of the Krebs. A gas bubbler is inserted into the tube to near the bottom and generates a stream of bubbles which by rising to the top of the tube draws the Krebs from the bottom of the beaker, so generating a circulation of Krebs which will act to hold the slices down on the net. The incubation chamber is placed in a heated water bath (a large water tank, 5-10 litres, heated with a standard aquarium heater is sufficient) and covered (e.g. with a Petri dish lid) to prevent evaporation.

The incubation chamber can be dismantled every night and reassembled next day with a new piece of cotton (standard white muslin is cheap enough that a meter of material bought at the local drapers shop will last for years!). However, if the chamber is rinsed with distilled water and then left to soak overnight in distilled water acidified with a few drops of HCl, then the same chamber can be used for several days at a time.

Immobilizing the slice in the recording chamber

It is necessary to immobilize the slice during recording so that no movements occur as a result of the solution flowing through the bath. Typical flow rates would be

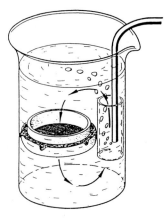

Fig. 1. A simple incubation chamber for maintaining brain slices. The chamber is made using a 100 ml beaker containing a cotton support which allows the slices to be held in a gentle circulation of oxygenated Krebs solution. The cotton support is made from standard cotton muslin stretched across two rings made from the top and bottom of a 35 mm Petri dish. This is then wedged halfway down the beaker using a stiff plastic tube which extends well below and a little above the cotton support. The tube should reach from near the bottom of the beaker to about 5 mm below the surface of the Krebs. When a bubbler is placed near the bottom of the tube the bubbles rise up the tube drawing solution with them and generating a current which flows down over the slices. (From Edwards & Konnerth, 1992).

between 1 ml and 3 ml per minute for bath volumes of less than 1 ml and stability is improved if the inflow and outflow are in separate chambers connected to the recording chamber by small, submerged passages. This has the disadvantage of tending to slow solution exchange around the slice so a good compromise is to have the outflow in a separate chamber and place the inflow on a ramp running directly into the recording chamber.

Several methods have been described for immobilizing the slice including the use of fibrin clots (Takahashi, 1978; Blanton *et al.* 1989) and pieces of netting. However, many people find that a grid (described in Edwards *et al.* 1989) made of flattened platinum wire with single nylon strands glued across it with cyanoacrylate glue works well for holding the slice firmly on the bottom of the recording chamber.

4. Visualizing cells in living brain slices

How healthy is the slice?

In the past the health of the slice was generally determined from physiological parameters such as resting membrane potential of impaled cells, size of the action potential, population spike etc. Comparison of these parameters measured *in vitro* with the same parameters measured *in vivo* suggests that healthy sliced brain tissue behaves in a remarkably similar way to the *in vivo* state. However, in the past only a few studies used high-resolution optics to allow cells in slices to be visualized

(Yamamoto, 1975; Takahashi, 1978; Llinas & Sugimori, 1980) so that a direct visual assessment could be made of the health of the sliced tissue.

One great advantage of visualizing the cells in slices is that it allows an immediate assessment of the health of the slice to be made and cells suitable for recording to be carefully picked. The first examination of a healthy slice under the microscope is a fantastic sight! In a hippocampal slice for example, lots of bright shiny cells should be visible with a variety of cell dendrites and different cell morphologies present. Compared to blindly inserting the electrode into the slice, a great deal of time can be saved by first picking out the good cells particularly, for example, if the cell of interest is a relatively rare interneurone. If the slice does not contain many bright cells, but instead is uniformly dark with many round opaque cells with clearly visible nuclei evident, then the slice should be discarded (see also Edwards & Konnerth, 1992 for a discussion of visually assessing the slice).

Labelling and identifying cells

It was partly as a result of the desire to record from identified cells (Takahashi, 1978) that the techniques for patch clamping visually identified cells in brain slices were developed (Edwards *et al.* 1989). Retrograde transport of fluorescent dyes or fluorescent beads (e.g. Takahashi, 1978; Katz *et al.* 1984; Gibb & Walmsley, 1987) has been used to allow subsequent identification of living cells in slices or following dissociation. However, these procedures can only be used where cells have a definite projection (e.g. motor neurones) and require expensive fluorescent optics on the microscope. Instead, cell bodies and parts of the dendritic tree can be easily observed using differential interference contrast (DIC) Nomarski or Hoffmann modulation optics. The identification of the cell to be studied then depends on the use of information about the local anatomy, and the size and morphology of the cells of interest. For most purposes this is sufficient to identify a cell clearly.

Microscope requirements

The particular brand of microscope used is not critical. Zeiss, Olympus and Nikon all make upright microscopes which can be used for visualizing cells in slices (Micro-Instruments in Oxford make a good customized microscope fitted with Nikon optics). Ideally the microscope should have a fixed stage so that focusing occurs without moving the preparation relative to the patch electrode. Although the standard Olympus BHS is not a fixed-stage microscope, it can easily be converted and Olympus will now do this conversion if requested. This is a cheap and satisfactory option. It is also important that the microscope is not mounted on rubber feet but instead is firmly fixed to the vibration isolation table or the electrode will crash into the slice every time the focus is adjusted! Use of a high numerical aperture (e.g. 0.75) ×40 water immersion objective on a standard upright microscope preferably fitted with Nomarski optics allows visualization of neurones and their dendrites with a resolution of about 1-2 μm, if the cell lies within 20 μm of the surface of the slice (looking from above). For Nomarski optics to be effective, however, the maximum slice thickness is around 300 μm. The thicker the slice, the more the light is scattered passing through the slice and the dimmer and lower the resolution of the image. On the other hand, it is more difficult to obtain healthy thin slices: slices 200-300 μm are usually a good compromise.

The choice of objective is a compromise between the need for high resolution (high numerical aperture) and the need for a reasonable working distance (at least 1.5 mm) to allow access to the surface of the slice with a normal patch electrode. The Zeiss ×40 achromat (numerical aperture 0.75, working distance 1.9 mm) fitted to the Zeiss Axioskop is a good example. In most countries the Zeiss Axioskop is considerably more expensive than the Olympus BHS fitted with the newly released Olympus 40× water immersion objective (NA 0.7, WD 3 mm). Unlike the Axioskop, the Olympus does not have infinity-corrected optics and so the new Zeiss and Olympus objectives are not interchangeable. There is also a Nikon 40× water immersion objective (NA 0.55, WD 2 mm) but, although this is a little cheaper, the image resolution seems to be not as good presumably because of the lower numerical aperture. The Zeiss Axioskop microscope gives excellent image quality. This may be partly because infinity-corrected optics are superior to standard optics (at least in principle) but could also be the result of a very stable condenser and Nomarski system combined with a very good light source. For patch clamping very small cells it may be an advantage to fit the microscope either with an octovar giving variable intermediate (1.0×, 1.25× and 1.6×) magnification or 16× eyepieces to give an overall magnification of more than 600×.

In principle, it might be expected that, when the water-immersion objective is in contact with the bath solution, a ground loop will occur because the objective will also be in electrical continuity with the rest of the microscope which is usually earthed. In practice we know of varying experiences on this where some objectives did, and some did not need insulating from the microscope, perhaps because some objectives are coated, which effectively insulates them anyway. If necessary, a solution is to manufacture an insulating collar to insert between objective and nose-piece.

It should be noted that for best results the numerical aperture of the condenser lens should always be as high or higher (0.9 for example) than that of the objective. This generally means that the working distance of the condenser will allow only a thin glass cover slip or glass base for the recording chamber, if the light from the condenser is to be focused properly on the surface of the slice (plastic chambers, although fine for phase contrast optics, destroy Nomarski imaging).

Whatever the precise optical arrangement it is essential for best results that good microscopic practice is followed (see e.g. Bradbury, 1989). In particular, good Köhler illumination must be set up with the condenser adjusted to focus the light source diaphragm exactly in the plane of the cells of interest. For DIC optics, the polarizers should be 90° to each other and the analyzer adjusted for optimum image quality. Secondary diaphragms in the condenser are then used to cut down the light entering the tissue and so improve the sharpness of the image.

It is often less tiring to view the image using a CCD camera in combination with a standard monitor. These are relatively cheap (e.g. from Radio Spares) and the smaller cameras (<300 g) are light enough to mount directly on top of the microscope trinocular head without applying too much weight to the microscope focusing mechanism. It is generally important to ensure that the camera is insulated from the microscope to avoid conducting interference into the patch clamp signal. When using a CCD camera, a better image may be achieved if the secondary diaphragms on the condenser are left open and the gain and contrast of the camera controller used to optimize the image, although this will tend to make the image down the eye pieces look very washed-out (Levis & Rae, 1992).

Dodt & Zieglgänsberger (1990) have described the use of Nomarski optics in combination with an infra-red filter placed in the normal light path to give infra-red DIC imaging of cells in brain slices. The infra-red image is then visualized with an infra-red-sensitive CCD camera (specialist infra-red CCD cameras are expensive but even an ordinary CCD camera is quite sensitive to infra-red light up to about 1000 nm wavelength) and the image can then be stored on video tape or on computer using a frame grabber. Analogue or digital image enhancement techniques can then be applied to the image. The improved resolution achieved with infra-red DIC may be partly due to reduced scattering of infra-red light during transmission through the slice. Although the infra-red imaging may add considerably to the cost of the microscope, it allows imaging much deeper in the slice and may be useful for specialist applications such as patching directly on to dendrites in slices (Stuart *et al.* 1993).

5. Recording from cells in slices

Electrodes

The electrodes used for patching cells in slices are fabricated in the normal way. Thick-walled glass and coating with Sylgard[R] are useful in minimizing the noise associated with the fact that the electrode is immersed quite deep in solution under the objective. For clamping large or fast currents where it is important to minimize the series resistance, it may be better to use thin-walled glass. The choice of glass can make a big difference with a thick-walled Aluminosilicate glass (e.g. Clark Electromedical SM150F 7.5) having a much lower noise than a thin-walled borosilicate glass (e.g. Clark Electromedical GC150TF 7.5). Rae & Levis (1992) discuss in detail a wide range of glass types for patch clamping.

Selecting a healthy cell

Selecting the best cell for patch clamping requires experience of the particular brain slice in use under the conditions presented by the way the microscope is adjusted. A good guide, however, is that the cells should be smooth with a clear outline and have a 'soft' appearance (see also Edwards & Konnerth, 1992). Cells that appear very shiny or 'hard' in appearance do not make seals easily and if observed for some time, appear to die gradually. Dead cells are opaque with visible nuclei and are often swollen and round. Although gigohm seals and single channel currents can be observed if a seal is made on a dead cell, these cells have no resting membrane potential and the recording is invariably lost on attempting to form the whole-cell configuration.

Patching cells under visual control

Cells are cleaned using a blunt patch pipette (tip diameter 3-10 μm according to the size of cell being cleaned). This pipette is inserted into the bath without filling and will fill with some Krebs by capillary action. The cleaning pipette is brought up close to the surface of the slice and then positive pressure used to produce a gentle stream of solution which will break up the surface of the slice over the cell of interest. After a few seconds of positive pressure, light suction can be used to remove the debris overlying the cell. The cleaning pipette is then discarded and a recording electrode filled and placed in the bath near the chosen cell. Finding the electrode under a water-immersion objective is often difficult at first: one method is to wind up the objective so that it is focused far above the slice, but still in solution. The electrode tip is then placed under the objective and moving from side to side, the point where the electrode cuts the light beam from the condenser is found. Looking down the objective, the electrode can then be found with only a small sideways movement of the electrode and then the electrode can be lowered vertically in full view until it is just above the slice.

Obtaining good seals on cells in slices involves essentially standard patch clamping procedures. All types of patch clamp configuration have been used in slices. The method is straight-forward. Positive pressure is applied to the back of the electrode. Then under visual control, the electrode is advanced until the tip is just in contact with the cell surface, detectable by a small increase in the electrode resistance. At this point the stream of solution from the electrode tip should be evident producing a dimple on the surface of the cell. On removal of the positive pressure the electrode resistance should increase a little more and then a small amount of suction applied to the back of the electrode should begin the sealing process. At this point it is often best to maintain gentle suction and wait. A gigohm seal may form immediately or may take some minutes. Sealing in slices seems to be generally much slower than with cultured or dissociated cells. Often, sealing is more successful if a negative voltage (around −50 mV) is applied to the electrode and this has the advantage that if the patch breaks through on seal formation then the cell membrane potential will be clamped near the resting potential (our usual procedure is to leave the pipette potential at 0 mV until a resistance of about 50-100 MΩ is achieved and then to depolarize the membrane by setting the holding potential to −50 mV).

Stimulating axons or cells in slices

Perhaps the single biggest advantage of being able to make patch clamp recordings from brain slices is that it allows synaptic transmission to be studied at synapses which are probably identical to those functioning *in vivo*. By using the patch clamp whole-cell recording configuration, the resolution of synaptic currents is increased by one to two orders of magnitude compared to what was possible with conventional intracellular recordings. In any whole-cell recording it is often possible to observe spontaneously occurring synaptic currents; most commonly $GABA_A$ receptor-mediated. The spontaneously occurring currents are due to the on-going firing of cells in the slice and are largely abolished by the application of tetrodotoxin to the slice, leaving only spontaneous miniature synaptic currents ('minis'). Since the origin of the spontaneous synaptic activity is generally unknown it is often preferable to stimulate a presynaptic cell or axon directly (either a single cell or axon, or a whole bundle of axons).

In order to try to stimulate a single presynaptic cell or axon a stimulating electrode (usually a standard patch electrode filled with Krebs) is pushed gently into the slice near the cell being recorded from and then short rectangular voltage pulses (e.g. 5-50 V in amplitude, 50-500 µs in duration) are applied to the electrode and the trace observed to see if a synaptic current occurs immediately after the stimulus artifact. The stimulating electrode is moved slowly through the slice until a connection is found or, if no currents are evoked, a different place in the slice is selected and the process repeated. Generally this procedure is more successful than the much harder method of obtaining a whole-cell recording on both pre- and post-synaptic neurones, although this is probably the only way to be sure that a single cell is producing the recorded input (and even a single cell may make multiple synapses on the post-synaptic neurone). An alternative strategy is to place the stimulating electrode on the surface of a nearby cell. If the presynaptic neurone also has a synaptic connection with this cell then it is possible to generate antidromic action potentials from there which will produce synaptic currents at the cell under the recording electrode. This strategy worked well with both inhibitory currents in hippocampal granule cells (Edwards *et al.* 1990) and with excitatory currents in visual cortex interneurones (Stern *et al.* 1992).

The alternative method of stimulation is to use a bipolar platinum stimulating electrode to stimulate a whole region of the slice or a whole fibre tract at once. This method may be much more convenient and reliable in some situations such as stimulating the Schaffer collaterals in a hippocampal slice. The advantages are that the electrodes probably don't need to be moved after placement on the slice and the stimulus artifact can be much smaller than when using a local electrode. The disadvantage is that the stimulus is not localized in any way and so many axons may contribute to the measured synaptic response, even in situations where the stimulus voltage has been adjusted to provide the smallest reliable input.

A problem can arise when stimulating if the stimulus artifact feeds into the clamp command voltage producing a voltage jump in the cell, but this can usually be avoided if the stimulus reference is connected to the common earth of the setup.

Blind patching

This method was originally developed in experiments on turtle cortex slices (Blanton *et al.* 1989). In marked contrast to the method described above, no attempt is made to visualize individual cells. Instead the electrode is slowly advanced ('blindly') into the slice while positive pressure is applied to the back of the electrode. The first contact of the electrode with the slice produces a change in resistance which is ignored and the electrode is advanced further into the slice. A cell is detected by an increase in electrode resistance and then the positive pressure is removed, suction applied to the electrode and a seal attempted in the usual way. The amazing thing about this procedure is that it works very well! It is now widely used, particularly in hippocampal slices where it is viewed as easier than standard intracellular recording with sharp microelectrodes. Stuart *et al.* (1993) used a very similar procedure to the blind patching method (except the patch electrode was moved under visual control) to make recordings from the dendrites of cells in slices.

The advantages of the blind patching method are (a) it requires no sophisticated microscopy (a dissecting microscope is used to guide the electrode onto the right area of the slice) and so it is cheaper to set up, (b) no cleaning of the slice is necessary before recording, so saving time, (c) the cell is not disrupted by the cleaning procedure which could conceivably damage synapses on the cell body and (d) cells deep in the slice can be recorded from. The disadvantage of this method of course is that the actual cell is not visually identified. Blind patching may also tend to give lower seal resistances and less stable series resistance so that the quality of the final recording may be lower.

6. Applications of patch clamp recording from cells in slices

Clearly, the key to the success of this technique is that it brings all the power of patch clamp procedures to bear on the study of almost intact nervous system. The main advantages over previous approaches are increased signal resolution, absence of enzyme treatment, manipulation of the intracellular environment and the ability to record from identified cells which are essentially in their *in vivo* state.

The increased signal resolution obtainable results from the fact that, when using the whole-cell recording configuration, the resistance and capacitance of neurones are often high enough and low enough respectively to allow resolution of currents as small as a few pA at a bandwidth of 1 kHz. The electrode resistance (i.e. the series resistance in whole-cell recordings) and capacitance are the key places where the signal-to-noise ratio and recording bandwidth can be degraded. The electrode capacitance is often high because the electrode may be immersed several millimetres in the solution in order to reach under a water-immersion objective, so use of thick-walled glass and careful coating of the electrode with a hydrophobic coat is useful. In addition, blocking unwanted synaptic activity and other conductances in the cell either with drugs or by ion substitution can greatly reduce the background noise in a recording. When recording synaptic currents it may be an advantage to use a patch clamp fitted with a capacitor feedback headstage since this headstage design allows larger transient currents to be recorded at high resolution than the traditional patch clamp with a resistor feedback in the headstage (see Levis & Rae, 1992 for discussion of the capacitor feedback headstage).

Using the whole-cell configuration, neurones can be rapidly filled with fluorescent dyes and their structure examined in living slices (Edwards et al. 1989). Access to the intracellular environment can be used to study second messenger pathways involved in signal transduction following synaptic transmission. Cells can be filled with Ca^{2+}-sensitive dyes and localized changes in intracellular Ca^{2+} concentration measured in response to synaptic transmission (Konnerth et al. 1992; Alford et al. 1993). It may also be possible to fill cells with 'caged' Ca^{2+} and use localized light flashes to produce discrete and localized changes in intracellular Ca^{2+}. Similar experiments are possible using other caged compounds such as caged ATP and caged IP_3 (see Chapter 15). In addition, patch clamping in slices is now being used to look at the properties of ion channels and receptors on identified cells at defined developmental stages of the nervous system (e.g. NMDA receptors: Hestrin, 1992b; Farrant et al. 1993).

In the future it is likely that more detailed and precise investigations of the mechanism of synaptic transmission will be made in which patch clamp techniques will be applied to recording from both pre- and post-synaptic neurones. For example, perfusing different fluorescent dyes into the pre- and post-synaptic neurones and using confocal laser-scanning microscopy (Edwards et al. 1989) may mean that the anatomy and function of synaptic connections can be studied together in a single experiment.

It is way beyond the scope of this article to mention even a small fraction of the present applications of patch clamp techniques to brain slice experiments and so here we will try to highlight a few situations where patch clamping in slices has been particularly useful.

Studies of synaptic transmission in slices

The advantage of studying synapses, receptors and ion channels in sliced brain tissue has been exploited in a variety of brain slice preparations but particularly, in the hippocampus (e.g. Edwards et al. 1990; Keller et al. 1991; Colquhoun et al. 1992), cerebellum (Konnerth, 1990; Farrant & Cull-Candy, 1991; Silver et al. 1992) and visual cortex (Stern et al. 1992; Hestrin, 1992a).

The properties of $GABA_A$ receptor-mediated inhibitory postsynaptic currents (IPSCs) were studied in hippocampal granule cells in order to investigate in detail the mechanism of inhibitory synaptic transmission (Edwards et al. 1990). The choice of this particular cell type was important in maximizing the resolution of the recordings. These cells have a small soma with long fine dendrites. The cell capacitance is therefore relatively small allowing wide-bandwidth voltage clamping and the cell input resistance is high (>1 $G\Omega$) giving good signal resolution when care is taken to block as many of the cell ionic conductances as possible. Inhibitory synaptic connections are on, or close to the cell soma and so there is a good space-clamp of the synaptic currents.

Figure 2A shows examples of miniature and stimulus-evoked IPSCs recorded from a hippocampal granule cell. The inhibitory synaptic currents were found to have fast risetimes (0.5 ms) and were of relatively small amplitude: miniature currents were around 10 pA (equivalent to the activation of only about 10 GABA$_A$ receptor channels at the peak) and evoked currents were 10-100 pA. Distributions of the amplitude of miniature and evoked currents could be described by the sum of several Gaussian components (Fig. 2B) with the peak of each component being a multiple of the predominant component in the miniature IPSC amplitude distribution. These results are not consistent with the idea that the quantal size (size of miniature currents) is determined by the amount of transmitter in each transmitter packet (as seems to occur at the neuromuscular junction). Instead, the results support the hypothesis of a different mechanism for central synaptic transmission where the quantal size is determined by the number of receptors on the postsynaptic membrane (see Edwards *et al.* 1990 for discussion).

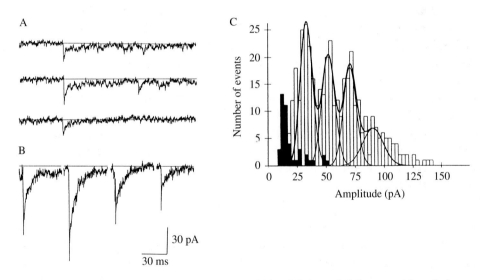

Fig. 2. Properties of miniature and stimulus-evoked IPSCs recorded from granule cells in hippocampal slices. (A) Three traces showing miniature IPSCs recorded in Krebs solution containing 1 μM tetrodotoxin. The traces are displayed with the peak of each current aligned. (B) Examples of four consecutive IPSCs evoked by stimulation of a nearby neurone in the slice. The stimulus was a rectangular voltage pulse of 5 V amplitude and 200 μs duration applied at a rate of 1 Hz. Traces were filtered at 2 kHz (−3 dB Bessel) and sampled at 10 kHz. Experimental details are given in Edwards *et al.* 1990. (C) Distributions of miniature and evoked IPSC amplitudes. The open bins show the distribution of the amplitudes of 428 evoked IPSCs in a cell voltage-clamped at −50 mV. The thick line superimposed on the histogram show the sum of 4 Gaussian functions which were fitted to the bins using a least-squares fitting routine. This was fitted over the amplitude range from 25-115 pA. The bin width is 3 pA. The Gaussian parameters fitted to the data were (mean±SD) 33±4.9 pA, 52±5.3 pA, 71±6.6 pA and 91±6.2 pA. The mean separation between the peaks was therefore 17.4 pA. The filled bins show the amplitude distribution of 43 miniature IPSCs recorded in the same cell in the presence of 0.5 μM TTX. The peak of the miniature IPSC distribution occurs at 13.8 pA. The background noise standard deviation was 2.8 pA. (Adapted from Edwards *et al.* 1990).

Single channel current recordings from cells in slices

A number of different labs now use patch clamping in brain slices as a means of obtaining single channel recordings (e.g. NMDA receptors, Gibb & Colquhoun, 1991; Edmonds & Colquhoun, 1992; Farrant *et al.* 1993; GABA$_A$ receptors, Edwards *et al.* 1990; glycine receptors, Takahashi & Momiyama, 1991). The rationale behind this approach is that the receptors and ion channels are presumably in the same location on the cell and in the same condition as they would be *in vivo*. This is potentially a powerful approach to elucidating the functional role of receptors and ion channels in central neurones since an understanding of the number, distribution and functional properties of receptors and ion channels is critical to understanding the processing of information at the neuronal level. At present, most of this information comes from recordings made from the cell body of neurones. However, it is quite likely that channel distribution and possibly functional properties are regulated according to location in the cell membrane and so techniques for recording from cell processes (Stuart *et al.* 1993) could be particularly useful in the future.

A second important reason for making single channel recordings from slices is that it allows information from *in situ* mRNA hybridization studies or antibody labelling studies to be used to pick cells expressing particular receptors or ion channel proteins. These results will contribute to determining the relationship between structure and function for different receptors.

Figure 3 shows examples of single NMDA receptor channel currents recorded from an outside-out patch excised from a CA1 cell in a hippocampal slice. The NMDA receptor is sensitive to nanomolar concentrations of glutamate and glycine and therefore when an outside-out patch containing NMDA receptors is isolated from a neurone in a slice, spontaneous channel activity is often evident due to background release of glutamate and glycine from the slice. In order to avoid this the patch is moved away from the slice towards the bath inflow and at the same time raised towards the surface of the bath reducing the depth of immersion of the electrode and so reducing the background noise in the recording.

Recordings like those shown in Fig. 3 were used to determine the properties of single NMDA receptor activations (Gibb & Colquhoun, 1991). Information from changes in the distribution of patch closed times (such as shown in Fig. 3B) with changes in glutamate concentration was used to identify which closed times and open times in the data record were likely to occur within a single receptor activation. Each activation was found to be composed of numerous openings and closings so that, although the distribution of open times (Fig. 3C) suggests a mean open time of around 3 ms, the length of each receptor activation will be several tens of ms. The results of these experiments led to the same conclusion as obtained from macroscopic experiments on NMDA receptor-mediated EPSCs and currents in outside-out patches: the time course of the NMDA receptor-mediated synaptic current is determined by the kinetics of the NMDA receptor activation (Hestrin *et al.* 1990; Lester *et al.* 1990).

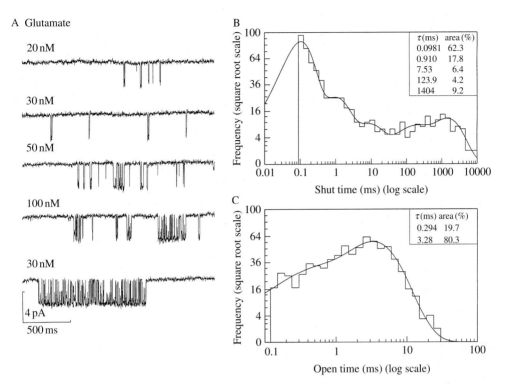

Fig. 3. (A) NMDA receptor channel. Examples of NMDA receptor channel openings recorded from a hippocampal CA1 cell outside-out patch in response to 4 different glutamate concentrations. Each trace shows a 2 second long period of activity. Each glutamate concentration was applied for between 120 and 200 seconds with 60 seconds wash between applications. The mean opening frequency was 3.48 s^{-1} at 20 nM and 19.9 s^{-1} at 100 nM. The lowest trace shows a clear example of the high activity periods which are occasionally observed at all glutamate concentrations. 1 µM glycine was present during all recordings. The patch membrane potential was −60 mV. (B) Distribution of 645 channel closed times recorded from an outside-out patch at a glutamate concentration of 30 nM. The mean measured closed time was 223 ms. The distribution is displayed using a log(t) transformation of the x-axis and a square root transformation of the bin frequencies (Sigworth & Sine, 1987) and fitted using the maximum likelihood method (Colquhoun & Sigworth, 1983; Colquhoun, Chapter 6) with 5 exponential components with time constants and relative amplitudes as shown. The fit predicts a total of 1061 gaps in the distribution with a distribution mean of 135 ms. The resolution for this record was 90 µs for closed times and 120 µs for open times. (C) The distribution of 767 openings from the same recording as analysed in B. The mean measured open time was 2.97 ms. The distribution was fitted with 2 exponential components and predicts that there were 847 openings in the distribution with a mean of 2.69 ms. (Adapted from Gibb & Colquhoun, 1991).

7. Conclusions

Although a relatively new technique, it is already clear that making patch clamp recordings from brain slices is an enormous advance in the study of brain function. The technique has allowed the application of all the patch clamp configurations to cells in brain slices and so many experimental approaches that are new to brain slices

are being used. Brain slices seem to be another major field of investigation that has
been revolutionized by the application of patch clamp techniques. It is fun to
speculate about which field patch clamping will be applied to next. As Fred Sigworth
has commented (Sigworth, 1986), 'The patch clamp is more useful than anyone
expected'!

References

AGHAJANIAN, G. K. & RASMUSSEN, K. (1989). Intracellular studies in the facial nucleus
illustrating a simple new method for obtaining viable motoneurones in adult rat brain slices. *Synapse*
3, 331-338.

ALFORD, S., FRENGUELLI, B. G., SCHOFIELD, G. J. & COLLINGRIDGE, G. L. (1993).
Characterization of Ca^{2+} signals induced in hippocampal CA1 neurones by the synaptic activation of
NMDA receptors. *J. Physiol.* **469**, 693-716.

ALGER, B. E., DHANJAL, S. S., DINGLEDINE, R., GARTHWAITE, J., HENDERSON, G., KING,
G. L., LIPTON, P., NORTH, A., SCHWARTZKROIN, P. A., SEARS, T. A., SEGAL, M.,
WHITTINGHAM, T. S. & WILLIAMS, J. (1984). Brain slice methods. In *Brain Slices* (ed. R.
Dingledine) pp. 381-437. New York: Plenum Press.

BARRES, B. A., KOROSHETZ, W. J., SCHWARTZ, K. J., CHUN, L. L. Y. & COREY, D. P.
(1990). Ion channel expression by white matter glia: the O-2A glial progenitor cell. *Neuron* **4**, 507-
524.

BLANTON, M. G., LO TURCO, J. J. & KRIEGSTEIN, A. R. (1989). Whole-cell recording from
neurones in slices of reptilian and mammalian cerebral cortex. *J. Neurosci. Meth.* **30**, 203-210.

BORMANN J., HAMILL, O. P. & SAKMANN, B. (1987). Mechanism of anion permeation through
channels gated by glycine and γ-aminobutyric acid in mouse cultured spinal neurones. *J. Physiol.* **385**,
234-286.

BRADBURY, S. (1989). An introduction to the optical microscope (Revised edition) *Royal
Microscopical Society Microscopy Handbooks* Oxford Science Publications, UK.

COLQUHOUN, D. & SIGWORTH, F. J. (1983). Fitting and statistical analysis of single channel
records. In *Single-channel Recording* (ed. B. Sakmann & E. Neher), pp 191-263. London: Plenum
Press.

COLQUHOUN, D., JONAS, P. & SAKMANN, B. (1992). Action of brief pulses of glutamate on
AMPA/kainate receptors in patches from different neurones of rat hippocampal slices. *J. Physiol.* **458**,
261-287.

CUELLO, A. C. & CARSON, S. (1983). Microdissection of fresh rat brain tissue slices. In *Brain
Microdissection Techniques* (ed. A. C. Cuello), pp 37-125. International Brain Research
Organisation.

CULL-CANDY, S. G. & OGDEN, D. C. (1985). Ion channels activated by L-glutamate and GABA in
cultured cerebellar neurones of the rat. *Proc. Roy. Soc. Lond.* B **244**, 367-373.

DODT, H.-U. & ZIEGLGANSBERGER, W. (1990). Visualizing unstained neurons in living brain
slices by infra-red DIC videomicroscopy. *Brain Res.* **537**, 333-336.

EDMONDS, B. & COLQUHOUN, D. (1992). Rapid decay of averaged single-channel NMDA receptor
activations recorded at low agonist concentration. *Proc. Roy. Soc. Lond.* B **250**, 279-286.

EDWARDS, F. A., KONNERTH, A., SAKMANN, B. & TAKAHASHI, T. (1989). A thin slice
preparation for patch clamp recordings from synaptically connected neurones of the mammalian
central nervous system. *Pflug. Arch. Eur. J. Physiol.* **414**, 600-612.

EDWARDS , F. A., KONNERTH. A. & SAKMANN, B. (1990). Quantal analysis of inhibitory synaptic
transmission in the dentate gyrus of rat hippocampal slices: a patch-clamp study. *J. Physiol.* **430**, 213-
249.

EDWARDS, F. A. & KONNERTH, A. (1992). Patch clamping cells in sliced tissue preparations.
Methods in Enzymology **207**, 208-222.

FARRANT, M., FELDMEYER, D., KOBYASHI, S., TAKAHASHI, T. & CULL-CANDY, S. G.
(1993). NMDA receptor channels of developing granule cells in rat thin cerebellar slices. *J. Physiol.*
467, 272P.

FARRANT, M. & CULL-CANDY, S. G. (1991). Excitatory amino acid receptor channels in Purkinje cells in thin cerebellar slices. *Proc. Roy. Soc. Lond.* B **244**, 179-184.

FENWICK, E. M., MARTY, A. & NEHER, E. (1983). Sodium and calcium channels in bovine chromaffin cells. *J. Physiol.* **331**, 599-635.

FERSTER, D. & JAGADEESH, B. (1992). EPSP-IPSP Interactions in cat visual cortex studied with *in vivo* whole-cell patch recording. *J. Neurosci.* **12**, 1262-1274.

FINKEL, A. S. & REDMAN, S. (1985). Theory and operation of a single microelectrode voltage clamp. *J. Neurosci. Meth.* **11**, 101-127.

FORSYTHE, I. D. & BARNES-DAVIES, M. (1993). The binaural auditory pathway: excitatory amino acid receptors mediate dual timecourse excitatory postsynaptic currents in the rat medial nucleus of the trapezoid body. *Proc. Roy. Soc. Lond.* B **251**, 151-157.

GIBB, A. J. & COLQUHOUN, D. (1991). Glutamate activation of a single NMDA receptor-channel produces a cluster of channel openings. *Proc. Roy. Soc. Lond.* B **243**, 39-45.

GIBB, A. J. & COLQUHOUN, D. (1992). Activation of NMDA receptors by L-glutamate in cells dissociated from adult rat hippocampus. *J. Physiol.* **456**, 143-179.

GIBB A. J. & WALMSLEY, B. (1987). A preparation for patch clamp studies of labelled, identified neurones from guinea pig spinal cord. *J. Neurosci. Meth.* **20**, 35-44.

GRAY, J. & JOHNSON, D. (1986). Rectification of single GABA-gated chloride currents in adult hippocampal neurones. *J. Neurophysiol.* **54,** 134-142.

HAMILL, O. P., MARTHY, A., NEHER, E., SAKMANN, B. & SIGWORTH, F. J. (1981). Improved patch-clamp techniques for high-resolution current recording from cells and cell-free membrane patches. *Pflug. Arch. Eur. J. Physiol.* **391**, 85-100.

HESTRIN, S., SAH, P. & NICOLL, R. A. (1990). Mechanisms generating the time course of dual component excitatory synaptic currents recorded in hippocampal slices. *Neuron* **5**, 247-253.

HESTRIN, S. (1992a). Activation and desensitization of glutamate-activated channels mediating fast excitatory synaptic currents in the visual cortex. *Neuron* **9**, 991-999.

HESTRIN, S. (1992b). Developmental regulation of NMDA receptor-mediated synaptic currents at a central synapse. *Nature* **357**, 686-689.

HUGUENARD, J. R., HAMILL, O. P. & PRINCE, D. A. (1988). Developmental changes in Na^+ conductances in rat neocortical neurones: appearance of a slowly inactivating component. *J. Neurophysiol.* **59**, 778-795.

KATZ, L. C., BURKHALTER, A. & DREYER, W. J. (1984). Fluorescent latex microspheres as a retrograde neuronal marker for *in vivo* and *in vitro* studies of visual cortex. *Nature* **310**, 498-500.

KAY, A. R. & WONG, R. K .S. (1986). Isolation of neurones suitable for patch-clamping from adult mammalian central nervous systems. *J. Neurosci. Meth.* **16,** 227-238.

KAY, A. R. & WONG, R. K. S. (1987). Calcium current activation kinetics in isolated pyramidal neurones of the CA1 region of the guinea-pig hippocampus. *J. Physiol.* **392**, 603-616.

KELLER, B. U., KONNERTH, A. & YARRI, Y. (1991). Patch clamp analysis of excitatory synaptic currents in granule cells of rat hippocampus. *J. Physiol.* **435**, 275-294.

KNOWLES, W. D. & SCHWARTZKROIN, P. A. (1981). Local circuit synaptic interactions in hippocampal brain slices. *J. Neurosci.* **1**, 318-322.

KOBAYASHI, S. & TAKAHASHI, T. (1993). Whole-cell properties of temperature-sensitive neurons in rat hypothalamic slices. *Proc. Roy. Soc. Lond.* B **252**, 89-94.

KONNERTH, A. (1990). Patch-clamping in slices of mammalian CNS. *Trends Neurosci.* **13**, 321-323.

KONNERTH, A., DRESSEN, J. & AUGUSTINE, G. J. (1992). Brief dendritic calcium signals initiate long-lasting synaptic depression in cerebellar Purkinje cells. *Proc. Nat. Acad. Sci. USA.* **89**, 7051-7055.

LANGMOEN, I. A. & ANDERSEN, P. (1981). The hippocampal slice *in vitro*. A description of the technique and some examples of the opportunities it offers. In *Electrophysiology of Isolated Mammalian CNS Preparations* (ed G. A. Kerkut & H. V. Wheal). New York: Academic press.

LESTER, R. A. J., CLEMENTS, J. D., WESTBROOK, G. L. & JAHR, C. E. (1990). Channel kinetics determine the time course of NMDA-receptor mediated synaptic currents. *Nature* **346**, 565-567.

LEVIS, J. L. & RAE, R. A. (1992). Constructing a patch clamp setup. In *Methods in Enzymology* **207**, 14-66.

LLINAS, R. & SUGIMORI, M. (1980). Electrophysiological properties of *in vitro* purkinje cell somata in mammalian cerebellar slices. *J. Physiol.* **305**, 171-195.

MADISON, D. V. (1991). Whole-cell voltage-clamp techniques applied to the study of synaptic function in hippocampal slices. In *Cellular Neurobiology. A Practical Approach* (ed J. Chad & H. Wheal). Oxford, UK: IRL Press.

MASON, A. J. R., NICOLL, A. & STRATFORD, K. (1991). Synaptic transmission between individual pyramidal neurons of the rat visual cortex *in vitro*. *J. Neurosci.* **11**, 72-84.

MILES, R. & WONG, R. K. S. (1984). Unitary inhibitory synaptic potentials in the guinea-pig hippocampus *in vitro*. *J. Physiol.* **356**, 97-114.

NUMANN, R. E. & WONG, R. K. S. (1984). Voltage-clamp study of GABA response desensitization in single pyramidal cells dissociated from the hippocampus of adult guinea pigs. *Neurosci. Lett.* **47**, 289-295.

NEHER, E. (1992). Ion channels for communication between and within cells. *Neuron* **8**, 605-612.

NOWAK, L., BREGESTOVSKI, P., ASCHER, P., HERBET, A. & PROCHIANTZ, A. (1984). Magnesium gates glutamate-activated channels in mouse central neurones. *Nature* **307**, 462-465.

RAE, R.A. & LEVIS, J.L. (1992). Glass technology for patch clamp electrodes. *Methods in Enzymology* **207**, 67-92.

SAII, P., GIBB, A. J. & GAGE, P. W. (1988). The sodium current underlying action potentials in guinea-pig hippocampal CA1 neurones. *J. Gen. Physiol.* **91**, 373-398.

SAKMANN, B. (1992). Elementary steps in synaptic transmission revealed by currents through single ion channels. *Neuron* **8**, 613-629.

SAYER, R. J., FRIEDLANDER, M. J. & REDMAN, S. J. (1990). The time course and amplitude of EPSPs evoked at synapses between pairs of CA3/CA1 neurons in the hippocampal slice. *J. Neurosci.* **10**, 826-836.

SCHARFMAN, H. E., KUNKEL, D. D. & SCHWARTZKROIN, P. A. (1990). Synaptic connections of dentate granule cells and hilar neurons: results of paired intracellular recordings and intracellular horseradish peroxidase injections. *Neurosci.* **37**, 693-707.

SIGWORTH, F. J. (1986). The patch clamp is more useful than anyone expected. *Fed. Proc.* **45** 2673-2677.

SIGWORTH, F. J. & SINE, S. (1987). Data transformations for improved display and fitting of single-channel dwell time histograms. *Biophysical J.* **52**, 1047-1054.

SILVER, R. A., TRAYNELIS, S. F. & CULL-CANDY, S. G. (1992). Rapid time-course miniature and evoked excitatory currents at cerebellar synapses *in situ*. *Nature* **355**, 163-166.

STERN, P., EDWARDS, F. A. & SAKMANN, B. (1992). Fast and slow components of unitary EPSCs on stellate cells elicited by focal stimulation in slices of rat visual cortex. *J. Physiol.* **449**, 247-278.

STUART, G. J., DODT, H.-U. & SAKMANN, B. (1993). Patch-clamp recordings from the soma and dendrites of neurones in brain slices using infra-red video microscopy. *Pflug. Arch. Eur. J. Physiol.* **433**, 511-518.

TAKAHASHI, T. (1978). Intracellular recording from visually identified motoneurones in rat spinal cord slices. *Proc. Roy. Soc. Lond.* B **202**, 417-421.

TAKAHASHI, T. (1992). The minimal inhibitory synaptic currents evoked in neonatal rat motoneurones. *J. Physiol.* **450**, 593-611.

TAKAHASHI, T. & MOMIYAMA, A. (1991). Single-channel currents underlying glycinergic inhibitory postsynaptic responses in spinal neurones. *Neuron* **7**, 965-969.

XING PEI, VOLGUSHEV, M., VIDYASAGAR, T. R. & CREUTZFELDT, O. D. (1991). Whole cell recordings and conductance measurements in cat visual cortex *in vivo*. *NeuroReport* **2**, 485-488.

YAMAMOTO, C. (1975). Recording of electrical activity from microscopically identified neurons of the mammalian brain. *Experentia* **31**, 309-311.

YOSHIMURA, M. & NISHI, S. (1993). Blind patch-clamp recordings from substantia gelatinosa neurons in adult rat spinal cord slices: pharmacological properties of synaptic currents. *Neuroscience* **53**, 519-526.

Chapter 11

Ion-sensitive microelectrodes

JUHA VOIPIO, MICHAEL PASTERNACK and KENNETH MACLEOD

1. Introduction

Electrophysiologists are generally interested in studying ion movements across cell membranes because such movements play critical roles in the function of the cells. Other chapters in this book elaborate on the techniques that are used to study fast electrical events related to ion transfer (e.g. single and two electrode voltage clamp). In this chapter we focus on a technique that allows the study of changes in ion activity at a cellular level. Ion-sensitive microelectrodes provide a means of directly assessing the extracellular or intracellular activity of an ion and for making prolonged measurements of these. It is the activity of an ion which is important because it, rather than the total or free concentration, determines, for example, the membrane potential, equilibrium potentials, or the thermodynamic conditions for ion transport mechanisms.

Other techniques used in intracellular ion measurements include optical indicators, radio-isotopic tracer and atomic absorption methods. The usefulness of the latter two is limited since these methods do not account solely for the activity of the ion in question, but give a measure of the total concentration of the ion including those sequestered in intracellular organelles and bound to intracellular buffers. In addition they cannot continuously monitor ionic changes.

A technique that could be compared with ion-sensitive electrodes is that using fluorescent indicators and one is often asked how the advantages and disadvantages of these two techniques compare. We have summarised the general features of these two methods below.

ION-SENSITIVE MICROELECTRODES

Advantages	Disadvantages
(1) Organelles excluded	(1) Need large cells (limit≈50×10 μm)
(2) Continuous measurement	(2) Specificity occasionally inadequate
(3) Two or more ions simultaneously	(3) Electrical interference
(4) Low cost and simple equipment	(4) Some degree of manipulative skill needed
(5) Calibration usually simple	(5) Response time of seconds

J. VOIPIO AND M. PASTERNACK, Department of Zoology, Division of Physiology, P.O. Box 17 (Arkadiankatu 7) SF-00014 University of Helsinki, Helsinki, FINLAND
K. MACLEOD, Department of Cardiac Medicine, National Heart & Lung Institute, University of London, Dovehouse Street, London SW3 6LY, UK

FLUORESCENT INDICATORS

Advantages	Disadvantages
(1) Response time of milliseconds	(1) May not exclude organelles
(2) Continuous measurement	(2) Calibration can be difficult
(3) Good specificity -though not always	(3) May alter intracellular buffering
(4) Can be used on small preparations	(4) Equipment costly
(5) Little manipulative skill needed	(5) Photo-toxicity
	(6) Light interference
	(7) Photo-bleaching
	(8) Good indicators for some important ions not yet available (e.g. Cl^-)

Let us now consider these in more detail.

Response times. Many changes in intracellular ion activities occur quite slowly - on a second to minute timescale - and the response times of ion-sensitive microelectrodes are well suited to record such changes. However, some intracellular ionic events occur much faster and it is the examination of such fast events which would benefit from fluorescent indicators.

Organelles. Ion-sensitive microelectrodes measure only activities of the cytoplasm as their tip size precludes proper and unruptured impalement of organelles. As discussed in the chapter on fluorescent indicators, a major problem of fluorescent indicators is that one of the main methods for their incorporation into cells can lead to the indicator becoming trapped in intracellular organelles. This makes it difficult to know what one is actually measuring: ion activity in the cytoplasm, organelle or a mixture of the two.

Continuous measurement. It is possible to record ionic changes inside cells with ion-sensitive microelectrodes for as long as one can maintain impalement of the cell (often many hours). The drawback of fluorescent indicators is that they bleach to greater or lesser extents and emission intensity of the dye can decrease over a period of time, which can preclude continuous recording. It is difficult, though not impossible, to measure two ions simultaneously with fluorescent dyes but this is more straightforward to do with ion-sensitive microelectrodes provided the cell under study can withstand the multiple impalements.

Equipment. The peripheral equipment required to use ion-sensitive microelectrodes is simple to make and cheap. Most electrophysiology laboratories will have a spare oscilloscope and a strip chart recorder. In contrast, the equipment for using fluorescent indicators can be an order of magnitude more expensive.

Cell size. The technique requires that two microelectrode tips should be in the cell at the same time and this becomes more difficult as cell size decreases. There are techniques that are partial solutions to this problem but cells around 50×10 μm in size are about the smallest on which we would contemplate using ion-sensitive microelectrodes. For small cells there are advantages in using fluorescent indicators.

Specificity. Many ion-sensitive microelectrodes are not uniquely selective for the ion one is trying to measure. Other ions cause interference and may combine with the true signal thus making it difficult to assess the exact change in activity of the ion one wishes to measure. In practice what this does is to set a detection limit for the electrode. Generally, fluorescent dyes are selective for the ion in question, but they can suffer from direct interference or changes in autofluorescence produced by, for example, pharmacological compounds.

Calibration. This is relatively straightforward for ion-sensitive electrodes, which can be calibrated separately, but can be difficult with fluorescent dyes.

Electrical interference. The sensitivity of ion-sensitive microelectrodes to electrical interference requires that they be well shielded. Fluorescent dyes obviously do not suffer from such problems but they must be used in the dark which can also put constraints on the design of the experimental area.

Buffering. Ion-sensitive microelectrodes themselves are inert measuring devices in that they do not add or remove anything from the cell under test. They do, however, require impalement of the cell membrane. Fluorescent dyes are not inert in the same way. Ca^{2+} indicators, for example, may signifcantly increase the Ca^{2+} buffering power of the cell.

2. Ion-sensitive electrode measurements

Ion activity and ion concentration

If we were to grind up a biological preparation and assess the calcium content of the tissue by atomic absorption spectrophotometry and then compare this value with one produced by an ion-sensitive microelectrode or fluorescent indicator we would find that the total calcium concentration would be of the order of millimoles per litre whilst the free intracellular calcium ion concentration would be 100 or so nanomoles per litre. Thus a huge percentage of the calcium is bound to moieties within the cell e.g. intracellular organelles and buffers. The differentiation between free ion concentration and total ion concentration is a simple matter. However, what is the difference between free ion concentration and activity? As another illustration consider a solution of 150 mM NaCl in water. The NaCl would fully ionise so the concentration of Na^+ ions ($[Na^+]$) in the solution would be 150 mM. However, the mutual electrostatic repulsion between the similarly charged species and attraction between anions and cations reduces their mobility and freedom especially as the concentration of the solution becomes higher. In other words, the ions exhibit non-ideal behaviour, which is not due to incomplete ionization but to the existance of inter-ionic forces. Thus the activity of the Na^+ ions in solution or their effective concentration is less than their total concentration. Only when the solute is infinitely diluted will its concentration equal its activity. It is the activity of the ion which is used in calculations related to thermodynamic processes. The constant that relates the ion's activity to concentration is the activity coefficient (γ_i) such that:

$$a_i = [i] \cdot \gamma_i \tag{1}$$

where a_i is the activity of the ion (i) and the square brackets denote free ion concentration (i.e. ions which are not bound). The activity coefficient is a correction factor for thermodynamic calculations. For a 150 mM solution of NaCl at 20°C, $\gamma_{Na}=0.75$. The activity of Na^+ in the solution is therefore 112.5 mM.

The inter-ionic forces are influenced by the number and valency of the ions in solution and so the activity coefficient will vary with the ionic strength of the solution and the valency of the ion. This departure from non-ideal behaviour was recognised by Debye and Hückel who produced a quantitative expression for activity coefficients in dilute (<0.01 M) solutions. At 25°C:

$$\log\gamma = -0.5091z^2\sqrt{I} \tag{2}$$

where z is the valency of the ion and I is the ionic strength of the solution, which is calculated as follows for n number of ionic species (i):

$$I = 0.5\sum_{k=1}^{n}[i_k]z_k^2 \tag{3}$$

In more concentrated solutions (>0.01 M but <0.1 M) expression (2) has to be modified so that:

$$\log\gamma \approx -A\left(\frac{\sqrt{I}}{1 + \sqrt{I}}\right) \tag{4}$$

where:

$$A = \frac{1.8246 \times 10^6}{(\varepsilon T)^{3/2}} \tag{5}$$

ε being the dielectric constant of the solvent (water = 80.1 at 20°C; 78.3 at 25°C) and T the absolute temperature.

One has a choice whether one expresses the voltages from the ion-sensitive microelectrodes in terms of free concentration or activity. The advantages and disadvantages of either way are reviewed by Thomas (1978) and Tsien (1983). Our own view is that stated by Tsien as follows: "When one says ... that the free calcium concentration in a cell was measured to be 1 µM, that really means that the calcium activity in the cell was the same as the calcium activity in a certain calibrating solution in which the other major constituents were considered to be similar to those of cytosol and which contained 1 µM of calcium ions not tightly bound to ligands". In other words we normally calibrate the electrodes in solutions of known concentration which are usually chosen to mimic the cytoplasm. When these electrodes are pushed into a cell we compare the voltages then obtained with the calibration curves. Provided the activity coefficient is the same inside the cell and outside, then the value that we read from the calibration will be the intracellular free ion concentration. We do not need to know the value of the activity coefficient. Because the ionic strength of the calibration solution is chosen to be nearly the same as that of the cytoplasm, then our value for the intracellular ion concentration does not get any better or become more exact if we multiply by γ. If, however, one is not prepared to accept that γ is similar on both sides of the cell membrane, then by knowing its value in your calibration solutions you can simply read off the intracellular ion activity from your curve.

One of the problems is in knowing with certainty the value of γ for the ion in question. Values can be found in the literature or they can be calculated using equations (3) and (4) above, but these will only yield an approximation which is good for monovalent ions but poorer for divalents. Mainly because ionic strengths are usually similar in calibrating and intracellular solutions it is acceptable to express results in free ion concentration. One advantage of expressing your results in activity is that this is precisely the quantity upon which biological processes depend. However, if you wish to compare an ion activity measurement with a biochemically derived one - say the K_m of a plasmalemmal exchange process - then you will need to convert your activity measurement back to concentration. The tendency in the literature is to use free ion concentration for divalents and activity for monovalents. This is largely because of uncertainty about γ for divalents. There is certainly great scope for confusion and it is largely a matter of preference but, whatever you decide to do, you should provide sufficient details of calibrations so that others know what you have done.

The Nernst relationship

Assume we have a membrane that is permeable to only one ion. If we separate, by this membrane, two solutions that have different activities of this ion, a net diffusion of the permeant ion will start to take place through the membrane. This flux gives rise to a net movement of electrical charge and therefore an electrical potential difference is rapidly generated across the membrane. At equilibrium, there is no net diffusion and the potential difference across the membrane is given by the well-known Nernst equation:

$$E = \frac{RT}{zF} \ln \frac{a_s}{a_f} \qquad (6)$$

where R is the gas constant, T is the absolute temperature, z is the valency of the ion, F is the Faraday constant and a_s and a_f are the activities of the permeant ion on the two sides of the membrane.

Ion-sensitive electrodes are devices with a selectively permeable membrane sealing a small volume of a solution with a constant activity of the permeant ion (a_f) (see Fig. 1). When such an electrode is exposed to a sample of a solution with an unknown activity of the permeant ion (a_s) the equilibrium potential which develops between the two solutions can be measured and used to calculate the activity of the permeant ion in the sample.

For practical measurements a closed circuit is set up as shown in Fig. 2A. In this system we have an ion-selective electrode which is measuring the activity (a) of an

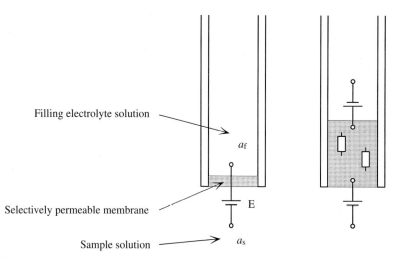

Fig. 1. Basic structure of an ion-sensitive electrode. Left: at equilibrium, the potential difference E, which exists across a selectively permeable membrane is given by the Nernst equation. a_f and a_s denote the activity of the permeant ion in the filling solution and in the sample solution respectively. Right: in practice, the selectively permeable membrane has a finite thickness and a potential difference is generated at each of the two membrane-aqueous solution interfaces. At the inner surface the potential difference can be assumed to be constant (but for its temperature sensitivity, see Section 3, *Temperature sensitivity and membrane column length*). Such a membrane column behaves in a manner analogous to a selectively permeable membrane displaying Nernstian characteristics, i.e. the two batteries can be lumped together and they generate E shown on the left.

ion (i) in solution. The potential difference (E_1) measured between the two electrodes in this solution (number 1) is:

$$E_1 = E_o + \frac{RT}{zF} \ln \frac{a_{i(1)}}{a_{i(f)}} \tag{7}$$

where E_o is the reference or offset potential (a constant consisting of several terms) and $a_{i(f)}$ is the activity of the ion in the filling solution of the electrode (which is also constant). Note that E_o includes the liquid junction potential of the reference electrode which, in the following discussion, is assumed to remain constant. If we now move the electrodes into a new solution (number 2) then E in this solution is:

$$E_2 = E_o + \frac{RT}{zF} \ln \frac{a_{i(2)}}{a_{i(f)}} \tag{8}$$

The difference in potential is dependent upon the ratio in ion activity between the two solutions because subtracting equation (7) from equation (8) yields:

$$E_{2-1} = \frac{RT}{zF} \ln \frac{a_{i(2)}}{a_{i(1)}} \tag{9}$$

$$= m \log \frac{a_{i(2)}}{a_{i(1)}}$$

where m is the slope of the relationship and at room temperature (20°C) equals (58/z) mV. Equations (9) are statements of the Nernst equation and what this means is that for a 10-fold change in monovalent ion activity one would expect a 58 mV change in potential.

Ion-sensitive microelectrodes are a miniaturisation of the system shown in Fig. 2A. In order to make intracellular ion measurements one must have two electrodes in the same cell (Fig. 2B). One electrode measures the reference potential while the other is the ion-sensitive device. Ion-sensitive microelectrodes are usually glass micropipettes plugged at the tip with an organic membrane solution. This medium does not provide electrical interactions for ions as water dipoles do in the water phase and therefore ions can only enter the membrane phase if they bind to specific carrier molecules (ionophores) in the membrane solution. These form

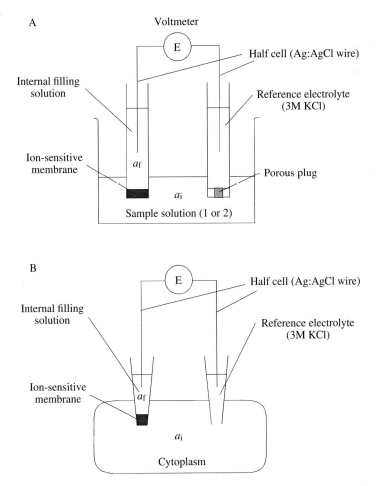

Fig. 2. Experimental set-up for ion-sensitive electrodes. The potential difference E, which exists across a selectively permeable membrane is measured by a voltmeter. a_f and a_i denote the activity of the permeant ion in the filling solution and in the sample solution respectively. For details, see text.

lipophilic complexes with ions and so promote the transfer of hydrophilic ions into and across a hydrophobic region. This then generates the potential across the membrane between the inner filling solution and the external solution being measured. Ion-sensitive microelectrodes will also record the membrane potential in addition to the intracellular ion changes. Thus an independent measure of membrane potential needs to be made (this is done by the separate reference electrode) and the voltage due to the membrane potential is then subtracted from the combined signal.

Detection limits and selectivity factors

The ion-sensitive electrode response tends to deviate from the Nernst relationship at low activities of the ion. The reason for this is that the total potential change is also governed by the presence of other "interfering" ions in the sample solution, which compete with the primary ion at the aqueous solution/ion-sensitive membrane interface. In mixed solutions the electrode response (E) is better described by the Nicolsky-Eisenman equation:

$$E = E_o + m\log(a_i + \Sigma K_{ij}^{pot}(a_j)^{z_i/z_j})$$ (10)

where j are the interfering ion(s) and K_{ij}^{pot} are the selectivity constants of the ion-selective electrode for the interfering ions. K_{ij}^{pot} or more simply, K_{ij} are a measure of the preference by the electrode for the interfering ion (j) relative to the primary ion (i) being detected. The potential measured by an ion-sensitive electrode will thus be a combination of an offset potential and the potential due to the primary ion and the interfering ions in solution. In practice, it is the logarithm of the selectivity constant which is usually given. As an example, consider the selectivity of a Na$^+$-sensitive electrode (made from the ETH 227 sensor and containing sodium tetraphenylborate) for K$^+$. The log K_{NaK} is quoted as −2.3 so K_{NaK} is 0.005. This means that the sensor is approximately 200 times more sensitive to Na$^+$ than to K$^+$. When log K_{ij} is given, a negative value means that the sensor is more selective for the primary ion and a positive value means that the sensor is more selective for the interfering ion. Therefore, the more selective sensors will have the more negative log K_{ij} values. When the selectivity of the sensor for the primary ion is the same as for the interfering ion, log K_{ij} will equal zero.

In equation (10) when:

$$a_i = \Sigma K_{ij}(a_j)^{z_i/z_j}$$ (11)

then the limit of detection for the electrode has been reached. This is about the lowest activity of the primary ion at which the electrode can discriminate; thereafter a decrease in primary ion activity will produce a progressively attenuated potential change because the influence of a constant background of interfering ion will predominate. It is then apparent why, at low activities of the primary ion, the slope of the relation between potential and primary ion activity (measured in a constant background of interfering ion) deviates from one which is Nernstian.

3. Practical electrode design and construction

Ion-sensitive glass electrodes

Ion-sensitive microelectrodes can be made using special glass with ion-sensitive properties. H^+-, Na^+- and K^+-sensitive glasses have been applied to physiological measurements. The development of a potential at the glass-solution interface of a glass pH-sensitive electrode is related to the transfer of protons into the glass in exchange for sodium ions passing into solution on the other side of the membrane. The glass behaves as a semi-permeable membrane, although the potential is developed by a different mechanism on each surface (see Bates, 1954). While some glass ion-sensitive microelectrodes are more selective than their liquid-membrane counterparts, practically their use is limited to the measurement of pH or Na^+. The main advantage of this type of ion-sensitive microelectrode, which is important in certain cases, is that they are very insensitive to substances that are known to interfere with the organic membrane of liquid membrane microelectrodes. The disadvantages with electrodes of this type are associated with tip size and response time and a lot of experience is required in their construction. The main difficulty in manufacturing microelectrodes of this type is sealing the tip of an ordinary micropipette with a small piece of ion-sensitive glass. For these reasons most ion-sensitive microelectrodes are now made from the liquid membrane cocktails. An interested reader can find a detailed description of ion-sensitive glass microelectrodes and different methods used to produce them in Thomas (1978). On the following pages, we concentrate on liquid-membrane microelectrodes.

Liquid-membrane microelectrodes

Liquid-membrane solutions generally have three components: the ion-selective compound or carrier (most often a neutral ligand), a membrane solvent or plasticizer in which the carrier is dissolved and a membrane additive (a lipophilic salt).

There are a variety of neutral carriers or ligands to bind (and hence detect) a variety of ions that are biologically important: Na^+, K^+, H^+, Ca^{2+}, Mg^{2+} and Cl^-. The carrier is dissolved in a solvent which must be non-polar (to reject hydrophilic ions and allow solubilization of lipophilic compounds) and be of moderate viscosity (to allow a microelectrode to fill easily). Additives produce significant improvements in selectivity and decrease membrane resistance and electrode response time. As an example, the components of a Ca^{2+}-sensitive membrane solution are shown in Fig. 3.

Single-barrelled microelectrodes

Single-barrelled liquid-membrane microelectrodes are the most widely used type of ion-sensitive microelectrodes. The detailed ways of making these electrodes vary between different laboratories, but basically the aims of all steps in the work are the same. In what follows, we have divided the manufacturing procedure into four main steps in the same order as they take place in practical work.

Pulling the micropipettes. Micropipettes suitable for ion-sensitive microelectrodes

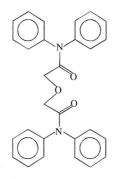

ETH 129 - Ca^{2+} ionophore

o-nitrophenyl octyl ether - membrane solvent

$$Na^+ \; B \left[\begin{array}{c} \end{array} \right]_4^-$$

Sodium tetraphenyl borate - membrane additive

Fig. 3. Components of a Ca^{2+}-sensitive membrane solution.

can be made by almost any puller. The shape at the extreme tip is not critical as it is bevelled away in most cases. We have, however, successfully used thin-walled tubing without any bevelling or breaking in crustacean preparations (e.g. Voipio *et al.* 1991). These electrodes have a small tip diameter combined with a reasonable electrode resistance, but their applicability is limited by easy breaking of the tips while impaling cells or tissue. We have also used thicker-walled tubing without bevelling in mammalian heart cells (MacLeod, 1989) but one has to be careful since such electrodes may suffer from poorer selectivity and sensitivity. Thick-walled borosilicate glass without filament (like GC150 from Clark Electromedical Instruments) is good for general use in short-column electrodes, but filamented glass is better if the pipettes are to be backfilled with membrane solution. The use of other glass materials (e.g. aluminosilicate) is not common despite speculation concerning non-specific cation permeability of hydrated borosilicate glass in the tip region (c.f. e.g. Tsien & Rink, 1981). For these reasons it is preferable to start with ordinary borosilicate tubing.

In our experience, the tubing rarely needs cleaning. Recommendations vary from avoidance of cleaning (Thomas, 1978) to very effective washing procedures (e.g. Tsien and Rink, 1980). If silanization fails, tubing may be simply soaked overnight in butanol, rinsed in distilled water and ethanol and finally dried at 200°C. Since pulling pipettes is not time consuming, it is best to pull pipettes on the day they will be used. Pipettes are mounted in a metal holder in a Petri dish with a lid to protect them from dust. Never use modelling wax or any sticky material e.g. plasticine or Blutack™ to hold the pipettes, since the remains of oily components on the glass may evaporate and prevent proper silanization. It is best to start with about 20 pulled micropipettes. After a few days of training, one should be able to obtain more than ten working ion-selective electrodes in a few hours.

Bevelling of the micropipette tip. Micropipettes are bevelled in order to obtain tips with the shape of a hypodermic needle. This results in a large tip opening without compromising sharpness. An increase in tip inner diameter decreases the resistance of an ion-sensitive microelectrode and improves electrode sensitivity and selectivity.

Bevelling techniques can be divided into two categories. In wet bevelling, a micropipette filled with an electrolyte solution is lowered towards a rotating surface covered with aluminium oxide particles or diamond dust (Brown & Flaming, 1975) or with loose particles as a "thick slurry" (Lederer *et al.* 1979). The surface is covered by an electrolyte solution, which makes it possible to control bevelling of the tip while monitoring electrode resistance. These methods have the disadvantage that pipettes must be filled for bevelling and therefore silanization should be undertaken first. In addition, impaling properties indicate that slurry-bevelled electrodes do not necessarily have sharp tips. We know of several laboratories where both wet and dry bevelling techniques are available and all prefer the latter at least in construction of ion-sensitive electrodes.

Our dry-bevelling technique (Kaila & Voipio, 1985) was originally developed to meet the needs of making ion-sensitive microelectrodes, but it has been successfully used also in making low-resistance microelectrodes for voltage clamping of cardiac (e.g. Kaila & Vaughan-Jones, 1987) and crustacean (e.g. Kaila & Voipio, 1987) preparations. With this method, tips with a 0.6 μm outer diameter at the base of the bevel still appeared to have a neatly bevelled shape when examined with scanning electron microscopy. The dry bevelling equipment is simple and is described in Section 4, *Microelectrode dry bevelling equipment.* A pipette to be bevelled is lowered with a micromanipulator until its tip touches a rotating bevelling surface at an angle of 30-45°. The total bevelling time is controlled by amplifying and listening to the noise originating from the shank of the pipette. The correct bevelling time is found by trial-and-error and it is seldom longer than a few seconds unless large diameter tips are desired. Each bevelled tip should be inspected with a high quality microscope.

If bevelling equipment is not available, one may try breaking the tips by pushing them against a piece of glass under a microscope. This often gives tips with a reasonable diameter and with sharp edges facilitating impalements (Thomas, 1978).

Silanization. The organic liquid membrane solution must be in tight contact with the glass wall of the electrode, otherwise the aqueous electrolyte solution will find a pathway along the luminal glass surface thus short-circuiting the ion-sensitive signal source. This can be avoided if the glass surface is made hydrophobic by silanization. Reactive silanes replace hydroxyl groups on the glass surface and bind to it with covalent bonds resulting in a monomolecular hydrophobic coating (for references, see Ammann, 1986). Vapour treatment with N,N-dimethyltrimethylsilylamine (TMSDMA; Fluka or Sigma) is a very effective and widely used silanization method. The disadvantages are that this compound evaporates rapidly and is extremely toxic. Therefore, the correct place for handling TMSDMA (including the placement of the oven used for silanization) is a fume-cupboard.

Silanization with TMSDMA is easy. Bevelled micropipettes are mounted horizontally on a metal holder in a Petri dish which is taken with its lid open to an oven. The pipettes are baked at 200°C for 15 min after which time 20-40 μl of TMSDMA is added and the lid of the dish immediately closed. Because TMSDMA evaporates rapidly even at room temperature, it is best added to the dish in a small open glass vial. After another 15 min the lid of the dish is opened and baking is continued for a few more minutes to let the remains of TMSDMA disappear. The micropipettes are now ready for filling. They may be stored in the oven at e.g. 110°C, but we usually keep them in a closed Petri dish on a lab bench for a day. Storage of pipettes for longer periods should be avoided. Some workers store pipettes in a hot oven or in a desiccator but despite keeping them dry you may find resilanization necessary if the pipettes have been stored for several days. The silanization temperature used in different laboratories ranges from 110 to 200°C, but we have found that, within these limits, higher temperatures give better results. Silanization may fail if some other use has lead to contamination of the oven by other organic substances which evaporate and stick to the inside surfaces of the oven. Cleaning the inside surfaces with a solvent and/or prolonged baking at a temperature much higher than that needed for silanization are tricks that usually help.

Filling. The method chosen for filling micropipettes with the liquid membrane and filling electrolyte solutions determines the length of the membrane column within electrodes. Since this is a critical factor concerning the temperature sensitivity of the finished ion-sensitive microelectrode, this matter will be discussed before going into the details of pipette filling.

(1) Temperature sensitivity and membrane column length. The electrical output of ion-sensitive electrodes is the sum of potentials at the two liquid membrane - aqueous solution interfaces (Fig. 1). These are the two energy barriers where the ion to be measured experiences a step change in electrochemical potential which is, along with other factors as outlined in section 2, a function of temperature. If the liquid membrane column within a microelectrode is so long that part of it remains above the surface level of a warmed experimental bath, changes in bath temperature or surface level can give rise to changes in the temperature gradient along the membrane column and therefore, cause changes in electrode output, thereby causing serious noise and errors in quantitative ion measurements (Vaughan-Jones & Kaila, 1986). For this reason, the column length should be shorter than the part of the electrode that is immersed in the experimental bath during measurements.

Obviously, the temperature sensitivity of long-column microelectrodes is much less of a problem if the experiment is performed at room temperature. However, they still suffer from two other minor drawbacks: backfilling consumes a lot of (often expensive) liquid membrane solution and capacitance compensation is more difficult since there is no low resistance connection for some distance up the shank. In spite of these problems, long-column electrodes are widely used because they are very easy to fill and they usually have a longer lifetime than their short-column counterparts.

(2) Filling long-column microelectrodes. Micropipettes are back-filled through the stem with a small sample of liquid membrane solution. If a filamented pipette is well silanized, it fills spontaneously up to its tip resulting in a column several millimetres long. Back pressure is required when using non-filamented pipettes. The rest of the pipette is then filled with an electrolyte solution and the electrode is ready for calibration.

(3) Filling short-column microelectrodes. The filling of a short-column microelectrode involves two steps: the whole pipette is first filled with an electrolyte solution and then a short column of a liquid membrane solution is taken up into its tip. If the pipettes are properly silanized, they should not show any self-filling properties with an aqueous solution. A pipette is first back-filled up to its shoulder by injection of an electrolyte solution (see below) and it is then coupled to a pressure source. When pressure is applied it is possible to see (provided light comes from a suitable angle) the air-solution interface moving along the shank while the air in the tip is first compressed and then is pushed out through the tip opening. If very little pressure is needed, one should suspect either poor silanization or a broken tip. Pressurised air or nitrogen are suitable gases, or more simply, a 10-20 ml syringe can be used if a very high pressure is not needed. To couple the pipette to the pressure source either use tightly fitting silicone tubing or a special adapter like a patch-clamp holder. Dental wax has been used to seal a copper capillary into the pipette stem. Whatever the connection is, never point the pressurised pipette at anyone - if it gets loose, it flies like a bullet!

Pipettes may be checked after filling by resistance measurement, which will reveal broken or blocked tips. At this stage (when the electrode contains its electrolyte filling solution only) the higher resistance values, compared with conventional microelectrodes of the same size, result from the lower conductivity of the internal filling solution. A DC current of 1 nA induces a voltage change of 1 mV/MΩ. In AC measurements, a low-frequency sine-wave signal is preferable. Problems with filling can occur if the tip tapers at a steep angle, i.e. its diameter changes rapidly. With this type of pipette tip the release of pressure after filling may be followed by air being taken up into the tip. Such a phenomenon is not seen if tips have a more tubular shape.

The filling electrolyte solution must contain the ion to be measured in addition to chloride, which is required for stable operation of the Ag:AgCl electrode. NaCl- and KCl-containing solutions are used in Na^+ and K^+ electrodes, respectively, as well as in Cl^- electrodes. Buffer solutions are used in pH and Ca^{2+} electrodes. Some examples of filling solutions are given in Table 1.

Sometimes bubbles appear in the filling solution within the shank of the

Table 1. *Typical examples of ion-sensitive microelectrodes as constructed in the authors' laboratories*

Measured ion	Liquid membrane	Filling solution (mmol.l^{-1})	Comments
H$^+$	Fluka 95291/95293	NaCl 100, Hepes 20, NaOH 10, (pH 7.5)	for pH 5.5-12
H$^+$	Fluka 95297	NaCl 100, Hepes 20, NaOH 10, (pH 7.5)	for pH 2-9
K$^+$	Fluka 60031/60398	NaCl 100, KCl 5	Valinomycin
Na$^+$	Fluka 71176	NaCl 100	
Na$^+$	Fluka 71178	NaCl 100	(see note 1)
NH$_4^+$	Fluka 09879	NH$_4$Cl 10 (see note 2)	
TeMA$^+$	Corning 477317	KCl 100 or TeMACl 150 (see note 3)	volume meas.
Ca^{2+}	Fluka 21048/21191	pCa 6 solution of Tsien & Rink (1981) or Hove-Madsen & Bers (1992)	ETH 1001±PVC
Ca^{2+}	Fluka 21196	pCa 6 solution of Tsien & Rink (1981) or Hove-Madsen & Bers (1992)	ETH 129
Mg^{2+}	Fluka 63085	KCl 100, MgCl$_2$ 5	
Cl$^-$	Fluka 24902	NaCl 100	
Cl$^-$	Corning 477913	NaCl 100	

TeMA$^+$=tetramethylammonium ion. Note 1: suitable for extracellular measurements due to higher selectivity against Ca^{2+}. Note 2: see Fresser, Moser & Mair (1991). Note 3: see p. 232 in Nicholson & Phillips.

microelectrode. Filamented tubing does not help in this case, since after silanization the self-filling properties with aqueous solutions are lost. If the bubble is initially small and it is attached to the inner glass wall, a fall in pressure during suction used to take up membrane solution will increase the bubble diameter until it breaks the continuity of the filling solution. Returning to normal pressure does not necessarily result in the reverse process since the hydrophobic glass surface provides no route for the filling solution to cross the gas phase. The resulting disk-shaped bubble appears as a transverse stripe on the shank and causes a break in electrical conductivity which is seen as an infinite electrode resistance. Such a bubble can usually be broken down by a gentle tap on the electrode or by pushing a thin fibre (e.g. the classical cats whisker or a hair) through it.

Small bubbles often consist of air that was originally dissolved in the filling solution. The reduction in pressure during suction or a slight increase in temperature during handling will favour bubble formation, since the solution becomes over-saturated with air. If bubbles occur frequently, it is a good idea to de-gas all filling electrolyte solutions now and then. This can be done either by shaking the solution under low pressure or at a slightly elevated temperature (about 40°C).

When a pipette has been filled with electrolyte solution, it is advisable first to dip its tip in distilled water before quickly immersing it in the liquid membrane solution. Sometimes, a sufficient amount of membrane solution is taken up by capillarity, but suction (by mouth or from a syringe or a suction pump) lasting from a few seconds to minutes is often required to obtain a proper column length. Keep the tip in membrane solution during the suction so that air is not taken into the pipette. The membrane

A

B

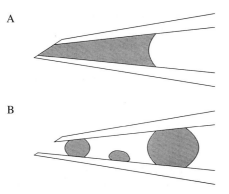

Fig. 4. (A) In a properly silanized micropipette, the liquid membrane solution forms a uniform column with a concave interface against the aqueous filling solution. (B) In a poorly silanized micropipette, the organic membrane solution is easily replaced by water at the hydrophilic glass surface which results in convex interfaces and, often, breaking of the membrane column.

column should appear as a 50-200 μm long continuous region ending in a concave surface against the filling electrolyte solution, which indicates proper silanization of the glass. A convex membrane-water interface, or breaking of the membrane phase into multiple sections, or its withdrawal into the shank are typical signs of bad silanization (see Fig. 4).

(4) Choice of membrane solution. Ready-to-use liquid membrane solutions - 'cocktails' - are commercially available for H^+, K^+, Li^+, Na^+, NH_4^+, Ca^{2+}, Mg^{2+} and Cl^-. Fluka Chemie AG publishes a separate catalogue entitled "Selectophore®, Ionophores for Ion Selective Electrodes and Optodes" which contains detailed information (including cited literature) on numerous products for ion-sensitive electrodes. Other sources for liquid membrane solutions are given in Table 2. For some ions there are several liquid membrane solutions available. Fluka supplies H^+ cocktails based on two different hydrogen ionophores. They differ in measuring range (pH 5.5-12 and 2-9) and electrode response time. Fluka also has different cocktails for Ca^{2+} microelectrodes. Those based on Calcium Ionophore I (ETH 1001) yield microelectrodes with a detection limit at or slightly below typical resting intracellular Ca^{2+} levels, but microelectrodes made using the cocktail containing Calcium Ionophore II (ETH 129) are able to record much lower Ca^{2+} levels and should be preferred for intracellular experiments. They also have two different valinomycin-based cocktails for K^+, which make highly selective low- and high-resistance microelectrodes, and two cocktails for Na^+, one of which is widely used but the other has better selectivity against Ca^{2+} and therefore it has been applied to extracellular measurements (see Table 1 and Coles & Orkand, 1985). A potassium ion-exchanger solution (Corning cat.no. 477317; or 5 mg potassium tetrakis p-chlorophenylborate in 0.1 ml 3-nitro-O-xylene, Alvarez-Leefmans *et al.* 1992; see also Ammann, 1986) is very selective to tetra-alkylammonium ions and is used in extra- and intracellular volume measurements. This technique, as well as a recently

developed CO_2 microelectrode, is briefly discussed below. In general, Cl^- liquid membrane electrodes are not ideal with respect to anion interference (bicarbonate, acetate, lactate and other intracellular anions). Fluka supplies a Cl^- cocktail (cat.no. 24902) with a reported selectivity coefficient of 0.03 against bicarbonate. In our experience, the widely used Cl^- exchanger (Corning cat.no. 477913) yields microelectrodes with a similar selectivity coefficient against bicarbonate of 0.03-0.04 (Kaila *et al.* 1989) which is much better than that reported previously (Baumgarten, 1981). The improvement is likely to be due to dry bevelling of the micropipettes used in the electrode construction. A membrane solution, originally published as a HCO_3^- sensor (Wise, 1973) but later shown to be sensitive to CO_3^{2-} (Herman & Rechnitz, 1974), has been used also in intracellular microelectrodes (Wietasch & Kraig, 1991). A nonspecific ion exchanger solution for reference electrodes (Thomas & Cohen, 1981) avoids problems related to electrolyte leakage into the cytoplasm, but such electrodes are now rarely used.

Electrode testing. Physiological experiments should never be made with electrodes that have not been properly tested. Different calibration methods are discussed in detail below (Section 5). An essential step in testing an ion-sensitive microelectrode is measurement of its resistance. This is done by passing a DC current of 1.0 or 0.1 pA through the electrode, which induces a voltage drop of 1.0 mV/GΩ or 0.1 mV/GΩ respectively, across the membrane column. Typical resistance values depend on tip

Table 2. *Sources for components used in the construction of ion-sensitive electrodes*

Components	Sources
Borosilicate, aluminosilicate and pH-sensitive glass, fused and theta-style capillaries, teflon-coated silver wire	Clark Electromedical Instruments, P.O. Box 8, Pangbourne, Reading, RG8 7HU, England. 0734 843888
Glass including thick-septum theta tubing, membrane solutions	World Precision Instruments, Inc., 175 Sarasota Central Blvd., Sarasota, Florida 34240, USA or World Precision Instuments Ltd., Astonbury Farm Business Centre, Aston, Stevenage, Hertfordshire, SG2 7EG, England
Liquid membrane solutions and components for practically all ions, silanizing agents	Fluka Chemie AG, CH-9470 Buchs, Switzerland or Fluka Chemicals Ltd., The Old Brickyard, New Road, Gillingham, Dorset, SP8 4JL, England. 0800 262300 or 0747 823097
A widely used Cl^- exchanger (cat. no. 477913) and a K^+ exchanger used in volume measurements (no. 477317)	Ciba Corning U.K., Colchester Road, Halstead, Essex, CO9 2DX, England. 0787 472461
High quality 100 GΩ resistors (type RX-1M). Useful when testing equipment	Victoreen Inc., 6000 Cochran Road, Solon, Ohio 44139, USA or LG Products Ltd., c/o ECl International, 17 Trident Industrial Estate, Blackthorne Road, Colnbrook, Slough, SL3 0AX, England. 0753 686667
Details of the construction of the dry beveller and of electrometer amplifiers	Information available from the author J.V. Fax +358 0 191 7301

diameter as well as on the liquid membrane solution being used and they range from a few GΩ to 100-200 GΩ.

Electrode holders and silver wires. Ion-sensitive microelectrodes should not be mounted on holders that might generate any pressure difference (positive or negative) between electrode interior and ambient air, since the membrane column within the tip is easily dislodged. Simple clamps attaching to the stem of a microelectrode give good results. Such clamps may be made of metal and be grounded without a risk of electrical leakage, since silanized glass has an exceedingly low surface conductivity under typical experimental conditions.

The AgCl coating on silver wires should be in perfect condition. We have obtained very good stability with wires chlorided by a DC current of +0.1-1.0 mA in 250 mM NaCl. Teflon coated wires are very useful, but only a few millimetres of teflon should be removed from each end, so that the chlorided area will be completely immersed in the electrode filling solution.

In lengthy experiments, water may evaporate from electrodes to such an extent that the activity of the electrolytes does not remain constant. This is, of course, seen as electrode drift, which can be avoided simply by sealing the upper end of each electrode. Evaporation of water is seldom a problem with ion-sensitive microelectrodes, but occurs sometimes in KCl-filled reference electrodes (affecting the electrode potential of their Ag:AgCl electrodes), where the filling solution has a tendency to creep up along the hydrophilic glass surface. To ensure stability of the potential in the bath, it should be grounded by a saturated KCl bridge with a conical tip (Strickholm, 1968).

Ion-sensitive surface electrodes

Net transmembrane ionic fluxes cause changes in intracellular ion activities, but they can also give rise to measurable ionic gradients within the unstirred water layer surrounding individual cells. Ion-sensitive surface microelectrodes, made using blunt

Table 3. *Steps in the construction of short-column single-barrelled liquid-membrane ion-sensitive microelectrodes*

0.*	Clean the tubing and de-gas the filling electrolyte solution
1.	Pull micropipettes with sub-micron tips
2.	Bevel the tips to the desired diameter
3.	Silanize the bevelled pipettes
3.1	Bake at 200°C for 15 min
3.2	Add 20-40 µl of TMSDMA in a glass vial, immediately close the Petri dish containing the pipettes and continue baking for 15 min
3.3	Open the dish and continue baking for 5 min
4.	Back-fill a pipette with the filling electrolyte solution
5.*	Measure the resistance of the electrode
6.	Take a column of membrane solution into the tip
7.	Look at the membrane column with a microscope
8.	Calibrate the electrode and measure its resistance

Less essential steps are indicated by an asterisk.

micropipettes in the shape of a patch pipette, can be gently pressed against a cell to measure changes in ion activity at the external surface (see also Vanheel *et al.* 1986). An example of the application of a surface electrode is given in Fig. 5. In this experiment, surface pH and intracellular pH were measured simultaneously in a crayfish muscle fibre. A prolonged application of γ-aminobutyric acid (GABA) gives rise to an extracellular surface alkalosis coupled to a fall in intracellular pH, both of which are due to a channel mediated efflux of bicarbonate ions (Kaila & Voipio, 1987). Surface electrodes have also been used in a double-barrelled configuration to measure changes in extracellular K^+ (Kline & Kupersmith, 1982) and depletion and accumulation of Ca^{2+} at the surface of cardiac muscle cells (Bers & MacLeod, 1986; MacLeod & Bers, 1987).

Double-barrelled electrodes

As explained in Section 2, *The Nernst relationship*, the ion-dependent signal is measured with respect to a reference electrode which will sense the electrical potential prevailing around its tip. For intracellular measurements it is important that both microelectrodes measure the same membrane potential and the same changes in membrane potential. This necessitates both microelectrodes being in the same cell but because of cell size, fragility or visual difficulties it is often not possible to make multiple impalements. When making extracellular electrode measurements significant voltage gradients can appear, especially within excitable tissues and if the

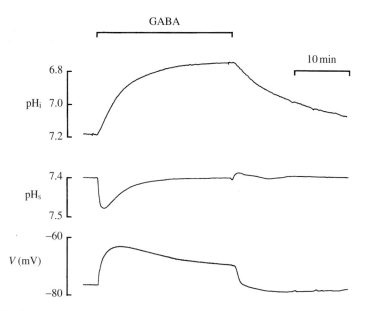

Fig. 5. Simultaneous measurement of intracellular pH (pH_i), extracellular surface pH (pH_s) and membrane potential (V) in a crayfish muscle fibre exposed to γ-aminobutyric acid (GABA, 100 μM). The experimental solutions were equilibrated with 5% CO_2 (30 mM HCO_3^-). The GABA-induced effects are due to a channel-mediated efflux of HCO_3^-. (Reproduced with permission from Kaila & Voipio, 1987).

reference electrode is not situated close to the ion-sensitive electrode, spurious changes in ion activity can be recorded. These difficulties can be overcome by using double-barrelled microelectrodes which contain the reference and the ion electrode in the same tip.

Several methods have been used to produce double-barrelled micropipettes. In the twist-and-pull method, two glass capillaries are first glued together and then twisted around each other whilst being pulled into micropipettes (see Zeuthen, 1980). An easier approach may be to use filamented and non-filamented capillaries which have been already fused together (e.g. 2GC150FS from Clark Electromedical Instruments). An eccentric positioning of two capillaries has also been used (Thomas, 1987). Theta-tubing is easy to pull and gives a spherical cross-section up to the tip, but other steps in making ion-electrodes are sometimes more tricky with them. Thick-septum theta tubing has a lower electrical coupling (for DC and AC signals) between the two barrels and it has been used also in making ion-electrodes. (Note: Bending theta-tubing easily causes cracks to form in the septum, which causes an inter-barrel electrical shunt). Making double-barrelled ion-sensitive microelectrodes differs from making single-barrelled electrodes in only one respect: only one of the two barrels should be silanized. If the tip of the reference barrel becomes hydrophobic, it is easily blocked by membrane solution. Therefore, the main objective is to obtain a specific silanization of one barrel only. Barrel-specific silanization has been done using wet silanizing methods and vapour treatments have been used in combination with sophisticated barrel-specific pressure/suction systems. However, good results are obtained by simply letting TMSDMA vapour evaporate into one barrel at room temperature and then moving the pipette to an oven. A simple holder, like the one shown in Fig. 6, makes it possible to silanize many pipettes simultaneously. TMSDMA is added to a Petri dish which is covered by the holder. Silane vapour is then allowed to flow into the non-filamented barrels for 5-15 min, thereafter the holder with the pipettes is rapidly moved into an oven at 200°C for 15 min.

If bevelling equipment is available, it is advisable to silanize the pipettes before bevelling. This will reduce the possibility of inter-barrel silane contamination at the

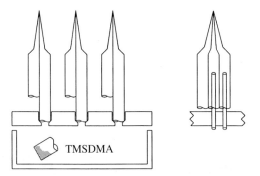

Fig. 6. Holder that allows selective silanization of individual barrels of double or triple-barrelled micropipettes. For details, see text.

tip. In addition, the opening of the reference barrel, which may still get silanized by vapour leaking from the other barrel, is then bevelled away after silanization.

If silanization has been successful, filling the pipettes is easy. Both barrels are first filled with their electrolyte solutions and the liquid membrane solution is then taken up into the ion-sensitive barrel by suction. The reference barrel will not become blocked with the membrane solution unless it has been contaminated by silane vapour. To check this, its resistance should be measured as the first step of electrode testing.

One should be cautious about the filling solution used in the reference barrel. Leakage from the tip makes a point source of ions at the very site of ion measurement. Just imagine the effect of 3 M KCl leaking out of the reference barrel of a double barrelled Cl^- microelectrode!

Mini-electrodes

Mini ion-sensitive electrodes (tip size $\geq$1-2 mm) are easy and cheap to make and can be very useful. An example is titration of the free Ca^{2+} concentration in calibration solutions (see Appendix, Section 7) and in physiological test solutions containing Cl^- substitutes which bind Ca^{2+}, such as glucuronate or gluconate. The relative amounts of constituents in liquid membrane solutions used in micro- and mini-electrodes are usually different. In addition, the solution has to be bound to a mechanically durable membrane matrix. PVC membranes are widely used, but silicone rubber membranes can also be made. Membrane compositions and instructions for making solvent polymeric membranes can be found in Ammann (1986) and in Fluka's Selectophore® catalogue. In brief, the liquid membrane components (usually a neutral ligand, a lipophilic salt and a plasticizer) are mixed in a small glass vial and dissolved in tetrahydrofuran (THF) together with high molecular weight polyvinylchloride (PVC). When all constituents are fully dissolved (this can often take 1-2 hrs) the solution is poured onto a glass-plate or glass Petri dish and the THF left to evaporate overnight. The membrane that forms upon evaporation of the THF can be stored for months. A mini-electrode is made by cutting out a piece of membrane, placing this over the end of a piece of PVC tubing and sealing the tubing and membrane together using a few drops of THF to dissolve the PVC. The tube is then filled with an appropriate electrolyte solution. In this state the finished electrode can be kept for anything from a week or so to a month depending on the membrane. Thereafter, the membrane may have deteriorated resulting in a decrease in the steepness of the calibration slope and a decrease in the detection limit.

Special applications of ion-sensitive microelectrodes

Volume measurements and related techniques

Extracellular (see e.g. Dietzel *et al.* 1980; Hablitz & Heinemann, 1989) as well as intracellular (Serve *et al.* 1988; Alvarez-Leefmans *et al.* 1992) volume changes can be detected and measured with ion-sensitive microelectrodes. If the extracellular fluid or cytoplasm contains an impermeable ion species, its activity will undergo transient changes upon transmembrane water movements. This basic idea has been applied to volume measurements using K^+ ion-exchanger microelectrodes (see above), which are far more sensitive to the membrane-impermeable tetramethylammonium ion than to the potassium ion. In addition to this, probe ions of different size have also been used to estimate the maximum pore size of unidentified permeability pathways, e.g. those activated by spreading depression in brain tissue (Nicholson & Kraig, 1981). The recovery of a transient change in the extracellular activity of an impermeable ion reflects volume changes and also its diffusion in the extracellular space. This has been used to examine the diffusional properties of the interstitial space in brain tissue (Nicholson & Phillips, 1981).

CO_2-sensitive microelectrodes

In general, changes in the partial pressure of CO_2 (P_{CO_2}) can result from cellular metabolism and from the operation of the CO_2/HCO_3^- buffer system. It appears that a frequent implicit assumption in physiological work done under *in vitro* conditions is a constancy of P_{CO_2}. However, this view may need re-evaluation. Direct evidence obtained with a recently developed CO_2-sensitive microelectrode clearly indicates that considerable changes in P_{CO_2} take place upon neuronal activity in rat hippocampal slices (≤ 400 µm) (Voipio & Kaila, 1993) that are kept in an interface-type chamber, which provides a route for gas exchange on both sides of the preparation. As shown in Fig. 7, surprisingly large changes in P_{CO_2} are induced by

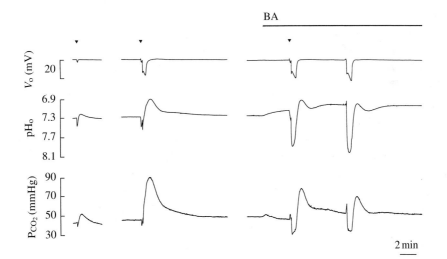

Fig. 7. Simultaneous recordings of changes in the interstitial partial pressure of CO_2 (P_{CO_2}), interstitial pH (pH_o) and extracellular potential (V_o) induced by neuronal activity (left) and by spreading depression (SD) in a rat hippocampal slice (stratum pyramidale of area CA1). Triangles (▼) indicate trains of stimuli applied to the Schaffer collaterals used to initiate SD (SDs are seen as pronounced negative deflections in V_o; one of the SDs occurred spontaneously). In the latter part of the experiment benzolamide (BA, 10 µM) was used to inhibit extracellular carbonic anhydrase activity. (Voipio, Paalasmaa, Taira & Kaila, unpublished observations.)

spreading depression in the same preparation (Voipio, Paalasmaa, Taira, Kaila, unpublished observations).

So far, all CO_2-sensitive electrodes have suffered from a slow response time of minutes rather than seconds. One of the electrode designs makes use of a membrane which is permeable to both H^+ and CO_2 (Niedrach, 1975; Funck *et al.* 1982). The pH affecting the inner surface of such a membrane depends on the external P_{CO_2} if the electrode is filled with an unbuffered solution. Therefore, the electrode output is a function of both pH and P_{CO_2}. In our novel CO_2-microelectrode (Voipio & Kaila, 1993), a Hepes-buffered solution and an unbuffered solution are used in a theta micropipette to obtain a pH-sensitive signal from one barrel and a P_{CO_2}-sensitive signal as the difference of the outputs of the two barrels. The membrane in both barrels consists of the same proton carrier solution. A response time of only a few seconds is obtained upon addition of carbonic anhydrase (CA) to the unbuffered filling solution which speeds up the equilibration of the CO_2 hydration reaction. By using PVC in the membrane solution, it is possible to produce very short (4-7 μm) yet mechanically stable membrane columns (a prerequisite for fast diffusion of CO_2 through the column). A useful membrane is obtained simply by using 14% (w/w) PVC in a H^+ cocktail (Fluka 95291) and by dissolving them in THF (Voipio & Kaila, 1993). The fast response time, stability, insensitivity to lactate and a P_{CO_2} measuring range from the level prevailing in nominally CO_2-free solutions to over 350 mmHg (unpublished observations) make the above microelectrodes suitable for a wide variety of applications.

Quantification of the pH signal that is measured simultaneously with P_{CO_2} requires a reference microelectrode if the potential around the tip is unknown. Clark Electromedical Instruments has started supplying a special triple-barrelled tubing (3GC150SF) which consists of one filamented and two non-filamented capillaries fused together. This material is advantageous in the construction of CO_2-microelectrodes, since the side-by-side positioning of the three pieces of tubing makes it possible, with the use of a microscope, to see the two membrane columns at the tip. The filling solution containing CA should be made daily from its three components (4.0 mg CA (C-7500, Sigma) + 1.0 ml 150 mM NaCl + 23.6 μl 0.1 N NaOH) and must not be shaken. The triple-barrelled micropipettes should be silanized before bevelling. Plastic tubing can be inserted into the barrels and sealed e.g. by modelling wax to obtain barrel-specific pressure/suction for filling.

Limitations imposed by electrical properties

The high resistance (10^{10}-10^{11} Ω) of ion-sensitive microelectrodes combined with a total capacitance of several picofarads means that their electrical time constant (τ) is generally between 0.1-1 sec (τ=RC). This results in a significant low-pass filtering of both electrical and ion signals if no attempt is made to compensate for the filtering. However, electrical compensation can greatly reduce the effect of long time constants. This matter is dealt with in detail in Section 4, *Differential amplifier and filtering*.

4. Equipment design and construction

Microelectrode dry bevelling equipment

As discussed above, a simple method for dry bevelling of micropipettes (Kaila & Voipio, 1985) has proven advantageous in the construction of ion-sensitive microelectrodes. The equipment consists of a rotating unit with a commercially available bevelling film, a detector coupled to the shank of a micropipette and a low-noise audio amplifier.

A rotating plate, which is free of vertical movements, can be made using two optically flat glass plates (Brown & Flaming, 1975). A piece of commercially available lapping film with 0.3 μm particle size (ILF-film 3M261X A/O 0.3 Mic; contact local subsidiary of 3M, or 3M, St. Paul, MN 55144-1000, USA) is used on top of the rotating unit, which is mounted on a rigid plate together with a micromanipulator. We cut the lapping film to a diameter of 8 cm and use a speed of 50-80 rotations per minute. A simple DC motor is mounted elastically to the system, in order to avoid mechanical coupling of its vibrations.

A cheap piezocrystal record-player pick-up can be used as the detector (moving magnet or moving coil pick-ups give a much weaker signal), but better sensitivity is obtained with a 70 mm long piezocrystal bimorph element (Philips order code 4322 0200823 is ideal, but the minimum order is 600 pieces!). A low-noise audio amplifier is easily built from standard components. Fig. 8 shows the circuit diagram of a simple amplifier, which can be used with headphones.

Electrometer amplifier design

Ion-sensitive microelectrodes have an extremely high resistance so they must be coupled to electrometer amplifiers that are designed to cause as small a load as

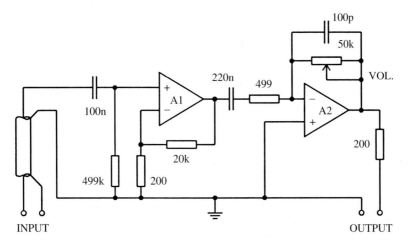

Fig. 8. Circuit diagram of a simple audio amplifier for headphones and a piezocrystal detector used in dry-bevelling of micropipettes. A1 is a low-noise op. amp. (e.g. LM 607), A2 is a general purpose op. amp. (e.g. LF 356). The circuit may be powered by two 9 V batteries or from a conventional ±15 VDC source.

possible to the signal source. The input impedance (Z_{in}) of an amplifier is usually expressed as the input resistance (R_{in}) from input to ground paralleled by the input capacitance (C_{in}). The effect of R_{in} is to short circuit the liquid membrane column of an ion-sensitive microelectrode. The input resistance of an amplifier is defined as the ratio of a given change in input voltage to the corresponding change in input current at steady state. Due to this relationship being often non-linear in electrometer circuits and the input current seldom vanishing with zero input voltage, the concept of input resistance is not very meaningful in this context. Instead, the performance of an electrometer amplifier is best described if an upper limit is given for the input bias current (I_b) throughout the whole input voltage range.

A common practice in electrophysiological laboratories is to build electrometer amplifiers instead of buying them. An operational amplifier with an I_b as low as possible is used as the input stage. Since the Ag:AgCl wire is frequently touched with fingers, the input must also tolerate some static discharge.

The best operational amplifier for use with ion-sensitive microelectrodes is no longer available. It is the ultra-low bias current, varactor-bridge operational amplifier AD311J from Analog Devices. In our amplifiers, I_b has stayed at about 6 fA (6×10^{-15}A) for years. However, other components that are both cheaper and smaller and have a reasonable I_b are available. The rather old AD515L (AD515AL) is available from both Analog Devices and Burr Brown and has a maximum input bias current of 75 fA. Its input is reliably protected by a series resistor (1 MΩ is probably sufficient but we have used resistors up to 100 MΩ). The AD549L (Analog Devices) is its improved version and has a typical I_b of 40 fA (60 fA max.). An OPA128LM (40 fA typ., 75 fA max., Burr Brown) may also be used. Other components exist, but do not try to combine an unprotected operational amplifier with an input protection circuit including ultra low-leakage diodes (e.g. FDH300 or FDH333) since this will result in an increase in I_b.

A small capacitor is used to couple an adjustable positive feedback signal to the amplifier input. This compensates for the effect of stray capacitances, which would otherwise reduce the input impedance to all AC signals (capacitance compensation, 'negative capacitance'). If a linear ramp is added to the feedback signal, a constant current flows to the amplifier input ($I = C \cdot \delta V/\delta t$) making it possible to measure the electrode resistance.

A poor lay-out can totally ruin the performance of an ambitiously designed electrometer circuit. Optimal performance cannot be achieved by using the often recommended guard loops around input lines on a printed circuit board. Instead, all connections of the input signal to the circuit board should be avoided and, if necessary, teflon stand-offs should be used for input wiring. Most BNC connectors use pure teflon as the insulator and are suitable as the input connector. Ordinary capacitors are too leaky to be used in coupling the capacitance compensation signal to amplifier input. Since the required capacitance value is only a few pF, it is preferable to construct a teflon or air insulated capacitor for this purpose.

The easiest way of measuring the I_b of an amplifier is to connect its input to ground via a low leakage capacitor. After instantaneously grounding the input, I_b starts to charge the capacitor causing a drift on the recorded signal, and I_b is obtained as $I_b = C \cdot \delta V/\delta t$. Fig. 9 shows a circuit diagram of an electrometer amplifier that allows capacitance compensation and measurement of the electrode resistance. For simplicity the circuit is based on standard operational amplifiers and a separate bias current compensation circuit is not included. Indeed, bias current with the AD 549L is low enough not to require compensation.

Differential amplifier and filtering

The gains of an electrometer amplifier, a reference electrode amplifier and a differential amplifier are easily adjusted to show no change at the differential output upon a common mode input signal (a test signal connected simultaneously to both inputs). However, the different time constants of ion-sensitive and conventional microelectrodes requires that the reference signal is slowed down

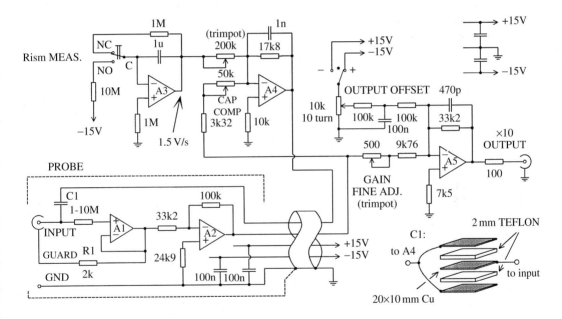

Fig. 9. Circuit diagram of an electrometer amplifier designed for use with ion-sensitive microelectrodes. The amplifier provides capacitance compensation, electrode resisitance (Rism) measurement and an output with offset and a gain of 10. The active guard signal can be obtained via R1 as shown, but it is advisable to replace this resistor by a voltage follower. The feedback capacitor C1 should be constructed from thin copper and 2 mm teflon plates sandwiched as shown in the inset. The 200 kΩ trimpot is adjusted to obtain a constant current of 1 pA through C1 for a few seconds when pressing the Rism MEAS. pushbutton. For the adjustment, note that a current of 1 pA induces a voltage change of 1 mV/1 GΩ (10 mV at output) when flowing through an electrode or a test resistor, or a linear ramp of $1~\text{mV.s}^{-1}$ ($10~\text{mV.s}^{-1}$ at output) if the input is connected to ground with a 1 nF capacitor and briefly short-circuited just before the measurement. A1 is preferably an AD 549L (Analog Devices), but an AD 515L, AD 515AL or an OPA 128LM (Burr Brown) may also be used. A2-A5 e.g. LF 356 (A2 must tolerate the capacitive load of the probe cable), resistors up to 1 MΩ with 1% tolerance.

before subtraction to mimic the frequency response of the ion electrode. A simple adjustable RC filter gives a reasonable approximation of the distributed resistance and stray capacitances of an ion-sensitive microelectrode. This kind of adjustment is important when recording from excitable cells and in voltage clamp experiments, where step changes in membrane potential are otherwise seen as large transient deflections on the differential ion signal. Suggested circuitry is shown in Fig. 10A.

The correct procedure to minimise transient deflections on the differential signal due to rapid membrane potential changes includes the following steps made in this order: (1) capacitance compensation adjustment of the ion-sensitive electrode, (2) slowing down of the reference signal and (3) low pass filtering of the differential signal. Fig. 10B shows the effect of these steps performed on an equivalent circuit mimicking an intracellular recording with step changes in membrane potential at a constant ion concentration.

Noise reduction

In practical work with ion-sensitive microelectrodes, capacitive coupling is the main source of noise. Stray capacitances from surrounding structures to the electrode and the Ag:AgCl wire mediate small currents, which flow through the high resistance of the membrane column, thereby giving rise to a voltage signal seen as noise in the recording. Capacitive current is usually given by $I = C \cdot \delta V/\delta t$ but it includes the assumption of a constant C. The general equation for capacitive current is obtained from $Q = CV$ and it is:

$$I = \left(C \cdot \frac{\delta V}{\delta t} \right) + \left(V \cdot \frac{\delta C}{\delta t} \right) \tag{12}$$

Therefore, capacitive coupling to ion-sensitive microelectrodes is minimised if: (1) cables etc. with AC (mains) signals are removed (to minimise $\delta V/\delta t$), (2) the total surface area of the wiring from the electrode to the amplifier input is as small as possible (to minimise C), (3) solutions have a steady surface level (pay particular attention to experimental bath design if the superfusate is to be pre-bubbled and if necessary, shield the waste line from the bath by a grounded metal tube and fix all wires and superfusion tubing so that they cannot move (to minimise $\delta C/\delta t$)), and (4) the lab has a proper antistatic floor coating (to minimise V associated with static charges on your clothes and skin). (Static charges can be a real problem if the relative humidity of air is very low. A portable humidifier helps in such cases).

The experiments should be performed within a Faraday cage which intercepts stray capacitances to external noise sources. The headstages of all microelectrode amplifiers should be positioned in the vicinity of the electrodes and ion-sensitive microelectrodes should be connected to their amplifiers with thin teflon-coated or non-insulated wires. (Some insulating materials generate a slowly declining noise current upon bending the cable.) Shielded cables even with the shield driven by an active guard signal should not be used to connect ion-sensitive microelectrodes, because the insulation resistance between the signal wire and the shield is not necessarily high enough. Direct capacitive coupling from a current-passing microelectrode can be reduced by wrapping a small piece of aluminium foil round it and grounding the foil.

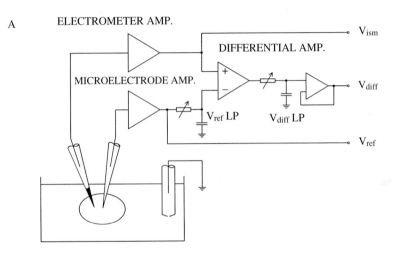

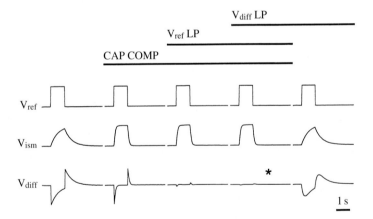

Fig. 10. (A) Circuitry for intracellular ion measurements. V_{diff}LP and V_{ref}LP indicate adjustable low-pass filters for the subtracted true ion signal (V_{diff}) and for the reference electrode signal (V_{ref}) respectively. The ion-sensitive microelectrode signal is V_{ism}. (B) Reduction of transient artifacts on the ion signal V_{diff}, caused by rapid changes in membrane potential in an intracellular recording, can be obtained by sequential adjustment of the capacitance compensation (CAP COMP) and of the two low-pass filters shown in A as described in the text (optimal response indicated by *). The measurements were made using an equivalent circuit with a 100 GΩ resistor as an ion-sensitive microelectrode and a 10 MΩ resistor as the reference electrode. Square pulses were connected as a common mode signal to both resistors to mimic step changes in membrane potential taking place at a constant intracellular ion activity. Note that low-pass filtering of the differential signal alone gives a poor result, as is evident from the trace on the right.

5. Practical electrode calibration

Empirical methods

There are various methods for calibrating ion-sensitive microelectrodes and the easiest and most practical from a biologists point of view are empirical techniques where no effort is made to ascertain the selectivity coefficients and there is no absolute requirement for knowledge of the activity coefficient (see Section 2, *Ion activity*). In addition to the extracellular solution which is used as a reference point one simply makes up solutions that are representative of the intracellular environment containing a fixed amount of the major interfering ion(s) likely to be encountered intracellularly and varying amounts of the primary ion. It is important to maintain constant ionic strength (so that γ is the same, for the reasons outlined in Section 2, *Ion activity*) and one way of achieving this is to make so-called "reciprocal dilutions" with a substituting ion of the same valency which does not interfere with the sensor. Calibration of the electrode should also be undertaken at the same temperature at which the experiment is performed because, as equation 9 shows, the slope obtained is dependent upon temperature. The reference point for calibration is obtained in the physiological solution which is used as a perfusate during impalements. The intracellular measurements obtained after such a calibration procedure are a measure of the intracellular free ion concentration (assuming γ is the same in the cytoplasm as in the calibration solutions). This method should be used with care for its approach assumes that the intracellular activity of the interfering ion does not change during the experiment. If it does, one has to find out by how much the interfering ion changes and calibrate using best and worst case solutions. The empirical approach to electrode calibration is outlined in Fig. 11.

Another way to calibrate electrodes is by a more formal approach involving calculation of the selectivity coefficient, K_{ij}. While this technique is certainly more rigorous it is probably of less benefit to the biologist than the physical chemist. There are two main reasons for this which stem from problems in the determination of K_{ij}. Firstly, the value of K_{ij} differs depending upon which method is used (see below); secondly, K_{ij} varies with the size of the ion-sensitive microelectrode tip (see e.g. Bers & Ellis, 1982) so it cannot be assumed to be constant from one electrode to the next.

Separate solution method

The potentials (E) are recorded from an electrode in single electrolyte solutions of varying activity of the primary ion (a_i) and are compared with potentials recorded in single electrolyte solutions of varying activity of the interfering ion (a_j). The selectivity coefficient is then calculated from the following equation:

$$K_{ij} = \frac{a_i}{a_j^{z_i/z_j}} \times 10^{(E_j-E_i)/m} \tag{13}$$

where m is the slope of the electrode (negative for anions).

Fixed interference method

With this method, potentials are recorded from an electrode in solutions of varying activity of the primary ion (a_i) and which also contain a fixed activity of the interfering ion (a_j). The selectivity coefficient is then obtained at the intercept of the asymptotes of the two linear portions of the calibration curve (i.e. at the detection limit):

$$K_{ij} = \frac{a_i}{a_j^{z_i/z_j}} \tag{14}$$

The empirical approach to electrode calibration is to be recommended but as we remarked earlier: however you calibrate, state how you did it!

6. Analysis of results

Ion fluxes

The net flux of an ion across the cell membrane can often be deduced by measuring the change in the intracellular activity of this ion over time, i.e. $\delta a_i/\delta t$. The net flux (J) is then defined as:

$$J = \frac{V{:}A}{\gamma_i} \cdot \left(-\frac{\delta a_i}{\delta t} \right) \tag{15}$$

where γ_i is the intracellular activity coefficient for the ion, V:A is the volume/surface area ratio of the cell. Flux is then a change in amount per area and time, commonly expressed as $mmol.min^{-1}.cm^{-2}$, and an outward flux has a positive value. Usually $\delta a_i/\delta t$ is taken as the instantaneous rate of change of a_i, which in a recording with an ion-selective electrode corresponds to the slope of the recorded trace at a given instant of time. As discussed above (Section 2, *Ion activity*) γ_i is usually assigned the value of the extracellular activity coefficient as a value for γ_i can normally not be obtained. The volume and/or surface area of a cell is often not known and so flux is expressed as a change in concentration per time ($mmol.l^{-1}.min^{-1}$), i.e.:

$$J = \frac{1}{\gamma_i} \cdot \left(-\frac{\delta a_i}{\delta t} \right) \tag{16}$$

A measured change in the intracellular ion activity can be converted to net flux *only* if it is known to be solely the result of transmembrane movement of the relevant ion and if the amount of intracellular buffering for this ion is known. It should be stressed that a change in the a_i can also result from, for example, a change in cell volume, or from movements of the ion into or out of subcellular compartments, like mitochondria or the nuclei.

In some cases it is possible to detect a transmembrane movement of an ion by means of an ion-sensitive electrode measuring a change in ion activity in an extracellular compartment (e.g. in brain slices) or at the outer surface of a cell membrane (see Fig. 5, Section 3). This type of measurement can be used to verify transmembrane movements of H^+, Ca^{2+} and K^+ as the transmembrane gradients and the physiological extracellular activities of these ions favour the detection of such movements with an ion-selective electrode. Convincing evidence for transmembrane fluxes of a given ion using ion-sensitive electrodes is often provided by the coupling of an intracellular and a surface ion measurement. An example of a simultaneous measurement of intracellular and extracellular pH is seen in Fig. 5.

Flux measurements made with ion-selective microelectrodes can be used for

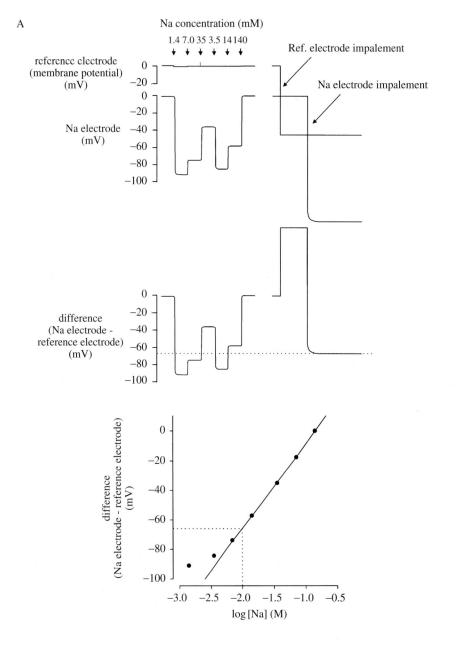

B

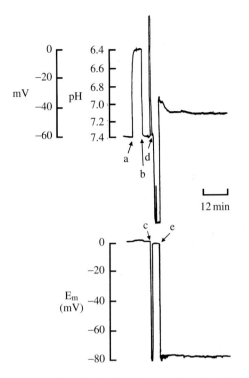

12 min

Fig. 11. (A) Representation of a typical calibration procedure for a Na$^+$-sensitive microelectrode. At the left side of the figure the reference electrode and the Na$^+$-sensitive microelectrode were exposed to a physiological saline including 140 mM NaCl and 4mM KCl and five other solutions containing various amounts of NaCl and KCl at a total concentration of 144 mM in addition to 1 mM MgCl$_2$, 10 mM glucose, 5 mM Pipes, 1 mM EGTA + 0.3 mM CaCl$_2$ to give a free [Ca^{2+}] of 0.1 μM, pH, 7.20. In this way, calibration for intracellular levels of Na$^+$ was carried out under conditions of constant ionic strength, with a high background of K$^+$ (the major interfering ion expected in the cytoplasm) and with values of pH and free [Ca^{2+}] also near those likely encountered in the cytoplasm. The calibration curve on the right was constructed from the values obtained from the difference signal. The solid line drawn through the points has a slope of 57 mV. Notice that at low concentrations of Na$^+$, the response of the electrode deviates from linearity as interference from K$^+$ begins to predominate. After the calibration was completed, the preparation was taken to the bath and superfused with a physiological saline. The reference electrode was pushed into the cell giving a value of −81mV for membrane potential. This produced an equal but opposite movement on the difference signal. Some minutes later, the Na$^+$-sensitive electrode was pushed into the cell. This produced a large deviation on the trace due to the electrode measuring (1) a much lower [Na$^+$] than outside the cell and (2) the membrane potential. The membrane potential is subtracted from the combined signal, producing, on the difference trace, a value of −64 mV which corresponds to an intracellular Na$^+$ concentration of 10.7 mM (dotted line). If $\gamma_{Na} = 0.75$ then intracellular Na$^+$ activity is about 8 mM. (B) Electrode calibration and subsequent impalement of a sheep cardiac Purkinje fibre with a pH-sensitive microelectrode. At point a, the solution bathing the tips of the reference (E_m) and pH-sensitive microelectrode was changed from a pH of 7.4 to a pH of 6.4. This 10-fold change in H$^+$ resulted in a 60 mV change in voltage from the pH electrode. At point b the solution was changed back to a pH of 7.4. Next, an impalement of the cell by the reference electrode was made (point c) and because the pH trace already has the reference potential subtracted (i.e. it is the difference) it moved an equal amount in the opposite direction. This impalement was soon lost while pushing in the pH electrode (point d). Finally, the reference electrode was put back into the cell (point e) which allows the intracellular pH (difference) to be read directly from the calibration. This gives an intracellular pH of about 7.15 and a membrane potential (E_m) of about −80 mV.

dissecting the ionic components of a current carried by more than one ion. An example of this is shown in Fig. 12. A crayfish muscle fibre is penetrated with conventional microelectrodes for measuring membrane voltage and injecting current, as well as with a Cl^--sensitive and a H^+-sensitive microelectrode. In a solution buffered with CO_2 and HCO_3^- the cell is clamped at different voltages and short pulses of GABA (γ-aminobutyric acid) are applied. The recording shows that the GABA-induced changes in intracellular chloride are dependent on the membrane voltage. Their direction reverses, by definition, at the chloride reversal potential, which is approximately at the resting membrane potential (RP). The current evoked by GABA is, however, not solely carried by chloride, since it reverses at a potential more positive to the chloride reversal potential. The other component of the GABA-induced current is now known to be HCO_3^-, to which GABA channels are also permeable. In the absence of HCO_3^- the GABA-induced current reverses at the chloride reversal potential.

Buffering power

The capability of a solution to "resist" changes in the activity of a given ion upon a

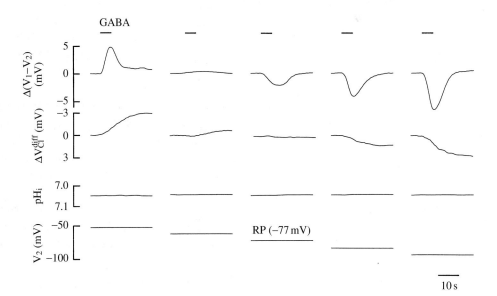

Fig. 12. Simultaneous measurement of the GABA-current and chloride reversal potentials in a crayfish muscle fibre. The muscle fibre is penetrated with five electrodes: three conventional ones for voltage-clamping in a three electrode voltage-clamp configuration, and two ion-sensitive electrodes for measuring changes in the intracellular chloride level (ΔV_{Cl}^{diff}) and the intracellular pH. $\Delta(V_1-V_2)$ is the difference in voltage seen by the electrode placed in the middle of the fibre (V_1) and that placed halfway between the middle and end (V_2). $\Delta(V_1-V_2)$ is proportional to the membrane current at the site of V_2. pH_i is measured to show that no changes in intracellular pH occur during the short GABA applications, as changes in pH_i would change the driving force for HCO_3^-. The chloride reversal potential is approximately at the resting membrane potential (RP; middle traces) where the intracellular Cl^- does not change upon the GABA application. The GABA-induced current (carried by both Cl^- and HCO_3^-) reverses at a more positive holding potential, where GABA induces a Cl^- influx. (Reproduced with permission from Kaila et al. 1989).

change in its concentration is called the buffering power of this solution for the ion. As stated above, a measured change in the ion activity cannot be converted to a flux unless the extent of buffering of this ion is known. Usually cells have appreciable buffering of H^+ and Ca^{2+} but other ions can be considered to be unbuffered. For this reason changes in K^+, Na^+ and Cl^- are usually directly converted to transmembrane fluxes, but in doing so one should bear in mind the possibility of cellular volume changes (see below).

Buffering of H^+

The H^+ buffering power (β) is defined as:

$$\beta = \frac{\Delta A}{-\Delta pH} \left(= \frac{\Delta B}{\Delta pH} \right) \tag{17}$$

where ΔA and ΔB are the changes in acid or base concentration respectively, which cause a change in pH (ΔpH). Intracellular H^+ buffering is usually divided into two categories: the intrinsic buffering power (β_i) which is attributable to intracellular proteins and the "CO_2 buffering power" (β_{CO_2}) which is due to the presence of CO_2 and bicarbonate (HCO_3^-). β_i is commonly between 10 and 50 $mmol.l^{-1}$ and is, in many cells, fairly constant within the physiological intracellular pH range. β_{CO_2} is the result of the familiar equilibrium between CO_2 and HCO_3^-:

$$CO_2 + H_2O \rightleftharpoons H_2CO_3 \rightleftharpoons H^+ + HCO_3^- \tag{18}$$

In an open buffer system, i.e. when CO_2 is held constant and the above reactions are fast (see below), it can be shown that $\beta_{CO_2} = 2.3[HCO_3^-]$ and hence β_{CO_2} is steeply dependent on pH. All buffering is additive, i.e. an increase in β_{CO_2}, for example, will add linearly to the total buffering power of the cell.

Equation 15 can now be rewritten for proton fluxes to give:

$$J^{H^+} = (\beta_i + \beta_{CO_2}) \cdot (V:A) \cdot \frac{\Delta pH}{\Delta t} \tag{19}$$

As pH is a logarithmic function, the above equation can also be rewritten replacing ΔpH with the voltage change recorded with a H^+-sensitive electrode (ΔV_{pH_i}):

$$J^{H^+} = (\beta_i + \beta_{CO_2}) \cdot (V:A) \cdot \left(-\frac{\Delta V_{pH_i}}{m\Delta t} \right) \tag{20}$$

where m is the slope of the electrode. Again the $V:A$ ratio is often omitted and the flux given as $mmol.l^{-1}.min^{-1}$.

One should note however, that rapid CO_2 buffering requires catalysis of the reversible hydration reaction of CO_2 to H_2CO_3, since this reaction is fairly slow in the absence of a catalyst (the half time is about 10-30 s at 21-37°C). In biological systems the catalyst for the reaction is the enzyme carbonic anhydrase (CA). Most cells

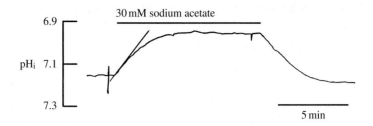

Fig. 13. Measurement of the intrinsic buffering power (β_i) of a crayfish muscle fibre. The intracellular pH (pH$_i$) is recorded with a H$^+$-selective microelectrode while the fibre is subjected to a solution containing 30 mM sodium acetate. The measurement gives a value for β_i of 53.2 mmol.l^{-1}. A line is drawn to indicate the initial slope of the change in pH$_i$ (ΔpH$_i$/Δt) used for calculating the initial acid flux into the cell (see text). (From an experiment made by the authors and a group of students in 1989 at the Plymouth workshop).

possess intracellular CA activity and some cells even have membrane-bound, extracellularly active CA (e.g. Kaila et al., 1990, 1992). In the absence of rapid catalysis, the CO_2 buffering becomes time-dependent and pH measurements may be more difficult to interpret quantitatively.

β_i can be estimated by introducing an acid or base load of known magnitude to the cell and measuring the evoked changes in intracellular pH in the absence of pH regulation. An example of the estimation of β_i of a crayfish muscle fibre using a membrane permeant weak acid is given in Fig. 13.

A crayfish muscle fibre is penetrated with two microelectrodes, a conventional electrode for measuring the membrane voltage and a H$^+$-sensitive microelectrode. In the crayfish muscle fibre the regulation of intracellular pH is almost totally dependent on HCO_3^-. When the cell is subjected to 30 mM of the Na$^+$-salt of acetic acid, sodium acetate (CH_3COONa), in a Hepes-buffered solution (pH 7.40) in the absence of CO_2 and HCO_3^-, the intracellular pH decreases with an exponential time course until it reaches a new, more acidic steady-state value. This is because the electrically neutral acetic acid is capable of penetrating the cell membrane causing an intracellular acidification. The acidification proceeds with an exponential time course until acetic acid has equilibrated across the membrane. The dissociation constant for an acid is defined as:

$$K_a = \frac{[H^+]\,[A^-]}{HA} \tag{21}$$

where HA is the non-dissociated acid and A^- is the acid anion. Assuming that the extracellular (K_a^o) and intracellular (K_a^i) dissociation constants are equal, one can write:

$$K_a^o = K_a^i = \frac{[H^+]_o\,[A^-]_o}{[HA]_o} = \frac{[H^+]_i\,[A^-]_i}{[HA]_i} \tag{22}$$

When *HA* is in equilibrium across the membrane, i.e. $[HA]_o = [HA]_i$ the equation can be solved for $[A^-]_i$:

$$[H^+]_o \, [A^-]_o = [H^+]_i \, [A^-]_i \Leftrightarrow$$

$$[A^-]_i = \frac{[H^+]_o}{[H^+]_i} \cdot [A^-]_o \Leftrightarrow$$

$$\log[A^-]_i = \log[H^+]_o - \log[H^+]_i + \log[A^-]_o \Leftrightarrow \qquad (23)$$

$$[A^-]_i = 10^{pH_i - pH_o} \cdot [A^-]_o$$

Since acetic acid is almost completely dissociated at physiological pH, $[A^-]_o$ can be considered equal to the total amount of acetate added, i.e. 30 mM. In this cell the intracellular pH change (ΔpH) is −0.2 pH units (from pH 7.15 to pH 6.95) and at this pH_i calculation of $[A^-]_i$ from equation 23 gives a value of 10.64 mM, which is equal to the intracellular acid load ($[H^+]_i$, or ΔA in equation 17). Fitting the values for ΔpH and $[H^+]_i$ to equation 17 gives β_i a value of 53.2 mmol.l^{-1}. The initial acid flux into the cell, J^{H^+}, can now be calculated using the measured value of β_i and the initial rate of change in pH_i ($\Delta pH_i/\Delta t = -0.08$.min^{-1}). Fitting these values into equation 19 gives a value of −4.26 mmol.l^{-1}.min^{-1}. The remarkable H$^+$ buffering power is seen in that the above measured changes in H$^+$ activity are in the submicromolar scale, whereas the acid load is expressed in mmol.l^{-1}!

In many cell types pH regulation involves several mechanisms and measurements of β_i with manipulations like the above may give incorrect results because of an attenuation of the evoked acidosis by the pH regulatory mechanism(s). It may be possible in some cells to use a weak base instead of a weak acid to circumvent the problem of pH regulation. However the cells may be equipped with a mechanism for pH regulation capable of intake of acid equivalents. An alternative possibility is to use a pharmacological inhibitor of the pH regulating mechanism(s).

Buffering of Ca^{2+}

Intracellular calcium ions are buffered in several ways: (1) Like H$^+$, Ca^{2+} can bind to intracellular proteins, (2) Ca^{2+} can be sequestered into endoplasmic or sarcoplasmic reticulum and cellular organelles, such as mitochondria, and (3) Ca^{2+} can bind to sites on the intracellular plasmalemma surface. Some of the buffering sites for Ca^{2+} may be shared with those for H$^+$ and so changes in intracellular pH may affect the levels of free Ca^{2+} concentration (often given as the logarithmic term pCa, akin to pH) and *vice versa*. The release and uptake of Ca^{2+} from and into internal stores, as well as the competing action of H$^+$ for Ca^{2+} buffering sites makes it difficult to assess the cellular Ca^{2+} buffering power.

The changes in intracellular pCa are often rapid and cannot be detected accurately with ion-sensitive electrodes. The response times of a Ca^{2+}-sensitive microelectrodes are, even with proper capacitance compensation (see Section 4, *Electrometer amplifier design* and *Differential amplifier and filtering*), in the order of a second. Ca^{2+}-sensitive electrodes can be used for the detection of *slow* changes in pCa (i.e.

those occurring over time courses from seconds to minutes). However, in the interpretation of Ca^{2+}-electrode measurements one should bear in mind that undetectable fast transients in pCa may be involved in the development of the slower changes, e.g. a rapid undetected increase in pCa may trigger an enhancement in intracellular Ca^{2+} storage and thus result in a decrease in pCa. In such a case the experimental manipulation would cause a biphasic change in pCa but only the slow decrease would be apparent to the experimenter.

Changes in cellular volume

A change in cellular volume will obviously result in a change in the activities of all intracellular ions, if the volume change is not parallelled with ion movements. On the other hand, fluxes that are not balanced with osmotically equivalent counterfluxes will change the cellular osmolarity and result in changes in cell volume. This may introduce errors in the quantitative analysis of ion-selective electrode recordings, unless the volume change is taken into account during the analysis.

The overall changes in intracellular ion activities produced by cotransport mechanisms, e.g. the K^+-Cl^--cotransport or the Na^+-K^+-Cl^- cotransport, may be complex, since water movements are perhaps more pronounced than those related to countertransport.

Fig. 14 depicts an example of the consequences of the activation of a Na^+-K^+-Cl^- cotransport. The stochiometry of this transport is assumed here to be 1 Na^+:1 K^+:2 Cl^-. In this example 5 mM of Na^+ and K^+ together with 10 mM of Cl^- are transported into the cell. Assuming no other ion movements, the end result of this transport is an increase of about 6.5% in the cell volume. The final ion activities are, due to this swelling, decreased by about 6% from those expected in the absence of water movements, resulting in the activities shown in the figure. As can be seen, intracellular Na^+ activity increases by 88% and intracellular Cl^- by 90% of the values expected with no swelling. What is perhaps more difficult to comprehend at first glance, is that K^+ activity *decreases* from its initial value. This is because the final activities of all solutes are decreased in the same proportion (6%) resulting in a larger absolute decrease in K^+ activity than in Na^+ or Cl^- activity.

In the flux equations given above the cellular volume to surface area ratio (V:A) is assumed to be constant. Obviously a change in this ratio will cause some error in the estimate of the flux, unless the flux is taken as the instantaneous change in ion activity and the V:A of the cell at this point in time is known.

At first glance a countertransporter mediating equal fluxes of ions in and out of the cell is not expected to induce volume changes. However, the activation of an equimolar countertransport may change cell volume if one of the transported ions is buffered. The activation of Na^+/H^+ exchange or Cl^-/HCO_3^- exchange can lead to changes in cell volume, since H^+ is buffered and changes in pH therefore do not contribute equally to the cellular osmolarity. HCO_3^- on the other hand may, together with H^+ (see equation 18), exit the cell as CO_2 and thus cause a decrease in intracellular osmolarity.

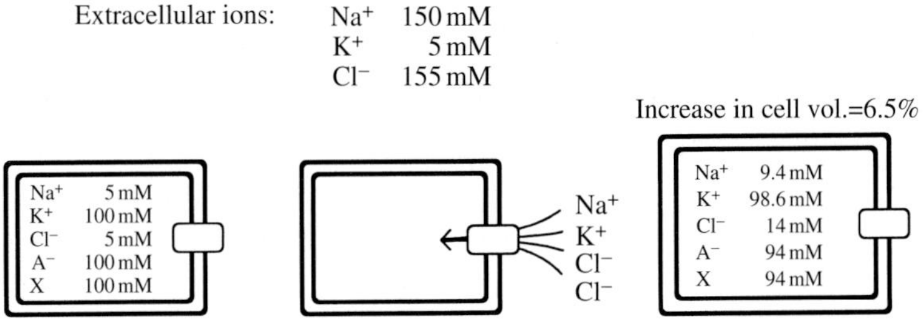

Extracellular ions: Na⁺   150 mM
                        K⁺   5 mM
                        Cl⁻   155 mM

Increase in cell vol.=6.5%

Na⁺	5 mM
K⁺	100 mM
Cl⁻	5 mM
A⁻	100 mM
X	100 mM

Na⁺
K⁺
Cl⁻
Cl⁻

Na⁺	9.4 mM
K⁺	98.6 mM
Cl⁻	14 mM
A⁻	94 mM
X	94 mM

Fig. 14. Effect of Na⁺:K⁺:Cl⁻ cotransport on cell volume and ion activities. Activation of a Na⁺:K⁺:Cl⁻ cotransport results in an increase of the cell volume, which attenuates the increase in intracellular ions. For further explanation, see text. (X, intracellular impermeant solutes; A⁻, intracellular impermeant anions).

7. Appendix: Ca^{2+}-sensitive electrodes and their use in measuring the affinity and purity of Ca buffers

One important use of Ca^{2+}-sensitive microelectrodes which is often overlooked is in the preparation of solutions of known Ca^{2+} concentration. These solutions can be used subsequently to, for example, calibrate Ca^{2+}-sensitive electrodes, fluorescent indicators or bathe skinned muscle preparations. Recall that the association constant (K_a) describes how well a calcium buffer (e.g. EGTA, NTA) binds Ca^{2+}:

$$K_a = \frac{[Ca]_b}{[Ca]_f \cdot [Ca\ buffer]_f} \tag{24}$$

where the subscripts b and f denote bound and free concentrations respectively. One of the difficulties in making up solutions of varying Ca^{2+} concentration is that association constants for calcium buffers are sensitive to ionic strength, temperature and can be particularly sensitive to pH (Bers & MacLeod, 1988; Harrison & Bers, 1987, 1989). Thus if any of these variables change then the free $[Ca^{2+}]$ in solution may be quite different from the expected value. The effect of pH on K_a should not be underestimated. It can be easy to make an error of 0.1 pH units particularly if the electrode is poorly calibrated and it is very worrying that many electrodes incorporating a ceramic porous plug can generate large liquid junction potentials when the ionic strength of the solution is altered which, in turn, can cause the pH measurement to be in error by an average of 0.2 pH units (Illingworth, 1981). Our advice when making pH measurements is to use a separate reference electrode with a flowing junction, calibrate carefully, check this with a known standard (i.e. a solution of 25 mM KH_2PO_4/25 mM Na_2HPO_4 which should give a pH of 6.865 at 25°C and when diluted 10 times with distilled water the pH should be 7.065) and calibrate often (every hour if you are using the electrode continuously). Not only is it important to be careful with consistency of the above variables, but also it is important to consider the purity of your sample of calcium buffer (Bers, 1982; Miller & Smith, 1984).

Variations in the purity can cause significant errors in the final free $[Ca^{2+}]$. It is important to realise to what extent an error in each of these will alter the final free $[Ca^{2+}]$. Table 4 shows for EGTA how a 5% change in purity, or a 0.2 unit error in pH or changing the temperature of the solution from 25 to 35°C alters the expected value of free $[Ca^{2+}]$. Making up solutions with accurate and precise Ca^{2+} concentrations is not difficult provided care is taken with weighing any calcium buffer, adding Ca^{2+} usually as $CaCl_2$ and making sure the solution pH is correct to within ±0.005 units (especially if using EGTA as the Ca^{2+} buffer). The pH of the solutions need not be as accurate if one uses BAPTA or dibromoBAPTA (Tsien, 1980) as the calcium buffer. These compounds exhibit good selectivity for Ca^{2+} over Mg^{2+} (like EGTA) but their affinities are much less sensitive to pH changes around physiological pH values (Tsien, 1980; Bers & MacLeod, 1986; Harrison & Bers, 1987).

Steps for making up Ca^{2+} calibration solutions:

(1) Calculate the binding constant for your buffer in the solution composition you will use

(2) Determine the purity of your batch of buffer

(3) Knowing the above, calculate the total $[Ca^{2+}]$ required to be added to your solutions using, for simple solutions equation (25) or, for more complex, multi-ligand solutions, a computer programme (Fabiato & Fabiato, 1979)

(4) Make up your solutions to about 90% of the required volumes weighing accurately and precisely the calcium buffer

(5) pH your solutions to pH 7.000 and add $CaCl_2$

(6) Re-pH to desired pH ± 0.005 unit

(7) Make up to near target volume and check pH - adjust if necessary and then make up to final volume.

Table 4. *The effects of altering purity, pH or temperature on free [Ca2+] in buffered solutions*

expected free $[Ca^{2+}]$		observed free $[Ca^{2+}]$ (µM)		
pCa	µM	Ca buffer 95% pure	pH error +0.2 unit	Δ temp. to 35°C
8.000	0.010	0.011	0.004	0.010
7.699	0.020	0.021	0.008	0.019
7.301	0.050	0.053	0.019	0.048
7.000	0.100	0.108	0.039	0.096
6.699	0.200	0.221	0.079	0.191
6.301	0.500	0.600	0.199	0.480
6.000	1.000	1.379	0.399	0.956
5.699	2.000	3.920	0.802	1.919
5.301	5.000	22.21	2.098	4.778
5.000	10.000	43.70	4.769	9.805

Ca buffer is EGTA (1 mM). K_a (expressed as $\log_{10}$ M^{-1}) of 6.45 is derived from constants in Martell & Smith (1974); 0.1 M ionic strength, pH 7.00, 25°C. pH sensitivity of $K_a = 0.2/0.1$ pH unit (see Bers & MacLeod, 1988). Temperature correction made using $\Delta H = 16.6$ kJ mol^{-1} (Harrison & Bers, 1987).

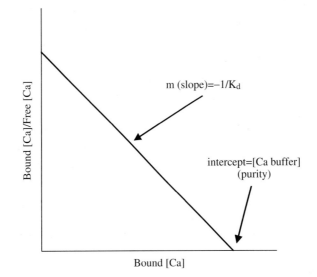

Fig. 15. Scatchard plot showing how K_a ($1/K_d$) and the purity of the Ca buffer are obtained from a plot of Bound [Ca]/ Free [Ca] against Bound [Ca].

Assuming that there is only one calcium buffer in your solutions then the total $[Ca^{2+}]$ can be calculated from the following equation:

$$[C_a]_t = \frac{[Ca\ buffer]_t + [Ca]_f + \dfrac{1}{K_a}}{1 + \dfrac{1}{K_a[Ca]_f}} \quad (25)$$

where the square brackets denote concentration of the species and the subscripts f and t denote free and total respectively. It is often useful to be able to calculate the free $[Ca^{2+}]$ knowing the total $[Ca^{2+}]$ added and this can be done by solving the quadratic:

$$[C_a]_f = \frac{\left([Ca]_t - [Ca\ buffer]_t - \dfrac{1}{K_a}\right) + \sqrt{\left([Ca\ buffer]_t - [Ca]_t + \dfrac{1}{K_a}\right)^2 + \dfrac{4[Ca]_t}{K_a}}}{2} \quad (26)$$

Ca^{2+}-sensitive electrodes can be used to calculate the binding constant of the buffer and its purity (Bers, 1982) although there are other methods (Miller & Smith, 1984; Smith & Miller, 1985). The procedure involves making an educated guess at K_a, assuming buffer purity is 100% and then making (say) 5-10 solutions aiming to obtain target pCa values of (say) 8.0, 7.5, 7.0, 6.5, 6.0, 5.5, 5.0, 4.5 and 4.0. You then measure the free $[Ca^{2+}]$ with a Ca^{2+}-sensitive electrode and calculate the corresponding bound [Ca^{2+}] in each solution. Once you have done this you will be able to produce a Scatchard-type plot of bound Ca/free Ca against bound Ca. The slope of this relationship gives you the affinity (K_a) and the intercept the purity of the buffer. Equilibrium binding is analysed by a Scatchard plot which is a modification of the Eadie-Scatchard rearrangement of Michaelis-Menten enzyme kinetics. Applying it to our needs yields the equation:

$$\frac{[Ca]_b}{[Ca]_f} = -\frac{1}{K_d}[Ca]_b + \frac{[Ca\ buffer]_{total}}{K_d} \tag{27}$$

where K_d is the <u>dissociation</u> constant for the Ca buffer, the square brackets denote concentration of the species and the subscripts b, f and t denote bound, free and total respectively. Thus a plot of the bound:free [Ca] against bound [Ca] is linear with a slope of $-1/K_d$ or simply $-K_a$ (see Fig. 15). The intercept with the x-axis gives the <u>total</u> concentration of Ca buffer in the solution. Assuming you weighed out the Ca buffer with a good degree of accuracy then this figure allows you to calculate the purity of the batch of EGTA or BAPTA etc. you are using. The advantage of the method is that you can establish the K_a for the calcium buffer in the solutions you wish to use and at the pH you desire.

Ca^{2+}-sensitive electrodes can be obtained commercially (Orion) or they can be made more cheaply yourself. The advantage of making them yourself is that you can make small minielectrodes (1 mm≤tip size≤5 mm) which are useful for measuring small amounts of solution. Good methods for making Ca^{2+} minielectrodes can be found in Bers (1982) or Hove-Madsen & Bers (1992), see also Section 3, *Mini-electrodes*.

Acknowledgements

Research in the authors' laboratories is supported by the Academy of Finland (J. V. and M. P.) and the Medical Research Council, British Heart Foundation and the Royal Society (K. M.). We thank Prof. K. Kaila for comments on the manuscript.

References

ALVAREZ-LEEFMANS, F. J., GAMIÑO, S. M. & REUSS, L. (1992). Cell volume changes upon sodium pump inhibition in Helix Aspersa neurones. *J. Physiol.* **458**, 603-619.
AMMANN, D. (1986). *Ion-selective Microelectrodes.* Berlin: SpringerVerlag.
BATES, R. G. (1954). *Electronic pH Determinations, Theory and Practice.* London: John Wiley & Sons.
BAUMGARTEN, C. M. (1981). An improved liquid ion exchanger for chloride ion-selective microelectrodes. *Am. J. Physiol.* **241**, C258-263.
BERS, D. M. (1982). A simple method for the accurate determination of free [Ca] in Ca-EGTA solutions. *Am. J. Physiol.* **242**, 404-408.
BERS, D. M. & ELLIS, D. (1982). Intracellular calcium and sodium activity in sheep heart Purkinje fibres. Effect of changes of external sodium and intracellular pH. *Pflugers Archiv* **393**, 171-178.
BERS, D. M. & MACLEOD, K. T. (1986). Cumulative depletions of extracellular calcium in rabbit ventricular muscle monitored with calcium-selective microelectrodes. *Circulation Res.* **58**, 769-782.
BERS, D. M. & MACLEOD, K. T. (1988). Calcium chelators and calcium ionophores. In *Handbook of Experimental Pharmacology*, (ed. P. F. Baker), pp. 491-507. Berlin Heidelberg: Springer-Verlag.

BROWN, K. T. & FLAMING, D. G. (1975). Instrumentation and technique for beveling fine micropipette electrodes. *Brain Res.* **86**, 172-180.

COLES, J. A. & ORKAND, R. K. (1985). Changes in sodium activity during light stimulation in photoreceptors, glia and extracellular space in drone retina. *J. Physiol.* **362**, 415-435.

DIETZEL, I., HEINEMANN, U., HOFMEIER, G. & LUX, H. G. (1980). Transient changes in the size of the extracellular space in the sensorimotor cortex of cats in relation to stimulus-induced changes in potassium concentration. *Exp. Brain Res.* **40**, 432-439.

FABIATO, A. & FABIATO, F. (1979). Calculator programs for computing the composition of the solutions containing multiple metals and ligands used for experiments in skinned muscle cells. *J. Physiol. (Paris)* **75**, 463-505.

FRESSER, F., MOSER, H. & MAIR, N. (1991). Intra- and extracellular use and evaluation of ammonium-selective microelectrodes. *J. Exp. Biol.* **157**, 227-241.

FUNCK, R. J. J., MORF, W. E., SCHULTHESS, P., AMMANN, D & SIMON, W. (1982). Bicarbonate-sensitive liquid membrane electrodes based on neutral carriers for hydrogen ions. *Analyt. Chem.* **54**, 423-429.

HABLITZ, J. J. & HEINEMANN, U. (1989). Alterations in the microenvironment during spreading depression associated with epileptiform activity in the immature neocortex. *Dev. Brain Res.* **46**, 243-252.

HARRISON, S. M. & BERS, D. M. (1987). The effect of temperature and ionic strength on the apparent Ca-affinity of EGTA and the analogous Ca-chelators BAPTA and dibromo-BAPTA. *Biochim. Biophys. Acta* **925**, 133-143.

HARRISON, S. M. & BERS, D. M. (1989). Correction of proton and Ca association constants of EGTA for temperature and ionic strength. *Am. J. Physiol.* **256**, C1250-C1256.

HERMAN, H. B. & RECHNITZ, G. A. (1974). Carbonate ion-selective membrane electrode. *Science* **184**, 1074-1075.

HOVE-MADSEN, L. & BERS, D. M. (1992). Indo-1 binding to protein in permeabilized ventricular myocytes alters its spectral and Ca binding properties. *Biophys. J.* **63**, 89-97.

ILLINGWORTH, J. A. (1981). A common source of error in pH measurements. *Biochem. J.* **195**, 259-262.

KAILA, K., PASTERNACK, M., SAARIKOSKI, J. & VOIPIO, J. (1989). Influence of GABA-gated bicarbonate conductance on potential, current and intracellular chloride in crayfish muscle fibres. *J. Physiol.* **416**, 161-181.

KAILA, K., PAALASMAA, P., TAIRA, T. & VOIPIO, J. (1992). pH transients due to monosynaptic activation of GABA$_A$ receptors in rat hippocampal slices. *Neuroreport* **3**, 105-108.

KAILA, K., SAARIKOSKI, J. & VOIPIO, J. (1990). Mechanism of action of GABA on intracellular pH and on surface pH in crayfish muscle fibres. *J. Physiol.* **427**, 241-260.

KAILA, K. & VAUGHAN-JONES, R. D. (1987). Influence of sodium-hydrogen exchange on intracellular pH, sodium and tension in sheep cardiac Purkinje fibres. *J. Physiol.* **416**, 161-181.

KAILA, K. & VOIPIO, J. (1985). A simple method for dry bevelling of micropipettes used in the construction of ion-selective microelectrodes. *J. Physiol.* **369**, 8P.

KAILA, K. & VOIPIO, J. (1987). Postsynaptic fall in intracellular pH induced by GABA-activated bicarbonate conductance. *Nature* **330**, 163-165.

KLINE, R. P. & KUPERSMITH, J. (1982). Effects of extracellular potassium accumulation and sodium pump activation on automatic canine Purkinje fibres. *J. Physiol.* **324**, 507-533.

LEDERER, W. J., SPINDLER, A. J. & EISNER, D. A. (1979). Thick slurry bevelling. *Pflügers Archiv* **381**, 287-288.

MACLEOD, K. T. & BERS, D. M. (1987). Effects of rest duration and ryanodine on changes of extracellular (Ca) in cardiac muscle from rabbits. *Am. J. Physiol.* **253**, C398-C407.

MACLEOD, K. T. (1989). Effects of hypoxia and metabolic inhibition on the intracellular sodium activity of mammalian ventricular muscle. *J. Physiol.* **416**, 455-468.

MARTELL, A. E. & SMITH, R. M. (1974). *Critical Stability Constants.* New York: Plenum Press.

MILLER, D. J. & SMITH, G. L. (1984). EGTA purity and the buffering of calcium ions in physiological solutions. *Am. J. Physiol.* **246**, C160-C166.

NICHOLSON, C. & KRAIG, R. P. (1981). The behavior of exctracellular ions during spreading depression. In: "*The Application of Ion Selective Microelectrodes*", (ed. T. Zeuthen), pp. 217-238. Amsterdam: Elsevier.

NICHOLSON, C. & PHILLIPS, J. M. (1981). Ion diffusion modified by tortuosity and volume fraction in the extracellular microenvironment of the rat cerebellum. *J. Physiol.* **321**, 225-257.

NIEDRACH, L. W. (1975). Bicarbonate ion electrode and sensor. U.S.Patent 3,898,147.

SERVE, G., ENDRES, W. & GRAFE, P. (1988). Continuous electrophysiological measurements of changes in cell volume of motoneurons in the isolated frog spinal cord. *Pflügers Archiv.* **411**, 410-415.

SMITH, G. L. & MILLER, D. J. (1985). Potentiometric measurements of stoichiometric and apparent affinity constants of EGTA for protons and divalent ions including calcium. *Biochim. Biophys. Acta* **839**, 287-299.

STRICKHOLM, A. (1968). Reduction of response time for potential insalt bridge reference electrodes for electrophysiology. *Nature* **217**, 80-81.

THOMAS, R. C. (1978). *Ion-sensitive Intracellular Microelectrodes, How to Make and Use Them.* London, New York & San Francisco: Academic Press.

THOMAS, R. C. (1987). Extracellular acidification at the surface of depolarized voltage-clamped snail neurones detected with eccentric combination pH microelectrodes. *Can. J. Physiol. Pharmac.* **65**, 1001-1005.

THOMAS, R. C. & COHEN, C. J. (1981). A liquid ion-exchanger alternative to KCl for filling intracellular reference microelectrodes. *Pflügers Archiv* **390**, 96-98.

TSIEN, R. Y. (1980). New calcium indicators and buffers with high selectivity against magnesium and protons: design, synthesis, and properties of prototype structures. *Biochemistry* **19**, 2396-2404.

TSIEN, R. Y. (1983). Intracellular measurements of ion activities. *Ann. Rev. Biophys. Bioengineering* **12**, 91-116.

TSIEN, R. Y. & RINK, T. J. (1980). Neutral carrier ion-selective microelectrodes for measurement of intracellular free calcium. *Biochim. Biophys. Acta* **599**, 623-638.

TSIEN, R. Y. & RINK, T. J. (1981). Ca^{2+}-selective electrodes: a novel PVC-gelled neutral carrier mixture compared with other currently available sensors. *J. Neurosci. Methods* **4**, 73-86.

VANHEEL, B., DE HEMPTINNE, A. & LEUSEN, I. (1986). Influence of surface pH on intracellular pH regulation in cardiac and skeletal muscle. *Am. J. Physiol.* **250**, C748-C760.

VAUGHAN-JONES, R. D. & KAILA, K. (1986) The sensitivity of liquid sensor, ion-selective microelectrodes to changes in temperature and solution level. *Pflügers Archiv* **406**, 641-644.

VOIPIO, J., PASTERNACK, M., RYDQVIST, B. & KAILA, K. (1991). Effect of γ-aminobutyric acid on intracellular pH in the crayfish stretch receptor neurone. *J. Exp. Biol.* **156**, 349-361.

VOIPIO, J. & KAILA, K. (1993). Interstitial P_{CO_2} and pH in rat hippocampal slices measured by means of a novel fast CO_2/H^+ sensitive microelectrode based on a PVC-gelled membrane. *Pflügers Archiv* **423**, 193-201.

WIETASCH, K. & KRAIG, R. P. (1991). Carbonic acid buffer species measured in real time with an intracellular microelectrode array. *Am. J. Physiol.* **261**, R760-R765.

WISE, W. M. (1973). Bicarbonate ion sensitive electrode. U.S. Patent 3,723,281.

ZEUTHEN, T. (1980). How to make and use double-barrelled ion sensitive microelectrodes. In: *Current Topics in Membranes and Transport* (eds. E. Boulpaep & G. Giebisch), Vol. 13, pp. 31-47. New York: Academic Press.

Chapter 12

Intracellular ion measurement with fluorescent indicators

MARK B. CANNELL and MARTIN V. THOMAS

1. Introduction

Most of this book deals with electrical recording techniques which, although powerful, cannot always be used to study the composition of the intracellular environment - particularly when small cells or populations of cells are the preparation of interest. Optical techniques are becoming an increasingly attractive alternative method because of their *apparent* non-invasive nature and ease of use. One area in which optical techniques have largely replaced other methods is the measurement of intracellular ion concentrations with fluorescent indicators. This chapter will provide an introduction to this increasingly important application and will also consider some of the basic features of optical versus electrical recording techniques.

Electrical recording necessarily measures the movement of electrical charge. Individual charges can neither be created nor destroyed, and the overall balance between positive and negative charge must remain within narrow limits, since only a small charge separation causes a large electrical potential difference (voltage). Electrophysiology revolves around the measurement and/or control of charge separation and, as described elsewhere, the separation of charge by ion selective membranes allows determination of the ion levels within the cell as well as the creation of the Nernstian membrane potential. Some of the problems associated with electrical measurements arise from the fact that energy must be extracted from a system in order to ascertain its properties. Although careful electronic design can minimise the extraction of energy from the system (cell), the measurement process must perturb the system. In contrast to the idea of extracting energy from the system to measure its properties, with optical techiques (such as fluorescence and absorbance measurements) the experimenter actually supplies energy to examine the system, by supplying photons to the system and recording the results of the interaction of the photons with the system. This has several important consequences that should be kept in mind: (1) Photons will interact with any molecule that has absorbance in the

M. B. CANNELL, Department of Pharmacology, St George's Hospital Medical School, Cranmer Terrace, London SW17 0RE, UK.
M. V. THOMAS, Cairn Research Ltd., Unit 3G Brents Shipyard Industrial Estate, Faversham, Kent, ME13 7DZ, UK.

wavelength range of the photon (thus molecules that are of no interest to the experimenter may also produce a signal that will contaminate the experimental record). (2) It is usually necessary to introduce special molecules into the cell so that specific cell functions can be examined (and the introduction of these "probes" will perturb the system). (3) The amount of material that is being examined with the photons may be small so that there will be quantal limitations in the signal to noise ratio that can be achieved. (4) Energy is being supplied to the system, which at the very least will generate heat. At the other extreme, the molecules that absorb the photons may be destroyed (cf. photolysis techniques). Put in more concrete terms for fluorescence measurements, these consequences take the form of: (1) Cell autofluorescence. (2) Difficulties associated with introducing molecules that bind the ions of interest (either by intracellular injection or allowing the molecule to diffuse into the cell). (3) The small intracellular volume and limited concentrations of indicator or ion that binds to the indicator produce "shot noise" in the experimental record (see subsequent discussion). (4) Photon-induced degradation of the indicator (bleaching) and cell damage.

Optical and electrical recording techniques share a fundamental property in that their signals are both quantised. Although for convenience we may regard electrical or optical signals as being continuous, they in fact consist of individual charges or photons, and the statistical variation in their rate of arrival constitutes a source of noise. In the case of electrical measurement, this particular form of noise is not usually significant. For example, even the very small currents that are recorded in patch-clamp experiments represent the flow of a large number of electrons; e.g. it takes the movement of about 6 million electrons per second to generate a current of only 1pA. However, the question of time resolution must also be considered, and the timescale of interest is often in the millisecond range or less. Over these intervals, only a few thousand charges will be moving, and so the quantal nature of the current becomes much more important, to the extent that it may become significant.

The quantal nature of optical signals can be observed more readily. For example, in a cell whose volume is 10 pl and resting calcium concentration is 100 nM there are about 600,000 free calcium ions. In order not to perturb the system too much a fluorescent indicator is chosen that binds to only 1% of the free ions, resulting in the need to detect about 6000 indicator molecules. Assuming that each molecule gave off 100 photons per second (at this rate the molecules would probably bleach very quickly!) and with a <u>very</u> good detection system with 10% efficiency, the photon arrival rate would be 60,000/s. This is two orders of magnitude lower than the flux of electrons associated with the 1pA current described above. For such low numbers of particles, statistical variation in the number detected becomes a major problem. Optical detection is a Poisson process, where the variance of the signal is equal to its mean, so that if a 10 ms time resolution is desired the best that might be obtained in this example would be a signal-to-noise ratio of about 20 (defining the signal-to-noise ratio as the mean signal divided by the standard deviation of the signal). Although at this high photon detection rate the signal-to-noise ratio is acceptable, poor optical design can easily reduce the detection efficiency to <1% and make the signal too noisy to be useful (at a 10 ms time resolution). Actually, most calcium indicators bind much more than 1% of the free calcium ions, which has the advantage of providing larger signals, but at the expense of greater disturbance to the cell by the measurement process.

The above calculation highlights some important areas for the measurement of ions with optical methods: (1) the importance of efficient optics and detector design; (2) the choice of affinity of the indicator for the ion of interest. (Note that further amplification of the signal will not improve the measurement as the signal-to-noise is limited by the quantal nature of the signal.) Paradoxically, it is the ease with which such weak optical signals can be recorded that may make them appear so noisy in comparison with electrical records.

2. General principles of fluorescence

Comparison with photoproteins and absorbance measurements

Fluorescence techniques are now very popular, and it is worth making a brief comparison with other older optical methods of intracellular ion concentration measurement. The first optical methods involved the measurement of dye absorbance (e.g. metallochromic indicators such as murexide, arsenazo or antipyrylazo) but this technique has several disadvantages. The main one is that cells are quite small so that the optical path along which absorbance is being meaured is quite short (< 1mm). This results in only a low level of absorbance for reasonable indicator concentrations, and the absorbance signal has to be separated from the signal due to light scattering as well as from intrinsic absorbance. In practice, the situation is greatly improved by measuring the absorbance changes differentially between a pair of wavelengths, which maximises the indicator-related changes and minimise the others. This allows small concentration changes to be measured with surprisingly high accuracy and time resolution (Thomas, 1982), but estimates of absolute concentration levels are much more uncertain unless in situ calibrations can be performed. On the other hand, since the source of the light is external rather than internal, the indicator is not consumed and so measurements can be carried out for quite long periods. Nevertheless, absorbance techniques have been mainly limited to large invertebrate cells or muscle fibres, since the average mammalian cell is at least 10 times smaller in linear dimensions, resulting in a considerable (approaching 100-fold) reduction in the absolute amplitudes of the recorded optical signals as well as in a 10-fold reduction in the absorbance, which may preclude useful measurements. Further description of absorbance measurements is given elsewhere (see Ashley & Campbell 1979; Thomas, 1991).

Photoproteins (such as aequorin) are enzymes (mol.wt. 22,000) that catalyse the oxidation of a bound prosthetic group (such as coelenterazine-for review see Blinks et al., 1976). The oxidation reaction gives off a photon (as do many oxidation reactions) and the catalysis is regulated by calcium. The principal disadvantages of this technique for measuring calcium arise from the difficulties associated with introducing the photoprotein into the cell. Furthermore, on average, less than one photon is emitted by each photoprotein molecule (typically, only one molecule in three releases a detectable photon). In addition, the rate of photon emission has a highly nonlinear calcium dependence, which means that photoproteins tend to overestimate the spatially averaged calcium concentrations. Despite these disadvantages, the technique has a major advantage over absorbance measurements in that the detected signal arises only from the injected photoprotein, so movement artifacts and light scattering are not a significant problem. The light emission at a given calcium concentration depends on the the amount of active photoprotein in the cell, but it is possible to take this into account by measuring the amount remaining in the preparation at the end of the experiment by using a non-ionic detergent to make the cell leaky to calcium. The signals obtained during the experiment can then be expressed as fractions of the amount of active indicator, and converted to calcium concentrations by relating them to in vitro calibration curves (Allen & Blinks, 1979). In some circumstances, a significant proportion of the photoporotein may be consumed during the experiment, but it is possible to correct for this, by expressing the luminescence at any given time as a fraction of the total luminescence remaing at that time. It was the success of the photoprotein technique that revolutionised the field of calcium metabolism in cardiac muscle cells (volume approx. 30 pl), but the difficulties associated with handling and injecting these sensitive calcium indicators have discouraged their widespread use. More recently, the prospect of expressing genetically engineered photoprotreins in situ (Rizzuto et al., 1983) could mean that the technique is on the verge of a renaissance (although the enzyme system that builds the coelenterazine has not yet been cloned and cells have to be incubated with it overnight).

Fluorescent indicators combine advantages from both of the above methods. The fluorescent indicators can be considered to be self luminous (and therefore provide the advantages of the photoproteins) without concern for loss of activity on exposure to calcium. Conventional organic synthesis allows the properties of the indicators to be modified (by, for example, adding dextran residues to increase molecular weight and reduce indicator compartmentalisation and excretion from the cell), and since each fluorescent molecule can emit many photons the signal to noise ratio is high. It is notable that calcium signals have been recorded from cells as small as platelets with the newer calcium indicators. The fluorescent molecule (or fluorochrome) is usually excited by the absorption of a photon. This raises the energy of the molecule to a new "singlet" state from which the molecule descends the vibrational ladder until a radiative transition takes place and the molecule returns to the ground state with the emission of a photon. Alternatively, the photon emitted may be reabsorbed or the excited state may be quenched by collision with another molecule. In any case, the number of emitted photons are somewhat less than the number of absorbed photons and the ratio between them is called the quantum efficiency (modern fluorochromes have quantum efficiencies of about 0.3). The energy of the emitted photon is ordinarily lower than that of the absorbed photon, so its wavelength is correspondingly longer. However, under conditions of very high light intensity, i.e.

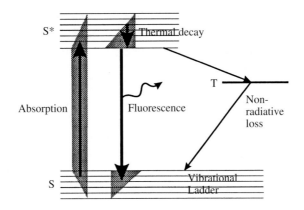

Fig. 1. An energy level diagram for fluorescence excitation. Note the presence of multiple vibrational levels around the principal energy levels S and S*.

pulsed laser illumination, it is possible for the indicator to absorb two long-wavelength photons in rapid succession, simulating the absorption of a single photon of half their wavelength. In the high energy state, the molecule is more likely to be oxidised and this will result in a loss of fluorescence (bleaching) but modern fluorochromes are also quite resistant to this effect. If the excitation and emission wavelengths are sufficiently far apart, the exciting light can be blocked from the detector with suitable filters so that the emission can be measured against a dark background. In practice, the background will not be completely dark, as other cell constituents will also fluoresce to some extent. This is known as autofluorescence, but at visible wavelengths its magnitude is relatively small and it is often ignored (not necessarily with good justification).

Fluorescence excitation and emission spectra

A potentially confusing aspect of fluorescence measurements is that there are not one but two sets of spectra to consider. First, some wavelengths will be more effective than others in exciting the fluorescence, and this dependence can be quantified by measuring the fluorescence excitation spectrum (there are multiple excitation wavelengths because of the large number of vibrational energy levels associated with the fluorochrome molecule). Second, the fluorescent light will be emitted over a range of wavelengths (due to the vibrational ladder associated with the excited and ground states - see Fig. 1). Ion sensitive probes are made by attaching groups that bind ions to the fluorescent part of the molecule. The binding of an ion alters the electronic configuration of the molecule and hence alters the fluorescence of the molecule. For example fluo-3 has a calcium co-ordination site based on the BAPTA molecule and the fluorescent group (based on fluorescene) is attached to one side of the BAPTA backbone. Calcium binding to fluo-3 draws electrons from the BAPTA rings, which in turn draw electrons from the rings of the fluorescene group and thereby increase the fluorescence of the molecule (this explanation is something of a simplification as it is really the resonance of the ring structures that is altered).

In order to measure ion levels with such a fluorescent indicator, all one has to do is measure fluorescence at a suitable wavelength (both for excitation and emission). However, the raw signal would not be quantitative because the absolute fluorescence will depend on (1) the concentration of indicator, (2) the volume of the cell (or path length which is being illuminated), (3) the intensity of the illumination, (4) the properties of the detection system, (5) the cell autofluorescence, and finally (6) the calcium concentration (which is the only variable of real interest). However, variables 1-5 can (in principle) be eliminated by recording the maximum fluorescence of the cell at saturating ion levels and minimum fluorescence in the prescence of a quenching ion (such as manganese) or in the absence of the ion at the end of the experiment (see below).

Spectral shifts and ratio measurements

It is also possible that binding of the target substance by the fluorescent molecule will cause a shift in the fluorescence spectra rather than a simple modulation as discussed above for fluo-3. In principle, this concept is straightforward enough, but consideration of it is complicated by the further possibility that either the fluorescence excitation spectrum or the fluorescence emission spectrum - or indeed both - may be shifted. For example, fura-2 and indo-1 also have a calcium coordination site based on the BAPTA molecule with a fluorescent group attached to it. The alteration of electronic structure on calcium binding has different effects on these molecules in that for fura-2 the <u>excitation</u> spectrum shifts to shorter wavelengths, while for indo-1 the <u>emission</u> spectrum shifts to shorter wavelengths (see Fig. 2). Such shifts are desirable because they allow the concentration of the ion of interest (in this example calcium) to be estimated from the relative levels of fluorescence measured at two different wavelengths. This technique is known as ratiometric fluorescence measurement and is described in detail below.

The most important point concerning these spectral shifts is that they must be large enough to be detected easily, ideally in a part of the spectrum that does not require very specialised detection equipment (i.e. it is easier to design instruments to work in the visible wavelength range than in the far ultraviolet or infrared range of wavelengths). Although one can argue about which type of shift is the best on theoretical grounds, one also requires a reasonable affinity and selectivity for the target substance, high fluorescence quantum yield, lack of biological side-effects, and molecular stability. These other factors all tend to be just as important, so in practice it is best to design instrumentation that is flexible enough to allow use of a range of fluorescent indicators.

From the measurement point of view, the case where the emission spectrum shifts is the simpler one to deal with. The consequence of a shift in the emission spectrum is that the fluorescence at some wavelengths will increase, whereas at other wavelengths it will decrease, as shown in Fig. 1B. Although it is useful for calibration, and some other purposes, to measure the complete spectra, the parameter that is of greatest interest in biological measurements is the time-dependence of the fluorescence change. In practice, therefore, one normally measures signals at just two

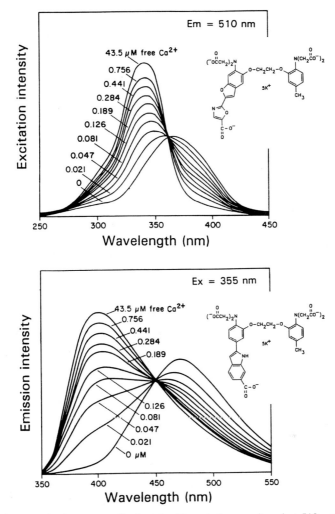

Fig 2. A shows excitation spectra for fura-2 with emission monitored at 510 nm. Note that with illumination at 340 nm the fluorescence increases with increasing calcium, while at 380 nm the fluorescence decreases with increasing calcium. B shows emission spectra for indo-1 with excitation at 355 nm. Note that as calcium increases the emission spectrum shifts to shorter wavelengths. The traces labelled F were obtained at 0.1 μM free calcium, a level comparable to resting levels in the cell. Data redrawn from Grynkiewicz et al., 1985.

wavelengths - or to be more precise, two waveBANDS - so that as many of the emitted photons as possible can contribute to the measurements to maximise the signal-to-noise ratio. Thus, all one needs to do is split the output signal and pass it to two photodetectors, preceded by optical filters of suitable characteristics to define their spectral sensitivity. The optical efficiency of the arrangement can be improved by using a dichroic mirror to split the output signal as this will maximise the signal to each detector (dichroic mirrors reflect light of some wavelengths and transmit the rest, so by placing one in the light path at an angle of 45 degrees, the emission light

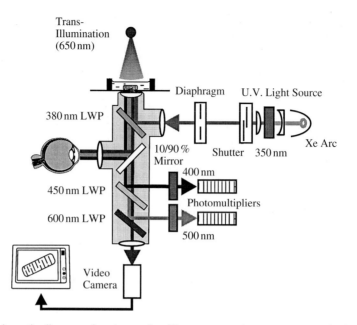

Fig. 3. A schematic diagram of an inverted epifluorescence microscope system for indo-1 fluorescence detection. For fura-2 this system would have to be modified to include some method for changing excitation wavelength. The LWP (long wave pass) mirrors are dichroic mirrors that split the light at the wavelengths indicated. Note that red light (650 nm) is used to give continuous illumination of the preparation, so that it can be observed by eye or via a camera.

can be steered to one detector in the straight-ahead position or to another one at 90 degrees according to its wavelength; see Fig. 3).

The case where the excitation spectrum shifts is somewhat more complicated, because the emission spectrum will (at least sometimes!) remain the same. To measure changes in the excitation spectrum, the individual excitation wavelengths must be supplied sequentially. This requires some form of optical chopping system, combined with some appropriate electronics on the detector side to separate out the fluorescence emission signals that have been detected for each of the excitation wavelengths. Clearly the time resolution of the system will depend on the chopping frequency, but other factors, such as the signal-to-noise ratio of the detected signals (which increases as the signal bandwidth increases), and the response times of the indicators, impose the ultimate limits, since it is relatively easy to perform the chopping on a millisecond timescale. Fortunately, the fluorescence lifetimes of all the indicators are very much shorter than this, so in practice there is no risk of cross-contamination between individual excitation wavelengths when operating at such speeds.

Multiple-excitation wavelength systems are more complicated, but in practice, much of the complexity is hidden from the user; indeed they may even appear simpler since only one detector is required. However, for a truly general-purpose instrument, the addition of a second photodetector allows dual-emission measurements to be made as well. It is also possible to make measurements using two or more indicators at the same time, if their excitation and/or emission spectra are sufficiently different. For example, simultaneous measurement of Ca^{2+} with fura-2 and pH with BCECF has been carried out successfully, since fura2 is excited at 340 and 380 nm, and BCECF is excited at 440 and 490 nm. These excitation wavelength pairs each have little effect on the other indicator, whereas they both emit in the 500-550 nm range and so they can be monitored with the same photodetector (see Fonteriz et al., 1991). However, the experimental difficulties inherent in using two indicators simultanoeusly should not be underestimated.

3. Detection of optical signals

Two basic types of photodetector are in general use. These are photomultipliers and solid-state photodiodes, and they are based on vacuum-tube and semiconductor technology respectively. This is one of the few areas in which semiconductor technology has <u>not</u> displaced vacuum tubes.

Photomultipliers

A photomultiplier consists of a evacuated glass tube with a photocathode, an anode and a series of about 10 intermediate electrodes known as dynodes between them (see Fig. 4A). The photocathode is so named because it is coated with a material which emits an electron in response to the absorbance of a photon. Although the energy carried by a photon at optical wavelengths is relatively small (about 2 eV), it is sufficient to liberate an electron from the materials coating the photocathode. The electron is accelerated towards the first dynode, which is held at a more positive potential (on the order of 100 V), so it gains an equivalent amount of energy in eV on the way. When the accelerated electron hits the first dynode it liberates further electrons, which are then accelerated towards the next dynode in the series. This process repeats through the series of dynodes (each of which is about 100 V more positive than its predecessor) until the anode is reached, where an electrically detectable pulse appears when the avalanche of electrons arrives. Thus the single electron liberated by the photon has been multipled by several million by the secondary emission throughout the dynode chain. (It should be noted that no earthed metal should contact the surface of the tube or emitted electrons will be attracted to the glass of the tube and a large increase in dark current will result). The pulse at the anode only lasts a few nanoseconds because the electron transit time down the tube is relatively constant. The numbers given here are necessarily approximate because they depend on the type of photomultiplier and on the total supply voltage, but the important point is that photomultipliers can give such enormous amplification that the output signal is well above the noise level of the subsequent electronics. An important caveat to varying the supply voltage to the photomultiplier (and hence its gain) is that the cathode to first dynode voltage should ideally be kept constant. In particular, if this voltage is reduced too much, the energy gained by the first emitted electron may

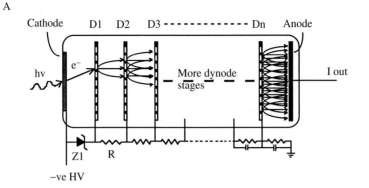

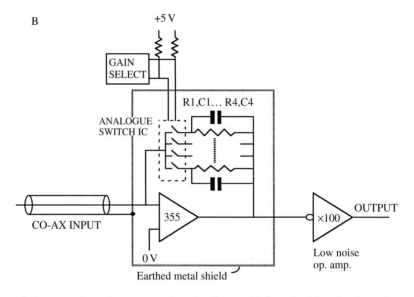

Fig. 4. A shows a schematic representation of a photomultiplier tube. Note that the voltage to the dynode voltages are set by a resistor chain, with the first dynode to cathode voltage stabilised with a zener diode. B is a schematic diagram of a current to voltage converter suitable for photomultiplier tube current measurements (see the Microlectrode Electronics chapter for further information). The gain is selected by an analog switch which could be controlled by a computer. The second ×100 inverting amplifier should be a low noise type with a bandwidth similar to that set by the combination of R1,C1..R4,C4 (the cut off frequency being $1/(2\pi RC)$. Alternatively the bandwidth of this amplifier could be computer controlled to filter the light signal as needed (see text). The gain of this circuit is: 100×R Volts/Amp, typical photomultiplier gains would be 0.1-1000 nA/V, corresponding to R= 1 MΩ–10 kΩ respectively.

prevent secondary emission, which would result in a serious reduction in detector efficiency. Thus if the experimenter wants to alter detector gain by altering the photomultiplier supply voltage, then the first dynode to cathode voltage should be set preferably by a zener diode rather than by a resistor in the divider chain (see Fig. 4A).

The requirement for a relatively high supply voltage (approx. 1 kV) for the photomultiplier is a slight inconvenience, but the usual practice is to run the photocathode at a negative potential so that the anode can be at ground potential. This makes the detection electronics much more straightforward. Detection can either be analogue (by measuring the average anode current) or digital (by counting the individual current pulses). The preferred configuration for analogue detection is to connect the anode to the input of a virtual ground amplifier, which acts as a current-to-voltage converter (see Fig. 4B). The other practical inconvenience is that photomultipliers tend to be physically large, but that may also be an advantage since there is no problem in directing all of the light onto the detecting area (10-25 mm diameter).

Photomultiplier technology has a number of variants, and the one most relevant to fluorescence is the image intensifier. This device uses a similar type of photoelectron multiplication, but spatial information is retained by preventing the electrons from wandering within the tube - either by using hollow fibres coated with a secondary emitter, with the accelerating voltage applied at the ends of the fibres (called a micro channel plate intensifier or second generation intensifier) or by using shaped electrodes within the tube to form an electrostatic lens (a first generation intensifier). The anode is replaced by a phosphorescent screen so that accelerated electrons hit the screen and produce light. Thus an intensified image appears at the end of the tube which may then be viewed directly (as in military image intensifier goggles) or else with a conventional TV camera. Just why such a device may be useful will become clear in the following discussion on photodiode technology.

It should be noted that photomultipliers must <u>always be protected from bright light</u> when powered, as permanent damage to the tube can occur if the anode current rises above the limit set by the manufacturer (although the high-voltage power supply may incorporate current-limiting circuitry to reduce this risk). Following exposure to bright light (even with the power off) the dark current of the tube may be increased (James, 1967), an effect that may decay over several hours if the tube has not been damaged. However, damage to the photocathode may permanently reduce the sensitivity of the tube, an effect that can only be determined if the experimenter has a light standard with which to test the tube. (A simple light standard can be made from a light emitting diode run from a stable current source. The intensity of the light source can be reduced by painting the surface of the diode with black paint and the diode should be mounted in a holder that fixes the geometry of the source.)

Photodiodes and CCD cameras

A photodiode is a single-point detector and is the solid state analogue of the vacuum photodiode. A CCD camera basically consists of an array of photodiodes, and although there are differences in detail, the underlying principles are similar, so the case of a single photodiode will be considered here. In these devices, the absorption of a photon causes charge separation (more specifically, the formation of an electron-hole pair). This process is highly efficient, and the quantum efficiency approaches unity for wavelengths close to the bandgap energy of the semiconductor, i.e. in the infrared at around 900 nm, but it is also high over most of the optical spectrum (in contrast, photomultipliers struggle to achieve much more than 25% quantum efficiency). If the diode is open-circuited, the charge separation will appear as a voltage across the capacitance of the device, or if it is shortcircuited by connection to a virtual ground amplifier, the charge separation can be measured directly as a current. This is the preferred method of operation for a single photodiode, since the response is linear over many orders of magnitude when measured in this way. However, unlike the photomultiplier currents, these photocurrents have not been amplified, so they must be measured against the thermal noise background of the device, plus that of the amplifier to which it is connected. In practice, the noise characteristics of the photodiode swamp those of the amplifier, since the primary noise source is the leakage resistance that appears in parallel with the device, which typically has a value of around 100 megohms for a photodiode of 1 mm^2 active area (see the application notes for the OPA101 in the Burr-Brown Data book, 1982, for a detailed analysis of this configuration). In a CCD camera, the leakage resistance per pixel should be at least 10,000 times higher, as the pixel area is correspondingly smaller, i.e. the linear dimensions are about 10 μm instead of 1 mm, which then places greater demands on the noise performance of the amplifier because the signal per pixel is correspondingly smaller. Since the current is measured by a current to voltage converter the noise limits are the same as those of the patch clamp amplifier (described elsewhere), with the same dependencies on source resistance and amplifier noise. For detailed treatment of noise see the next section, but *assuming* that it is possible to detect about 1 pA photocurrent at 1 kHz in a small photodiode (<1 mm^2), this represents a lower limit for detection of about 6,000,000 photons/s. Since this implies 6000 photons per measurement bin (a photomultiplier might give no more than one spurious count per bin at this bandwidth) the superiority of the photomultiplier is obvious (although the reduced quantum efficiency would result in an ouput of about 10^6 counts/s the photomultiplier output would be almost all signal (signal to noise ratio = 1000) whereas the diode output would be almost all noise). One way that the "sensitivity" of silicon diodes can be improved is by allowing the charge to accumulate over a period of time until it is large enough to be reliably read out. However, to allow long integration times the detector must be cooled to prevent thermal energy from destroying the electron-hole pairs as well as contributing to Johnson noise (see below) (astronomers use liquid nitrogen cooling and sometimes integration times of more than an hour!). In arrays of photodiodes (which make an imaging sensor), it is impractical to connect each pixel diode to its own amplifier, and instead the charge separations are accumulated within each pixel diode and then sent in turn to a single amplifier via a shift register arrangement (hence the term Charge Coupled Device for this type of photodetector). Unfortunately, the charge readout and reset process introduces additional noise. Nevertheless, total detector noise levels of only a few photoelectrons per pixel can be achieved from the best CCD cameras currently available, and there may still be some scope for further improvement.

To extend the light range over which CCD cameras operate a microchannel intensifier (see above) can be optically coupled to the sensor (by lenses or with a coherent fibre optic plug). Such intensified CCD (or ICCD) cameras have become very popular for fluorescence imaging. However the dynamic range of these cameras (at a given gain) is not as large as photomultipliers, and saturation effects are easily observed. The experimenter should therefore ensure that the camera gain is adjusted so that the majority of the image field is captured within the linear part of the camera response - a precaution that applies to both images when ratio imaging. Put another way, the average ratio in a field should be near unity, a condition that can be obtained by altering the intensity of excitation wavelengths appropriately.

In summary therefore, although photomultipliers have lower quantum efficiencies, they allow the detection of individual photons (albeit one in 3 or 4), which is not possible with silicon technology. Silicon devices are only useful if the total number of photons received within the measurement period exceeds a limit set by the read out electronics, so that the photomultiplier with its large detection area remains a superior choice for the majority of low light level fluorescence measurements.

Noise in detectors

In this section we must distinguish between THERMAL (Johnson) noise which is generated by the random movement of electron in conductors and SHOT noise which is due to the statistical variation in the rate of arrival of particles (photons or electrons). This is important because in photodiodes, both shot noise and thermal noise will contribute to the signal whereas in photomultipliers the enormous current amplification that occurs in the dynode chain ensures that only the shot noise is important.

Since noise generation is a random process, the individual noise sources add their contributions in quadrature, i.e. the total noise is the square root of the sum of the squares of the individual components. This also applies to the bandwidth over which the noise is measured, so if the noise is equal at all frequencies, it increases with the square root of the bandwidth. Therefore noise is typically specified in units of volts (or amps) per Hz. The thermal noise also varies with the square root of the resistance. The equation for the noise voltage (V_n) across a resistance R is:

$$V_n = \sqrt{4KTR\Delta F}$$

(where K is Boltzmann's constant (1.38×10^{-23} J per degree), T is absolute temperature and F is the bandwidth in Hz). Assuming a temperature of 25°C we can write:

$$V_n/\sqrt{Hz} = 1.28 \times 10^{-10}\sqrt{R}$$

Thermal noise is usually expressed as a voltage but we are more interested in current noise (I_n) which from Ohm's law is given by:

$$I_n/\sqrt{Hz} = 1.28 \times 10^{-10}/\sqrt{R}$$

This is to be compared to the shot noise, which is given by:

$$I_n = \sqrt{2eI\Delta F}$$

where e is the charge on an electron (1.6×10^{-19} coulombs) so

$$I_n/\sqrt{Hz} = 5.66 \times 10^{-10}\sqrt{I}$$

The two components will be equal when:

$$1.28 \times 10^{-10}/\sqrt{R} = 5.66 \times 10^{-10}\sqrt{I}$$

i.e. when

$$I = 1/19.64R$$

Note that this relationship is independent of the measurement bandwidth, since both noise components increase with the square root of the bandwidth. Thus for the shot noise current (which increases with the square root of the current, i.e. light intensity) to equal the thermal noise in a 100 megohm resistance of our typical photodiode, the current itself must be:

$$I = 5.09 \times 10^{-10} \text{ A or } 3.18 \times 10^9 \text{ (photo)electrons per second.}$$

It is therefore clear that such a photodiode is no match for the photomultiplier when it comes to detecting low photon fluxes.

Actually this discussion is relevant to any current measurement where all the thermal noise (including amplifier noise) can be modelled as a single resistance. This

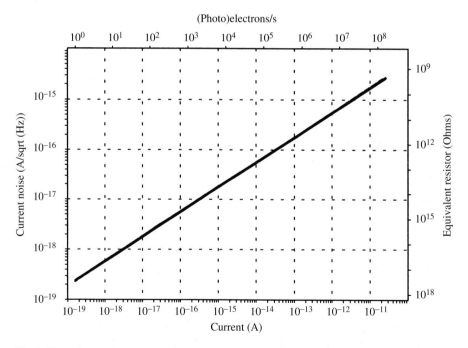

Fig 5. The relationship between current flow and shot noise at 25°C. The top of the graph shows the electron flow rate and the value of the resistor that gives the same noise level is shown at the right axis. Note the large resistor values that would be required to reproduce shot noise.

is illustrated in Fig. 5 where the current is shown on the x-axis both as a current (bottom) and as an electron flow (top) and the associated noise is shown on the y-axis. The resistor that would give the equivalent noise is shown on the right. Thus if the background noise from a photomultiplier is modelled as an equivalent resistance (this is perfectly legal) we see that at a typical value of 100 counts per second, the noise is equivalent to that from a resistor of about 3×10^{15} Ω. For a good photomultiplier, cooled to 0°C, a noise level of 1 count per second can be achieved, giving an equivalent resistance of about 3×10^{17} Ω. These figures make the gigaohm resistances used in patch clamping look quite pedestrian, and serve to emphasise just how sensitive optical measurements with photomultipliers really are!

In a CCD image sensor, the pixel area (photodiode) may be much smaller and a resistance of 10^{12} Ω per pixel might be realised. However, this does not reflect a real improvement in "sensitivity" because the light flux is spread over the whole sensor area and so the number of photons arriving at each pixel falls (in direct proportion to the pixel area). However, if the experimenter reduces the magnification of the image (keeping other factors constant) the sensitivity will improve (because the photon flux incident on each pixel will increase). Thus the image magnification should be kept as low as possible, consistent with the desired spatial resolution, to maximise signal to noise in a fluorescence image.

As noted earlier, cooling helps improve CCD performance. This arises because (1) the Johnson noise is proportional to the square root of the absolute temperature (this is not a large factor unless liquid nitrogen temperatures are approached) and (2) the resistance of the pixel elements increases as the temperature is reduced. This increase in resistance also allows the image to be acquired over a longer period, since it reduces the rate at which charges leak away. Taking advantage of this effect to reduce the measurement bandwidth will reduce the noise still further. The other important noise sources are the noise in the detecting amplifier and the charge shift/read out electronics, in which the noise will be reduced by reducing the bandwidth. These effects make cooled CCD cameras very suitable for long term observation of dim objects but are rather less well suited to the relatively rapid changes in fluorescence encountered in biophysics. Despite the advantages of cooling CCD sensors, they will not be able to detect single photons.

Despite these criticisms, it is important to note that for shot noise levels, above thermal levels, photodiode detectors will be superior to photomultipliers, because their quantum efficiency is much higher. For example, in absorbance measurements photodiodes are the detectors of choice, and one of the authors (M.V.T.) found that for absorbance measurements on molluscan neurones, the shot noise was indeed the dominant noise source in a detector circuit using a 1mm photodiode.

In conclusion, shot noise will always limit the signal to noise ratio for a perfect detector, thus there can be no substitute for efficient collecting optics. In practice it may be easier to improve collection efficiency rather than squeeze the last drop of performance from the detector system. Nevertheless, the correct choice of detector and its proper implementation can be crucial for experimental success.

Analogue detection or photon counting?

There tends to be a certain amount of confusion concerning photomultiplier signal detection methods, particularly with regard to the possible advantages of the photon counting technique over analogue detection. This section gives a detailed comparison between the two methods, so that the advantages and drawbacks of each can be fully appreciated. However, for readers who just want to know which method is the better one, the answer (in the authors' opinion) is that for typical applications analogue circuitry is perfectly adequate (and is simpler than photon counting). For analogue signal detection, one just measures the average anode current, whereas photon counting uses a discriminator circuit, which provides an output pulse when the input current rises above a threshold value. Since the current pulses are not instantaneous, they can only be recognised reliably as single pulses when the interval between them is large compared with their individual duration. As the light level increases, the pulses become closer together until they can no longer be individually recognised, and the detection efficiency will fall. Analogue circuitry does not exhibit this problem as the anode current rises in proportion to the light intensity until tube saturation effects occur. Furthermore, the light level at which tube saturation starts can be increased by reducing the supply voltage (but see caveat above). Photon counting is thus possible only for relatively low light levels, whereas analogue detection is possible at any light level.

For very low light levels, photon counting does has definite advantages, since (1) the output pulses are of constant amplitude and duration (2) the output signal in the absence of light is lower because of the rejection of (i) thermally generated, (ii) cosmic ray and radioactive decay generated events. At a given tube operating voltage, the amplitude of a current pulse caused by a photon arriving at the photocathode occurs within a relatively narrow range, so the discriminator threshold can be set just below it (in practice, it may be more convenient to set the current pulse amplitude to match the discriminator threshold, by varying the photomultiplier supply voltage - but see above). When this is done, the discriminator rejects most of the thermally generated pulses (thermionic emission occurs from the photocathode and dynodes; however only thermally emitted electrons arising at the photocathode will cause a full size output pulse whereas those occurring within the dynode chain cause smaller pulses because the number of amplification stages is reduced). In addition, extra large pulses resulting from cosmic rays or radioactive decay particles hitting the tube may be rejected by the discriminator (or else are simply reported as normal photon pulses). While the rejection of these spurious pulses might seem a real advantage, in practice the number of large pulses is small (1 per second) and the quite large number of thermionic pulses does not cause as large a dark current as might be expected since they are of reduced amplitude (see Sharpe, 1964). As a practical example, one of the authors (M.B.C.) uses EMI 9789 tubes, which have a dark current of about 0.1 nA. The same tubes used in a typical fluorescence experiment give a signal of about 50 nA showing that the tube dark current is negligible.

For biological <u>luminescence</u> measurements, where only a few photons may be detected per second, the use of photon counting may be considered essential, but it should be noted that greater improvements may be made by building a cooled housing for the photomultiplier tube (cf. Cannell & Allen, 1983) to reduce the dark current. Only when this has been done is it worth contemplating the extra complexities of photon counting.

Photon counting can offer an improvement for signals in the biological fluorescence range, but the effect is quite small. It arises because there is a statistical variation in the amplitude of the current pulses arriving at the anode, which results mainly from the variation in the number of electrons that are emitted by the first dynode as a result of the impact of the photoelectron (the same effect occurs at all the other dynodes too, but what happens at the first dynode is the predominant source of the variation). For analogue detection, this variation causes an additional noise component, but for well-designed tubes the effect should increase the total noise by no more than about 10-20%. The relative size of the variation can be reduced by applying a higher accelerating voltage to the first dynode, causing each photoelectron to emit a larger number of electrons from it, and the voltage divider networks supplied with commercially available tubes usually incorporate such an arrangement.

This improvement may be worth having, although one has to consider the loss of information that can occur with photon counting on account of the dead time. The main problem with photon counting is that two coincident or closely spaced pulses cannot be resolved as separate events, whereas they would both be recorded by analogue detection. The minimum interpulse interval for both pulses to be recognised by photon counting is known as the dead time, and it is normally set to a known figure by design, so that a statistical correction can be applied to the data to compensate for the undetected pulses. The correction is simple. If we detect N pulses in time T, and the dead time is given by t, then the count will actually have been obtained in a time $T-Nt$ instead of T, so to compensate for the lost pulses we can estimate the true count N' to be $N' = N \times T/(T-Nt)$. However, we cannot compensate for the loss of signal-to-noise arising from the loss of those pulses. Typically t is in the range of 100-500 nanoseconds, so to take a typical example with t = 200 nsec and N = 100,000 per second, then the true count N' will be underestimated by about 2%. This is clearly not very significant, but for count frequencies approaching 1 MHz the losses become more serious, to the extent that analogue detection would be the better choice on theoretical as well as practical grounds.

To offer a similar word of caution on the analogue side, it is quite easy to devise an analogue detection system that performs nowhere near as well as it should, and claims that photon counting is far superior may well have been based on comparisons with such systems. The best method of analogue detection is the direct analogue equivalent of photon counting, i.e. to use an integrator to store the total charge that arrives within a given interval. At the end of the measurement period, the integrator output is digitised for data processing by computer, and/or stored in a sample-and-hold amplifier to give an analogue output that remains at a steady level until the next measurement is made. The integrator output is then set to zero in preparation for making the next measurement. A simpler technique is to use a current-to-voltage converter, with its bandwidth limited by a capacitor across the feedback resistor. This can also be viewed as a "leaky integrator", in which the continuous discharge pathway presented by the resistor removes the need for discontinuously discharging the capacitor (see Fig. 4). However, as with the any other analogue signal, if it is to be digitised, then a sampling rate appropriate to the signal bandwidth must be chosen. Too low a sampling rate will result both in loss of information (i.e. reduced signal-to-noise ratio) and in the generation of additional noise components via the phenomenon of aliasing. The integration method method allows the integation time (and hence the effective bandwidth) to be altered in step with the sampling rate so that the bandwidth is always appropriate for the sampling frequency. An additional advantage is that in multiple-excitation systems, the integration period can be made to correspond exactly to the period during which a particular light filter is in the excitation path. This eliminates the possibility of contamination by previous signals from other filter wavelengths.

To summarise, photon counting is only really needed if the experimenter anticipates detecting very low light signals (i.e. those that are not visible to the dark adapted eye). It may actually <u>degrade</u> the signal if the discriminator is improperly adjusted or if large signals are to be measured. Analogue circuitry is simpler and is more likely to perform at or near its theoretical limits. It is certainly the preferred method for those who wish to construct their own equipment.

4. Instrumentation

Light source

Many of the currently available fluorescent indicators need to be excited in the near-ultraviolet, e.g. around 360 nm for the Calcium indicator indo-1, and 340-380 nm for the calcium indicator fura-2 and the sodium indicator SBFI. These requirements make fluorescence instrumentation somewhat more expensive, as incandescent light sources (e.g. quartz-halogen) do not give sufficient light at these wavelengths, so an arc source is all but essential. The almost universal choice is xenon, as this gives a very uniform output spectrum, which extends between ultraviolet and infrared wavelengths. Mercury arc lamps are less suitable for this application, because their spectra are very irregular, giving huge outputs at some wavelengths and very little at others (see Fig. 6). However, this distinction is not usually relevant to the choice of light source, as most arc sources will take either type of bulb.

Arc sources produce high-intensity light with reasonable efficiency, but stability of the arc source with time is not as good as that of incandescent lamps. In addition, there is a problem with these sources that arises from the arc "wandering" between the arc electrodes. Since the arc colour and intensity are not constant over the area that the arc encompasses, quite large variations in illumination intensity (up to 40%) may result from using optical systems that depend on imaging the arc on the specimen (Levi, 1968). While this problem can be overcome by obtaining a feedback signal from the illumination to control arc current, some care should be taken to ensure that the arc has not wandered so far that the arc lamp is excessively overdriven. In comparison, incandescent light sources are far more stable and are easily stabilised to less than 1% intensity variation.

Selection of illumination wavelength may be performed with interference filters or with grating monochromators (note that coloured glass filters do not have a rapid enough roll-off of transmission efficiency to be generally useful). While interference filters are relatively inexpensive and readily available, they have problems in their application that are not present in monochromators. The first is that attention must be paid to the incident energy load that is presented to the filter. Interference filters should not be used to select illumination wavelength in such a way that an appreciable heat load is placed on the filter; typical interferences filters will be permanently damaged in situations where their temperature rises to more than about 70°C. In practice this means that an interference filter should not be placed directly in front of an arc lamp source. Instead, unwanted light should be first absorbed with coloured glass filters or, better still, reflected from the optical path with dichroic mirrors (often called "hot" or "cold" mirrors and commercially available for this purpose). Interference filters are usually constructed so that their front surface reflects as much as possible of the unwanted radiation, which greatly reduces the amount of energy that they absorb, and so it is particularly important that they are installed in the correct orientation. Water filters are good at absorbing unwanted infrared light, but the white light that passes through them may still be intense enough to cause damage

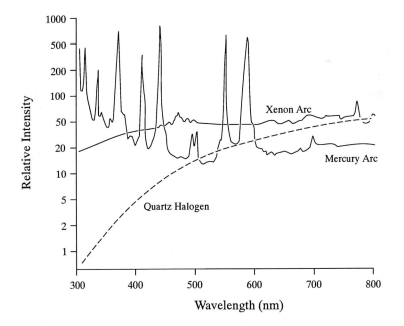

Fig. 6. Comparison of typical spectra of xenon, mercury and quartz halogen light sources. Note that the halogen source is the most uniform but the output drops quite rapidly in the violet and is not useful for UV illumination. The xenon arc source has a smoother spectrum than the mercury arc, but the sharp peaks of the mercury spectrum can be useful if they occur at desired wavelengths.

to the interference filter with time. A second problem arises from the pass-band of an interference filter being dependent on the angle of incident illumination, although for light within 0-10° of the optical axis the effect is relatively small. In practice this means that the light passing through the filter should be reasonably well collimated to achieve the manufacturer's specifications for the filter. Since the pass-band of the filter shifts to shorter wavelengths with increasing angle of illumination, the filter may be "tuned" to the desired wavelength by placing it at an angle to the optical path, but collimation becomes more important under these conditions, since the variation in the pass-band wavelength with angle of incidence increases considerably as the (average) angle is increased. Typical filters will shift to about 98% of the nominal pass band wavelength at an incident angle of 15°. Where the light is not collimated, this will result in a broadening of the pass band of the filter and an effective shift of the pass band to shorter wavelengths.

A growing number of indicators require excitation at around 490 nm (e.g. fluo-3 for calcium), for which an incandescent source would be perfectly satisfactory. Most of these indicators were introduced for confocal microscopy applications, where a laser must be used to obtain sufficient excitation intensity, and the argon laser at 494 nm is a good choice. Unfortunately, these indicators are generally not ratiometric, so they are not quite so attractive for other applications.

The other problem with the near-UV requirement is that ordinary glass has a very sharp transmission cut-off at around 350 nm, so all lenses in the optical excitation pathway must be of quartz or silica in order to allow fluorescence excitation at 340 nm. In practice, this does not seem to be a problem now that microscope manufacturers recognise the need for such lenses, but it does still rule out some very nice microscope objectives (e.g. some plan apochromats). Objectives that would otherwise be marginal or unsuitable can sometimes give satisfactory results if the excitation wavelength can be increased slightly, e.g. use of 350 nm rather than 340 nm for short-wavelength excitation of fura-2 will give more signal at the expense of a smaller range of ratios.

The optical system

The most commonly used optical system is the inverted microscope. This is particularly convenient for cell cultures or single-cell preparations, as individual cells can be illuminated and viewed from below, while preserving accessibility from above for simultaneous use of other techniques such as voltage recording or patch-clamping (see Fig. 3). In both this and the conventional (upright) microscope configuration, the fluorescence excitation light is supplied through a side port. A dichroic mirror situated behind the objective reflects the excitation light through the objective, which focuses the light onto the cell, thereby acting as the condenser lens. The objective also captures the emitted light, and the dichroic mirror is selected so that it transmits rather than reflects this light, so that it passes through the rest of the microscope. The numerical aperture (NA) describes the ability of the objective to capture light and is equal to **n.sin**Θ where Θ is the half angle of the cone of light collected by the objective and **n** the refractive index of the medium between the specimen and the lens. Most biological specimens are immersed in saline for which **n**=1.3. This represents the upper limit for the effective numerical aperture of the objective, so that even though higher numerical aperture objectives are made, they will be limited by the refractive index of the bathing medium. The amount of light collected (assuming the entire object is visible within the field) is given by the solid angle fraction $(1-\cos\Theta')/2)$, where Θ' is the angle of the cone of light from the specimen. Θ' and Θ are related by Snell's law: $\sin\Theta'/\sin\Theta = n'/n$ where n and n′ are the refractive indexes of the medium containing the specimen and the medium between the objective lens and the specimen medium. These equations suggest that a 1.4 NA objective will collect about 25% of the emitted light whereas a 0.8 NA objective will collect about 7%.

Clearly one should use as high a numerical aperture lens as possible for efficient detection and illumination. In fact the objective lens may be the most critical component for imaging and the experimenter would be advised to compare different lenses even if they have similar NA as the presence of field stops within the lens may reduce the light collection further. Some form of optical switching system is incorporated to allow light to be sent to the eyepieces for visual observation or to one or other output ports for photometry and/or camera viewing. For photometric measurements, it is also usual to provide an adjustable diaphragm, together with some method of viewing it, to select the area of view that is being sent to the photomultiplier(s). A particularly elegant way of doing this is to use deep red transmission illumination, to allow the selected area to be viewed with an inexpensive CCD camera. Since CCD cameras are very sensitive in the deep red/near-infrared (640-900 nm), the rejection of these wavelengths by the filters used to shape the fluorescence emission waveband is quite straightforward. Thus it is possible to view the preparation while recording very low levels of fluorescence.

A practical tip that can avoid a great deal of uncertainty over whether the equipment is working properly is that if a fluorescence signal is large enough to give a satisfactory recording, it should also be bright enough to be observed with the dark-adapted eye. This can be particularly informative when adjusting the conditions to give the best results when the ester loading technique is used to introduce the indicator into the cells.

For relatively large cells, for measurements in situ, or for experiments in cuvettes, a pair of fibre optic probes can be used to carry the excitation and emission light, or alternatively a single probe can be used in conjunction with a dichroic mirror to separate the two signals.

Detector arrangements

Continuously spinning filter wheels are a convenient (and commercially available) way of changing wavelength. These systems are usually available with all the necessary electronics to extract the relevant signals. They often allow up to six or eight filters to be used and the extra filter positions can be used to make measurements at additional wavelengths, and/or (as often happens) where the fluorescence from one wavelength is relatively weak, two or more filters can be provided for that wavelength and the individual outputs can be combined electronically.

5. Using fluorescent indicators

Fluorescence is a rapidly developing field, and many further developments are likely to occur during the lifetime of this edition of the book. This section will therefore concentrate on the principles underlying the use of the fluorescent indicators, and the reader should consult current publications for the most up-to-date information on individual indicators. A useful source of additional information is the catalogue and regular updates published by Molecular Probes (see the references for their address).

Introducing the indicators into cells

The most reliable method of indicator loading is by direct injection through a microelectrode (Cannell et al., 1987, 1988). The injection technique is discussed elsewhere in this volume, so little needs to be said about it here, but of course it is limited to those cells that are large enough for application of such methods. However, indicator can also be loaded through the electrode when the cell-attached patch recording technique is used, which allows relatively small cells to studied.

Once introduced into cells, the fluorescent indicators generally remain there for a reasonable period, allowing stable recordings to be made. Even if there is some loss of indicator over time, the effects may be compensated for by the ratiometric measurement method. However, there are cases in which loss of indicator is a problem, and this effect can be reduced by injecting the indicator in a dextran-linked form. Linkage of the indicators to the dextran polymer does not in general seem to affect their properties significantly but inhibits cellular transport of the indicator (see the Molecular Probes catalogue for further information and for availability of these forms).

A particularly attractive feature of many of the fluorescent indicators is the possibility of introducing them into cells by the ester loading technique. This technique has extended fluorescence measurements into the realm of very small cells such as blood platelets. The fluorescent indicators are highly impermeant on account of being multiply charged at neutral pH, but on most indicators these charges are all carried on carboxyl groups. By esterifying these groups, an uncharged derivative of the indicator can be produced (Tsien, 1981). This derivative is not an active indicator, but it is sufficiently lipophilic to permeate biological membranes and thereby enter cells. Inside cells, the derivative is converted to the active indicator by the action of intrinsic esterase enzymes. Since the active indicator is impermeant, the effect is to cause accumulation of this form in the cells, so only a low concentration of the ester need be present in the external medium. This is useful, because the esters are highly insoluble (in fact they need to be dissolved in an appropriate carrier solvent before addition to the medium). Some care must be taken with this step, and the incubation conditions need to be adjusted carefully in order to achieve satisfactory loading of the cells. There is also the risk that the indicator will be loaded into other cell compartments as well as into the cytosol (see below). A further potential problem is that the type of ester that needs to be used (acetoxymethyl) liberates formaldehyde as a hydrolysis product, although serious toxicity problems have not been reported.

The major problem with the ester loading technique is that the experimenter has little direct control over where the indicator ends up in the cell. The ester will enter all intracellular compartments and the active indicator concentration in each compartment will depend on the relative esterase activity. Thus the endoplasmic reticulum and mitochondria will also contain indicator that can confound interpretation of the signals. An additional problem is that de-esterification may be incomplete so that a fluorescent intermediate, which is not ion sensitive, may be produced (e.g. Highsmith et al., 1986). The magnitude of these effects can only be ascertained with careful control experiments, and at the very least the fluorescence from the cell should be examined under a microscope to ensure that it is relatively uniform.

A related problem is that some cells seem to be very good at clearing their cytoplasm of the indicator. In endothelial cells for example, after about an hour, the cell fluorescence appears punctate and mitochondria are clearly visible (M. B. Cannell, unpublished observations). Whether this represents simply the removal of cytoplasmic indicator or accumulation of indicator in the punctate regions is unclear at this time. In any case, considerable caution should be applied to the interpretation of signals from the cells loaded with the ester form of the indicator. (It should be noted that during the Plymouth Microelectrode Workshop we routinely observe that cells loaded with the ester give smaller ratio changes when depolarised to 0 mV than when the patch pipette loading method is used).

In summary, direct injection is always preferable to ester loading if there is a choice. However, good results can be obtained with the ester loading technique in many cases, provided adequate control experiments are performed. The ester loading technique may be the only route to take if cells cannot be loaded by patch pipette and it also has the advantage that many cells can be loaded at the same time allowing experiments to be performed on cell suspensions in cuvettes.

A few words on the handling of the indicators is in order here. The fluorescent indicators are all subject to oxidation during storage and will lose activity in a few days if exposed to light and air at room temperature. (It is for this reason that the indicators are supplied in ampoules sealed under argon or nitrogen). It is best to make up the acetoxymethyl esters in dry DMSO and split the indicator into a number of ampoules each of which contains enough indicator for a single experiment. The ampoules can then be frozen and this will avoid repeated freeze/thawing cycles. A similar procedure should be used for the free acid form of the indicator except that instead of DMSO a suitable buffer solution should be used (e.g. 140 mM KCl). The free acid form of the indicator should never be exposed to metal (such as stainless steel syringe needles). On the day of the experiment the aliquot of indicator is dissolved in about 1 ml of intracellular solution and placed in a 1 ml polyethylene tuberculin syringe, the end of which has been drawn into a fine capillary (to allow it to pass down inside the electrode barrel). The syringe is used to place about 50-100 μl at the shoulder of the electrode, allowing a reasonable number of electrodes to be used before the syringe needs to be refilled. The experimenter should ensure that the small volume placed in the electrode is contacted by the silver wire of the electrode holder. Since the tip of the electrode is about 5 mm from the silver wire, diffusion of "silver contaminated" indicator should not be a problem. Finally, positive pressure must be maintained on the back of the electrode at all times before forming a giga-seal or else the experimenter will find that the indicator will be seriously diluted at the tip of the electrode (quite apart from the problem of contamination of the intracellular solution by bathing solution!). After breaking into the cell, dialysis of the cell may take 5-20 minutes, but the speed of dialysis can be markedly improved by applying brief pulses of positive pressure to the back of the electrode to "inject" the indicator (considerable care and experience is needed for this technique). The extent of dialysis can be determined from the magnitude of the resting fluorescence signal, and experienced patch clamp electrophysiologists may be amazed at how slow dialysis can be!

Interpretation of measurements

Although fluorescent indicators have proved to be extremely powerful, it is important to bear in mind that they share (as with all other recording techniques) the possibility of giving inaccurate measurements and/or of influencing whatever is being measured. In particular, it should be appreciated that the indicators work by reversibly binding to the target, so by definition they have a buffering action. The extent of the buffering will depend on the concentration of the indicator relative to the free concentration of the target ion as well as the affinity of the indicator, so it will be most significant for those ions whose free concentrations are relatively low, i.e. Ca and protons. Fortunately, the natural cell buffering capacity for these ions reduces the effect of indicator buffering. Nevertheless, the amplitudes and time courses of ion changes may well be significantly altered by the increased buffering, so one really needs to carry out experiments over a range of indicator concentrations to demonstrate that the results are reasonably independent of the indicator concentration. The binding equation is:

$$\text{fraction bound} = \frac{K_b[X]}{1+K_b[X]}$$

where

$$K_b = \frac{1}{K_d}$$

As an example, consider an experiment using fura-2 at 100 μM. Assuming a resting level of Ca and K_d of 150 nM, the fraction bound $= 0.5$. Hence, 50 μM Ca is bound to the dye and 0.15 μM is free. Thus the buffering power (B) = bound/free = 333.

Using another dye such as Furaptra: $K_d = 44$ μM and using 500 μM dye (very bright) gives B=10. These values can be compared to endogenous buffering powers of 75 for immobile buffers and 20 for mobile buffers (Neher & Augustine, 1992), so that it is clear that the high affinity indicators can swamp the intrinsic buffers at usable concentrations. This will result in a reduction in the amplitude of calcium transients as well as a slowing in their time course.

Another problem is that the kinetics of the fluorescence change do not reflect the kinetics of the underlying calcium transient. This arises from (1) saturation effects in indicator response and (2) the kinetics of dye binding. While the first problem can be circumvented by converting the dye signal to ion concentrations (but see below), workers often assume that, because the dye:calcium stoichiometry is 1:1, the dye is linear and that the time course of fluorescence change should reflect the time course of calcium change. However this is not the case. From the equation:

$$\frac{dF}{dt} = \frac{d[X]}{dt} \cdot \frac{dF}{d[X]}$$

($dF/d[X]$ is the slope of the relationship between F and [X] -the ligand concentration) it is clear that we must consider the properties of the relationship between F and [X]. Initially, assume that the dye kinetics are so fast that the dye reaction is at equilibrium at all times (see below) and the binding is 1:1. Fig. 7 shows the relationship between F and $dF/d[X]$ as a function of $[X]/K_d$ (the ligand concentration divided by the affinity of the indicator for the ligand). It is clear that $dF/d[X]$ is considerably less than unity at [X] > 0.1Kd. This is the range within which most indicators are used, so that we

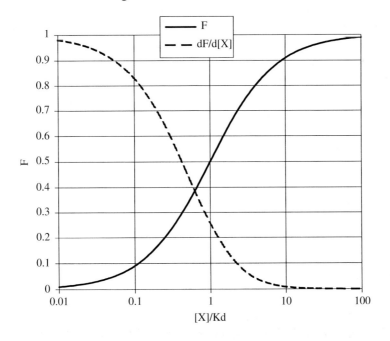

Fig. 7. The relationship between fluorescence (F) and ligand concentration ([X]) for a 1:1 stoichiometry. The ligand concentration is divided by its Kd; at the Kd 50% is bound. Note that the slope of the relationship between F and [X] is considerably less than unity around the Kd.

cannot take the time course of F as being the same as the time course of [X]. The simplest way around this is to convert F (or the ratio measurement) to [X] with a calibration curve.

However, dF/d[X] is also a function of time as the kinetics of the dye are limited. For the reaction:

$$D + X \underset{k_{on}}{\overset{k_{off}}{\rightleftharpoons}} DX$$

If the ion concentration undergoes a step change to [X] the indicator signal will change exponentially to a new level with a rate constant of (k_{on}[X] + k_{off}). For example, the kinetics of the fura-2 and indo-1 reactions with calcium have been investigated in stopped-flow experiments (Jackson et al., 1987). The rate of calcium binding by fura-2 is 2.5-6.5×10^8 M^{-1} s^{-1} and the rate of dissociation is 84 s^{-1} at 20°C. Indo-1 is slightly faster, with the rate constants of 0.5-1.0×10^9 M^{-1} s^{-1} and 130 s^{-1}, respectively. As might be expected, the rate of calcium binding is probably diffusion limited, while the lower affinity of indo-1 is reflected in the slightly higher rate of calcium dissociation. Thus at a level of 200 nM free calcium, the apparent fura-2 reaction will be about 180 s^{-1}. While this response speed is clearly superior to that of calcium-selective microelectrode and at least as fast as aequorin, it is not so fast that possible kinetic distortion of the calcium signal can be ignored. It is notable that Baylor and Hollingworth (1988) reported that the fura-2 kinetics appeared to be considerably slower in single frog muscle fibres than measurements of Jackson et al. (1987) would suggest. Baylor and Hollingworth (1988) compared the antipyrylazo absorbance signal to the fura-2 signal and suggested that the fura-2 signal might be explained if the fura-2 off rate were about 25 s^{-1}, a value three to four times lower than expected from *in vitro* experiments. The exact effect of the intracellular environment on the properties of fura-2 and indo-1 are still unknown. However, both Williams et al. (1985) and Baylor & Hollingworth (1988) reported that the fluorescence spectrum of fura-2 inside single smooth muscle cells was very similar to the spectrum obtained in a cuvette. In addition, although Williams et al. (1985) were loading the cells with the esterified form of fura-2 (which may have suffered from partial de-esterification problems - see above), they found that the calibration curve obtained by ionomycin exposure (to control intracellular calcium) agreed reasonably well with that obtained in a cuvette. It is notable that Williams et al. (1985) and Weir et al. (1987) found that the maximum ratio obtained from the intracellular dye appeared to be slightly less than that obtained in calibrating solutions, which suggests that the intracellular milieu alters the properties of the dye. A reduced coefficient for diffusion of fura-2 was observed by Baylor & Hollingworth (1988), an effect also found by Timmerman and Ashley (1986) in barnacle muscle. Fluorescence anisotropy also suggested some dye immobilisation (Baylor & Hollingworth, 1988). These results may all be due to the dye binding to some intracellular constituent, but whether such binding seriously alters the calcium sensitivity of the dye remains unclear.

Another potential difficulty concerns the fact that any indicator will give best resolution over a fairly narrow ion concentration range, and attempts to use one indicator to cover all experimental situations is likely to give inaccurate results. For example, the early fluorescent indicators for Ca ions were devised primarily to measure resting ion concentrations in small cells. The greatest accuracy for such measurements is obtained when the dissociation constant of the Ca-indicator complex is similar to the free Ca concentration, since the indicator fluorescence will be midway between its values at the concentration extremes. Unfortunately, this is not the best situation for measurement of transient concentration changes.

An indicator with a dissociation constant of a few hundred nM will be driven towards saturation by a Ca transient which locally elevates the free Ca concentration into the micromolar range, i.e. the slope of the Ca/fluorescence relationship will be progressively reduced as the Ca concentration increases. Although concentrations in this range can still be estimated if the Ca concentration is uniform throughout the measurement area, in practice this will not be the case. Instead, as the Ca diffuses away from the release sites, the total fluorescence signal will increase even though the average Ca concentration remains the same, because the reduction in fluorescence in the areas near the release sites is less than the increase in fluorescence in the areas further away. For accurate transient measurements, one should really use an indicator with a dissociation constant that is no lower than the highest concentration transient, but unfortunately this reduces the signal at resting ion levels.

Calibration

Although the calibration of fluorescent indicator recordings are, in general, easier than for photoproteins (which are highly non-linear) and for absorbance indicators (where absolute calibrations are difficult and interference effects from other ions tend to be greater), calibration can still be problematic. Ion concentrations can be estimated from fluorescence recordings by reference to *in vitro* calibration solutions, but such methods should be supplemented by measurements in which defined ion concentrations are imposed on the cell, to see if equivalent results are obtained (e.g. a known Ca concentration could be imposed by using an appropriate buffer solution outside the cell, and permeabilising the membrane with a Ca ionophore such as ionomycin, or perhaps an ion-sensitive electrode could be used). Unless such a verification can be made, calibrations should be taken only as a guide. This is a particularly important point, since calibration with reference to *in vitro* solutions is very straightforward, so it is always tempting to convert the results to concentrations without paying too much regard to possible errors.

For fluorescence measurements made at a single wavelength, the free ion concentration [I] is related to the fluorescence F by:

$$[I] = K_d(F-F_{min})/(F_{max}-F)$$

where F_{min} and F_{max} are respectively the fluorescence levels at zero and saturating ion concentrations, and K_d is the dissociation of the ion-indicator complex.

For fluorescence measurements made at a pair of wavelengths using ratiometric indicators, the free ion concentration [I] is related to the fluorescence ratio R by the analogous equation:

$$[I] = K_d.S.(R-R_{min})/(R_{max}-R)$$

where S is a scaling factor given by the fluorescence at the denominator wavelength of R at zero ion concentration, divided by the fluorescence at a saturating ion concentration (see Grynkiewicz, Poenie and Tsien (1985) for derivation of this equation).

For single-wavelength indicators, F_{max} and F_{min} have to be estimated for each experiment because they depend on the indicator concentration, whereas for ratiometric measurements the parameters R_{min}, R_{max} and S (in theory) need to be measured only once on the experimental setup because by dividing the signals at two wavelengths the concentration terms in the equation are cancelled (although regular checks of these parameters are recommended). However, the values calculated from these equations need to be treated with some caution. Apart from the obvious risk that the K_d may be somewhat different in the cell from that in the calibration solution, there is also the possibility that the fluorescence properties of the indicator may be different in the cell, and such effects have been observed for several indicators (see above).

These equations also assume that the indicator is sufficiently selective that formation of complexes with other ions does not occur to any significant extent. Although this may be a questionable assumption, the fluorescent indicators generally perform quite well in this respect. This is particularly true for Ca indicators based on the BAPTA molecule, e.g. fura-2 and indo-1 (Grynkiewicz, Poenie and Tsien, 1985), which are more selective against Mg and protons than are the metallochromic Ca indicators, where competing effects from such ions pose a major calibration problem (Thomas, 1982, 1991). Where interference effects from other ions are significant, the simplest approach is to calibrate the indicator in a solution that contains appropriate concentrations of those ions. However, such calibrations will only apply so long as the concentrations of the interfering ions in the cell do not change significantly during the experimental procedures.

As discussed above, when the ester loading technique is used, additional problems occur. First, there is the risk that the indicator may not be completely de-esterified, giving rise to a variety of indicator species with different fluorescence and ion-binding characteristics, and thereby introducing further calibration errors. Second, de-esterification may occur in other cell compartments as well as in the cytosol (particularly in the endoplasmic reticulum), where the ion concentrations may be very different. In some cases this can cause entirely different results compared with those obtained by injection of free indicator.

References

ALLEN, D. G. & BLINKS, J. R. (1979). The interpretation of light signals from aequorin-injected skeletal and cardiac muscle cells: a new methods of calibration. In *Detection & Measurement of free Ca²⁺ in cells*. (Ed. C. C. Ashley & A. K. Campbell). Elsevier/North Holland.

BLINKS, J. R., PRENDEGAST, F. G. & ALLEN, D. G. (1976). Photoproteins as biological calcium indicators. *Pharm. Rev* **28**, 1-28.

BAYLOR, S. M. & HOLLINGWORTH, S. (1988). Fura-2 calcium transients in frog skeletal muscle fibres. *J. Physiol.* **403**, 151-192.

CANNELL, M. B. & ALLEN, D. G. (1983). A photomultiplier assembly for the detection of low light levels. *Pflugers Arch.* **398**, 165-168.

CANNELL, M. B., BERLIN, J. R. & LEDERER, W. J. (1987). Effect of membrane potential changes on the calcium transient in single rat cardiac muscle cells. *Science* **238**, 1419-1423.

CANNELL, M. B., BERLIN, J. R. & LEDERER, W. J. (1988). Intracellular calcium in cardiac myocytes: Calcium transients measured using fluorescence imaging. In *Cell calcium and control of membrane transport.* (Ed. L. J. Mandel & D. C. Eaton). Chapt. 13. pp 201-214. New York: Rockefeller University Press.

FONTERIZ, R. I., SANCHEZ, A., MOLLINEDO, F., COLLADO-ESCOBAR, D. AND GARCIA-SANCHO, J. (1991). The role of intracellular acidification in calcium mobilization in human neutrophils. *Biochim. Biophys. Acta* **1093**, 1-6.

GRYNKIEWICZ, G., POENIE, M., AND TSIEN, R. Y. (1985). A new generation of Ca^{2+} indicators with greatly improved fluorescence properties. *J. Biol. Chem.* **260**, 3440-3450.

JACKSON, A. P., TIMMERMAN, M. P., BAGSHAW, C. R. & ASHLEY, C. C. (1987). The kinetics of calcium binding to fura-2. *FEBS Lett.* **216**, 35-39.

JAMES, J. F. (1967). On the use of a photomultiplier as a photon counter. *Monthly Not. Roy. Astronom. Soc.* **137**, 15-23.

LEVI, L. (1968). *Applied Optics.* Volumes 1 & 2. New York: Wiley.

MOLECULAR PROBES CATALOGUE (1993). *Handbook of Fluorescent Probes and Research Chemicals.* Molecular Probes Inc. Eugene Oregon, USA.

RIZZUTO, R., SIMPSON, A. W. M., BRINI, M. & POZZAN, T. (1983). Rapid changes in mitochondrial Ca^{2+} revealed by specifically targeted recombinant aequorin. *Nature* **358**, 325-327.

SHARPE, J. (1964). *Dark Current in Photomultipliers.* EMI Electronic Ltd. Document ref no. R/P021.

THOMAS, M. V. (1982). *Techniques In Calcium Research.* London: Academic Press.

THOMAS, M. V. (1991). Metallochromic indicators. In *Cellular Calcium: A Practical Approach,* (ed. J. G. McCormack and P. H. Cobbold) Oxford: IRL Press.

TIMMERMAN, M. P. & ASHLEY, C. C. (1986). Fura-2 diffusion and its use as an indicator of transient free calcium changes in single striated muscle cells. *FEBS Lett.* **209**, 1-8.

TSIEN, R. Y. (1981). A non-disruptive technique for loading calcium buffers and indicators into cells. *Nature* **290**, 527-528.

WIER, W. G., CANNELL, M. B., BERLIN, J. R., MARBAN, E. & LEDERER, W. J. (1987). Fura-2 fluorescence imaging reveals cellular and sub-cellular heterogeneity of $[Ca^{2+}]_i$ in single heart cells. *Science* **235**, 325-328.

Chapter 13

Microelectrode techniques in plant cells and microorganisms

C. BROWNLEE

1. Introduction

This chapter will review recent progress in the application of microelectrode techniques to the study of a range of cell physiological problems in plant cells and microorganisms. Plant cells present a special set of problems for electrophysiology and other microelectrode methods, related principally to cell structure (Fig. 1). The past five years have been witness to an great expansion in the application of microelectrode techniques and great progress towards overcoming some of the most significant problems facing plant electrophysiologists, particularly those associated with the presence of the cell wall and intracellular compartmentalization. Consequently, a wide range of plant cell types are now amenable to investigation using modern electrophysiological techniques and this is reflected by the rapidly increasing literature pertaining to plant cell electrophysiology. Virtually all techniques applied to animal cells are now possible with many plant cell types. Here I will outline some of the most significant recent advances in plant cell electrophysiology and discuss problems which remain to be overcome, indicating some of the likely advances in the near future.

2. Voltage clamp

While conventional voltage clamp of large plant cells, particularly the giant algae, using two or three electrodes, has provided the foundations for much of the current plant electrophysiology, (see e.g. Findlay, 1961; Kishimoto, 1961; Beilby, 1982; Lunevsky et al. 1983; Gradmann, 1978), technological developments in voltage and patch clamp have allowed more wide-ranging studies. One of the most significant advances in conventional two-electrode voltage clamp studies has been the application of double-barrelled electrodes to single small cells of higher plants. In this configuration, one barrel of the electrode measures voltage while current is injected through the other barrel. This is perhaps best exemplified by the work of Blatt and co-workers on the stomatal guard cell (e.g. Blatt, 1987, 1991a,b, 1992). The stomatal

Marine Biological Association of the UK, The Laboratory, Citadel Hill, Plymouth PL1 2PB, UK.

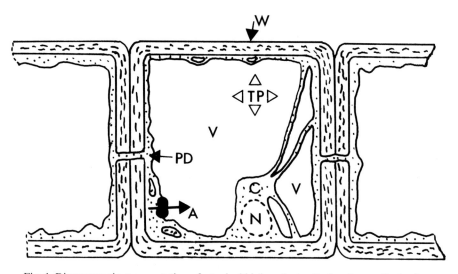

Fig. 1. Diagrammatic representation of a typical higher plant cell, showing particular features relevant to the application of microelectrode methods. W, cell wall; C, cytoplasm; V, vacuole; A, anion transporters in vacuolar and organelle membranes; PD, intercellular electrical continuity via plasmodesmata; TP, turgor pressure.

guard cell is probably the most intensively studied higher plant cell from an electrophysiological aspect. In addition to playing a major role in plant-water relations, stomatal guard cells show very clear responses (stomatal opening or closure, corresponding to changes in cell shape resulting from changes in ion and water fluxes across the plasma membrane (Mansfield *et al.* 1990)) to well-defined signals, including plant hormones (abscisic acid, auxin), light and CO_2. In the studies of Blatt and co-workers, electrodes were filled with 200 mM K^+-acetate to minimise the effects of Cl^- leakage into the relatively small cytoplasmic volume. Typically, steady-state I/V relations were obtained using bipolar staircase voltage clamp protocols. This work has concentrated largely on currents through inward- and outward-rectifying K^+ channels involved in the response to abscisic acid, particularly their control by pH and Ca^{2+} (e.g. Blatt, 1990; 1992; Blatt & Armstrong, 1993). This method has been extended further to allow loading of the cells with caged compounds (caged Ins(1,4,5)P3 and caged ATP) which were introduced through the recording barrel of the electrode (Blatt *et al.* 1990). Advantages of the double-barrelled voltage clamp method over patch clamping of plant protoplasts (see below) are that it can be used *in situ* on intact cells inside their cell walls and probably reflects a more physiologically relevant situation. Complications arise where the cells are electrically coupled to other cells (which is the case for many plant cell types), restricting the application of this technique. Other disadvantages of this method are that it is often technically difficult to impale the cytoplasm of small highly vacuolated plant cells without damage.

The single electrode discontinuous voltage clamp (see Chapter 2) has been applied to eggs of the marine alga, *Fucus serratus* (Taylor & Brownlee, 1993). With this method, a single electrode performs both the voltage recording and current injection tasks with a rapid switching frequency, determined by the time constants of the electrode (τe) and cell membrane (τc), allowing any voltage associated with charging of the electrode capacitance during the current injection cycle to decay before membrane potential is sampled. The relatively large size (~70 μm diameter) of *Fucus* eggs ensured a sufficiently high τc despite a relatively low input resistance (100 MΩ). Electrode switching frequency was optimised by bevelling the tip of the electrode (Kaila & Voipa, 1985) and dipping in mineral oil to minimize stray capacitance. Limitations on the range of the voltage clamp potentials that could be used arose from the relatively small τc/τe ratio (typically around 150), resulting in clamp error or oscillations induced by extreme command voltages (Finkel & Redman, 1984). Nevertheless, this study provided evidence for the existence of a voltage-activated inward Ca^{2+} current and a slowly-activating outward K^+ current which were postulated to be involved in the generation of the fertilization potential (Taylor & Brownlee, 1993). As far as we know, this technique has not been applied to other plant cell types, though the relatively high input resistance of many plant cells may render them suitable where other voltage clamp methods may not be feasible.

3. Patch clamp

Plasma membrane

The application of the patch clamp technique to plant cells has undergone a major expansion, significantly improving our understanding of the nature and roles of ion transport in plant cells, relating to nutrient transport, osmoregulation, and intracellular signalling. (For recent reviews on plant cell ion channels and pumps, see Brownlee & Sanders, 1992; Hedrich & Schroeder, 1989; Schroeder & Thuleau, 1991, Tester, 1990). The primary, and most crucial step in patch clamping plant cells is the removal of the cell wall to reveal a clean plasma membrane on which giga-ohm seals can be obtained with a patch pipette. To date, most patch clamp studies on plant cells have utilized protoplasts in which the cell wall has been removed with enzyme cocktails based primarily on cellulase. A variety of enzyme methods have been developed for the production of viable plant cell protoplasts. Methods vary considerably between different cells and tissue types. A major problem with protoplast production is the isolation of particular identified cell types from tissue which may contain several different kinds of cell. Isolation of a particular cell type may involve initial dissection of the tissue, followed by preferential disruption of the unwanted cells and selective enzymatic digestion of the walls of the cells of interest. A good example of the successful isolation of a particular protoplast type is the preparation of stomatal guard cell protoplasts from leaf tissue (Kruse *et al.* 1989). The technique essentially involves mechanical removal of the epidermal tissue containing the guard cells, enzyme treatment to liberate the protoplasts, followed by centrifugation for removal of debris and purification of protoplast fractions. Details of the methods for production of a variety of guard cell protoplasts suitable for patch clamping are given by Raschke & Hedrich (1989). Separation techniques of this kind, however are only applicable to a few cell types and the often drastic treatments involved may have unwanted side effects, such as alteration of the plasma membrane properties by the enzyme treatment. An indication of the physiologically altered state of protoplasts comes from the few measurements of protoplast membrane potential that have been made. For example, tobacco mesophyll protoplast membrane potential has been measured at around -10 to -15 mV (Venis et al. 1992), whereas membrane potentials from most mesophyll cells are very negative (-100 to -150 mV), reflecting the activity of the plasma membrane proton pump (e.g. Senn & Goldsmith, 1988). The depolarized state of the protoplast membrane may reflect, in part, increased leakage arising from electrode impalement, though this would tend to produce variable membrane potentials, depending on the quality of impalement, which is not generally observed. The other possibility is that the activity of the proton pump is reduced in protoplasts for reasons related to the protoplast isolation procedure. This uncertainty awaits resolution. Enzymatically produced protoplasts also lack the plasma membrane/extracellular matrix connections as a result of wall removal. The cell wall performs many functions which may have significant bearing on the physiology of the cell. For example, plasma membrane-wall connections are known to be essential for polar axis fixation in *Fucus* zygotes (Kropf et al. 1988). Loss of turgor and non-physiological osmotic environment are unavoidable factors in the application of patch clamp techniques to most plant cells.

The success rate of giga seal formation in plant protoplasts prepared by standard enzyme methods varies but is generally low compared with most animal cells. This probably reflects combinations of the lack of cytoskeletal rigidity, incomplete wall removal or protoplast cleaning, and regeneration of the wall. Fairley & Walker (1989) showed that inhibition of wall regeneration in maize protoplasts did not improve electrode-membrane sealing rates. Significant improvements in seal rates have, however, been produced by modifications of the enzyme treatment, particularly by reducing the enzyme treatment to a minimum. Elzenga et al. (1991) used a 5 minute enzyme treatment to weaken the cell wall, followed by osmotic adjustment to make the protoplasts swell and pop out of the weakened wall. The method was shown to be successful in several cell types (Table 1). Giga seal (>10 MΩ) formation was obtained in $>40\%$ of attempts, usually within 2 minutes. A similar approach has been used to obtain improved seal rates from root cell protoplasts (Vogelzang & Prins, 1992). In this case, protoplasts were released by application of mechanical pressure.

The patch clamp technique has been used in a variety of animal cells to measure capacitance changes in the plasma membrane related to exocytosis (e.g. Lindau & Neher, 1988). So far, there is only one published report of the application of this technique to plant cells (Zorec & Tester, 1992). In that study, capacitance of protoplasts from barley aleurone cells was shown to increase when cells were dialysed through the patch pipette with high (1 μM) Ca^{2+}. Whole cell patch clamp of enzymatically produced protoplasts has also been used in combination with ratio photometry using the fluorescent Ca^{2+}-sensitive dye, fura-2 (Schroeder & Hagiwara,

Table 1. *Success rate of giga-ohm seal formation on protoplasts prepared by brief enzyme treatment*

Protoplast source	Total seals	R>10GΩ	1GΩ>R >1GΩ	Fail/WC	Isolates
Pisum sativum, leaf epidermis	123	67	32	24	29
P. sativum stem epidermis	98	42	31	25	27
Phaseolus vulgaris leaf mesophyll	15	10	3	2	2
Avena sativa coleoptile cortex	15	6	5	4	2
Arabidopsis thaliana cotyledon mesophyll	15	9	5	1	2

Numbers indicate seal formation with resistance >10GΩ; resistance between 1 and 10GΩ; and attempts which resulted in no seal formation or where the patch ruptured during seal formation i.e. went whole cell (fail/WC). "Isolates" represents the number of different protoplast batches used. (Reproduced, with permission from Elzenga *et al.* (1991).)

1990). In this study, dye was introduced into the cell via the patch clamp electrode. Transient elevations of cytosolic Ca^{2+} were observed with simultaneous increases in inward current after the application of abscisic acid.

Mechanical removal of the cell wall presents an alternative to enzymatic treatments for plasma membrane access. This was initially used in giant cells of *Chara australis*, using a knife cannula or fine forceps to cut the wall (Laver, 1991) The patch pipette could be inserted through the hole cut in the cell wall. Development of a microsurgical technique using a U.V. laser arose out of a need to maintain the polarity of rhizoid cells of *Fucus* in studies of the distribution of channel activity during polarized growth (Taylor & Brownlee, 1992). In recent years, the use of lasers as microsurgical tools has become established (see Berns *et al.* (1991); Greulich & Weber (1992); Weber & Greulich, (1992) for recent reviews). The most useful laser for plant cell microsurgery is a pulsed nitrogen laser such as a 337 VSL or 337 ND (Laser Science Inc., U.S.A.). This can be passed via appropriate optics through the fluorescence port of a microscope and focussed to a very small spot by a U.V.-conducting objective. We routinely use a 40, 1.3. N.A. objective (Nikon). The power density of the focussed spot is sufficient to ablate selected regions of the cell wall. Prior osmotic shrinkage of the cytoplasm allows the plasma membrane to be withdrawn from the cell wall during ablation. Careful reflation of the cytosol by adjusting the osmotic potential of the bathing medium results in extrusion of varying amounts of plasma membrane-bound cytoplasm which can be accessed by a patch electrode (Fig. 2). To date, this method has allowed rapid access to the plasma membrane of a growing *Fucus* rhizoid, giving a rate of seal formation (40-80% for 5-15 GΩ seals), not possible with enzymatic protoplasting techniques. Laser microsurgery has also recently been used to expose the plasma membrane of root hairs of *Medicago* (Kurkdjian *et al.* 1993). The high success rate of sealing observed in *Fucus* suggests that the technique may be useful as an alternative for plant cell types that are difficult to patch clamp following enzymatic wall removal. Laser microsurgery should also enable studies of other systems where ion channel activity is important in relation to polarity, such as fungal hyphae, root hairs and pollen tubes. It should also be possible to patch clamp membranes of identified cells which are difficult to isolate using bulk protoplasting techniques. Our own experience (unpublished) indicates that laser microsurgery can produce high quality protoplasts from cells types as varied as fungal hyphae and stomatal guard cells.

Organelles

Significant progress has now been made in patch clamp studies of the vacuolar membrane of a variety of plant cells. Vacuolar isolation techniques are relatively straightforward and generally involve washing the surface of a freshly cut tissue slice with buffer to liberate the exposed vacuoles (Coyaud *et al.* 1987; Keller & Hedrich, 1992). Vacuoles can also be produced from protoplasts (e.g. Ping *et al.* 1992). Whole-vacuole and single channel recordings have been obtained from a variety of channel types. These include K^+, Cl^- and Ca^{2+} channels (e.g. Alexandre & Lassalles, 1990; Hedrich & Neher, 1987; Johannes *et al.* 1992; Pantoja *et al.* 1992; Ping *et al.* 1992). The success of vacuolar patch clamp recordings is dependent in part on obtaining appropriate osmotic conditions for both bath and electrode filling solutions. Improvement in signal-to-noise ratio has been reported by dipping the pipette in a 30% silane:70% carbon tetrachloride solution after filling, presumably by forming a hydrophobic surface at the pipette shank (Alexandre & Lassalles, 1990).

Patch clamp recordings have been made from giant chloroplasts liberated from osmotically shocked protoplasts of *Peperomia metallica* leaf cells (Schonknecht *et al.* 1988; Keller & Hedrich, 1992). This treatment caused osmotic swelling of the chloroplast and rupturing of the thylakoid envelope, exposing thylakoid membrane blebs which could be patch clamped and allowed studies of the ion fluxes associated with the pH gradient across the thylakoid membrane.

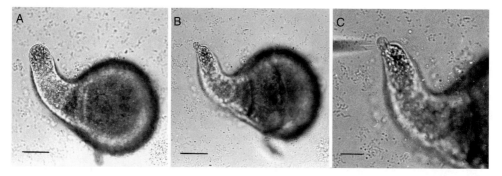

Fig. 2. Localized patch clamping from a *Fucus* rhizoid cell using U.V.-laser microsurgery. (A) 24-h *F. serratus* zygote with developing rhizoid. Scale bar: 10 μm. (B) After laser ablation of the cell wall and reflation of the cytoplasm. Scale bar: 10 μm. (C) Exposed plasma membrane and patch clamp electrode. Scale bar: 5 μm. (Reproduced, with permission from Taylor & Brownlee (1992).)

Patch clamp and microorganisms

Progress in patch clamp studies of microbial membranes has been reviewed recently (Siami *et al.* 1992). The protozoan, *Paramecium* can be induced to form plasma membrane blisters under conditions of low Ca^{2+} and shear which are accessible for patch clamping. Ion channels have also been recorded in isolated *Paramecium* cilia (Siama *et al.* 1992). Developments in the production of protoplasts from yeast has enabled detailed patch clamp studies of both the plasma membrane and tonoplast. Bertl & Slayman (1990) and Bertl *et al.* (1992), for example, used a tetraploid strain of *Saccharomyces cerevisiae* (YCC78) as starting material for protoplasts from which both vacuolar and plasma membrane channels have been characterized. Both voltage- and Ca^{2+}-dependent cation channels were found in the plasma membrane and tonoplast. Methods for the production of yeast protoplasts (spheroplasts) are described in detail by Siami *et al.* (1992).

Giant spheroplasts can also be produced from bacteria, such as *E. coli* and *B. subtilis* (Siami *et al.* 1992). Various treatments have been utilised to effect wall loosening and osmotic swelling of the underlying protoplast. These include cephalexin or U.V. treatment to produce unseptated filaments, followed by lysozyme treatment and osmotic swelling. Magnesium chloride coupled with cephalexin has also been used to produce giant cells. Mutants lacking certain cell wall components, or osmotically sensitive mutants have also been used (Siami *et al.* 1992; Criado & Keller, 1987). Fusion of bacterial membrane vesicles with liposomes followed by unilamellar blister formation in the presence of 20 mM $MgCl_2$ has yielded recordings from a variety of reconstituted channels, including a mechanosensitive channel, a voltage-sensitive channel and a cation-selective channel (Delcour *et al.* 1989a,b).

Protoplasts have also been obtained from fungal hyphae, allowing patch clamp recordings and preliminary characterisation of channel distribution between protoplasts isolated from different regions of the hypha of *Saprolegnia ferax* (Garril *et al.* 1992a). Problems were encountered in obtaining high resistance seals, limiting studies to cell-attached recording mode. However, both whole cell and excised patch recordings were made from protoplasts of the fungus, *Uromyces* (Zhou *et al.* 1991), though Garril *et al.* (1992b) have questioned the origin of the membrane in these studies, suggesting that it may be tonoplast. It is likely that the application of laser microsurgery will improve the quality of giga-seal formation on the fungal plasma membrane. Our preliminary experiments show that high resistance seals (>10 GΩ) can be obtained relatively easily with protoplasts extruded from *Phytophora* following laser microsurgery of the cell wall (A. Taylor, P. Whiting, D. Sanders & C. Brownlee, unpublished observations).

Wall-free mutants of the unicellular green alga, *Chlamydomonas* have been used in a patch clamp study of the rhodopsin photoreceptor current (Harz & Hegemann, 1991; Harz *et al.* 1992). In these experiments whole cells were gently sucked into fire-polished pipettes, forming seals with resistances up to 250 MΩ, allowing cell-attached recordings from a relatively large membrane area, though higher resistance seals were not achieved.

4. Microinjection

Studies employing microinjection based on both iontophoresis and pressure are now widespread in plants. Examples include microinjection of fluorescent antibodies for microtubules and actin into living cells (Cleary *et al.* 1992), caged compounds (Blatt *et al.* 1990; Gilroy *et al.* 1990) and Ca^{2+} indicators (e.g. Brownlee & Pulsford, 1988; Gilroy *et al.* 1990; McAinsh *et al.* 1990; Miller *et al.* 1992; Rathore *et al.* 1991) Microinjection has proven to be essential in several cell types for the use of fluorescent dyes for monitoring cytoplasmic Ca^{2+} and pH. Plant cells generally do not take up and hydrolyse the acetoxymethyl esters of these dyes which are commonly used in many animal cell studies. Even when this does occur, compartmentalization of the free acid form of the dye into vacuoles and intracellular vesicles and loss from the cell across the plasma membrane is frequent (Brownlee & Pulsford, 1988; Bush & Jones, 1987; reviewed by Read *et al.* 1992). Circumvention of these problems has recently been achieved by the use of fluorescent dyes linked to high relative molecular mass (M_r) dextran (10,000 or more: Miller *et al.* 1992). This usually necessitates the use of pressure and may be problematic in highly turgid plant cells. Reduction of the cell turgor pressure was found to facilitate injection and reduce cell damage during microinjection of mixtures of the Ca^{2+} indicator, Calcium Green-dextran and the pH indicator SNARF-dextran into *Fucus* zygotes (Berger & Brownlee, 1993). In this study it was found that 10,000 M_r Calcium Green-dextran, though excluded from intracellular vacuoles and vesicles and retained in the cytoplasm, did enter the nucleus. 70,000 M_r Calcium Green-dextran, however, was excluded from the nucleus, except during cell division when nuclear envelope breakdown occurred. Using these long wavelength-excitable dextran-linked dyes enabled the acquisition of confocal sections of Ca^{2+} distribution, followed over a period of 3 days, without any significant loss of dye signal or compartmentalization. It is possible to microinject 10,000 M_r dextran dyes iontophoretically using positive current pulses. Negative current, however, causes preferential injection of any non-dextran linked dye molecules present as impurities and should be avoided.

5. The pressure probe

An intracellular electrode technique which appears to be uniquely applied to plant cells is the pressure probe. This was developed by Zimmerman *et al.* (1969) for the study of the large intracellular (turgor) pressures which are a fundamental feature of the water relations of walled plant cells. Following its initial application to giant algal cells, the technique has been adapted for use with much smaller cells of higher plants (Husken *et al.* 1978). An oil-filled micropipette is connected to a microsyringe and a pressure transducer. Impalement of a cell results in displacement of the oil meniscus in the tip of the micropipette which can be observed microscopically. The pressure required to return the meniscus to the pipette tip is a measure of the internal hydrostatic pressure of the cell. The technique has been used in several studies of water and solute relations of plant cells (e.g. Malone & Tomos, 1990, 1992; Pritchard & Tomos, 1993; Rygol *et al.* 1993; Tomos *et al.* 1992). Modifications of this technique include turgor clamp (Murphy & Smith, 1989; Zhu & Boyer, 1992) which allowed investigation of growth under carefully controlled conditions of cell turgor, and precise pressure injection of fluorescent probes along with simultaneous monitoring of turgor (Oparka *et al.* 1991). A further modification of this technique allows rapid sampling of vacuolar contents of higher plant cells (Leigh & Tomos, 1993; Malone *et al.* 1989, 1991). This has allowed analysis of vacuolar ion concentrations (using X-ray microanalysis), osmotic potential (freezing point method) and metabolites from a variety of identified cell types (see Tomos *et al.* (1993) for experimental details).

6. Ion-selective electrodes

Despite the emergence of fluorescent dyes for measuring intracellular ion concentrations, work with ion-selective microelectrodes has continued to give important insights into a range of processes, particularly those involving Ca^{2+} and pH, though nitrate-selective electrodes have been used in both giant algal and higher plant cells (Miller & Zhen, 1991; Zhen *et al.* 1991). The advantages of ion-selective microelectodes continue to be their relative ease of manufacture and low cost. Provided attention is paid to calibration, they can provide accurate measurements of ion concentration in selected compartments. Improvements in sensor (particularly for Ca^{2+}) and electrode fabrication has largely overcome problems of sensitivity and sensor displacement due to turgor pressure. Problems still exist in re-calibration of Ca^{2+} electrodes after intracellular recording (e.g. Amtmann *et al.* 1992) and in their low sensitivity at physiological free Ca^{2+} concentrations and tip diameters small enough to penetrate most plant cells (Felle, 1993). This, together with the fact that the response of a Ca^{2+} electrode is only half of that of a monovalent ion-selective electrode has limited their use in plant cells (see Felle (1993) for review on the use of Ca^{2+}-selective microelectrodes in plants). In contrast, pH-selective microelectrodes have found considerably wider application. They do not appear to suffer from significant calibration problems and the sensitivity and response time is significantly better than for Ca^{2+} electrodes. Felle (1993) reviews the use of pH microelectrodes in plants and provides a general overview of their role in elucidation of pH relations of plant cells. An additional problem with the use of ion selective electrodes is the determination of the location of the electrode tips in the cytosol or vacuole. This has been discussed by Felle (1993), Felle & Bertl (1986), Miller & Sanders (1987) and Miller & Zhen (1991).

An extracellular vibrating Ca^{2+}-selective electrode has been developed by Kuhtreiber & Jaffe (1990). This electrode contains a Ca^{2+}-selective sensor in the tip and oscillates between two fixed points close to the surface of a cell converting small gradients of Ca^{2+} into voltage differences. Provided external $[Ca^{2+}]$ is low, high sensitivity and low noise are achieved by averaging the oscillating voltage signal, allowing the detection and quantification of small Ca^{2+} gradients resulting from Ca^{2+} influx around the surface of a variety of plant and animal cells (Kuhtreiber & Jaffe, 1990; Schiefelbein *et al.* 1992).

References

ALEXANDRE, J., LASSALLES, J. P. & KADO, R. T. (1990). Opening of Ca^{2+} channels in isolated red beet vacuole membrane by inositol 1,4,5-trisphosphate. *Nature Lond,* **343**, 567-570.

ALEXANDRE, J. & LASSALLES, J. P. (1990). Effect of D-myo-inositol 1,4,5, trisphosphate on the electrical properties of the red beet vacuolar membrane. *Plant Physiol.* **93**, 837-840.

AMTMANN, A., KLIEBER, H-G. & GRADMANN, D. (1992). Cytoplasmic free calcium in the marine alga *Acetabularia*: measurement with Ca^{2+}-selective microelectrodes and kinetic analysis. *J. Exp. Bot.* **43**, 875-885.

BEILBY, M. J. (1982). Chloride channels in *Chara. Phil. Trans. Roy. Soc. Lond.* **299**, 435-455.

BERGER, F. & BROWNLEE, C. (1993). Ratio confocal imaging of free cytoplasmic calcium gradients in polarizing and polarized *Fucus* zygotes. *Zygote* **1**, 9-15.

BERNS, M. W., WRIGHT, W. H. & WIEGAND-STEUBING, R. (1991). Laser microbeam as a tool in cell biology. *Int. Rev. Cytol.* **129**, 1-44.

BERTL, A., GRADMANN, D. & SLAYMAN, C. L. (1992). Calcium- and voltage-dependent ion channels in *Saccharomyces cerevisiae. Phil. Trans Roy. Soc. Lond. B.* **338**, 63-72.

BERTL, A. & SLAYMAN, C. L. (1990). Cation-selective channels in the vacuolar membrane of *Saccharomyces*: dependence on calcium, redox state and voltage. *Proc. Natl. Acad. Sci. U.S.A.* **87**, 7824-7828.

BLATT, M. R. (1987). Electrical characteristics of stomatal guard cells: the contribution of ATP-dependent "electrogenic" transport revealed by current-voltage and difference-current-voltage analysis. *J. Membr. Biol.* **98**, 257-274.

BLATT, M. R. (1990). Potassium channel currents in intact stomatal guard cells: rapid enhancement by abscisic acid. *Planta* **180**, 445-455.

BLATT, M. R. (1991a). Ion channel gating in plants: physiological implications and integration for stomatal function. *J. Membr. Biol.* **124**, 95-112.

BLATT, M. R. (1991b). A primer in plant electrophysiological methods. In *Methods in Plant Biochemistry* (ed. K. Hostettmann), pp281-321 London: Academic Press.

BLATT, M. R. (1992). K^+ channels of stomatal guard cells. Characteristics of the inward rectifier and its control by pH. *J. Gen. Physiol.* **99**, 615-644.

BLATT, M. R. & ARMSTRONG, F. (1993). K^+ channels of stomatal guard cells: abscisic acid-evoked control of the K^+ inward-rectifier by cytoplasmic pH. *Planta* (in press).

BLATT, M. R., THIEL, G. & TRENTHAM, D. R. (1990). Reversible inactivation of K^+ channels of *Vicia* stomatal guard cells following the photolysis of caged inositol (1,4,5) tris phosphate. *Nature Lond* **346**, 766-769.

BROWNLEE, C. & SANDERS, D. (eds). (1992). *Calcium Channels In Plants. Phil Trans. Roy. Soc. Lond B.* **338**, 112 pp.

BROWNLEE, C. & PULSFORD, A. L. (1988). Visualization of the cytoplasmic free Ca^{2+} gradient in growing rhizoid cells of *Fucus serrartus. J. Cell Sci.* **91**, 249-256.

BUSH, D. S. & JONES, R. L. (1987). Measurement of cytoplasmic calcium in aleurone protoplasts using indo-1 and fura-2. *Cell Calcium* **8**, 455-472.

CLEARY, A. L., GUNNING, B. E. S., WASTENEYS, G. O. & HEPLER, P. K. (1992). Microtubule and F-actin dynamics at the division site in living *Tradescantia* stamen hair cells. *J. Cell Sci.* **103**, 977-788.

COYAUD, L., KURKDJIAN, A., KADO, R. T. & HEDRICH, R. (1987). Ion channels and ATP-driven pumps involved in ion transport across the plasma membrane. *Biochim. Biophys. Acta.* **902**, 263-268.

CRIADO, M. & KELLER, U. B. (1987). A membrane fusion strategy for single channel recordings of membranes usually non-accessible to patch clamp pipette electrodes. *FEBS Lett.* **224**, 172-176.

DELCOUR, A. H., MARTINAC, B., ADLER, J. & KUNG, C. (1989a). Modified reconstitution method used in patch clamp studies of *Escherichia coli* ion channels. *Biophys. J.* **56**, 631-636.

DELCOUR, A. H., MARTINAC, B., ADLER, J. & KUNG, C. (1989b). Voltage-sensitive ion channel of *Escherichia coli. J. Membr. Biol.* **112**, 267-275.

ELZENGA, J. T. M., KELLER, C. P. & VAN VOLKENBURG, E. (1991). Patch clamping protoplasts from vascular plants. *Plant Physiol.* **97**, 1573-1575.

FAIRLEY, K. A. & WALKER, N. A. (1989). Patch clamping corn protoplasts. Gigaseal formation is not improved by congo red inhibition of cell wall regeneration. *Protoplasma* **153**, 111-116.

FELLE, H. (1993). Ion-selective microelectrodes: their use and importance in modern plant cell biology. *Bot Act.* **106**, 5-12.

FELLE, H. & BERTL, A. (1986). The fabrication of H^+-selective liquid membrane microelectrodes for use in plant cells. *J. Exp. Bot.* **37**, 1416-1428.

FINDLAY, G. P. (1961). Voltage clamp experiments in *Nitella. Nature Lond.* **191**, 9-19.

FINKEL, A. S. & REDMAN, S. J. (1984). Theory and operation of a single microelectrode voltage clamp. *J. Neurosci. Meth.* **11**, 101-127.

GARRIL, A., LEW, R. & BRENT-HEATH, I. (1992a). Stretch-activated Ca^{2+} and Ca^{2+}-activated K^+ channels in the hyphal tip plasma membrane of the oomycete *Saprolegnia ferax. J. Cell Sci.* **101**, 721-730.

GARRIL, A. LEW, R. & BRENT HEATH, I. (1992b). Measuring mechanosensitive channels in *Uromyces. Science* **256**, 1335-1336.

GILROY, S., READ, N. D., & TREWAVAS, A. J. (1990). Elevation of cytoplasmic calcium by caged calcium or caged inositol triphosphate initiates stomatal closure. *Nature Lond.* **346**, 769-771.

GRADMANN, D. (1978). Green light (550 nm) inhibits electrogenic Cl^- pump in the *Acetabularia* membrane by permeability increase for the carrier ion. *J. Membr. Biol.* **44**, 1-24.

GREULICH, K-O. & WEBER, G. (1992). The light microscope on its way from an analytical to a preparative tool. *J. Microsc.* **167**, 127-151.

HARZ, H. & HEGEMANN, P. (1991). Rhodopsin-regulated calcium currents in *Chlamydomonas. Nature Lond.* **351**, 489-491.

HARZ, H., NONNEGASSER, C. & HEGEMANN, P. (1992). The photoreceptor current of the green alga, *Chlamydomonas. Phil. Trans. Roy. Soc. Lond. B.* **338**, 39-52.

HEDRICH, R. & SCHROEDER, J. I. (1989). The physiology of ion channels and pumps in higher plants. *Annu. Rev. Plant Physiol. Plant Molec. Biol.* **40**, 539-569.

HEDRICH, R. & NEHER, E. (1987). Cytoplasmic calcium regulates voltage-dependent ion channels in plant vacuoles. *Nature* **329**, 833-835.

HUSKEN, D., STEUDLE, E. & ZIMMERMANN, U. (1978). Pressure probe technique for measuring water relations of cells in higher plants. *Plant Physiol.* **61**, 155-163.

JOHANNES, E., BROSNAN, J. M. & SANDERS, D. (1992). Calcium channels in vacuolar membrane of plants: multiple pathways for intracellular calcium mobilization. *Phil. Trans. Roy. Soc. Lond. B* **338**, 105-112.

KAILA, K & VOIPA, J. (1985). A simple method for dry bevelling of micropippettes used in the construction of ion-selective microelectrodes. *J. Physiol.* **369**, 8.

KELLER, B. U. & HEDRICH, R. (1992). Patch clamp techniques to study ion channels from organelles. *Meth. Enzymol.* **207**, 673-681.

KISHIMOTO. U. (1961). Current-voltage relations in *Nitella. Biol. Bull.* **121**, 370-371.

KROPF, D. L., KLOAREG, B. & QUATRONO, R. S. (1988). Cell wall is required for fixation of the embryonic axis in *Fucus* zygotes. *Science* **239**, 187-190.

KRUSE, T., TALLMAN, G. & ZEIGER, E. (1989). Isolation of guard cell protoplasts from mechanically prepared epidermis of *Vicia faba* leaves. *Plant Physiol.* **90**, 1382-1386.

KUHTREIBER., W & JAFFE, L. F. (1990). Detection of extracellular calcium gradients with a calcium-specific vibrating electrode. *J. Cell Biol.* **100**, 1565-1573.

KURKDJIAN, A., LEITZ, G., MANIGAULT, P., HARIM, A. & GREULICH, K-O. (1993). Non-enzymatic access to the plasma membrane of Medicago root hairs by laser microsurgery. *J. Cell Sci.* **105**, 263-268.

LAVER, D. R. (1991). A surgical method for accessing the plasma membrane of *Chara australis. Protoplasma* **161**, 79-84.

LEIGH, R. A. & TOMOS, A. D. (1993). Ion distribution in cereal leaves: pathways and mechanisms. *Phil. Trans. Roy. Soc. Lond. B.* (in press).

LINDAU, M. & NEHER, E. (1988). Patch clamp techniques for time-resolved capacitance measurements in single cells. *Pflugers Arch.* **411**, 137-146.

LUNEVSKY, V.Z., ZHERELOVA, O.M., VASTRIKOV, I.Y. & BERETOVSKY, G.N. (1983). Excitation of Characeae cell membranes as a result of activation of calcium and chloride channels. *J. Membr. Biol.* **72**, 43-58.

MALONE, M. & TOMOS, A. D. (1990). A simple pressure-probe method for the determination of volume in higher-plant cells. *Planta* **182**, 199-203.

MALONE, M. & TOMOS, A. D. (1992). Measurment of gradients of water potential in elongating pea stem by pressure probe and picolitre osmometry. *J. Exp. Bot.* **43**, 1325-1331.

MALONE, M., LEIGH, R. A., & TOMOS, A. D. (1989). Extraction and analyisis of sap from individual wheat leaf cells: the effect of sampling speed on the osmotic pressure of extracted sap. *Plant Cell Envir.* **12**, 919-926.

MALONE, M. LEIGH, R. A. & TOMOS, A. D. (1991). Concentrations of vacuolar inorganic ions in individual cells of intact wheat leaf epidermis. *J. Exp. Bot.* **42**, 305-309.

MANSFIELD, T. A., HETHERINGTON, A. M. & ATKINSON, C. J. (1990). Some current aspects of stomatal physiology. *Annu. Rev. Plant Physiol. Plant Molec. Biol.* **41**, 55-75.

MCAINSH, M. R., BROWNLEE, C. & HETHERINGTON, A. M. (1990). Abscisic acid-induced elevation of guard cell cytosolic calcium precedes stomatal closure. *Nature* **343**, 186-188.

MILLER, A. J. & SANDERS, D. (1987). Depletion of cytosolic free calcium induced by photosynthesis. *Nature* **343**, 186-188.

MILLER, A. J. & ZHEN, R-G. (1991). Measurement of intracellular nitrate concentrations in *Chara* using nitrate-selective microelectrodes. *Planta* **184**, 47-52.

MILLER, D. D., CALLAHAM, D. A., GROSS, D. J. & HEPLER, P. K. (1992). Free calcium gradient in growing pollen tubes of *Lilium*. *J. Cell Sci.* **101**, 7-12.

MURPHY, R. & SMITH, J. A. C. (1989). Pressure-clamp experiments on plant cells using a manually operated pressure probe. In *Plant Membrane Transport: the Current Position* (Ed. J. Dainty, M.I. de Michelis, E. Marre & F. Rasi-Caldogno), pp. 577-578. Elsevier, Amsterdam.

OPARKA, K. J., MURPHY, R., DERRICK, P. M., PRIOR, D. A. M. & SMITH, J. A. C. (1991). Modification of the pressure probe technique permits controlled intracellular microinjection of fluorescent probes. *J. Cell Sci.* **98**, 539-544.

PANTOJA, O., GELLI, A. & BLUMWALD, E. (1992). Voltage-dependent calcium channels in plant vacuoles. *Science* **255**, 1567-1570.

PING, Z., YABE, I. & MUTO, S. (1992). Identification of K^+, Cl^- and Ca^{2+} channels in the vacuolar membrane of tobacco cell suspension cultures. *Protoplasma* **171**, 7-18.

PRITCHARD J. & TOMOS, D. (1993). Correlating biophysical and biochemical control of root cell expansion. In *Water Deficits: Plant Responses from Cell to Community* (Eds J.A.C Smith & H. Griffiths), Oxford, Bios. (in press).

RASCHKE, K. & HEDRICH, R. (1989). Patch clamp measurements on isolated guard cell protoplasts and vacuoles. *Meth. Enzymol.* **174**, 312-330.

RATHORE, K. S., CORK, R. J. & ROBINSON, K. R. (1991). A cytoplasmic gradient of Ca^{2+} that is correlated with the growth of lily pollen tubes. *Dev. Biol.* **148**, 612-619.

READ, N. D., ALLAN, W. T. G., KNIGHT, H., KNIGHT, M. R., MALHO, R., RUSSELL, A., SHACKLOCK, P.S. & TREWAVAS, A.J. (1992). Imaging and measurement of cytosolic free calcium in plant and fungal cells. *J. Microsc.* **166**, 57-86.

RYGOL, J., PRITCHARD, J., ZHU, J. J., TOMOS, A. D. & ZIMMERMANN, U. (1993). Transpiration induces radial turgor pressure gradients in wheat and maize roots. *Plant Physiol.* (in press).

SCHIEFELBEIN, J. W., SHIPLEY, A. & ROWSE, P. (1992). Calcium influx at the tip of growing root hair cells of *Arabidopsis thaliana*. *Planta* **187**, 455-459.

SCHONKNOECHT, G., HEDRICH, R., JUNGE, W. & RASCHLE, K. (1988). A voltage-dependent chloride channel in the photosynthetic membrane of a higher plant. *Nature* **336**, 589-592.

SCHROEDER, J. I. & HAGIWARA, S. (1990). Repetitive increases in cytosolic Ca^{2+} of guard cells by abscisic acid activation of nonselective Ca^{2+}-permeable channels. *Proc. Natl. Acad. Sci. U.S.A.* **87**, 9305-9309.

SCHROEDER, J. I. & THULEAU, P. (1991). Ca^{2+} channels in higher plant cells. *Plant Cell* **3**, 555-559.

SENN, A. P. & GOLDSMITH, M. H. M. (1988). Regulation of electrogenic proton pumping by auxin and fusicoccin as related to the growth of *Avena* coleoptiles. *Plant Physiol.* **88**, 131-138

SIAMI, Y., MARTINAC, B., DELCOUR, A. H., MINORSKY, P. V., GUSTIN, M. C., CULBERTSON, M. R., ADLER, J. & KUNG, C. (1992). Patch clamp studies of microbial ion channels. *Meth. Enzymol.* **207**, 681-691.

TAYLOR, A. R. & BROWNLEE, C. (1992). Localized patch clamping of plasma membrane of a polarized plant cell: laser microsurgery of the *Fucus spiralis* cell wall. *Plant Physiol.* **99**, 1686-1688.

TAYLOR, A. R. AND BROWNLEE, C. (1993). Calcium and potassium channels in the *Fucus* egg. *Planta* **189**, 109-119.

TESTER, M. (1990). Plant ion channels: whole cell and single channel studies. *New Phytol.* **114**, 305-340.

TOMOS, A. D., LEIGH, R. A., HINDE, P., RICHARDSON, P. & WILLIAMS J. H. H. (1992). Measuring water and solute relations in single cells *in situ*. *Curr. Topics in Plant Biochem. Physiol.* **11**, 168-177.

TOMOS, D., HINDE, P., RICHARDSON, P., PRITCHARD, J. & FRICKE, W. (1993). Microsampling and measurements of solutes in single cells. In *Plant Cell Biology - a practical Approach*. (Ed. N. Harris & K. Oparka). Oxford, IRL Press (in press).

TYERMAN, S. D. (1992). Anion channels in plants. *Annu. Rev. Plant Physiol. Plant Molec. Biol.* **43**, 351-373.

VENIS, M. A., NAPIER, R. M., BARBIER-BRYGOO, H., MAUREL, C., PERROT-RECHENMANN, H. & GUERN, J. (1992). Antibodies to a peptide from the maize auxin-binding protein have auxin agonist activity. *Proc. Natl. Acad. Sci. U.S.A.* **89**, 7208-7212.

VOLGELZANG, S. A. & PRINS, H. B. A. (1992). Plasmalemma patch clamp experiments in plant root cells: procedure for fast isolation of protoplasts with minimal exposure to cell wall degrading enzymes. *Protoplasma* **171**, 104-109

WEBER, G. & GREULICH, K-O. (1992). Manipulation of cells, organelles, and genomes by laser microbeam optical trap. *Int. Rev. Cytol.* **133**, 1-41.

ZHEN, R-G., KOYRO, H-W., LEIGH, R. A., TOMOS, A. D. & MILLER, A.J. (1991). Compartmental nitrate concentrations in barley root cells measured with nitrate-selective microelectrodes and by single-cell sap sampling. *Planta* **185**, 356-361.

ZHOU, X-L., STUMPF, M. A., HOCH, H. C. & KUNG, C. (1991). A mechanosensitive channel in whole cells and in membrane patches of the fungus *Uromyces*. *Science* **253**, 1415-1417.

ZHU, G. L. & BOYER, J. S. (1992). Enlargement in *Chara* studied with a turgor clamp. *Plant Physiol.* **100**, 2071-2080.

ZIMMERMANN, U., RAEDE, H. & STEUDLE, E. (1969). Kontinuierliche Druckmessung in Pflannzenzellen. *Naturewiss.* **56**, 634.

ZOREC, R. & TESTER, M. (1992). Cytoplasmic calcium stimulates exocytosis in a plant secretory cell. *Biophys. J.* **63**, 864-867.

Chapter 14

Techniques for dye injection and cell labelling

PETER MOBBS, DAVID BECKER, RODDY WILLIAMSON,
MICHAEL BATE and ANNE WARNER

1. Introduction

The introduction of compounds into cells via iontophoresis or pressure injection from micropipettes is a powerful technique of wide application in modern biology. The many uses to which this technique can be put include:

(i) Cell identification following electrophysiological recording.

(ii) Delineation of cellular architecture in anatomical studies.

(iii) Tracing neuronal pathways.

(iv) Identification of cell progeny in lineage studies.

(v) Investigations of the transfer of molecules from one cell to another via gap junctions or other routes.

(vi) The introduction of genetic material that affect protein synthesis or gene expression.

(vii) The measurement of intracellular ion concentrations, for example pH or calcium ion.

This chapter describes the techniques used to inject cells and focuses upon the design of experiments for some common applications of these methods. In the final sections, we offer sample protocols and advice on the necessary equipment.

The basic methods for cell injection are similar whatever the compound to be used. This chapter concentrates on techniques that involve iontophoresis or pressure injection using intracellular micropipettes while section 9 describes some other routes by which compounds can introduced into cells. For each application described below, we concentrate upon the factors that influence the

The modifications made to this Chapter between the first and second editions have been driven by experience at successive annual Workshops. We would like to thank David Shepherd, who shared the teaching for two years, for his valuable tips that have become part of the practical advice offered here.

PETER MOBBS, Department of Physiology, University College London, Gower St., London WC1E 6BT, UK
DAVID BECKER AND ANNE WARNER, Department of Anatomy and Developmental Biology, University College London, Gower St., London WC1E 6BT, UK
RODDY WILLIAMSON, The Laboratory, Citadel Hill, Plymouth, PL1 2PB, UK
MICHAEL BATE, Department of Zoology, University of Cambridge, Downing St., Cambridge CB2 3EJ, UK

choice of the compound to inject, since this is usually the factor most crucial to success.

2. Microinjection methods

Manufacturing micropipettes

Pipettes for intracellular microinjection can be produced on any standard microelectrode puller. The best pipettes generally have the following characteristics: (a) a relatively short shank (b) a relatively large tip diameter. The latter is frequently a limitation because, for successful penetration of small cells without damage, the tip diameter also must be small. When the diameter of the tip is small then both the iontophoresis and pressure injection of compounds is impeded, the former by the charge on the glass and the electrical resistance of the tip and the latter by the tip's resistance to bulk flow of solution. Several different types of glass are available for the production of micropipettes. A number of manufacturers (see appendix B) provide suitable capillaries with a variety of outside diameters, with thick or thin walls, with and without internal filaments, made from soda or borosilicate glass. Pipettes made from thick-wall borosilicate glass are usually the most robust and useful for penetrating tough tissue. However, thin-wall glass has the advantage that the channel through the tip is usually larger, and thus the resistance is lower, for any given tip size. The characteristics of micropipettes for use in microinjection experiments can sometimes be improved by bevelling (see Chapter 11). Soda-glass is somewhat less fragile than borosilicate glass but is difficult to pull to fine tips, it has been dropped from some supplier's lists. No matter what the theoretical expectations, the best electrodes to use are those that work!

Pipette filling

Modern micropipette glass incorporates an internal 'filament' (actually a second narrow capillary). The filament increases the capillarity of pipettes so that fluid is drawn into the tip. This characteristic can be exploited to enable very small volumes of fluid to be loaded into the pipette tip, which is useful where the compound to be injected is expensive. Solutions can be introduced into the back of the pipette either by immersion or by bringing into contact with a drop of fluid. The volume drawn into the tip depends upon its diameter. Pipettes with tips of 1 μm will draw up about 100 nl and those of 5 μm will fill with about 1 μl of fluid. Coarse pipettes can be filled by sucking fluid directly through the tip. Electrical connections to pipettes in which only the tip is filled can usually be effected simply by sticking a wire into the pipette lumen. The presence of a thin trail of electrolyte along the outside of the internal filament provides the necessary path for current flow. It is advisable to centrifuge all solutions before use to remove material that may block the tip.

Iontophoresis

Iontophoresis involves the ejection of a substance from a pipette by the application of current. The polarity of the ejection current employed depends on the net charge on the substance to be injected (negative pulses are used to eject negatively charged molecules). Most modern microelectrode amplifiers are equipped with a current pump that can be used to provide an iontophoretic current that is, within limits, independent of the electrode resistance (see Chapters 1 and 16). If only a simple amplifier is available, or the current pump is unable to provide sufficient voltage to drive the required current through the electrode tip, then it is possible to use a battery and a current limiting resistor as a current source. If a battery is employed then the headstage of the amplifier should be switched out of the circuit when the battery is connected. Obviously the current provided by this crude arrangement will be governed by Ohm's Law. The current applied to a cell should be as small as is consistent with the introduction of sufficient of the compound into the cell. In all events the voltage produced by the passage of the iontophoretic current must be limited (to say +100 to −100 mV) to avoid damage to the cell membrane.

Continuous application of current should be avoided since it often causes the electrode tip to block. This block can sometimes be relieved by reversing the polarity of the current for a short time. However, once an electrode shows signs of block the trend is usually irreversible and the pipette should be discarded. Often the best strategy is to employ short duration current pulses of alternating polarity. Whatever the form of the pulse, small currents for long periods are usually more successful than high currents for shorter times. To recognise electrode block and standardise procedures, it is essential to monitor the *current* flow through the electrode. It is not sufficient simply to monitor the voltage applied to the electrode! If the amplifier employed does not have a current monitor then a simple one can be improvised by measuring the voltage drop across a resistor in the earth return circuit. The membrane potential of the cell should be measured during electrode insertion, before switching to current injection. It is sensible to check the condition of the cell by measuring its resting potential at intervals during iontophoresis. Such measurements are simplified by using a bridge amplifier (see Chapters 1 and 16) that enables the membrane voltage to be monitored continuously during current passing experiments. For a detailed discussion of the circuits for current injection and current monitoring see Purves (1981).

A useful technique for achieving bulk flow from the electrode tip is to cause high frequency oscillations of the voltage across the electrode resistance. This is achieved by pressing the 'buzz' or 'zap' buttons present on some amplifiers. The effect of these can be imitated by turning up the capacity compensation control, found on nearly all microelectrode amplifiers, to the point at which the electrode voltage oscillates (termed 'ringing').

In theory the amount of a substance ejected from the pipette during an iontophoretic pulse can be estimated from a consideration of its transport number (Purves, 1981). In practice, these estimates are highly unreliable and the transport number is often unknown for the compound employed.

Pressure injection

Pressure ejection is the method of choice for the injection of neutral molecules and those of low iontophoretic mobility. Commercial pressure injection devices are available (see list of suppliers) that enable the application of calibrated pressure pulses to the back end of the injection pipette. Essentially a pressure injection system consists of a gas cylinder connected, via a timing circuit, a solenoid-operated valve and a pressure regulator, to a side-arm pipette holder. Commercial equipment is expensive, but a home-made rig can be simply made from the components listed above. The timing circuit can be replaced by a manually operated switch. Take care to ensure that the connections and tubing are safe at the pressures employed and that the pipette is firmly held within the holder. The pressure and timing of the pulse can be roughly established by measuring the diameter of a drop expelled from the pipette tip into a bath of liquid paraffin. However, this method frequently over-estimates the back-pressure from the cytoplasm and quantification of pressure injection is often as uncertain as in iontophoresis.

Patch-pipettes

Many substances can be introduced into cells from patch-pipettes while recording in the whole-cell mode. The concentration that a compound reaches within the cell during whole-cell recording is equal to that within the patch-pipette solution. Thus for most dyes and labels the concentrations to employ are a fraction of those used in iontophoresis or pressure injection experiments. For example, Lucifer Yellow CH incorporated into the patch-pipette solution at 1 mg ml^{-1} will produce intense fluorescence of the cell (40 mg ml^{-1} is used in sharp electrodes for iontophoresis; Fig. 1B).

3. Techniques for visualizing cells

Visualizing cells prior to injection

In order to inject a cell you must be able to guide your micropipette toward it. There are three techniques available to aid in the steering of electrodes:

(a) Stereotaxic movements combined with continuous electrical recording (mainly used for penetration of cells in brain nuclei).

(b) Visual guidance using white light and interference contrast optics to visualize the cell and identify targets.

(c) Visual guidance using cells prelabelled with fluorescent dyes as the target.

In solid tissue, whatever technique is chosen to guide the electrode, the target must lie along initial trajectory of the electrode. Manipulation out of this axis will break the electrode.

1. *Stereotaxis.* This method requires that you know precisely where your target cells lie even though you can not see them. Such information is sometimes available from stereotactic atlases. Micromanipulators can be roughly calibrated to give depth measurements but errors always arise as a result of tissue distortion during electrode penetration. The identification of the target cells can sometimes be achieved through knowledge of their electrical properties or synaptic connections,

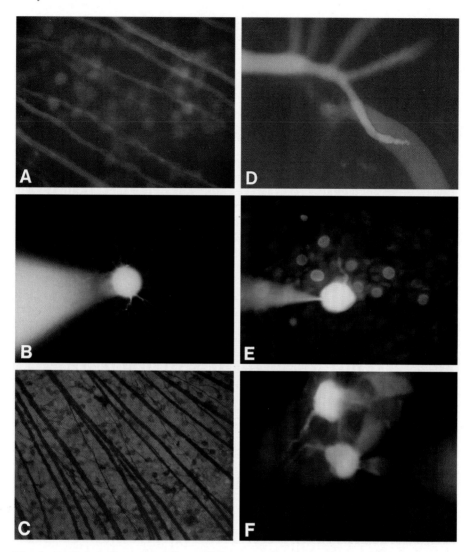

Fig. 1. (A-C) Labelling of ganglion cells in the salamander retina. (A) The living retina prelabelled with FITC-HRP viewed in the fluorescence microscope. The label has been introduced into the axons of the ganglion cells from the cut end of the optic nerve. Ganglion cell axon bundles and cell bodies are clearly visible. (B) Using the FITC fluorescence as a guide a patch-pipette has been attached to a labelled ganglion cell body. After entering whole-cell mode Lucifer Yellow CH included in the patch-pipette solution has diffused into the cell and brightly labels the cell body and dendrites (the FITC florescence is not visible in this photograph because the film has been exposed to best show the LY labelled cell). (C) An area of the same retina as in A reacted with diamino-benzidine to give a permanent preparation in which all the ganglion cells are visible. The retinal ganglion cell somata are 10 μm in diameter. (D) A living whole-mount preparation of the squid stellate ganglion in which the pre-and post-synaptic elements of the giant synapse have been labelled with Lucifer Yellow CH and sulphrhodamine 101 respectively. The axons are approximately 0.5 mm in diameter. Only the medial of the 5 postsynaptic giant fibres was injected and staining of the presynaptic elements of the other 4 giant synapses can be seen. (E) An autonomic ganglion prelabelled by immersion in Fast Blue and Diamidino Yellow. The ganglion has been fixed and one of the cells selected for injection with Lucifer Yellow CH via a microelectrode. Cell bodies 15 μm in diameter. (F) Cells within the same autonomic ganglion labelled via microelectrodes with sulphrhodamine 101 (red), Lucifer Yellow CH and Cascade Blue. The sulphrhodamine labelled cell is coupled to two others (presumably by gap junctions) and these are more faintly stained. Cell bodies 15 μm in diameter.

for example by the response to current injection or stimulation of a peripheral nerve.

2. *Interference contrast optics.* Phase contrast and differential interference contrast techniques (Nomarski) are good for visualising living cells. Phase contrast is useful for cells in tissue culture but does not work well for tissue slices. Nomarski optics provide high resolution and can be used to provide effective optical sections of transparent tissue. The more recently introduced Hoffman optics are cheaper than Nomarski optics and are useful for viewing tissue slices because they provide a greater depth of field.

3. *Prelabelling with a fluorophore.* There are two approaches to the prelabelling of cells to identify them as targets for subsequent microinjection experiments. Cells can either be bathed in a dye that becomes internalized (Fig. 1E), or labelled by retrograde transport of a marker from their axons (Fig. 1A). Whilst some dyes are either actively taken into cells or simply diffuse across the membrane others only enter if the membrane is disrupted by osmotic shock or through exposure to dimethyl sulphoxide. Whatever the method of prelabelling, the choice of the label is crucial to success. Ideally the label should be visible under the same filter set as the dye used in subsequent injection experiments and the intensity of the prelabel's fluorescence should not mask that of the injected fluorophore. Since the prelabel may remain inside the cell for an extended period, it is important that it is non-toxic.

Retrograde labelling of neurons via their axonal projections is an extremely useful means of identifying populations of cells that project to particular targets. Fast blue and diamidino yellow are amongst the most popular of the labels available for this purpose. Fast blue labels the cell cytoplasm and diamidino yellow stains the nucleus (Fig. 1E). Both pass rapidly across the cell membrane and can be used to label cells from their axon terminals or from cut axons. General labelling of all the cells in a tissue can be achieved by bathing in a dilute solution of the dyes. Both of these dyes work well on formaldehyde-fixed tissue. Target cells identified with these prelabel dyes can subsequently be injected with Lucifer Yellow, carboxyfluorescein or Cascade Blue which are visible with the same filter set (Fig. 1E).

Some fluorophores with useful properties are neither taken up nor transported by cells. However, they can be made into useful labels through conjugation to lectins, dextrans or plastic microspheres. Lectins bind to sugar moieties on the cell membrane, are brought into the cell through endocytosis and transported. Dextrans can also be conjugated to most fluorophores. Plastic microspheres can be coupled to fluorescent molecules. They are available in a variety of materials and sizes. Applied to damaged axons they are taken up by and retrogradely transported. Microspheres are visible in the electron microscope.

Visualizing labelled cells

The object of many microinjection experiments is to render the cell under study visible by introduction of a label. The majority of such labels are either fluorescent or can be processed to produce a coloured reaction product. Below we describe the techniques for visualizing and recording the results of cell labelling experiments.

Often labelled cells can be visualised without any histological processing and some labels can be used to follow changes in cell morphology that occur over extended periods of time (Purves *et al.* 1986). Methods for the fixation of tissue and the histological processing of tissue containing labels are given later.

Fluorescent labels are excited by light at one wavelength and emit light at another longer wavelength. The user must choose the excitation and emission filters most suitable to their application (see appendix A). It is convenient to have the microscope used for positioning the electrode equipped with a light source and filters capable of exciting the label. This allows the user to determine the endpoint of the injection experiment by observation. Many of the labels in common use are excited by far blue or UV light. The tungsten or quartz halogen bulbs found in most microscope illuminators do not provide much light at these wavelengths and an additional mercury or xenon light source is required. Most manufacturers provide some convenient means for switching between the white and UV light sources. If this switching mechanism is to be used during the course of a labelling experiment, it is imperative that it operates without vibration if the microelectrode is to remain in the cell under study.

The factors involved in the choice of the optics and light sources for fluorescence microscopy are complex. The short discussion below is offered as an introduction that may be supplemented by consulting some of excellent free literature provided by major manufacturers (see for example the booklets offered by Zeiss, Lieca and Nikon). Mercury lamps are cheaper than xenon lamps. However, the emission spectrum of a xenon lamp is relatively continuous throughout the UV and visible spectrum while that of mercury lamps consists of a series of sharp peaks (emission lines). With mercury lamps, it is important to ensure that a line exists at a wavelength appropriate to the dye in use. Most modern fluorescence microscopes employ epi-illumination, a system in which the light used to excite the dye is focused on the specimen through the same objective used to view the light emitted by the dye.

The choice of objective is critical in fluorescence microscopy. Quartz objectives pass much more short wavelength light than those made from glass. However, quartz objectives are expensive and unnecessary for use with dyes excited by light in the visible and near UV regions of the spectrum. It is crucial that the objective has a high numerical aperture (NA) since both the intensity of the light focused on the specimen and the light gathering power of the lens increase with the square of the aperture. An objective with an NA of 1.0 will yield 16 times as much light as a 0.5 NA lens. High NA objectives have shorter working distance and need an immersion medium - water, oil or glycerol (for UV). For injection of cells in thick preparations on an upright microscope water immersion lenses are preferable to those that work in air because they have a greater NA and there is no optical distortion due to meniscus effects of the micropipette on the bath surface. On the other hand, very long working distance air electrodes can be covenient, if optically inferior. Two particularly useful lenses are Zeiss ×40 0.75 NA W water immersion and the Nikon ×40 ELWD air (NA 0.5) with correction collar. Intensity of fluorescent light also depends upon the magnification. It decreases as the square of the magnification: a ×10 eyepiece produce an image of

25% the intensity of an image formed by a ×5 eyepiece. Low magnification eyepieces are therefore preferable for visual observation.

Fluorescent images can be recorded on film or by analog or digital video techniques. There are many black and white, colour print and transparency films suitable for recording fluorescence images. Generally a film of high speed and acceptable grain should be chosen. Colour films of speed greater than 400 ASA tend to be too grainy, however, black and white films such as Kodak's TMAX give excellent results even at 2400 ASA (must be developed in TMAX developer). In normal photography, the reciprocity law applies and the total amount of exposure is given by the product of the luminance and the exposure time. Thus an exposure of 1/60th of a second at f8 is the same as for 1/30th at f11. With dim objects the reciprocity law fails to predict the exposure and the exposure time has to be increased. Most film manufacturers provide a guide to the performance of their films at low light intensities. In practice it is often better simply to take several exposures of increasing duration starting with the exposure time indicated by the meter on the camera.

The advent of cheaper video cameras that operate at low light intensities has opened up the possibility of recording fluorescent images either on video-tape or in digital form on a computer. Digital image recording has the advantage of allowing complex analysis of an image.

Labels that result in a coloured or opaque reaction product are much simpler to photograph than those labelled with fluorescent compounds. No special equipment is required.

4. Labelling cells for subsequent identification and for determination of overall cell architecture

Dyes injected for these purposes should have the following properties: (a) they should be visible, either immediately or after chemical reaction; (b) they should remain in the injected cell, either because they are too large to move across the cell membrane and through gap junctions or because they are strongly bound by the cytoplasm; (c) they should not be toxic, although this requirement can be relaxed if the tissue is to be processed immediately after the cell has been injected; (d) they should be stable and not break down to give products with different properties; (e) they should withstand histological processing. In practice, property (e) is the most difficult to achieve.

Six classes of compound are used for this purpose:

1. *Inherently fluorescent molecules and those tagged with a fluorescent probe.* Lucifer Yellow (MW 457) and carboxyfluorescein (MW 376) are the most popular fluorescent compounds for determining overall cellular architecture. However, they are far from ideal for this purpose. Both pass through gap junctions (see below) and carboxyfluorescein cannot be fixed. Lucifer Yellow withstands fixation well but as with all other dyes some fluorescence intensity is lost. Passage through gap junctions

can be prevented by conjugation of the fluorophore to dextrans. Dextrans (MWs 3000-70000) can be coupled to fluorescein, rhodamine isothiocyanate or Texas Red. They can be prepared in the laboratory (see Gimlich & Braun, 1985) or purchased commercially (Molecular Probes, 48-49 Pitchford Avenue, Eugene, Oregon, OR97402-9144 USA). Cascade Blue and sulphrhodamine 101 are also useful for determining cellular architecture and extend the range of colours available for double marking experiments. For examples of multiple labelling see Fig. 1D,F.

Advantages:
  Can be pressure injected or iontophoresed.
  Can be seen in living cells with appropriate fluorescent illumination.
  Are not toxic provided the amount injected is kept fairly low.
  Do not break down.
  Will withstand routine fixation and embedding techniques, provided the fixative or mountant does not generate auto-fluorescence. Glutaraldehyde fixation, for example, must be avoided. Many commercial mountants, such as DPX, are unsuitable for this reason. Mountants that are designed to reduce fading can now be obtained (e.g. Citifluor, City University, London).

Disadvantages:
  Limit of detection determined by threshold of fluorescence. Detection levels can be improved by electronic image intensification.
  Fluorescence fades under continuous illumination. This can be reduced by using anti-fade mountants.
  Fluorescein fades particularly fast, but is more fluorescent than rhodamine or Texas Red.
  Sometimes become incorporated into cellular organelles with time, making fluorescence particulate.
  Margin between visible not toxic, and visible but toxic is narrow.

2. *The carbocyanine dyes.* Octadecyl(C_{18})-indocarbocyanine (DiI) and oxycarbocyanine (DiO) (MWs 934 and 882) are highly fluorescent lipophilic compounds. They dissolve in, and diffuse throughout, the lipids of the plasma membrane. They are not toxic and they have been reported to remain in the cell membrane for up to one year (Kuffler, 1990). They will also diffuse along membranes in lightly fixed tissue. In the absence of any sites of membrane fusion the carbocyanines label single cells. The diffusion rate for these compounds is slow (about 6 mm/day, slower in fixed tissue), however, carbocyanines with unsaturated alkyl chain segments (FAST-DiI and FAST-DiO) exhibit accelerated diffusion rates. The polyunsaturated "DiASP" compounds (N-4(4-dilinoleylaminostyryl)-N-methylpyridinium iodide and related molecules) (MW~800) are also reported to diffuse more rapidly. Because the carbocyanines are insoluble in water they must either be pressure injected into cells in solution in DMSO or alcohol or applied to the cell membrane in which they rapidly dissolve.

DiI and DiO can be visualized by fluorescence microscopy. DiI has similar excitation properties to rhodamine, excited by green it fluoresces red. DiO is similar to fluorescein in that it is excited by blue light and produces green fluorescence. DiAsp has a broad excitation spectrum and fluoresces orange. These dyes can be converted into a permanent reaction product via the Maranto reaction (Maranto, 1982) in which the singlet oxygen released by illumination is used to oxidise diamino-benzidine (DAB).

Advantages:

They are not toxic and can remain in the cell membrane without harm over several years.

Disadvantages:

Not water soluble.
They tend to fade quickly particularly in laser scanning confocal microscopy.
Long diffusion times.
Can only be pressure injected.

3. *Enzymes such as horse radish peroxidase.* Horse radish peroxidase (HRP) is reacted with diamino-benzidine or other chromogens to generate a product visible in the light or electron microscope. There are many protocols for developing HRP (see Mesulam, 1982 and Heimer & Robards, 1981 for a selection). Widely used in studies in the central nervous system. The injection of enzymes can also be used to kill individual cells (e.g. pronase). This is potentially useful in lineage and regeneration studies.

Advantages:

Can be pressure injected or iontophoresed.
Not toxic.
Remains within the injected cell, provided the preparation is free from micro-peroxidases. Will cross synapses, which can be useful when tracing pathways.
Does not break down.
Good visibility.
Reaction product visible in the electron microscope.

Disadvantages:

Can only be seen after reaction product produced. However, by using a fluorescent peroxidase conjugate, such as RITC-peroxidase (Sigma P5031), an indication of the staining can be obtained during the fill period (see Fig. 1A-C).
Can get reaction product from endogenous peroxidases, so method has to be modified if this is likely to be a problem.
The penetration of chromogen into tissue is rather poor (about 100 μm), so that whole mounts or slices have to be below this thickness.

Much of the enzyme activity is lost on fixation. If possible the material is best fixed after reaction.

4. *Biocytin*. A recently introduced intracellular marker (Horikawa & Armstrong, 1988) comprising a highly soluble conjugate of biotin and lysine (MW 372.48) that has a high binding affinity for avidin. The injected biocytin is visualised by attaching a label to avidin, e.g. a fluorescent label such as FITC or rhodamine, or a chromogenic enzyme such as HRP. Suitable avidin conjugates are widely available (e.g. Sigma, Vector Labs.). A small molecular weight biotin compound, biotinamide (MW 286), is also available (Neurobiotin, Vector Labs, 16 Wulfric Square, Bretton, Peterborough PE3 8RF, UK) and may be easier to inject (Kita & Armstrong, 1991).

Advantages:
Highly soluble in aqueous solutions.
Can be pressure injected or iontophoresed.
Low toxicity.
Does not break down.
Good fluorescent, visible light, or electron microscopic visibility after avidin reaction.

Disadvantages:
Can only be seen after avidin reaction.
Reaction penetration limited to about 100 μm even with detergents or surfactants so tissue may have to be sectioned.
Some ultrastructural degradation from penetration agents.
Can pass between coupled cells.
Occurs naturally in trace amounts.

5. *Heavy metals such as cobalt and nickel*. The metal is precipitated with ammonium sulphide or hydrogen sulphide. The sensitivity can be improved by intensification with silver (Pearse, 1968; Bacon & Altman, 1977). Double labelling can be achieved by using different metals in the same preparation followed by precipitation with rubeanic acid (Quicke & Brace, 1979); this results in precipitates of different colours depending on the metal, e.g. cobalt = yellow, nickel = blue, copper = olive.

Heavy metal complexes, such as lead EDTA (Turin, 1977) can be suitable in cells that are not linked to their neighbours by gap junctions (see later section). In principle, it is possible to prepare a range of heavy metal complexes of different sizes so long as the complex is firmly held, so that there is no free metal or anion which might be toxic, and the metal has a much higher affinity for sulphide than for the anion used to make the complex. This is essential to ensure precipitation of the metal out of the complex. The advantage of a heavy metal complex is that the complex can be much less toxic than the heavy metal itself and may be much easier to eject from the pipette. However, some metal sulphides will re-dissolve if the precipitant (usually ammonium sulphide) contains polysulphides. Freshly prepared solutions saturated with H_2S do not suffer from polysulphide formation.

(i) Cobalt and nickel

Advantages:

Can be iontophoresed or pressure injected.
Strongly bound to cytoplasm, therefore retained in the cell despite small size.
Good visibility after reaction, very good after intensification.
Very good in whole mount, because the sulphide precipitation step permeabilizes the cells. Electron opaque, so product visible in the electron microscope.
Will withstand fixation.
Multiple labelling possible with rubeanic acid precipitation.

Disadvantages:

Electrodes liable to block and require a high current for a long period to eject sufficient cobalt. Nickel filled electrodes suffer less from this.
Treatment with sulphide compounds interferes with cytological appearance.
Toxic, therefore only suitable when precipitated immediately after injection.

(ii) Lead EDTA (as an example of a heavy metal complex).

Advantages:

Very easy to inject iontophoretically or by pressure.
Not toxic; very well tolerated by cells.
Electron opaque.

Disadvantages:

Moves easily through gap junctions therefore not suitable if the cell is linked to others by electrical synapses or gap junctions.
Requires intensification to improve sensitivity.
Sulphide treatment spoils cytology.

6. *Compounds tagged with radioactive label and then visualized with autoradiography.* In principle any suitable molecule can be labelled. Large proteins can be tagged with ^{125}I, but a tritium or carbon label is preferable if autoradiography is to be used. Proline has been useful in studies of the central nervous system; like Horse Radish Peroxidase, proline is transported trans-synaptically and can therefore trace extensive, interlinked neuronal pathways.

Advantages:

Not toxic provided total radioactivity kept low.
Compound normally present in the cell can be used.
Permanent preparation, no fading.
If appropriate compound (usually one that is not normally found within cells, e.g. deoxyglucose) is chosen, no breakdown.
Very good for tissue cultured cells, because no sectioning required.

Disadvantages:

Label only withstands fixation if compound is bound to cell contents.

*If a naturally occurring molecule is used, it may be broken down by cell
    metabolism.*
Can only be used on tissue sections.
Potential delay in obtaining results introduced by autoradiography.

5. Identifying the progeny of the labelled cell (lineage tracing)

This technique is used extensively in developmental biology, as a way of analysing
the prospective fate of a cell and its progeny at different stages of development. The
technique also can reveal the extent to which the progeny of the injected cell remain
as coherent clones and so provide valuable information on the degree to which cells
mix during development. The most important factor when selecting a suitable
compound as a lineage label is the degree to which the label is diluted during cell
division and growth. Cell labelling by injection is, therefore, often only suitable at
certain stages of development, when cell division and growth are relatively slow.
This method of determining lineage has been most successful in the amphibian and
leech embryos, because early development in these animals involves reduction in
the size of each cell without extensive growth, so that the cytoplasm of the egg is
gradually partitioned into smaller and smaller units. In species such as the mouse,
where the embryo arises from a very small number of cells formed during the early
cleavages (most of which contribute to extraembryonic structures), there is
extensive growth and cell division causing dilution of the label. In this situation,
other methods that rely on cell autonomous labels, such as differences in enzymes,
have been more successful (see Gardner, 1985). The incorporation of
self-replication defective retro-viruses into the genome of a host, as yet
undifferentiated, cell is proving useful as a way of providing a cell autonomous
label.

   1. *Fluorescently labelled compounds such as labelled dextrans.* The overall
properties of these compounds are dealt with above. The specific advantage for cell
lineage studies is that fluorescent compounds can be observed in living cells,
provided the level of illumination is kept low. This means that the way in which the
fluorescent cells are distributed in the embryo can be followed sequentially in living
specimens. In order to avoid damage from illumination, it is sensible to use an image
intensifying system. Dilution is not a serious problem in the amphibian and leech
embryos and these compounds remain at an analyzable level for two or three days of
development. They have been applied also to studies of the zebra fish embryo, an
increasingly popular model system for the study of developmental processes.

   2. *Horse Radish Peroxidase (HRP).* This has been used extensively in the
amphibian embryo (e.g. Jacobson & Hirose, 1978) and more recently in *Drosophila*
(Technau & Campos Ortega, 1985). Its advantages and disadvantages are as above. In
amphibia the degree of dilution is much the same as for lysinated Dextrans. Some
caution should be exercised since there have been reports (see Serras & Biggelaar,

1987) showing that HRP can induce an exocytotic/endocytotic cycle, which causes artefactual transfer of HRP and any other compound injected with it.

3. *Radioactively labelled molecules that are incorporated in DNA and/or RNA.*

Advantages:

Label is permanent.

Will withstand fixation.

As long as the precursor is available to be incorporated into DNA or RNA the label will not be diluted out.

Disadvantages:

Precursors to DNA and RNA, such as small nucleotides, may not be restricted to the injected cell.

Labelled breakdown products may not be restricted to the injected cell.

Once all available label is incorporated, dilution occurs at each division.

Levels of radioactivity, and therefore concentration of precursor, have to be kept low to reduce radiation damage. This exacerbates the dilution problem.

Only usable on sections or very thin whole mounts.

May be long interval between experiment and obtaining results.

4. *Labelled proteins, which are usually tagged radioactively.*

Advantages:

Foreign protein can be used, so reducing likelihood of breakdown.

Disadvantages:

May require special fixative to ensure the protein will withstand histological processing.

Foreign protein may be toxic, or be handled by cell metabolism in an unpredictable way.

Naturally occurring proteins may be broken down into small metabolites which could leave the cell.

5. *Incorporation of viral or foreign DNA and recognition of the products of expression of the foreign genes.* This technique is expanding rapidly, and has been successfully used in tracing the lineage of some cells in the vertebrate nervous system. Lineage studies have depended on deficient retro-viruses, modified so that they can no longer spontaneously replicate and can therefore transfect only one cell. Since single copies only are incorporated into the host cell, the virus may insert only into one copy of cellular DNA and so replicate within 50% rather than 100% of the progeny of the transfected cell. This complication often is inadequately recognized. Transfection of single cells frequently is achieved by the injection of virus into the extracellular fluid (such as the cerebro-spinal fluid) at very low concentration. This technique relies on dilution by the CSF to reduce infection level and thus requires careful controls to ensure that single clones are chosen for analysis.

The generation of chimeric embryos also has given information on lineage. In this case either a single cell, or group of cells, may be injected into the blastocoel cavity of the mammalian embryo, become incorporated into the embryonic and extra-embryonic lineages, and are then recognized at specified intervals after injection. Alternatively aggregation chimaeras of whole embryos can be made. There are a number of strategies for recognizing the foreign cell(s) and the progeny. These include: (i) incorporation of genetic material that puts expression of, for example, the enzyme alkaline phosphatase under the control of the promoter for the foreign gene and (ii) molecular recognition (by *in situ* hybridization) of DNA specific to the injected cells.

Advantages:

Can be injected by iontophoresis (because of overall charge on DNA and RNA) or by pressure injection.

Label is autonomous and is amplified at each cell division. This is undoubtedly the major advantage of the approach because it eliminates the problems associated with dilution at each cell division.

If the appropriate gene is selected the product will be retained within the cell.

Disadvantages:

In order to make the product of gene expression visible some reaction step is likely to be required.

The label is unlikely to withstand fixation so that frozen sections, or permeabilization of the labelled cells may be necessary before reaction.

Expression of the foreign or viral gene may not be uniform throughout all the progeny of the injected cell because of difficulties with transcription or translation, or because expression of the foreign gene is subject to controls on gene expression exerted during development, which may be tissue or product specific. The site at which the foreign DNA is incorporated into the host genome cannot be controlled; this may lead to aberrant expression patterns and/or differentiation (see below).

6. Studying cell-cell communication

One of the commonest uses of dye injection is to determine the ability of cells to communicate with each other. The experiments may require simple determination of the presence or absence of cell-cell communication, or may be directed towards determining the size range of molecules that can be exchanged. This section considers the problems involved in determining direct cell-cell communication; that is, specifically, the exchange of small molecules from one cell to the next without recourse to the extracellular space, through the morphologically identified structure, the gap junction. The properties of gap junctions are discussed extensively in the literature; a useful start may be obtained by examining recent reviews (e.g. Seminars in Cell Biology Ed: Gilula, 1992). Transfer from cell to cell also can occur through the extracellular space, as with molecules like HRP, which can cross synapses, probably because they are successfully exocytosed and then endocytosed by adjacent cells.

The requirements of suitable molecules for tracing pathways of gap junctional communication are necessarily very different from those associated with lineage studies or labelling for subsequent identification. The *size* and *charge* of the injected molecules is of importance, because this will determine whether the molecule moves from one cell to the next. The most sensitive way of recognizing communication through gap junctions is to examine the spread from one cell to the next of injected current, where the voltage change induced in neighbours of the injected cell by injection of a current pulse reflects the ability of small ions to move through gap junctions. Because gap junctions allow the transfer of a range of small molecules (MWs generally less than 1000) in addition to small ions, the injection of dyes allows the upper limits of gap junction permeability to be explored. When working near the cut-off limit, dye transfer is the most useful technique, because it can reveal relatively small differences in permeability. It is important to recognize that the lower limit of available methods for detection of the selected compound can determine whether transfer from one cell to the next is recognized. Dye transfer is, therefore, inherently less sensitive than electrophysiological methods and failure to observe transfer may be a reflection of the detection method, rather than the permeability of gap junctions.

The major requirements when selecting compounds to examine the permeability of gap junctions are: (a) the compound should be visible at the time of injection; (b) it should be freely diffusible in the cytoplasm, so that transfer from cell to cell is not limited by binding; (c) preferably the compound should withstand fixation, so that the distribution can be examined in greater detail at the end of the experiment, possibly in sections; (d) it should not be toxic; (e) it should not influence intracellular pH, intracellular free calcium or intracellular cyclic AMP because the permeability of gap junctions is sensitive to pH, Ca^{2+} and cyclic AMP; (f) it should not influence the properties of the junction itself; (g) ideally the size and charge of the molecule should be known. In practice the molecular weight is often used as an indicator of size, because the degree of hydration and shape of the injected molecule are not available; (h) the injected compound should not be able to cross the surface membrane of the cell, so that entry into cells cannot take place if dye leaks into the extracellular space from the pipette or from damaged cells.

A large number of compounds have been used to trace the degree and pattern of cell-cell communication through gap junctions. Few of these compounds possess all the desirable characteristics. A useful discussion of the approach to synthesizing compounds with the appropriate properties can be found in Stewart (1978) and Stewart & Feder (1985). However, good chemists with an interest in generating suitable compounds are in short supply and most workers have proceeded on a trial and error basis. The reagents currently in most common use are:

1. *Fluorescein and 6-carboxy fluorescein.* Low molecular weight (fluorescein: 332) highly fluorescent compounds.

Advantages:

Easily injected iontophoretically or by pressure.

High quantum yield on excitation, so that low levels of dye can be detected easily.

Not toxic.

Not bound to cellular components.

Disadvantages:

Will not withstand fixation, so can only be used in live preparations.

Can cross cell surface membranes, although 6-carboxy-fluorescein is better in this respect.

2. *Lucifer Yellow* (MW 457). Two versions of this dye were originally available: CH and VS. Most published papers use the CH form; when no indication is given it is likely that the CH form has been used. A detailed description of the properties of these dyes is given in the two Stewart references (see above), which also indicate the variety of purposes for which Lucifer Yellow may be used. Lucifer Yellow, introduced in 1978, remains probably the most popular dye currently in use. It has proved a useful dye for developmental studies because the transfer of LY seems to be particularly sensitive to regional differences in gap junction properties so that its transfer can be restricted even when electrical coupling and the transfer of other molecules is not (e.g. Warner & Lawrence, 1982; there are now many examples in the literature). Several new forms of Lucifer Yellow are now available (see the Molecular Probes catalogue for details).

Advantages:

Easily injected by iontophoresis or pressure.

Highly fluorescent.

Diffuses through the cell rapidly, although it does become bound to cell contents and particularly nuclei with time.

Not toxic.

Withstands fixation, provided formalin or formaldehyde fixative used.

Permanent preparations can be made with an antibody to Lucifer Yellow (Taghert et al. 1982).

Will react with diamino-benzidine in the presence of irradiating light to give an electron dense product (Maranto, 1982).

It can be injected into cells after weak formaldehyde fixation. The prefixation technique can be useful for examining the structure of small cells that are liable to excessive damage by penetration of the electrode when alive.

Disadvantages:

Forms an insoluble precipitate with potassium so that electrodes must be backfilled with lithium chloride when iontophoresing. This is not a problem for short term experiments, but can lead to problems when injecting early embryos because lithium is extremely teratogenic at low intracellular concentrations. The potassium salt of Lucifer Yellow can be obtained from Molecular Probes, but it is much less resistant to fixation than the lithium salt.

Electrodes tend to block during iontophoresis, probably because Lucifer in the electrode tip is precipitated by potassium ion from the cytoplasm. The block can be temporarily relieved by applying depolarizing pulses.

Binding to cell components means that Lucifer is only available for transfer to adjacent cells for short periods of time, so that its distribution at longer times is not simply a reflection of the presence of gap junctions.

Some fading on irradiation.

Some loss of dye on fixation.

Illumination for long times leads to damage from singlet oxygen (but see below).

Other uses for Lucifer Yellow: The damage induced by over-irradiation provides a useful way of precisely killing a single cell.

3. *Tetramethyl Rhodamine Isothiocyanate (TRITC)/sulphrhodamine.* Both dyes are excited by green and emit red light. TRITC (MW 444) is poorly fluorescent, toxic and strongly bound, and therefore not a dye of choice, but can be useful if another label is to be used simultaneously. Sulphrhodamine 101 (MW 607) is a better choice, it has a high quantum yield and fades only slowly. It does not pass through gap junctions as fast as Lucifer Yellow or Cascade Blue but can be useful in multiple-label studies.

4. *Cascade Blue.* A relatively new dye available from Molecular Probes which shares many of the properties of Lucifer Yellow. Versions are available with MWs between ~600 and 700. These dyes are excited by near-UV light and fluoresce blue with a high quantum yield. They can be usefully combined with Lucifer Yellow in multiple-labelling experiments. Remains visible within fixed tissue.

5. *Biocytin/Neurobiotin* (see earlier for properties). These small molecules (MW 372/286) have proved useful as tracers of gap junctions because they are sufficiently small to give a diffusion pattern much closer to that predicted by electrical coupling studies than observed with Lucifer Yellow. This has proved particularly advantageous in the central nervous system, where both biocytin and Neurobiotin have revealed extensive networks of coupled cells, many times larger than seen with Lucifer Yellow (Vaney, 1991, Peinado et al., 1993).

6. *Heavy metal complexes.* For example lead EDTA, potassium argentocyanate (Turin, 1977). It must be possible to precipitate the metal out of the complex with hydrogen sulphide and ammonium sulphide. It is also important to ensure that traces of free metal or chelating anion are not present in the preparation. By choosing appropriate metal complexes, molecules of a wide range of molecular weight and dimensions can be generated. These compounds have not yet been widely used.

Advantages:
Easily injected iontophoretically or with pressure.
Not toxic, if the compound has been chosen with care.
Good sensitivity after intensification.
Permanent preparation, which does not fade.

Disadvantages:

Can only be visualized after chemical reaction.

Artefact can result from the presence of polysulphides in ammonium sulphide, which can redissolve the precipitated metal sulphide. This can lead to an over-estimate of the distribution of, and a false pattern for, the injected compound.

6. *Sugars and small peptides (and other small molecules) coupled to a fluorescent label such as fluorescein or rhodamine.* This approach allows the range of molecules that can be tested for the ability to pass through gap junctions to be greatly extended (see Simpson *et al.* 1977). However, considerable care must be taken to ensure that the label is not split off by metabolism and also that the test molecule is not broken down to smaller metabolites.

7. Achieving functional 'knock-out', ectopic expression and the generation of transgenics

The techniques of intracellular microinjection form the basis of a number of new and important approaches and methods for the analysis of cellular function. A full description is inappropriate here but they are introduced in order to demonstrate that skills obtained when learning intracellular injection can be translated immediately to exploit a variety of new technologies.

One way of determining the functional contribution of a particular molecule or mechanism is to neutralize its function by injection of an antibody or anti-sense RNA/oligonucleotides. This is a potentially powerful approach since, in principle, it allows direct demonstration of a specific function. Antibodies can be injected with pressure; however, "ringing" the electrode is also a rapid and efficient means of introduction. Careful selection of pipettes (for glass, tip size and shape) can improve the success rate. It can be helpful to include a low concentration of fluorescent dye to confirm that the injection has been successful. To be convincing, such experiments require a set of adequate controls (pre-immune serum, other antibodies, IgGs and, preferably, Fab fragments), which can make them time consuming and labour intensive. Nevertheless their considerable power makes them extremely informative.

The intracellular injection of RNA and/or oligonucleotides (both sense and antisense) to achieve over-expression, ectopic expression or functional knockout is now used widely as a method of exploring the functional role of genes and molecules. The *Xenopus* oocyte has proved to be a useful expression system, whether for expression of crude RNA extracts (e.g. the early experiments on the properties of neurotransmitters) or for expression of pure, *in vitro* synthesized RNA transcripts (e.g. Parke *et al.* 1993: a recent example from a very large literature). The large size of the oocyte (1 mm in diameter) allows substantial volumes to be ejected under pressure from relatively large tipped pipettes (up to 10 μm). Relatively unsophisticated (and therefore inexpensive) equipment is adequate; a simple manipulator and a dissecting microscope will suffice. The difficulties relate to the need to culture the oocytes to allow expression levels to build up, which requires sterile injections and good quality oocytes. Failure to obtain expression may reflect oocyte quality rather than difficulties with injected RNA. There are concerns about certain aspects of such experiments (e.g. whether the injection of foreign RNA induces inappropriate expression of endogenous genes and whether down-stream signalling cascades activated by, for example, exogenously expressed 5-HT, reflect the properties of the oocyte or that of the signalling mechanism in the system from which the RNA is drawn). Both concerns are known to be valid in some circumstances, but not in others.

For developmental studies, ectopic expression, achieved by injection of sense RNA into a cell where the protein product is not found normally, also has proved illuminating.

Knock out, by injection of antisense RNA or oligonucleotides, also can be achieved by intracellular injection. Again, the problems relate to the injected material and the way it is handled by the injected cell (and its progeny) rather than the injection itself.

A high quality microinjection set-up, based on a compound microscope and good manipulators, also can be used for the generation of transgenic mice, where the gene of choice is injected into the nucleus. The tricks associated with nuclear injection relate primarily to the preparation and it will generally be necessary to ensure sterility, since injected embryos must be returned to foster mothers to continue development. Nevertheless, the skills associated with microinjection of dyes and molecules into small cells (such as using electronic oscillation to eject molecules of DNA) are not often the province of mammalian developmental biologists; any one competent in microinjection should be capable of acquiring the basic technology relatively quickly.

8. The injection of indicators and buffers for ions

Indicators

A wide range of molecules are available that can act as fluorescent indicators of the concentrations of ions inside cells. Molecules are available to measure the concentration of most ions of biological interest (see the Molecular Probes catalogue). The mostly widely employed compounds are those used to measure free calcium levels (aequorin, Quin-2, Fura-2: Blinks *et al.* 1982; see Chapter on fluorescent indicators) and pH (e.g. BCECF, Rink *et al.* 1982) . The use of these dyes in imaging and microspectrofluorimetry is the subject of Chapter 12 in this volume and will not be covered here.

Indicator reagents can be introduced into single cells by injection through micropipettes and the underlying principles are the same as for other compounds. The difficulties are generated by methods of detection and measurement, which are beyond the scope of this chapter.

Buffers

Similar comments apply to the injection of buffers such as BAPTA and EGTA. Buffers allow the experimenter to set the intracellular level of the ion of interest and can be important when exploring the role of particular ions in, for example, controlling the permeability of gap junctions or controlling current flow through ion channels (e.g. calcium currents in invertebrate neurones).

Note: BAPTA is far superior to EGTA as a calcium buffer because its ability to complex calcium is not pH sensitive making calculation of the free calcium concentration much easier (see Chapter 11), and it has faster binding kinetics.

9. Other methods for introducing compounds into cells

For the sake of completeness we finish with a brief summary of methods other than pressure injection and iontophoresis that can be used to load cells with reagents. All these methods are directed towards loading cells in large numbers, rather than singly.

(i) A membrane soluble derivative of the chosen compound, which is converted by metabolism into an insoluble form is used, so that the compound can enter, but not leave, the cells. Usually an ester of the chosen compound is used. Esters cross the cell membrane rapidly and are then acted on by intracellular esterases. This method has been used to load cells with fluorescein (fluorescein diacetate) and Quin-2 and Fura-2 (acetoxymethyl esters; see Chapter 12). The hydrolysis of the ester inside the cell also generates hydrogen ions, so that a small fall in intracellular pH is inevitable. It is important to ensure that other products of the hydrolysis are not toxic.

Fluorescein diacetate can be used as a vital dye because fluorescein liberated inside the cell will only cross membranes of damaged cells, rendering intact cells fluorescent at FITC wavelengths, and can be combined with ethidium bromide staining of nuclei of the dead cells (fluorescent at rhodamine wavelengths).

(ii) The compound is dissolved in a reagent such as DMSO which permeabilizes the cell so that quite large molecules can gain entry. On return to normal solution the compound is trapped inside the cell. Used in prelabelling techniques (see above).

(iii) The cells are permeabilized transiently by osmotic shock. The exact sequence of changes in osmotic pressure that is most effective (i.e. from high to low or vice versa) depends on the cells being used. Low permeability is restored on return to normal osmotic strength.

(iv) Artificial endocytosis. Lipid vesicles are loaded with the substance to be incorporated into the cell. These vesicles then fuse with the cell membrane, releasing their contents to the cell interior (see Spandidos & Wilkie, 1984).

(v) Electroporation (e.g. Potter *et al.* 1984). Brief, high voltage shocks allow molecules to enter through holes made in the membrane by the electric field. This method is used routinely to incorporate DNA into cells (as when transforming cell lines). Voltages of about 4000 V cm^{-1} are required. Because of the high voltages used, this method is potentially hazardous and should not be attempted without advice from someone who is already experienced in its use.

(vi) Retrograde and anterograde labelling of neurones. These methods are widely used for tracing anatomical pathways and are covered, for example, in Helmer & Robards (1981). Some methods rely on the uptake of the tracer by damaged neurones others upon spontaneous endocytosis (see below). In the periphery the chosen axon is cut and sucked up into a pipette containing the label. The compound (HRP, Wheat germ agglutinin either fluorescently tagged or complexed with HRP, cobalt chloride and radioactively labelled proline have all been widely used) then enters the neurone and is transported back to the cell body. The transport of the label can be enhanced by applying a standing voltage to the cut end. In the central nervous system, the marker is injected fairly crudely into the brain or spinal cord and is taken up by damaged cells close to the injection site. The degree to which the pathway is traced depends on the time allowed for axonal transport.

(vii) Spontaneous endocytosis. Some compounds (particularly lectins) are avidly taken up by cells via an endocytotic pathway. Dye can be applied in an Agar pellet or in a small piece of gelatine sponge. The compound enters the lysozomes and provides a relatively permanent label so long as it is not broken down. Substances that are relatively toxic when directly injected are well tolerated by cells if allowed to enter by this natural pathway (e.g. tetra-methyl rhodamine, Texas Red). Cells labelled with TRITC have been used to examine the commitment of embryonic cells in *Xenopus laevis* (Heasman *et al.* 1985). Texas Red has been used to follow the outgrowth of neurones from retinal ganglion cells during the establishment of retino-tectal connections in *Xenopus* (O'Rourke & Fraser, 1986).

(viii) Scrape labelling. Cells are damaged by scraping! Compounds present in the external solution enter the cell during the period before the cell membrane reseals (El-fouly *et al.* 1987). Not a method of choice when studying gap junctions although many authors do so.

10. Sample protocols

Intracellular injection of HRP

(i) Fill microelectrode with solution of HRP. There are a number of different recipes in the literature. We have used 4% HRP in 0.2 M KCl.

An alternative is to use 4% HRP in 0.2 M Tris and 0.2 M KCl at pH 7.4. Some authors use slightly less HRP, slightly more KCl and may or may not add Tris. The type of HRP used also varies. Type II can be used, but many people prefer Type VI. Type VI is a single isozyme of HRP, type II a mixture. If you wish the preparation to survive for long periods of time after injection then type VI is probably the best.

(ii) Inject cell with positive going pulses of about 10 nA and 0.5 sec duration and a frequency of about 1 Hz for 15-20 min. You may see the cell swell as it fills.

(iii) When filling cells with fine processes it often helps to leave the preparation for 20 to 40 min after the end of injection to allow the HRP to diffuse. If the preparation will stand it, put it in the fridge.

(iv) Fix the preparation for about half an hour. 4% glutaraldehyde in saline or 0.1 M phosphate buffer for 15 to 20 min gives acceptable results.

There are various opinions in the literature concerning the concentration of fixative to employ. Some authors suggest fix as low as 0.8% glutaraldehyde. It may be important to use a low level of fixative to prevent inactivation of the peroxidase. In general the right amount and time of fixation are determined for each preparation by trial and error.

(v) Wash the preparation thoroughly in buffer.

(vi) Transfer preparation to 0.5 mg ml^{-1} diaminobenzidine (DAB) in 0.1 M phosphate buffer for 10 min. This allows the DAB to penetrate.

(vii) Add one or two drops of 1% hydrogen peroxide ml^{-1} of DAB solution, to the solution bathing the preparation. Watch the reaction. When the brown colour is fully developed wash in buffer.

(vii) Dehydrate in alcohol and clear for permanent preparation.

(ix) There are many alternative methods for HRP in the literature. Most of them work so it probably doesn't matter which you use. The main variant lies in the reagent used to visualise the HRP.

Many of the chromagens are potentially carcinogenic.

Some protocols recommend including 0.02% cobalt chloride and 0.02% nickel ammonium sulphide in the incubation medium (Adams, 1981). This can help to intensify the reaction product.

Biocytin injection technique

(i) Fill electrode with a 2-4% solution of biocytin or Neurobiotin (Vector Labs) in 2 M potassium acetate.

(ii) Inject with 1 nA depolarising pulses for up to 10 minutes.

(iii) Fix in 4% paraformaldehyde in 0.1 M phosphate buffer at pH 7.4 for at least 2 hours, depending on tissue thickness.

(iv) Rinse in buffer and, where necessary, slice tissue into sections (<100 μm).

(v) Wash well in buffer containing 0.4% Triton X-100 (or Tween 20) for at least 2 hours.

(vi) Incubate in 'ABC' (Streptavidin-biotinylated-HRP) complex (Vector Labs.) in Triton/ buffer for at least 2 hours; some protocols recommend 24 hours. (A variety of alternative avidin complexes, including a number of fluorescent conjugates are also available.)

(vii) Rinse well with 3 changes of Triton / buffer.

(viii) React with freshly mixed 0.05% diaminobenzidine (Caution, DAB is a potential carcinogen) and 0.003% hydrogen peroxide in Triton / buffer.

(ix) Wash well in buffer, if appropriate, mount sections, dehydrate in an alcohol series and clear in Histoclear or methyl salicylate

Cobalt injection technique

(i) Fill electrode with either 6% cobalt hexamine or 100 mM cobaltous chloride. Cobalt hexamine is best.

(ii) Inject with positive going pulses, 1-10 nA for about 10-15 min. For very small cells a shorter time will probably be enough.

(iii) Rinse preparation with bathing medium.

(iv) Either add a few drops of ammonium sulphide to the bath and observe appearance of dark brown precipitate, or saturate the bathing medium with hydrogen sulphide gas and then add to preparation; a dark brown precipitate of cobalt sulphide will appear.

Although both work, hydrogen sulphide is preferable if available. It produces a finer grain precipitate and does not contain polysulphides. Polysulphides can cause the sulphide-metal complex to re-dissolve.

(v) Fixation: The intensification procedure that follows works well with a glacial acetic acid/ethanol mix made up as:

1 part glacial acetic acid

4 parts 70% ethanol

The method will also work satisfactorily with glutaraldehyde fixation, provided the preparation is adequately washed. If a formaldehyde- or formalin-based fixative is essential, then the preparation should be washed extensively with chloral hydrate before proceeding to intensification.

For small preparations half an hour should be sufficient.

Method for intensification of a metal precipitate

There are two main physical developer methods of intensification, one based on Timm's solution, using gum Arabic as a protective colloid (e.g. Bacon & Altman, 1977), and a second using tungsto-silicic acid as the protective colloid (e.g. Szekely & Gallyas, 1975). Both methods give good results but the former is carried out in the dark at 60°C, whereas the latter, illustrated below, can be done at room temperature in the light.

We are indebted to Barry Roberts for introducing us to this intensification method.

(i) Wash preparation in distilled water for 15 min.

(ii) Incubate in 2% sodium tungstate - 10 min for sections, half an hour for whole mounts.

(iii) Place in intensification solution in Petri dish. Intensification solution should be freshly prepared.

(iv) Observe under microscope until tissue begins to discolour (2-10 min).

(v) Rinse with 3 changes of distilled water.

(vi) For permanent preparation, dehydrate through graded alcohols and then clear in xylene or methyl salicylate.

If the tissue is over-intensified, it can be worth attempting a partial de-intensification using Farmer's photographic reducer method (Pitman, 1979).

Intensification solution (for step iii)

A	distilled water	355 ml
	1% Triton X-100	15 ml
	sodium acetate 3H$_2$O	1.5 gm
	glacial acetic acid	30 ml
	silver nitrate	0.5 gm

This solution can be kept in the fridge until a silver precipitate begins to appear.

B   5% sodium tungstate
C   0.25% ascorbic acid in distilled water.

Make immediately before use.
Mix A, B, C in proportions 8A:1B:1C, freshly prepared. For 40 ml of solution take 32 ml A:4 ml B:4 ml C.

APPENDIX A

Table of absorption and emission maxima for some common fluorophores

Compound	MW	ABS (nm)	EM (nm)	Zeiss filter set
bis-Benzimide (Hoechst 33258)	534	365	480	01, 02
bis-Benzimide (Hoechst 33342)	562	355	465	01, 02
Carboxyfluorescein	376	492	516	09, 10, 16, 17, 23
Cascade Blue	607	375/400	410	02, 05, 18, 21, 30
DAPI	457	347	458	01, 02, 18
DiASP	787	491	613	09, 10, 16, 17
DiI/Fast DiI	934	550	565	14, 15, 23
DiO/Fast DiO	882	484	501	09, 10, 16, 17, 23
Diamidino Yellow	NA	~365	~480	05, 18
Ethidium bromide	394	526	605	14, 15
Fast Blue	NA	~365	~480	05, 18
Fluorescein (FITC)	389	495	519	09, 10, 16, 17, 23
Lucifer Yellow	453	428	535	05, 06, 18
Propidium iodide	668	536	617	14, 15
Rhodamine (TRITC)	444	544	570	14, 15, 23
Sulphrhodamine 101	607	~586	607	00, 14, 15, 23
Texas Red	625	589	615	00, 14, 15, 23

It is worth experimenting with different filter combinations to obtain the best result for any particular application.

ABS, absorbance max; EM, emission max; NA, not available.

APPENDIX B

Equipment

1. Electrode glass. From many suppliers, including:

Clarke Electromedical Ltd, P.O. Box 8, Pangbourne, Reading, RG8 7HU, UK. Clarke also supply pollers, electrophysiological equipment and act as agents for a number of manufacturers. A helpful firm.

Glass Company of America, Bargaintown, New Jersey, USA.

2. Electrophysiological equipment. No special requirements related to injection. Simple amplifiers can be constructed at low cost (see circuits in Purves, 1981). High quality amplifiers are available from:

Axon Instruments Inc., 1101 Chess Drive, Foster City, CA 94404 USA

Digitimer Ltd., 37 Hydeway, Welwyn Garden City, AL7 3BE, UK

World Precision Instruments, Astonbury Farm Business Centre, Unit J, Aston, Stevenage, Herts SG2 7EG UK (Obtainable from Clarke).

3. Pressure Injection. Ready made devices are available from:

General Valve Corporation, East Hanover, New Jersey, USA. (Picospritzer 11., cheap).

Eppendorf Geratebau, P.O. Box 630324, 2000 Hamburg 63, F.D.R. (not cheap).

Alternatively devices can be made using Agla (micrometer driven) syringe, plastic tubing and liquid paraffin which is cheap and messy, or using a gas cylinder and a solenoid operated tap; available from General Valve Corp., (above) and from:

RS Components Ltd, P.O. Box 99, Corby, NN17 9RS.

4. Optical equipment. A fluorescence microscope (preferably epifluorescence) is an absolute essential for many injection experiments and potentially expensive.

A dissecting microscope may be used for injection, but a compound microscope with a fixed stage and head focussing, or an inverted microscope (for cultured or dissociated cells), extends the range of injectable cells downwards to about 5 µm. Ideally the fluorescence head should be fitted to the microscope used for injection. Zeiss, Nikon and Leitz microscopes are available to order with a fixed stage.

Advice and a wide range of microscopes are available from:

Micro Instruments (Oxford) Ltd, 18 Nanborough Park, Long Hanborough, Oxford OX7 2LH, UK. They manufacture a fixed stage micromanipulation microscope.

Leica UK Ltd (Leitz), Davy Avenue, Knowlhill, Milton Keynes, MK5 8LB, UK

Nikon UK Ltd, Instrument Division, Haybrook, Halesfield, Telsford, TF7 4EW, UK

Carl Zeiss (Oberkochen) Ltd, PO Box 78, Woodfield Road, Welwyn Garden City, AL7 1LU, UK

References

N.B. Additional references, which are not quoted in the text, are included in this list for information.

ADAMS, J. C. (1981). Heavy metal intensification of DAB-based HRP reaction product. *Journal of Histochemistry and Cytochemistry*, **29**, 775.
AGHAJANIAN, G. K. & VANDERMAELEN, C. P. (1982). Intracellular identification of central noradrenergic and serotonergic neurons by a new double labelling procedure. *J. Neurosci.* **2**, 1786-1792.

BACON, J. P. & ALTMAN, J. S. (1977). A silver intensification method for cobalt-filled neurones in whole mount preparations. *Brain Research*, **138**, 359-363.

BENNETT, M. V. L. & SPRAY, D. C. (1985). *Gap Junctions*. Cold Spring Harbour Symposium. Cold Spring Harbour Press.

BLENNERHASSET, M. & CAVENEY, S. (1984). Separation of developmental compartments by a cell type with reduced junctional permeability. *Nature, Lond.* **309**, 361-364.

BLINKS, J. R., WIER, W. G., HESS, P. & PRENDERGAST, F. G. (1982). Measurement of Ca^{2+} concentrations in living cells. *Prog. Biophys. molec. Biol.* **40**, 1-114.

COLMAN, A. (1984). Translation of Eukaryotic messenger RNA in oocytes. In *Transcription and Translation: a practical approach*. London and Washington: IRL Press.

EL-FOULY, M. H. , TROSKO, J. E. & CHANG, C.-C. (1987). Scrape loading and dye transfer. *Exp. Cell Res.* **168**, 422.

GARDNER, R. L. (1985). Clonal analysis of early mammalian development. *Phil. Trans. Roy. Soc. Lond.* B **312**, 163-178.

GILULA, N. B. (ed) (1993). Gap junctional communication. *Seminars in Cell Biology* Volume 3. Academic Press.

GIMLICH, R. L. & BRAUN, J . (1985). Improved fluorescent compounds for tracing cell lineage. *Devl Biol.* **109**, 509-514.

GOODWIN, P. B. & ERWEE, M. G. (1985). Intercellular transport studied by microinjection methods. In *Botanical Microscopy* (ed. A. W. Robards), pp. 335-358. Oxford: Oxford University Press.

HEASMAN, J., SNAPE, A., SMITH, J. C., HOLWILL, S. & WYLIE, C. C. (1985). Cell lineage and commitment in early amphibian development. *Phil. Trans. Roy. Soc. Lond.* B **312**, 145-152.

HEIMER, L. & ROBARDS, M. J. (1981). *Neuro-Anatomical Tracing Methods*. New York and London: Plenum Press.

HORIKAWA, K. & ARMSTRONG, W. E. (1988). A versatile means of intracellular labeling: injection of biocytin and its detection with avidin conjugates. *J. Neurosci. Methods* **25**, 1-11.

JACOBSON, M. & HIROSE, G. (1978). Origin of the retina from both sides of the embryonic brain: a contribution to the problem of crossing over at the optic chiasma. *Science* **202**, 637-639.

KATER, S. & NICHOLSON, C. (1979). *Intracellular Staining Methods in Neurobiology*. Berlin: Springer Verlag.

KITA, A. & ARMSTRONG, W. (1991). A biotin-containing compound N-(2-aminoethyl) biotinamide for intracellular labeling and neuronal tracing studies: comparison with biocytin. *J. Neurosci. Methods* **37**, 141-150.

KIMMEL, C. B. & WAGA, R. M. (1986). Tissue specific cell lineages originate in the gastrula of the zebra fish. *Science* **231**, 365-368.

KUFFLER, D. P. (1990). Long-term survival and sprouting in culture by motoneurons isolated from the spinal cord of adult frogs. *J. Comp. Neurol.* **302**, 729.

MARANTO, A. R. (1982). Neuronal mapping: A photooxidation reaction makes Lucifer Yellow useful for electron microscopy. *Science* **217**, 953-955.

MESULAM, M-M. (ed.) (1982). Tracing neural connections with Horse Radish Peroxidase. *IBRO Handbook Series*: *Methods in Neurosciences*. Wiley.

O'ROURKE, N. & FRASER, S. (1986). Dynamic aspects of retinotectal map formation as revealed by a vital dye fiber tracing technique. *Dev. Biol.* **114**, 265-276.

PAPKE, R. L., DUVOISIN, R. M. & HEINEMANN, S. F. (1993). The amino terminal half of the nicotinic B-subunit extracellular domain regulates the kinetics of inhibition by neuronal bungarotoxin. *Proc. Roy. Soc. (Lond.)* B **252**, 141-148.

PEARSE, A. (1968). *Histochemistry*. London: Churchill.

PIENADO, A., YUSTE, R. & HATZ, L. C. (1993). Extensive dye coupling between rat neocortical neurons during the period of circuit formation. *Neuron* **10**, 103-114.

PITMAN, R. M. (1979). Block intensification of neurons stained with cobalt sulphide: a method for destaining and enhanced contrast. *J. Exp. Biol.* **78**, 295-297.

POTTER, H., WEIR, L. & LEDER, P. (1984). Enhancer dependent expression of human K immunoglobin genes introduced into mouse pre-B lymphocytes by electroporation. *Proc. natn. Acad. Sci. U.S.A.* **81**, 7161-7163.

PURVES, D., HADLEY, R. D. & VOYVODIC, J. T. (1986). Dynamic changes in the dendritic geometry of individual neurons visualised over periods of three months in the superior cervical ganglion of living mice. *J. Neurosci.* **6**, 1051-1060.

PURVES, R. D. (1981). *Micro-electrode Methods for Intracellular Recording and Ionophoresis.* London: Academic Press.

QUICKE, D. L. J. & BRACE, R. C. (1979). Differential staining of cobalt- and nickel-filled neurones using rubeanic acid. *J. Microsc.* **15**l, 161-163.

RINK, T. J., TSIEN, R. Y. & POZZAN, T. (1982). Cytoplasmic pH and free Mg^{2+} in lymphocytes. *J. Cell Biol.* **95**, 189-196.

SERRAS, F. & VAN DEN BIGGELAAR, J. A. M. (1987). Is a mosaic embryo also a mosaic of communication compartments. *Dev. Biol.* **120**, 132-138.

SIMPSON, I., ROSE, B. & LOEWENSTEIN, W. R. (1977). Size limit of molecules permeating the junctional membrane channels. *Science* **195**, 294-296.

SPANDIDOS, D. A. & WILKIE, N. M. (1984). Expression of exogenous DNA in mammalian cells. In *Transcription and Translation: a practical approach* (ed. B. D. Hames & S. J. Higgins). Oxford and Washington: IRL Press.

STEWART, W. W. (1978). Functional coupling between cells as revealed by dye-coupling with a highly fluorescent napthalamide tracer. *Cell* **14**, 741-759.

STEWART, W. W. & FEDER, N. (1985). Lucifer dyes as biological tracers: a review. In *Cellular and Molecular Control of Direct Cell-Cell Interactions*, *NATO Life Sciences*, vol. 99, pp. 297-312. London and New York: Plenum Press.

SZÉKELY, G. & GALLYAS, F. (1975). Intensification of cobaltous sulphide precipitate in frog nervous tissue. *Acta biol. Acad. Sci. hung.* **26**, 176-188.

TAGHERT, P. H., BASTIANI, M. J., HO, R. K. & GOODMAN, C. S. (1982). Guidance of pioneer growth cones: Filopodial contacts and coupling revealed with an antibody to Lucifer Yellow. *Devl Biol.* **94**, 391-399.

TECHNAU, G. M. & CAMPOS ORTEGA, J. A. (1985). Fate mapping in wild type Drosophila melanogaster. 11. Injections of horse radish peroxidase in cells of the early gastrula. *Roux. Arch. Devl Biol.* **194**, 196-212.

THOMAS, J., BASTIANI, M., BATE, C. M. & GOODMAN, C. (1984). From grasshopper to Drosophila: a common plan for neural development. *Nature, Lond.* **310**, 203-206.

TURIN, L. (1977). New Probes for studying intercellular permeability. *J. Physiol., Lond.* **269**, 6P.

VANEY, D. I. (1991). Many diverse types of retinal neurons show tracer coupling when injected with biocytin or Neurobiotin. *Neuroscience Letters* **125**, 187-190.

WARNER, A. E. & LAWRENCE, P. A. (1982). Permeability of gap junctions at the segment border; in insect epidermis. *Cell* **28**, 243-252.

WEISBLAT, D. A., SAWYER, R. T. & STENT, G. (1978). Cell lineage analysis by intracellular injection of a tracer enzyme. *Science* **202**, 1295-1298.

WILLIAMS, D. A., FOGARTY, K. E., TSIEN, R. Y. & FAY, F. S. (1985). Calcium gradients in single smooth muscle cells revealed by the digital imaging microscope using Fura-2. *Nature, Lond.* **318**, 558-561.

Chapter 15
Flash photolysis of caged compounds

ALISON M. GURNEY

1. Introduction

Photolabile 'caged' compounds are biologically inert precursors of active molecules, which when irradiated with a pulse of light free the active species at its site of action (see Fig. 1). Speed is the principal advantage offered by these probes; photochemical reactions can be very fast, with release of the active species often complete within a millisecond. The method can thus be used to study the kinetics and concentration dependence of reactions at sites that are not otherwise accessible. A number of processes can potentially distort the time course of responses to receptor ligands. These include diffusion of the ligand to the receptor, receptor desensitisation and breakdown and removal of the ligand by tissue enzymes, all of which can be substantial during the onset of a response when application is slow. These problems are by-passed with photolabile probes, which can be pre-equilibrated with the tissue in the inactive form, flash photolysis producing a pulse of agonist at a known concentration on demand.

Photolabile probes can be used to study intracellular pathways, overcoming the barrier formed by the cell membrane to the application of intracellular mediators or modulators. Access of the caged molecule can be by membrane permeabilisation, by microinjection or, most effectively, by perfusion from a patch pipette in whole-cell patch-clamp recording. It is important for quantitative studies that the intracellular concentration of the caged molecule is known. Depending on the size of the caged molecule, the cell and the patch pipette, it can take many minutes for exchange of the cell contents in whole-cell recording to reach equilibrium (Pusch & Neher, 1988).

Two further advantages offered by photolabile probes are the relative ease with which the intensity and the area of the light spot can be varied. By varying the intensity of the activating light, it is possible to vary the extent of photolysis and hence the concentration of the active species released at its site of action. Sufficient intensity for photolysis can be achieved in light spots large enough to encompass an intact tissue, for example a strip of muscle, or a population of cells. On the other hand, light can also be focussed to a small spot, allowing irradiation of a single cell or even part of a cell. Thus it should be possible to apply flash photolysis to study responses to

Department of Pharmacology, United Medical & Dental Schools, St Thomas's Hospital, Lambeth Palace Road, London, SE1 7EH, UK.

localised release of important signalling molecules in specialised regions of the cell, and to compare this with the effects of release imposed throughout the cell.

Studies in the nineteen seventies and early eighties were limited by the availability of only a few 'caged' molecules, such as caged ATP and photoisomerisable, bisquaternary, cholinergic ligands (Lester & Nerbonne, 1982). Although photolabile cyclic nucleotide derivatives were introduced around the same time as caged ATP (Engels & Schlaeger, 1977), it was another six years before their photochemistry was exploited to rapidly activate a biological pathway (Nargeot *et al.*, 1983). Since then there has been an acceleration in the design and development of new caged compounds, with improved photochemical and biological properties, and this has given rise to a growing interest in flash photolysis techniques. A wide range of molecules, specifically designed for flash photolysis studies with biological preparations, is now available and more are under development (Adams & Tsien, 1993). These include photolabile precursors for intracellular second messengers, such as cyclic nucleotides, inositol trisphosphate and calcium, and modulators of intracellular pathways, such as ATP, GTP, GTPγS and caged calcium chelators. There are also caged neurotransmitters such as glutamate, GABA, noradrenaline and serotonin, caged local hormones such as nitric oxide and arachadonic acid, as well as other receptor ligands such as carbachol and phenylephrine. Table 1 lists some of the photochemical properties of probes that are currently available from commercial sources. Methods developed for many more probes have been described (see Walker, 1991). Molecular Probes sell a kit that may be used to synthesise caged phosphate esters and they also offer a custom synthesis service. The available caged compounds may be used for studies on a wide variety of biological systems.

2. The light induced reactions

Light acts as the trigger to initiate a series of reactions. The first step is the absorption by the caged compound of a photon of energy (E),

$$E = h\nu, \tag{1}$$

where h is Planck's constant and ν is the frequency of the light. As a result of absorbing the photon, the caged molecule (M) gains energy and is thereby promoted to an excited state (M^*). This step, which occurs on a time scale of < 1 ns, can be represented as:

$$M + h\nu \rightarrow M^*. \tag{2}$$

M^* is a new chemical species that is highly reactive. After absorbing the photon, M^* can either lose its extra energy and decay back to M or it can proceed to form stable products ($M^* \rightarrow$ products). The reactions leading to the formation of photoproducts are known as the dark reactions, because once M^* has been formed they will procede in the absence of light. The dark reactions are usually relatively slow and vary widely in time course, from microseconds to hundreds of milliseconds, and in efficiency.

Table 1. *Properties and sources of commercially available caged compounds*

Caged compound	Source	Structure	ε (M^{-1}cm^{-1}×10^{-3})	φ	release half time (ms)	*Useful concentration range
Intracellular probes						
ATP	C, M	NPE	17.5 at 260 nm	0.63	7	5-20 mM
	M	DMNPE	5.3 at 354 nm			
ADP	M	NPE	19.6 at 260 nm		8.7	1-20 mM
GTP	C	NPE	~16.9 at 255 nm			1-10 mM
GTP-γ-S	M	NB				1-10 mM
	M	DMNB	4.8 at 355 nm			
	M	NPE	15.4 at 260 nm	0.35	6	
	M	DMNPE	3.6 at 362 nm			
GDP-β-S	M	DMNPE	~16.9 at 260 nm			1-10 mM
cyclic AMP	C, M	NPE	20 at 259 nm			50-500 µM
	M	DMNB	4 at 350 nm		<5	
cyclic GMP	C, M	NPE				50-500 µM
Inositol 1,4,5-trisphosphate	C	NPE	4.2 at 200 nm	0.65	3	1-500 µM
Inorganic phosphate	M	NPE	4.2 at 260 nm	0.54	<0.1	up to 20 mM
Calcium:						
nitr-5	C	see text	see Table 2			0.2 - 10 mM
nitr-7	C					
DM-nitrophen	C					
calcium chelator:						
diazo-2	M					
Extracellular probes						
carbachol	C, M	NPE	5.2 at 262 nm	0.29	0.07	50 -100 µM
	M	CNB	5.2 at 266 nm	0.8	0.04	
adrenaline	M	DMNB				50-500 µM
noradrenaline	M	DMNBOC				50-500 µM
dopamine	M	DMNBOC				50-500 µM
isoprenaline	M	NB				50-500 µM
propranolol	M	NB				100 -500 µM
serotonin	M	DMNBOC				50-500 µM
glutamate	M	DMNB	5.7 at 347 nm			up to 20 mM
MK-801	M	DMNBOC				100-500 µM
aspartate	M	DMNB	6.0 at 347 nm			50-500 µM
GABA	M	DMNB	5.4 at 347 nm			50-500 µM
glycine	M	DMNB	5.7 at 345 nm			50-500 µM
arachidonic acid	M	DMNB				
nitric oxide	M,A, 1	K$_2$Ru(NO)Cl$_5$	0.56 at 320 nm	0.06	<5	0.5-50 µM

C: Calbiochem Novabiochem, 3 Heathcoat Building, Highfields Science Park, University Boulevard, Nottingham NG7 2QJ. M: Molecular Probes, Inc., 4849 Pitchford Avenue, Eugene, Oregon 97402-9144 USA. A: Alfa, Johnson Matthey, Orchard Road, Royston, Hertfordshire, SG8 5HE. 1: See *Biophys. J.* **64**, p. A190, Bettache *et al.* (1993).

NB: 2-nitrobenzyl. DMNB: 4,5-dimethoxy-2-nitrobenzyl. NPE: 1(2-nitrophenyl) ethyl. DMNPE: 4,5-dimethoxy-1(2-nitrophenyl) ethyl. CNB: α-carboxy-2-nitrobenzyl. DMNBO:C 4,5-dimethoxy-2-nitrobenzyloxycarbonyl

* Many of these compounds have not been widely tested. The concentration ranges suggested are based on experience with the better tested probes.

The mechanism of photolysis of caged ATP has been studied in detail and is reviewed in McCray & Trentham (1989). One of the stable products is the biologically active molecule. The other products should ideally be biologically inert.

Since only light that is absorbed by the caged molecule can trigger photochemical reactions, the higher the proportion of molecules absorbing the light, the more product will be formed. Absorbance (A) of a caged molecule in solution is determined by

$$A = \varepsilon c l \tag{3}$$

where ε and c are the extinction coefficient and concentration of the molecule respectively, and l is the length of the light path through the solution. Thus for efficient photolysis, ε should be high at wavelengths triggering the photochemical reactions. On the other hand, a high extinction coefficient can lead to poor and uneven photolysis in thick preparations, such as muscles of 100 μm or so diameter or the large neurones (> 200 μm) found in several molluscs, because the molecules at the front surface may absorb much of the light before it reaches the centre of the

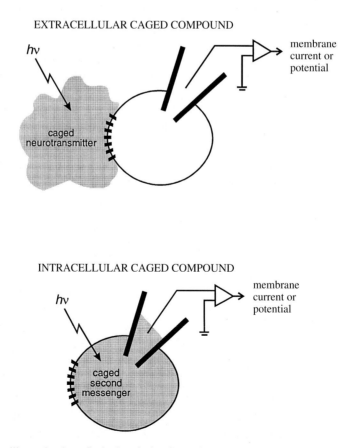

Fig. 1. Cartoon illustrating how flash photolysis of caged compounds can be used to activate biological pathways.

tissue. Photolysis of most of the currently available caged compounds requires wavelengths in the near-UV region of the spectrum, with a peak in efficiency at around 320-360 nm. Wavelengths below 300 nm cause damage to cells and are therefore filtered out of the activating light. Most caged compounds absorb little above 500 nm, so at the wavelengths of normal, incandescent room lighting, most caged compounds are fairly stable. This means that experiments do not need to be carried out in a dark room. It is good practice, however, to minimise exposure of caged compounds to light at all wavelengths. Experimental solutions should be prepared under dimmed light and be kept in light-tight containers or containers wrapped in foil. Most mammalian cells are sufficiently transparent in the near-UV region that they contribute little to the absorbance of the exciting light. Thus photolysis can be efficiently effected in isolated cells and in many thin tissue preparations. Some cells are pigmented, for example cell bodies found in the nervous system of the marine mollusc *Aplysia*. In this case the cells may contribute to absorbance and reduce the amount of effective light reaching the caged molecules (Tsien & Zucker, 1986; Nerbonne & Gurney, 1987). This can interfere with calibration of the amount of active molecule released from a probe, although, if the absorbance properties of the pigment are known, it can be allowed for in the calculations. The influence of cell absorbance and preparation thickness on the photolytic efficiency of a flash are discussed in detail in Landó & Zucker (1989), along with calculations of the volume-average photolysis of nitr-5 in an *Aplysia* neurone, where the cytoplasm has an absorbance coefficient (εc) of 25 cm^{-1} at 360 nm. By comparison rat cerebellar slices have an absorbance coefficient of 10 cm^{-1} at 320 nm, giving 26% attenuation of a flash over 300 μm (Khodakhah & Ogden, 1993).

Efficiency of photolysis is not only determined by ε. It is not sufficient that the caged compound absorbs light well, but each time a photon is absorbed it should be likely to result in formation of the photoproduct. A measure of the effectiveness of an absorbed photon is given by the quantum yield (φ), which is defined as

$$\varphi = \frac{\text{product molecules formed}}{\text{photons absorbed}}. \tag{4}$$

Thus less light would be required to trigger the release of active molecules from caged compounds with a high extinction coefficient and a high quantum yield. The commercially available probes all have sufficiently high extinction coefficients and quantum yields to permit concentration changes in the physiological range to be produced with flashlamps or lasers.

The proportion of molecules photolysed by a single flash may also be influenced by the lifetime of M^* relative to the flash duration. With some caged molecules, a single flash produces a greater percent conversion than is predicted by the quantum yield. For example, caged cyclic nucleotides have very low quantum yields compared with caged ATP, but photolysis produced by a 1 ms flash is only two-fold greater with caged ATP (Wootton & Trentham, 1989). The discrepancy can be explained if the excited intermediate formed by irradiating cyclic nucleotides is shorter lived than that formed from caged ATP, with the result that more excitations occur during the flash with the former compounds, thereby amplifying the response.

3. Structure and photochemistry

The currently available caged compounds have the general structure shown in Fig. 2A. Light sensitivity is conferred by the *o*-nitrobenzyl moiety and the simplest caged compound is the *o*-nitrobenzyl ester of the biologically active molecule. Photolysis occurs when irradiation cleaves the precursor at the benzyl carbon to release the active species. Released with the active molecule are a proton and a nitroso by-product, which is usually either an aldehyde or a ketone. These too could have biological effects that must be prevented or controlled as discussed below.

The photochemical properties of caged compounds are modified by varying the nature of the substituents at the benzyl carbon (R_1) and at positions R_2 and R_3 in Fig. 2A. For example, adding methoxy groups at positions R_2 and R_3 causes a red shift of the absorption maximum of caged compounds, thereby improving light absorption in the 300-360 nm range. The rate and efficiency of the photochemical reaction, as well as the biological activity of the precursor and the photolysis by-products, are also influenced. Unfortunately, there is no general rule that can predict how these properties will be affected, because they are also influenced by the nature of the molecule being caged. Thus methoxy groups at positions R_2 and R_3 improve the speed and efficiency of photorelease from caged cyclic nucleotides (Nerbonne, 1986), but slow release from caged ATP and caged phosphate (Wootton & Trentham, 1989). Similarly, a methyl group at R_1 accelerates photorelease from caged carbachol (Milburn et al, 1989) and caged ATP (Kaplan *et al.*, 1978) while making dimethoxy *o*-nitrobenzyl cyclic nucleotides unstable in aqueous solution (Wootton & Trentham, 1989).

Photolabile cation chelators

Caged compounds can be used to control the concentration of cations inside or outside cells. There are several molecules available that can act either as caged calcium ions (nitr-5, DM-nitrophen) or caged calcium ion chelators (diazo-2). All of these agents incorporate an *o*-nitrobenzyl group (Fig. 2B), the photolysis of which alters the affinity of the molecule toward Ca^{2+}.

The nitr-5 family of compounds (Adams *et al.*, 1988), including diazo-2 (Adams *et al.*, 1989), was designed around BAPTA, a Ca^{2+} chelator (K_D=110 nM) with high selectivity for Ca^{2+} over Mg^{2+}. This selectivity is retained in the photolabile probes, both before and after photolysis, and like BAPTA, their Ca^{2+} affinity shows little dependence on pH above pH7. Absorption of a photon by these compounds results in structural rearrangement to a form with altered Ca^{2+} affinity. In the case of nitr-5, photolysis induces a forty-fold loss of affinity for Ca^{2+}. The result is a net release of Ca^{2+} and H_2O is released as a by-product. Diazo-2 gains affinity for Ca^{2+} upon photolysis and therefore reduces free Ca^{2+}. This change is accompanied by the release of N_2 gas. There is a structurally related molecule, diazo-3, which can provide a suitable control for experiments with the BAPTA-derived molecules. It has similar photochemical properties but has little affinity for Ca^{2+} before or after photolysis, so the photochemical reaction does not produce a change in Ca^{2+} concentration.

DM-nitrophen (Kaplan & Ellis-Davies, 1988) is derived from the chelator EDTA. It binds Ca^{2+} ions tightly ($K_D = 5$ nM), but also has a significant affinity for Mg^{2+} (K_D

A.

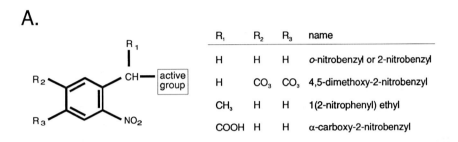

R_1	R_2	R_3	name
H	H	H	*o*-nitrobenzyl or 2-nitrobenzyl
H	CO_3	CO_3	4,5-dimethoxy-2-nitrobenzyl
CH_3	H	H	1(2-nitrophenyl) ethyl
COOH	H	H	α-carboxy-2-nitrobenzyl

B.

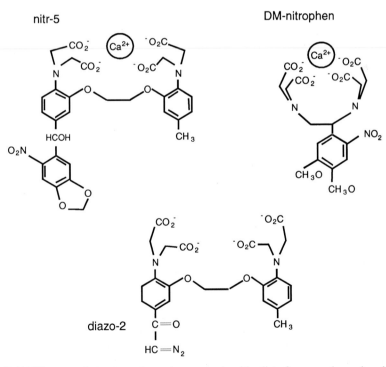

Fig. 2. (A) The general structure of caged compounds with a list of commonly used analogues, showing the variation in constituents at positions R_1, R_2 and R_3. (B) Structures of caged Ca^{2+} (nitr-5 and DM-nitrophen) and a caged Ca^{2+} chelator (diazo-2).

= 2.5 μM). When irradiated, the DM-nitrophen molecule is cleaved to yield two photoproducts, both with negligible affinity for Ca^{2+}. The change of affinity is large ($K_D > 3$ mM after photolysis), enabling millimolar changes in Ca^{2+} concentration to be produced.

DM-nitrophen and nitr-5 each have distinct advantages and disadvantages for use as Ca^{2+} donors. Nitr-5, but not DM-nitrophen at present, can be loaded into cells non-invasively using the membrane permeant acetoxymethyl ester form. Once this has entered cells, it is cleaved by intracellular enzymes to release free nitr-5, enabling responses to a rapid rise in intracellular Ca^{2+} to be studied in unperturbed cells. Both molecules release Ca^{2+} rapidly when irradiated, with submillisecond time constants. Following release, the free Ca^{2+} re-equilibrates with unphotolysed chelator. In the case of nitr-5, re-equilibration is rapid (around 2 μs) and faster than the rate of photolysis, whereas released Ca^{2+} re-equilibrates slowly with unphotolysed DM-nitrophen. The consequences of this are discussed in detail by Zucker (1993). Essentially it means that Ca^{2+} released by photolysis of nitr-5 reaches a steady level in an essentially step-like fashion, whereas partial photolysis of DM-nitrophen produces a pulse of Ca^{2+} that peaks in less than 1 ms, but then declines to a steady level over the next several milliseconds. Thus DM-nitrophen can be used to generate spikes of intracellular Ca^{2+}.

Nitr-5 and DM-nitrophen also differ in their affinities for Ca^{2+} and Mg^{2+}, the dependence of these affinities on pH, and the light-induced change in affinity. The low selectivity of DM-nitrophen for Ca^{2+} over Mg^{2+} means that at the millimolar concentrations of Mg^{2+} present inside cells, a significant fraction of DM-nitrophen would be present as the Mg^{2+} complex, and photolysis would release a mixture of Ca^{2+} and Mg^{2+}. Although this is a disadvantage for studying Ca^{2+} actions in intact cells, it means that DM-nitrophen can be exploited to selectively release Mg^{2+} (O'Rourke *et al.*, 1992). DM-nitrophen also has the disadvantage of a high pH sensitivity.

The favourable properties of DM-nitrophen, as compared with nitr-5, are its higher Ca^{2+} affinity and much greater change in affinity when irradiated. At resting cellular levels of Ca^{2+}, a greater proportion of DM-nitrophen would be bound to Ca^{2+} and larger changes in Ca^{2+} concentration could be produced by photolysis. The Ca^{2+} affinities of nitr-5 before ($K_D = 145$ nM) and after ($K_D = 6$ μM) photolysis are in the physiological range of Ca^{2+} concentrations, so nitr-5 will affect cell Ca^{2+} buffering.

Table 2. *Properties of commercially available photolabile Ca^{2+} chelators*

	K_D for Ca^{2+} binding (μM)				K_D for Mg^{2+} binding (mM)	
	pre photolysis	*post photolysis*	photolysis rate (s^{-1})	quantum yield	*pre photolysis*	*post photolysis*
nitr-5	0.145	6	4000	0.03-0.1	8.5	8
nitr-7	0.048	~2	556	~0.1	6	
DM-nitrophen	0.005	2000	3000	0.18	0.005	3
diazo-2	2.2	0.073	>2000			

In whole-cell, patch-clamp experiments, nitr-5 is usually the dominant Ca^{2+} buffer in the cell. Provided the extent of photolysis is known (see below), the free Ca^{2+} concentration before and after a flash can be calculated using the programme given in Gurney (1991). Landó and Zucker (1989) also discuss calculations of free Ca^{2+} produced in cells by photolysis of nitr-5, which incorporate the Ca^{2+} buffering capacity of cytoplasm.

Relevant properties of the photolabile Ca^{2+} probes are listed in Table 2 for comparison.

4. Sources of error in the use of caged compounds

The main sources of error in flash photolysis experiments with caged compounds are:

1. the unphotolysed caged compound itself has pharmacological activity in the tissue;

2. the by-products of photolysis are biologically active;

3. the flash itself triggers a response in the absence of a caged compound;

4. release of the active molecule limits the time course of the response.

Ideally, the caged compound and its photolysis products other than the one of biological interest should be inert, and flashes in the 300-400 nm wavelength range should cause no measurable response in the absence of the probe. This has usually been found to be the case, although it is important to test the effects of the 'caged' precursor, photolysis by-products and light in the biological system being studied.

Effects of the precursor

As already noted, nitr-5 can alter the Ca^{2+} buffering capacity of cells, both before and after photolysis, and this property is frequently exploited to control the intracellular Ca^{2+} concentration. Other caged compounds are also known to influence cellular activity prior to photolysis. For example, the nitrophenylethyl analogue of caged ATP was found to bind to the same site on the Na/K ATPase as free ATP, albeit with lower affinity (Forbush, 1984). It has also been found to bind to myosin (Dantzig *et al.*, 1989) and to cause partial blockade of ATP-sensitive K^+ channels in cardiac (Nichols et al, 1990) and smooth (Gurney, 1993) muscle. Nitrobenzyl acetate is a photolabile proton donor, which in some cells is cleaved by intracellular enzymes to lower pH without photolysis (Spray *et al.*, 1984). The nitrobenzyl ester of cyclic AMP spontaneously generates cyclic AMP inside cells in a similar way (Korth & Engels, 1979), whereas the dimethoxy nitrobenzyl ester, which for a number of reasons is the preferred analogue for photolysis experiments, is more stable (Nerbonne *et al.*, 1984). We have also found differences in the basal activity of the nitrophenylethyl and the dimethoxy nitrobenzyl derivatives of cyclic GMP in vascular muscle (Gurney, 1993). An early version of caged carbachol, the nitrophenylethyl derivative, was found to be active at both nicotinic and muscarinic acetylcholine receptors in the absence of photolysis (Walker *et al.*, 1986). This activity was subsequently eliminated by introducing a carboxylate group into the benzyl carbon to make N-(α-carboxy-2-nitrobenzyl) carbachol (Milburn *et al.*, 1989). Three isomeric forms of caged IP3, resulting from esterification of each of the three phosphate groups, have been isolated and they each have different properties. The 4-phosphate ester is inactive before photolysis, but IP3 caged on the 1-phosphate induces intracellular Ca^{2+} release in smooth muscle while the 5-phosphate ester inhibits IP3 3-kinase (Walker *et al.*, 1989). Commercial sources of caged IP3 contain a mixture of the 4- and 5-phosphate esters.

Effects of the photolysis by-products

As indicated earlier, photolysis of most caged compounds results in the release of a proton and a nitroso by-product as well as the biologically active molecule. Changes in pH caused by proton release can be suppressed by strongly buffering the experimental solution. The nitroso by-products present a more difficult problem, since they are reactive towards sulphydryl groups on proteins. Reactivity can be reduced by derivatising the benzyl carbon (McCray et al, 1980) with, for example, a methyl group, as in the nitrophenyl ethyl derivatives ($R_1 = CH_3$ in Fig. 2A). Even so, the nitrosoacetophenone produced alongside photolysis of nitrophenylethyl ATP inactivates the Na/K ATPase of erythrocyte ghosts (Kaplan et al, 1978). Such effects can, however, be minimised by adding a hydrophilic thiol such as glutathione or dithiothreitol to the experimental solution, since these agents protect sulphhydryl and amino groups on cell proteins by inactivating the nitrosoketone photoproduct (Kaplan *et al.*, 1978). Fortunately, most receptor ligands and intracellular messengers work in the micromolar concentration range, so effective concentrations of the biologically interesting molecule can be produced with relatively small amounts of the by-product. The photoproducts are only likely to cause a problem when high concentrations of a caged compound are photolysed, for example, when millimolar concentrations of ATP are needed to activate muscle contraction.

Effects of the photolysis by-products are not always easy to control for. Parallel experiments could be performed using a structurally related caged compound that undergoes similar photochemical reactions with the same efficiency, but does not release the molecule of interest. For example, caged inorganic phosphate could be used as a control for caged ATP or GTP, as could diazo-3 be used as a control for diazo-2 or nitr-5. Another approach is to add excess of the biologically active molecule to the caged compound solution to pre-activate the pathway of interest prior to photolysis. Since under these conditions the same photoproducts will be generated, any remaining response to photolysis is probably due to the by-products of the photochemical reaction. These kinds of experiment would also control for biological effects of the intermediate, M^*, which, although it is highly reactive, is sufficiently short-lived to have minimal effect.

Direct effects of the activating light

The application of caged compounds is clearly complicated in tissues that show sensitivity to the activating light. There are only a few biological tissues, such as photoreceptors, that are expected to respond to light at 300-400 nm. Thus in most tissues studied so far, the light pulses used to photolyse caged compounds have not been found to have effects unless the caged compound was present. Surprisingly, though, vascular smooth muscle is sensitive to near-UV light (Furchgott *et al.*, 1955), such that flashes of the intensity required to photolyse caged compounds can cause the muscle to relax and can alter the activity of membrane channels (Gurney, 1993). The relaxation results from light-induced activation of the cytosolic guanylyl cyclase (Karlsson *et al.*, 1984; Wigilius *et al.*, 1990) and it can be blocked by inhibitors of the enzyme, such as haemoglobin. The relaxant effect of light is only observed when the muscle has been precontracted by exposure to a vasoconstrictor or elevated extracellular K^+ concentrations. In common with other agents that stimulate guanylyl cyclase in vascular muscle, light has no effect on basal tension, even although intracellular cyclic GMP concentrations would be raised. This highlights the need to test for light-induced effects thoroughly; it does not follow that because light does not appear to produce a response that it has not altered the biochemistry of the cell in a way that could interfere with the response of interest. Nitric oxide, an activator of the cytosolic guanylyl cyclase is increasingly being shown to have widespread biological effects (Moncada *et al.*, 1991), suggesting that this enzyme could be important in a number of tissues. Thus potential effects of light should be seriously considered.

The Xenon flashlamps used for photolysis produce a wide spectrum of light, including long wavelengths that could potentially heat the experimental preparation and induce a response. These wavelengths have rarely been found to cause problems, probably because there is usually a sufficient depth of fluid covering the preparation for these wavelengths to be absorbed and dissipated before they reach the preparation. These wavelengths can be filtered out of the activating light if they do cause a problem.

Slow photolysis

Although the potential speed offered by photolysis is one of the main reasons for developing caged compounds, it should not be assumed that photorelease is always fast. The rates of the photochemical reactions vary enormously among different caged compounds, and among different analogues of the same caged compound. For example, ATP is released from nitrophenylethyl ATP with a time constant of 12 ms, compared with 55 ms from the dimethoxy nitrophenylethyl analogue at 20°C and pH 7 (Wootton & Trentham, 1989). A new analogue of caged ATP, a benzoin ester, was recently described that photolyses more rapidly, with a time constant less than 10 µs (Trentham *et al.*, 1992). The currently available forms of caged glutamate (Corrie *et al.*, 1993) and caged phenylephrine (Walker & Trentham, 1988) photolyse slowly, with time constants of around 50 and 200 ms respectively, although promising new analogues that photolyse more rapidly are being developed (reviewed in Adams & Tsien, 1993). The slow release from these compounds is due to slow dark reactions, which appear to involve multiple steps. Such slow release may be acceptable for some experiments. However, when the principal interest is in the kinetics of a biological process, then it is important to ensure that photorelease from the probe proceeds with sufficient speed that it will not limit the time course of the response.

5. Equipment

An intense light source, with output in the near UV, is the only specialised equipment needed for experiments with caged compounds. This usually takes the form of a laser or a xenon flashlamp, both of which can provide high intensity UV light concentrated into a brief pulse. Lasers commonly used include the frequency-doubled ruby laser, which produces an intense, 200 mJ pulse at 347 nm in 50 ns, and a cheaper, lower energy nitrogen laser (200 µJ at 337 nm), with which sufficient intensity can be achieved by focussing through a microscope objective. Lasers have the advantage of producing very brief (ns) pulses of monochromatic light at sufficient intensity to cause photolysis. Xenon flashlamps produce a broad spectrum, from 250 to 1500 nm, the pulse usually lasting around 1 ms. Flashlamps are less expensive than UV lasers, easier to maintain and are somewhat easier to incorporate into a microelectrode setup. They function by discharging a high-voltage capacitor across a short-arc flash tube that is filled with xenon gas. A 12 kV trigger pulse ionises the gas, which then emits light as it conducts. The intensity of the light emitted is determined by the energy discharged through the lamp upon ignition, i.e. the charge on the capacitor. This can be adjusted to vary the light intensity and extent of photolysis. The arc is generated between the two electrodes in the bulb, which are separated by a few mm. Thus the arc essentially forms a point source, the light from which is collected and focussed with quartz lenses as illustrated in Fig. 3A, or with an elliptical mirror (Fig. 3B). The overall energy output is about 3-fold higher when an elliptical mirror is used, although the light is focussed into a larger diameter spot (8-10 mm compared with ~3 mm). This configuration is well suited for use with larger preparations, for example intact muscle strips. The energy density is about 60-70 % of that produced using the more usual configuration with a short focal length (25 mm). Flashlamps designed for photolysis experiments can be obtained from Hi-Tech Scientific Ltd., Salisbury, the U.K. distributers for Dr. Rapp, Optoelektronik, Hamburg, Germany. See Rapp and Guth (1988) for a technical description of these lamps. A flashlamp is also available from Chadwick Helmuth, El Monte, California (Strobex model 238). However, it is not supplied with a lamp housing or focussing optics, which can be obtained from an optical supplies company.

Filters are placed in the light path to narrow the spectrum and remove wavelengths <300 nm. One that is often used is the Schott UG 11 bandpass filter, which shows peak transmission (~80%) at 320 nm, but also has a smaller transmission peak in the near infrared region of the spectrum, which can be a problem. For this reason, the Hoya U350 is sometimes preferred. After filtering, the total output of the lamp between 300 and 400 nm can be 200 mJ. The Rapp lamp photolyses 50% of caged ATP molecules when fired directly through a UG11 filter from 4 cm, equivalent to an energy density of about 400 mJ cm^{-2}. In order to increase the intensity of effective

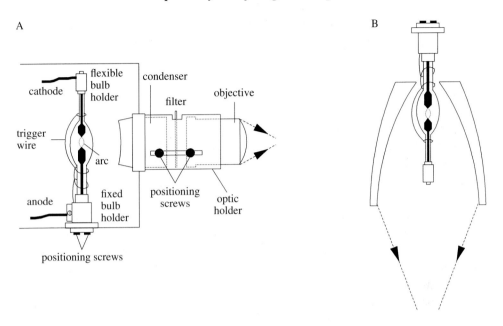

Fig. 3. (A) Diagram of the optical arrangement inside the lamp housing of a standard Rapp flashlamp, illustrating the use of a quartz condensor and objective to focus the light spot. From Gurney (1991). (B) An ellipsoid mirror arrangement to focus the light.

light reaching a preparation, and hence the extent of photolysis caused by a single flash, we frequently use broader spectrum, cut-off filters that only remove wavelengths <300 nm.

Several problems are encountered when incorporating a flashlamp into an electrophysiology set up. One is the well documented photoelectric effect, which occurs when light hits a metalic surface, causing electrons to be ejected and current to flow. Thus any metal in the light path, such as AgCl electrodes, will introduce an artifactual current when the light is flashed. This can be prevented by covering the electrodes and keeping them out of the light path. Currents could also be introduced by stray light hitting electrodes and input connector pins, although we have never found this to be a problem. Bigger problems are presented by the discharge current (2,000 A) and the 12 kV pulse that triggers the flash. These can generate an enormous electrical artefact, which can saturate amplifiers and they are not easily shielded. The recharging of the capacitors is also apparent sometimes as a ripple in the current record immediately following a flash. In addition, there is an audible thumping noise when the lamp discharges, which is associated with movement of the lamp in its housing. This produces mechanical artefacts and can sometimes result in loss of microelectrode impalement or a membrane-pipette seal. To minimise these effects it is desirable to place the lamp as far away from the electrodes as possible, and preferably outside the Faraday cage. This can be done if the light is directed onto a preparation through a light guide (Fig. 4). Fibre optics do not transmit enough UV light to be useful, unless they are made of very expensive quartz. Liquid light guides

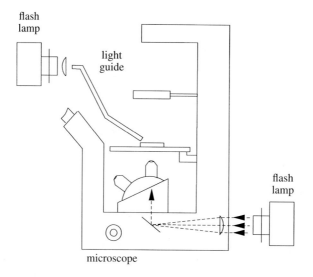

Fig. 4. Light can be focused directly onto a preparation, or directed through a liquid light guide or the optical path of a microscope. From Gurney (1993).

transmit well from 270 to 720 nm, although the efficiency of transmission depends on the length and diameter of the light guide. With a 1 m guide, maximum transmittance is about 80%, but it falls off fairly steeply below 400 nm, such that at 300 nm only 40% is transmitted. The loss of light in this region is unacceptably high if the light guide is used in combination with a UG11 filter. We prefer to remove short wavelengths by placing a borosilicate glass coverslip, about 0.1 mm thick (no. 0 thickness; BDH, Poole, Dorset), at the input to the light guide. The coverslip cuts off sharply below 320 nm, but transmits greater than 90% at longer wavelengths.

The light from a flashlamp can also be directed onto cells *via* the optical path of a microscope, most easily through the epifluorescence port (Fig. 4). In theory this should be the optimal configuration because the light is focussed into an extremely small spot. It is important, however, to use a microscope that transmits well in the near UV; most microscopes do not. In addition the objective should have a high numerical aperture to let as much light through as possible. The simplest arrangement is to place the flashlamp condensor directly in front of the epifluorescence port and focus it using a 200 mm quartz lens onto a dichroic mirror, which is selected to reflect maximally below 500 nm. The dichroic mirror is placed in the light path below the objective, and set at a 45% angle so that short wavelengths are reflected up through the objective. With a Nikon fluor 40×, n.a. 0.85 lens in an Olympus IMT 2 microscope, we can photolyse >5% of caged ATP with a single flash. That particular microscope is limited by a glass lens in the light path, which we have not yet been able to replace. Using a similar micrcope with improved optics should permit greater photolysis. The main difficulty with this configuration is calibrating the amount of photolysis, because the irradiated spot is so small (~400 μm diam).

6. Calibration

For the photosensitive probes to provide useful biological information, it is important that the concentration change resulting from photolysis can be determined accurately. This is usually done by flashing a drop of caged compound solution and measuring the concentration of the active molecule generated or the amount of caged compound destroyed. I find this the hardest part of the experiment, because to ensure that all of the drop is irradiated during photolysis, it must be smaller than the light spot that is focussed on to it. It is a particular problem when the light flash is focussed through microscope optics. Since the droplets are very small they can evaporate quickly, particularly when flashed. To prevent this the droplets can be covered with a thin film of mineral oil. After flashing, the droplets are diluted to provide a sufficient volume for analysis. Depending on the properties of the caged compound and the released species, photolysis can be measured from absorbance changes of the solution or by high performance liquid chromatography (HPLC). The volume of the flashed droplets can be small. To measure photolysis through a microscope, we inject <0.2 µl of test solution into a blob of mineral oil and then retrieve it after photolysis. These volumes are difficult to measure accurately, so it is useful to add a marker molecule to the solution, whose concentration is known and does not change with photolysis. For example, for calibrating the photolysis of caged ATP the marker could be adenosine monophosphate, because it has a different retention time on HPLC to ATP and caged ATP.

The photolysis of a caged compound can also be monitored by measuring the pH change that accompanies the release of the biologically active molecule. This method is particularly well suited to calibrating photolysis in the small volumes of solution flashed through a microscope objective. Several indicators are available for the direct measurement of pH changes. Khodakhah & Ogden (1993) used the fluorescent indicator, BCECF (2′,7′-bis(2-carboxyethyl)-5(6)-carboxyfluorescein), with micro-spectrofluorimetry, to calibrate the photolysis of caged ATP and IP$_3$. They measured 52% photolysis of caged ATP by a single, full-power flash (~100 mJ) at 300-350 nm in 1 ms. The pH change could alternatively be measured using the absorbance indicators phenol red or bromophenol blue. In theory, the changes in Ca^{2+} concentration produced by photolysis of caged calcium or caged Ca^{2+} chelators can be simultaneously measured with Ca^{2+} indicators, either in test droplets of solution or by incorporating the indicator into a cell along with the caged compound. Fluorescent Ca^{2+} indicators may be used to monitor changes in cell Ca^{2+}. However, their use is inadvisable when quantitative measurements of Ca^{2+} concentration are needed. Recent experiments have shown that the presence of caged Ca^{2+} compounds in a mixture with fluorescent Ca^{2+} indicators changes the fluorescent properties of the indicators (Zucker, 1992). The excitation spectra of the indicators fura-2 and fluo-3 were distorted by the presence of photolabile chelators, due to differential absorbance of the excitation light and Ca^{2+}-dependent fluorescence by the photolabile chelators. It may be possible to exploit the inherent fluorescence of nitr-5 to monitor the Ca^{2+} concentration changes produced by photolysis. The absorbance of caged Ca^{2+}

molecules can also be monitored to calibrate their photolysis (Gurney, 1991), since the absorbance spectra change markedly upon photolysis.

It is not always necessary to calibrate different caged compounds separately. If the quantum yield of the caged compound is known, then caged ATP (φ=0.63) may be used as a 'standard' to calibrate its photolysis in a particular experimental setup. As noted earlier, the extent of photolysis produced by a flash depends on the number of excitations during the flash as well as φ. Nevertheless, in practice it is found that with most caged compounds, the amount of flash-induced photolysis relative to that of caged ATP is equal to the ratio of their quantum yields (Walker *et al.*, 1989). Caged cyclic nucleotides appear to be the exceptions (Wootton & Trentham, 1989).

7. Further reading

A number of reviews of caged compounds have appeared in the last few years, which cover their chemistry and applications (Nerbonne, 1986; Gurney & Lester, 1987; Kaplan & Somlyo, 1989; McCray & Trentham, 1989; Kao & Adams, 1992; Gurney, 1993; Adams & Tsien, 1993; Corrie & Trentham, 1993) including several chapters in the 1990 volume of Annual Review of Physiology. In addition there are a number of technical papers outlining the methodology (Gurney, 1991; 1993), including methods for synthesising caged compounds and measuring their photolysis (Walker, 1991; Walker *et al.*, 1989).

References

ADAMS, S. R., KAO, J. P. Y., GRYNKIEWICZ, G., MINTA, A. & TSIEN, R. Y. (1988). Biologically useful chelators that release Ca^{2+} upon illumination. *J. Am. Chem. Soc.*, **110**, 3212-3220.

ADAMS, S. R., KAO, J. P. Y. & TSIEN, R. Y. (1989). Biologically useful chelators that take up Ca^{2+} upon illumination. *J. Am. Chem. Soc.*, **111**, 7957-7968.

ADAMS, S. R. & TSIEN, R. Y. (1993). Controlling cell chemistry with caged compounds. *Ann. Rev. Physiol.* **55**, 755-784.

CORRIE, J. E. T. & TRENTHAM, D. R. (1993). Caged nucleotides and neurotransmitters pp 243-305 in *'Bioorganic Photochemistry'* Vol 2 (H. Morrison, ed) Wiley, NY.

CORRIE, J. E. T., DESANTIS, A., KATAYAMA, Y., KHODAKHAH, K., MESSENGER, J. B., OGDEN, D. C. & TRENTHAM, D. R. (1993). Postsynaptic activation at the squid giant synapse by photolytic release of l-glutamate from a 'caged' l-glutamate. *J. Physiol.* **465**, 1-8.

DANTZIG, J. A., GOLDMAN, Y. E., LUTTMAN, M. L., TRENTHAM, D. R. & WOODWARD, S. K. A. (1989). Binding of caged ATP diastereoisomers to rigor cross-bridges in glycerol-extracted fibres of rabbit psoas muscle. *J. Physiol.* **418**, 61P.

ENGELS, J. & SCHLAEGER, E. J. (1977). Synthesis, structure and reactivity of adenosine cyclic 3′,5′-phosphate benzyl triesters. *J. Med. Chem.* **20**, 907-911.

FORBUSH, B. III (1984). Na^+ movement in a single turnover of the Na pump. *Proc. Natl. Acad. Sci. USA* **81**, 5310-5314.

FURCHGOTT, R.F., EHRREICH, S. J. & GREENBLATT, E. (1955). The photoactivated relaxation of smooth muscle of rabbit aorta. *J. Gen. Physiol.* **44**, 499-519.

GURNEY, A. M. (1991). Photolabile calcium buffers to selectively activate calcium-dependent processes. In *Cellular Neurobiology. A Practical Approach.* (ed. J. Chad and H. Wheal). pp 153-177. IRL Press, Oxford.

GURNEY, A. M. (1993). Photolabile caged compounds. In *Fluorescent and Luminescent Probes for Biological Activity. A Practical Guide to Technology for Quantitative Real-Time Analysis.* (ed. W.T. Mason) pp 335-48. Academic Press, London.

GURNEY, A. M. & LESTER, H. A. (1987). Light-flash physiology with synthetic photosensitive compounds. *Physiol. Rev.* **67**, 583-617.

KAO, J. P. Y. & ADAMS, S. R. (1992). Photosensitive caged compounds: design properties, and biological applications. In *Optical Microscopy: New Technologies and Applications.* (ed. B. Herman & J. L. Lemasters), pp. 27-85. Academic Press, New York.

KAPLAN, J. H. & ELLIS-DAVIES, G. C. R. (1988). Photolabile chelators for the rapid photorelease of divalent cations. *Proc. Natl. Acad. Sci. USA* **85**, 6571-5.

KAPLAN, J. H., FORBUSH, B. & HOFFMAN, J. H. (1978). Rapid photolytic release of adenosine 5'-triphosphate from a protected analogue: utilization by the Na:K pump of human red blood cell ghosts. *Biochemistry* **17**, 1929-35.

KAPLAN, J. H. & SOMLYO, A. P. (1989). Flash photolysis of caged compounds: new tools for cellular physiology. *Trends Neurosci.* **12**, 54-59.

KARLSSON, J. O. G., AXELSSON, K. L. & ANDERSSON, R. G. G. (1984). Effects of ultraviolet radiation on the tension and the cyclic GMP level of bovine mesenteric arteries. *Life Sci.* **34**, 1555-1563.

KHODAKHAH, K. & OGDEN, D. (1993). Functional heterogeneity of calcium release by inositol trisphosphate in single Purkinje neurones, cultured cerebellar astrocytes, and peripheral tissues. *Proc. Natl. Acad. Sci. USA* **90**, 4976-4980.

KORTH, M. & ENGELS, J. (1979). The effects of adenosine- and guanosine 3',5'-phosphoric acid benzyl esters on guinea-pig ventricular myocardium. *Naunyn-Schmiedeberg's Arch. Pharmacol.* **310**, 103-111.

LLANDÓ, L. & ZUCKER. R. S. (1989). "Caged calcium" in *Aplysia* pacemaker neurones *J. Gen. Physiol.* **93**, 1017-1060.

LESTER, H. A. & NERBONNE, J. M. (1982). Physiological and pharmacological manipulations with light flashes. *Ann. Rev. Biophys. Bioeng.* **11**, 151-175.

MCCRAY, J. A., HERBETTE, L., KIHARA, T. & TRENTHAM, D. R. (1980). A new approach to time-resolved studies of ATP-requiring biological systems: laser flash photolysis of caged ATP. *Proc. Natl. Acad. Sci. USA* **77**, 7237-7241.

MCCRAY, J. A. & TRENTHAM, D. R. (1989). Properties and uses of photoreactive caged compounds. *Ann. Rev. Biophys. Biophys. Chem.* **18**, 239-270.

MILBURN, T., MATSUBARA, N., BILLINGTON, A. P., UDGAONKAR, J. B., WALKER, J. W., CARPENTER, B. K., WEBB, W. W., MARQUE, J., DENK, W., MCCRAY, J. A. & HESS, G. P. (1989). Synthesis, photochemistry, and biological activity of a caged photolabile acetylcholine receptor ligand. *Biochemistry* **28**, 49-55.

MONCADA, S., PALMER, R. M. J. & HIGGS, E. A. (1991). Nitric oxide: physiology, pathophysiology and pharmacology. *Pharmacol. Rev.* **43**, 109-142.

NARGEOT, J., NERBONNE, J. M., ENGELS, J. & LESTER, H. A. (1983). Time course of the increase in the myocardial slow inward current after a photochemically generated concentration jump of intracellular cAMP. *Proc. Natl. Acad. Sci. USA* **80**, 2395-2399.

NERBONNE, J. M. (1986). Design and application of photolabile intracellular probes. In *Optical Methods in Cell Physiology* (Soc. Gen. Physiol. Ser.), (ed. P. De Weer & B. Salzberg) Wiley, New York, pp 417-445.

NERBONNE, J. M., RICHARD, S., NARGEOT, J. & LESTER, H. A. (1984). New photoactivatable cyclic nucleotides produce intracellular jumps in cyclic AMP and cyclic GMP concentrations. *Nature* **310**, 74-76.

NERBONNE, J. M. & GURNEY, A. M. (1987). Blockade of Ca^{2+} and K^+ currents in bag cell neurones of Aplysia californica by dihydropyridine Ca^{2+} antagonists. *J. Neurosci.* **7**, 882-893.

NICHOLS, C. G., NIGGLI, E. & LEDERER, W. J. (1990). Modulation of ATP-sensitive potassium channel activity by flash-photolysis of 'caged-ATP' in rat heart cells. *Pflügers Archiv.* **415**, 510-512.

O'ROURKE, B., BACKX, P. H. & MARBAN, E. (1992). Phosphorylation-independent modulation of L-type calcium channels by magnesium-nucleotide complexes. *Science* **257**, 245-8.

PUSCH, M. & NEHER, E. (1988). Rates of diffusional exchange between small cells and a measuring patch pipette. *Pflügers Archiv.* **411**, 204-11.

RAPP, G. & GUTH, K. (1988). A low cost high intensity flash device for photolysis experiments *Pflügers Archiv.* **411**, 200-203.

SPRAY, D. C., NERBONNE, J. M., CAMPOS DE CARVALHO, A., HARRIS, A. L. & BENNET, M. V. L. (1984). Substituted benzyl acetates: a new class of compounds that reduce gap junctional conductances by cytoplasmic acidification. *J. Cell Biol.* **99**, 174-179.

TRENTHAM, D. R., CORRIE, J. E. T. & REID, G. P. (1992). A new caged ATP with rapid photolysis kinetics. *Biophys. J.* **61**, A295

TSIEN, R. Y. & ZUCKER, R. S. (1986). Control of cytoplasmic calcium with photolabile tetracarboxylate 2-nitrobenzhydrol chelators. *Biophys. J.* **50**, 843-853.

WALKER, J. W., FEENEY, J. & TRENTHAM, D. R. (1989). Photolabile precursors of inositol phosphates. Preparation and properties of 1-(2-nitrophenyl)ethyl esters of *myo*-inositol 1,4,5-trisphosphate. *Biochemistry* **28**, 3272-3280.

WALKER, J. W., MCCRAY, J. A. & HESS, G. P. (1986). Photolabile protecting groups for an acetylcholine receptor ligand. Synthesis and photochemistry of a new class of *o*-nitrobenzyl derivatives and their effects on receptor function. *Biochemistry* **25**, 1799-1805.

WALKER, J. W., REID, G. P. & TRENTHAM, D. R. (1989). Synthesis and properties of caged nucleotides. *Methods Enzym.* **172**, 288-301.

WALKER, J. W. & TRENTHAM, D. R. (1988). Caged phenylephrine: synthesis and photochemical properties. *Biophysical J.* **53**, 596a.

WALKER, J. W. (1991). Caged molecules activated by light. In *Cellular Neurobiology. A Practical Approach.* (ed. J. Chad and H. Wheal). pp 179-203. Oxford: IRL Press.

WIGILIUS, I. M., AXELSSON, K. L., ANDERSSON, R. G. G., KARLSSON, J. O. G. & ODMAN, S. (1990). Effects of sodium nitrite on ultraviolet light-induced relaxation and ultraviolet light-dependent activation of guanylate cyclase in bovine mesenteric arteries. *Biochem. Biophys. Res. Comm.* **169**, 129-135.

WOOTTON, J. F. & TRENTHAM, D. R. (1989). "Caged" compounds to probe the dynamics of cellular processes: synthesis and properties of some novel photosensitive P-2-nitrobenzyl esters of nucleotides. *NATO ASI series C*, **272**, 277-96.

ZUCKER, R. S. (1992). Effects of photolabile calcium chelators on fluorescent calcium indicators. *Cell Calcium* 13, 29-40.

ZUCKER, R. S. (1993). The calcium concentration clamp: spikes and reversible pulses using the photolabile chelator DM-nitrophen. *Cell Calcium* **14**, 87-100.

Chapter 16
Microelectrode electronics

DAVID OGDEN

1. Introduction

These notes are intended to provide an introduction to the electronics of microelectrode and patch clamp amplifiers. How much electronics do you need to use a physiological amplifier? Enough to know how much distortion is introduced by the measurement. This means (1) testing the response to an input that simulates the physiological signal, (2) calibrating the gain and frequency response, and (3) knowing the errors that might arise from limited performance. This 'black box' approach to instruments is the minimum needed and requires a knowledge of the basic principles of electronic circuits. It is fine until something goes wrong or a special requirement arises which prompts a look inside to see how things work and whether a modification can be made.

It is worthwhile taking a practical course in e.g. medical electronics if one is available and working in the electronics workshop for a period to learn soldering from an expert. For those interested in making circuits, applications are given below of operational amplifiers in circuits which may be useful for signal processing and can be built relatively easily and cheaply. Building operational amplifier circuits is a very good way of learning the basics of electronics - the amplifiers commonly used are inexpensive enough to permit a degree of trial and error and standard printed circuit boards are available.

A knowledge of the properties and jargon of low-pass filters is necessary for survival. These are introduced and their use prior to digitizing data for computer analysis is discussed.

Many topics have not been included and for these, and wider coverage of topics introduced here, a list of books and articles for further reading and reference is appended.

Current flow in resistors and capacitors

Current, units Amps, is the rate at which charge (measured in Coulombs) flows at a point in a circuit. The driving potential, measured in Volts, is the energy of each unit of charge, Joules/Coulomb, and is analogous to pressure in a gas or concentration in solution. The conversion between chemical concentration of charged particles and electrical quantities of charge is by the Faraday, about 96500 Coulombs/mole of univalent ion, so 96.5 nA of current flowing in a solution is carried by a flux of 1 picomole of univalent ions/s.

National Institute for Medical Research, Mill Hill, London NW7 1AA, UK

Resistance and conductance. Charge flows through a wire, a solution or other conductive media by the movement of charged particles, electrons or ions, against resistance imposed by random thermal motion. The reciprocal of resistance is conductance.

Current flow in a resistor is proportional to the voltage applied across the terminals. Resistance is measured as 1 Ohm, Ω = 1 Volt/Amp. Conductance, Amp/Volt, is measured in Siemens, S = 1/Ohm.

Capacitance. Charges accumulate where two conductors are in close contact (a capacitor) and at different voltages. Energy is stored by polarisation of the medium (the dielectric) between the conductors. The charge accumulated is proportional to the voltage applied, and charges move into and out of the conductors when the voltage changes.

Current flow in a capacitor is proportional to the *rate of change* of voltage, Amps = capacitance $\times$ dV/dt. As a consequence the presence of capacitance modifies the timecourse of potential with respect to current flow.

The unit of capacitance measures the accumulation of charge for 1 volt change of potential, 1 Farad = 1 Coulomb/Volt.

Electrical models of the properties of cells and tissues comprise networks of resistors and capacitors, the former representing paths for current along the core and through ion channels in the surface membrane and the latter capacitative flow across the nonconducting lipid bilayer. No charges (ions) physically cross the membrane capacitor, but flow in the adjacent solution as the membrane potential fluctuates.

Time course of capacitor charging. The single most important circuit for an electrophysiologist is the charging, or discharge, of a capacitor through a resistor. For a voltage V applied to a resistor and capacitor in series, the voltage measured across the capacitor, V_C, can be derived as follows.

Assume (1) that the applied voltage can supply enough current (i.e. has negligable internal resistance) and (2) that the voltage measurement draws no current from the circuit.

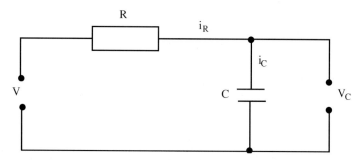

Current flowing into the capacitor is supplied via the resistor, so currents flowing *into* the junction (defined positive) of R and C, where the voltage across the capacitor is measured, are

$$i_C + i_R = 0.$$

The currents (1) through the resistor and (2) into the capacitor are

$$(1) \qquad i_R = (V - V_C)/R \text{ and}$$

$$(2) \qquad i_C = -C.dV_C/dt$$

(NB. i_C flows into the junction for dV/dt negative)

$$(V - V_C) - R.C.dV_C/dt = 0.$$

If V changes abruptly from 0 to V′ at time $t = 0$ then the solution for the time course of V_C is

$$V_C = V'(1 - e^{-t/R.C})$$

a rising exponential with final value $V_C = V'$. For the reverse change of V to 0, so V_C discharges from V′ to 0 the timecourse is

$$V_C = V'.e^{-t/R.C}.$$

Both the timecourse of charging and discharge are determined by the product of the resistance and capacitance and this product is known as the *time constant* of the circuit. It has dimensions of time (Ohms . Farads = seconds) and is the time taken to discharge to $e^{-1} = 0.38$ of the initial value or charge to $(1-e^{-1}) = 0.62$ of the final value.

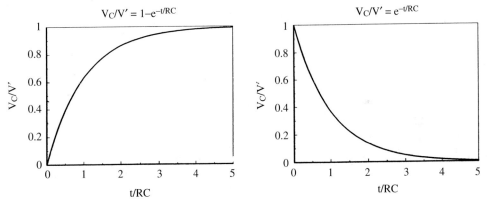

Ideal circuit elements

It is useful initially to consider circuit elements with perfect properties and to take account of practical limitations or secondary properties at a more detailed level of design, analysis or testing. As an example, the electrical properties of microelectrodes for some purposes may be represented electrically as a resistance, but have capacitance, an inherent tip potential and generate noise as well, all of which may be important under some conditions.

Resistors

Generally, current in resistors is simply proportional to potential difference (1 Ohm, $\Omega = 1$ Volt/1 Amp, V/A).

However, there are important practical considerations :

(1) large currents generate heat ($P = I^2R$ Watts). Most resistors are rated at 0.25 or 0.5 Watt.

(2) Resistors have capacitance across the terminals (~0.1-1 pF) which may be important with high resistances (>10 MΩ) and fast voltage changes (e.g. a step of potential) because current will flow through the capacitance as the voltage changes quickly to its new value, producing an initial spike of current.

(3) Voltage noise in resistors has a component (Johnson noise) which increases with the value of resistance, plus an additional component that depends on the resistor composition and voltage difference applied. The rms (standard deviation) of Johnson noise is $V(rms)=(4kTf_cR)^{0.5}$ (k is Boltzmanns constant 1.36×10^{-9} Joule/degree, T temperature °Kelvin, R resistance, Ω, and f_c the bandwidth, Hz).

(4) Moisture or dirt/grease may conduct appreciable current across high value resistors (>100 MΩ).

(5) Commonly, resistor tolerances are 5% or 2% - more precise values can be selected (with a digital multimeter, DMM) or obtained with 2 resistors in series or parallel.

Capacitors

The charge, Q (Coulombs), accumulated on the plates of a capacitor is proportional to the potential difference, V, between the plates, Q=C.V, where the capacitance C has dimensions of Farads (F). The energy difference between the plates due to the potential difference (Volts=Joules/Coulomb) is absorbed in the insulating dielectric separating the plates. No current flows between the plates of an ideal capacitor, but if the voltage changes then current, i, flows into the plates as the charge accumulated changes (Amp=Coulomb/s). Practically, some types of capacitor, such as the large value electrolytic types used in power supplies, may have small leakage currents across the plates. Some types distort rapidly changing signals due to the poor properties of the dielectric, and should not be used where fast signals are encountered. Tolerances are usually 20%. Precise measurement of capacitance is much more difficult than the measurement of resistance.

2. Voltage measurement

The circuit below represents the generation and measurement of a potential and can be split into 3 sections:

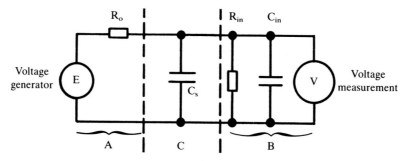

A. *Source of voltage.* The potential V is developed across the output of a voltage generator represented by an ideal voltage source E in series with a small output resistance R_o. Although E produces a constant voltage even if very large currents are generated, the output voltage V is reduced by an amount $(E-V)=IR_o$; this is the situation in real voltage generators, where R_o may represent the internal resistance of a battery or the output impedance of an amplifier. These are normally small ($<10 \Omega$) and only become important at relatively large currents (>100 mA). However, the same considerations apply to microelectrode recording, where E may represent the cell membrane potential and R_o the electrode resistance ($> 10 M\Omega$), so large errors $(E-V)$ may result from small currents (>10 pA).

B. *The measuring circuit* consists of 3 elements, an ideal voltmeter, which draws no current from the circuit and responds instantly to potential changes at the input, a resistance R_{in} to account for current flowing in the input in response to the input voltage, and a capacitance C_{in}. The current drawn by the input is V/R_{in} and determines the error in measuring E, $(E-V) = R_o.V/R_{in}$, so $V/E = R_{in}/(R_o+R_{in})$. The input resistance, R_{in}, of an oscilloscope amplifier is often 1 MΩ or 10 MΩ, of a microelectrode amplifier $10^{12} \Omega$ and of a pH meter $10^{14} \Omega$. It is clearly important that $R_{in} \gg R_o$ to minimize the error in potential measurements. The input capacitance contributes to slowing the response to a change of V, to an extent that depends on C_{in} and R_{in}; the output for a step input is an exponential of time constant $\tau = C_{in}.R_{in}R_o/(R_{in} + R_o)$ (see below).

C. *The connection* between voltage generator and measuring instrument usually involves a wire of low resistance, but often with important stray capacitance to ground. In the case of screened cable (with braided copper shield connected to earth) this amounts to 100-200 pF/m and may restrict the speed of transmitted signals if it is inseries with a large output impedence. A second effect of this capacitance may be to produce instability in the output of an operational amplifier, requiring insertion of a resistor of 20-50 Ω at R_o to limit the current flowing into the capacitance in fast signals. In the case of a microelectrode or other high resistance source, the stray capacitance should be kept as small as possible by using short connections between the electrode and amplifier input, as discussed below.

3. Rules for circuit analysis

There are two basic rules; (1) currents flowing into a node sum to zero, so in the example shown

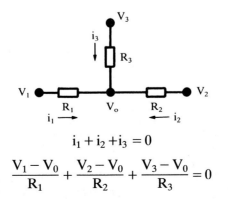

$$i_1 + i_2 + i_3 = 0$$

$$\frac{V_1 - V_0}{R_1} + \frac{V_2 - V_0}{R_2} + \frac{V_3 - V_0}{R_3} = 0$$

(2) the voltage between two nodes is the same *via* all connecting pathways. Complex circuits can often be simplified by applying circuit theory to produce equivalent circuits for analysis. As an example, the circuit

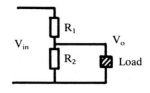

can be simplified to

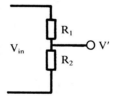

Equivalent

The value of R′ is given by the ratio of the open circuit voltage (i.e. load disconnected), V′, to the short circuit current (i.e. zero resistance load) I′. Thus

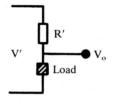

Open circuit

$$V' = V_{in} \frac{R_2}{R_1 + R_2}$$

Short circuit

$$I' = V_{in}/R_1$$

and

$$R' = V'/I' = \frac{R_1 R_2}{R_1 + R_2}$$

For example, in the circuit used to describe voltage generation and measurement above, the load can be represented by the parallel capacitors $C_{in}+C_s=C$. The equivalent potential and resistance are given by

$$V' = E\frac{R_{in}}{R_{in} + R_o} \quad \text{and} \quad R' = \frac{R_{in}R_o}{R_{in} + R_o}$$

and the charging time constant by $\tau=R'C$. The time course of the potential measured, $V(t)$, following a step change of E at $t=0$, is given by

$$V(t) = V'[1 - e^{-t/R'C}]$$

4. Operational amplifiers

These provide a convenient and inexpensive means of processing analogue signals and may also be suitable for use as input stages, in voltage clamp amplifiers and for current generation and measurement. The most common type has two 'differential' (i.e. A–B) inputs and a single output and is represented by

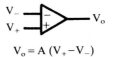

$$V_o = A(V_+ - V_-)$$

where (+) is the *non-inverting* input, i.e. the output has the same polarity as (+), and (−) is the *inverting* input, for which the output has opposite polarity. The output voltage is proportional to the difference, $(V_+ - V_-)$, of the input voltages. A few microvolts potential difference between the inputs is sufficient to cause the output to change by several volts, so the proportionality constant, or *open loop gain* A, is very large, typically more than 10^5.

Negative feedback

The effect of connecting the output to (−), to produce negative feedback, is to minimize the voltage difference between (−) and (+) inputs as follows.

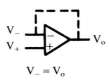

$$V_- = V_o$$

If the open loop gain is A, then $V_O=A(V_+-V_-)$ and, since $V_-=V_O$, $V_O=V_+A/(1+A)$. Therefore, if A is large, $V_O \approx V_+$ and $V_- \approx V_+$, i.e. to a good approximation the output follows the input, V_+, and voltages on (+) and (−) are equal. By applying only a proportion, say $1/x$, of V_O to V_-, a circuit with gain=x is produced. In this case $V_O=A(V_+-V_O/x)$ which can be rearranged to give

$$V_O = xV_+/(1 + x/A)$$

so, provided x/A is much smaller than 1,

$$V_O \approx xV_+ \text{ and } V_- = V_+A/(x + A) \approx V_+$$

In this way, operational amplifiers can be used to give amplification of differing characteristics by modifying the feedback from V_O to V_-.

This kind of analysis can be applied to voltage clamp circuits, in which V_+ is the command potential and V_- the output of the amplifier which monitors membrane potential; the clamp amplifier works to make these equal with gains of 500-5000. However, in this case the cell, the microelectrodes and membrane potential amplifier are included in the negative feedback circuit so the factor x is a complex function of frequency and may become large, particularly at high frequencies, producing poor voltage control and instability.

Amplifier circuits

An approach to building circuits is to suppose initially that amplifiers have ideal characteristics as follows:

(1) Infinite gain ($A > 10^6$) so that circuit gain can be set by external components.
(2) Very high input impedence, so that current flow into the inputs is negligible.
(3) Wide frequency response with no phase changes.
(4) Very low output impedence.
(5) Zero voltage and current offsets at the inputs, so zero input voltage gives zero output voltage. Some basic circuits will be introduced with these properties in mind before considering the deviations from ideal behaviour usually encountered. The two basic configurations have the input signal applied either to the inverting input or to the non-inverting input. In each circuit shown current flowing into the nodes N sums to zero.

Inverting amplifier

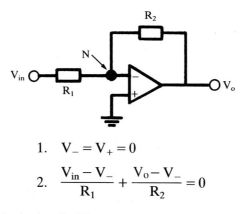

1. $V_- = V_+ = 0$

2. $\dfrac{V_{in} - V_-}{R_1} + \dfrac{V_o - V_-}{R_2} = 0$

Rearranging and substituting for V_-

$$\frac{V_o}{V_{in}} = \frac{-R_2}{R_1}$$

Non-inverting amplifier

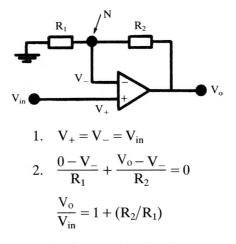

1. $V_+ = V_- = V_{in}$

2. $\dfrac{0 - V_-}{R_1} + \dfrac{V_o - V_-}{R_2} = 0$

$$\dfrac{V_o}{V_{in}} = 1 + (R_2/R_1)$$

The gains V_o/V_{in} of these two circuits were obtained in two steps
(1) $V_+ = V_-$ i.e. open loop gain A is large.
(2) Sum of currents into the node N is zero with none entering the amplifier inputs.

A number of useful circuits stem from the inverting amplifier. The point V_- is known as virtual ground since it is at the same potential as V_+ i.e. 0 V in the illustration (provided A is large) and the input resistance seen by the signal at V_{in} is R_1. If $R_1=0$ then a *current to voltage amplifier* results since the input current is equal to the feedback current ie. $-V_O/R_2$.

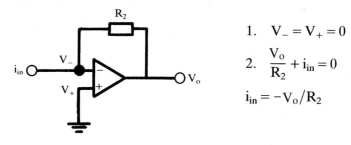

1. $V_- = V_+ = 0$

2. $\dfrac{V_o}{R_2} + i_{in} = 0$

$i_{in} = -V_o/R_2$

Summing amplifier

A number of inputs may be summed with differing gains as follows.

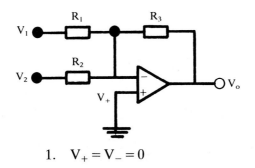

1. $V_+ = V_- = 0$

2. $\dfrac{V_1}{R_1} + \dfrac{V_2}{R_2} + \dfrac{V_o}{R_3} = 0$

$$V_o = -V_1 \dfrac{R_3}{R_1} - V_2 \dfrac{R_3}{R_2}$$

Gain for input V_1 $\dfrac{V_o}{V_1} = -\dfrac{R_3}{R_1}$; for V_2 $\dfrac{V_o}{V_2} = -\dfrac{R_3}{R_2}$

Integrator

The mean level of the input voltage over time may be estimated by integration with a capacitor as the feedback element.

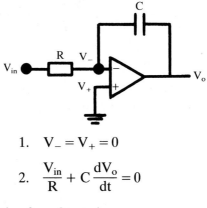

1. $V_- = V_+ = 0$

2. $\dfrac{V_{in}}{R} + C \dfrac{dV_o}{dt} = 0$

Rearranging and integrating from 0 to t gives

$$V_o(t) = \dfrac{-1}{RC} \int_o^t V_{in}\, dt$$

Integrators usually require a variable steady offset voltage at V_1 to zero the input initially and a 'reset' switch to discharge C to zero at the end of a measurement.

Differentiator

The first time derivative of a signal may be obtained by reversing the positions of R and C.

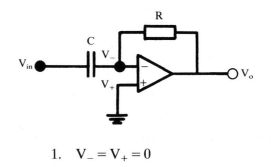

1. $V_- = V_+ = 0$

$$2. \quad \frac{CdV_{in}}{dt} + \frac{V_o}{R} = 0$$

$$V_o = -RC\frac{dV_{in}}{dt}$$

Non-inverting amplifiers

These are used mainly as buffers from a high resistance voltage scource, e.g. a potentiometer, to provide a low output resistance to drive the subsequent circuitry. The most common gain used is 1, i.e. as a voltage follower

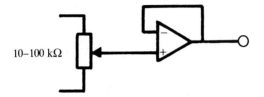

$10\text{--}100\ k\Omega$

but gains of 10 or more may be used as described above. If good quality operational amplifiers are used, the voltage follower configuration may be used in microelectrode amplifiers.

Differential amplifiers

In this case signals are applied at both (+) and (−) inputs and the difference signal $V_1 - V_2$ is required, sometimes with a gain factor. The following circuit uses a single operational amplifier, although for fine tuning of gain and rejection of signals common to both V_1 and V_2 ('common mode rejection'), a circuit with two or more amplifiers may be preferable.

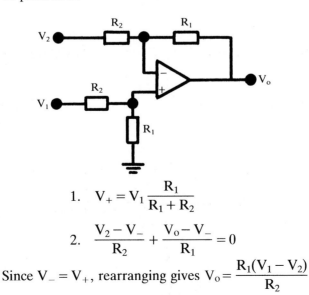

$$1. \quad V_+ = V_1 \frac{R_1}{R_1 + R_2}$$

$$2. \quad \frac{V_2 - V_-}{R_2} + \frac{V_o - V_-}{R_1} = 0$$

Since $V_- = V_+$, rearranging gives $V_o = \dfrac{R_1(V_1 - V_2)}{R_2}$

It can be seen that the gain and common mode rejection (the V_o obtained with $V_1=V_2$) of this circuit requires accurately matched values of resistors R_1 and of resistors R_2.

Non-ideal characteristics

(1) The open loop gain (A) is high at low frequencies, usually about 10^5 at 0 Hz, but declines markedly as frequency increases, falling to 1 at about 10^5-10^7 Hz, depending on the amplifier. The product of gain and bandwidth is a characteristic sometimes specified. Good frequency response can only be obtained at low circuit gain, x, since the condition A/x»1 has to be maintained over a wide frequency range. Gain-bandwidth product is often 100 kHz-1 MHz but may be lower e.g. the 'standard' 741 has only 10 kHz. It is usually better to realize high gains with 2 or more sequential amplifiers of low («100) gain if good frequency response is required.

(2) Stability. Phase changes occur at high frequencies which may result in instability due to positive feedback of these frequencies at gains >1. External compensation or a small feedback capacitor may be required to reduce the gain at high frequencies.

(3) Input resistance and leakage currents. The input impedence and leakage currents depend on the type of transistor junction used at the input. Bipolar inputs have 1-2 MΩ impedence and 0.1-10 nA leakage current. The μA741 and NE5534 are commonly used bipolar types. Junction field effect transistor (JFET) inputs may have 10^{11}-10^{13} Ω impedence and 1-100 pA leakage current e.g. LF356, BB3523. Other inputs e.g. MOSFET or varactor diode inputs may have 10^{13}-10^{15} Ω impedence.

(4) Offset voltage of 0.5-2 mV referred to the input is usually present and can usually be adjusted by an external potentiometer to give zero output.

(5) Noise and drift are usually acceptable but if a critical application is needed, such as a microelectrode input stage, more expensive versions of standard amplifiers selected for low noise and low drift should be used.

Some practical suggestions

It is straightforward and inexpensive to make basic circuits for signal processing, even if some integrated circuits are destroyed in the process. If possible, use preformed printed circuit boards, e.g. from RS™, which reduce the possibility of wiring errors. Always 'decouple' each amplifier from interference in the supply lines by 1-10 μF tantalum (NB polarity of tantalum capacitors) and 0.1 μF ceramic capacitors (for removing high frequency interference) from +15 V and −15 V to the common of the power supply. Use a terminating resistor of 50-100 Ω on the output if a screened cable is used. Make sure that the ground lines to the input and feedback circuits of the amplifier do not carry currents to the common of the power supply, or other sources of large currents, by use of a parallel earthing pattern to a common point.

5. Current measurement

Current flow in a circuit is measured as the voltage difference across a known value resistor. In microelectrode experiments the current to ground from the preparation bath is often required. Measurement by the voltage across a resistor e.g. 50 kΩ connected in the ground from the bath is unsatisfactory, mainly because the potential of the bath then varies with the current and the sensitivity is low (50 μV/nA).

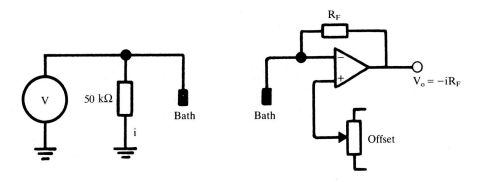

A better arrangement is the virtual earth circuit discussed above, which has the advantage that the bath potential is clamped at a constant level (set by the offset circuit), provided currents are not so large that polarization of the bath electrode occurs. The sensitivity is now set by the value of R_F, $V_O = -iR_F$, without affecting the bath potential, and practically may be 1 or 10 mV/nA (R_F=1 MΩ or 10 MΩ).

A second use of the virtual ground circuit is to avoid changes of bath potential resulting from polarization of the ground electrode with large current flow (>1 mA). The current is supplied via the feedback arm to a large surface area platinum or AgCl electrode and the bath potential monitored with zero current flow at the inverting input by means of a stable AgCl electrode. It should be noted that the bath electrodes form part of the feedback circuit and the output V_O is therefore influenced by polarization and not useful for current measurement directly.

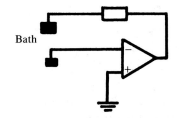

Current measurement within a circuit is also achieved with an inverting amplifier. A commonly used circuit is

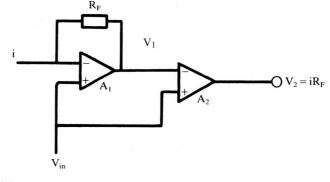

$$(1)\ V_- = V_{in} \quad (2)\ iR_F = V_{in} - V_1 \quad (3)\ V_2 = V_{in} - V_1 = iR_F$$

where the second stage is a differential amplifier (A2, resistors omitted) used to subtract V_{in} from the output of the current to voltage amplifier A1. This arrangement is used in the patch clamp amplifier and sometimes in the current injection and monitoring part of voltage clamp circuits.

Injection of constant current

A source of constant current pulses is useful for ionophoresis, dye injection and determining the passive electrical properties of cells. A 'constant current source' ideally injects constant current into even a very high and fluctuating load resistance such as a microelectrode. An approximation to this may be achieved with a large voltage (e.g. 100 V) and large resistance (R_s=100 MΩ–1 GΩ) in series with the microelectrode (R_e=10–100 MΩ). In order to give constant current, $R_s \gg R_e$.

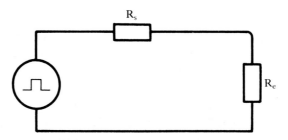

This arrangement works for small currents. With large currents, a capacitative transient may occur due to conduction of the rapidly rising edges of a rectangular pulse over the parallel capacitance of the large value series resistor.

Constant current can be generated electronically by clamping the potential across a resistor in the current path by means of operational amplifiers. A schematic diagram of a circuit given by Purves (1981, 1983) is as follows.

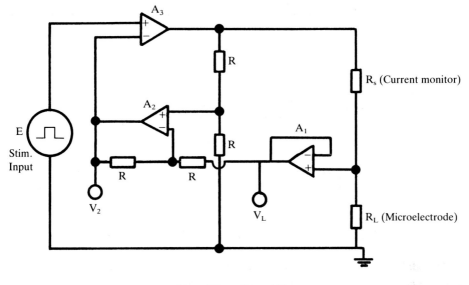

$$V_2 = IR_s \quad V_L = IR_L$$

E is the output of a stimulator, and resistor R_s is of known value (10 or 100 Ω) used to monitor current through R_L, which is a variable load (e.g. a microelectrode). Amplifiers A1 and A2 form a differential amplifier of unity gain to monitor the voltage across R_s; A1 acts as a high impedence buffer drawing no current from the injection path through R_s and R_L. V_2, the output of A2, gives the current through R_L as $I=V_2/R_s$. This signal is also compared at A3 with the command E, A3 acting to maintain $V_-=V_+$, keeping $V_2=E$ and therefore $I/R_s=E$, even if the load resistance, R_L (microelectrode + cell) changes. The potential at the microelectrode R_L can be monitored with the output of A1.

6. Filters

Purpose

Filters are used to remove unwanted high or low frequency components of a signal usually to improve the signal/noise ratio. The most common applications are (1) to remove high frequency noise arising in the recording instruments, microelectrodes or radio interference. (2) To prevent aliasing of digitized signals when sampling into a computer or other digital equipment. (3) To remove the steady DC component and low frequencies when recording low amplitude fast events superimposed on a steady membrane potential or current. It is important to remember that in addition to generating unwanted noise all electronic apparatus has some degree of inbuilt filtering.

Properties of filters and an explanation of filter jargon

Low pass filters progressively reduce the amplitude of high frequency signals; low frequencies below a specified value are passed unattenuated. These are particularly useful in removing high frequency instrument noise from relatively low frequency biological signals.

High pass filters produce attentuation of low frequencies below a specified value. Steady signals give zero amplitude output. The a.c. switch on oscilloscope amplifiers produces high pass filtering below about 1 Hz.

Band pass filters attenuate frequencies above and below specified values.

Filter characteristics are represented (1) in terms of the amplitude response to sine wave inputs of different frequencies and (2) the time course of the response to a step input.

(1) The POWER SPECTRUM of a signal is the power generated in a nominal 1 Ohm resistor at different frequencies plotted against frequency, usually on log-log coordinates. Power spectra are most often plotted as the ratio of output power to input power P_{out}/P_{in}. Fig. 1 shows the power spectrum of white noise (equal power at all frequencies) as a straight line parallel to the abscissa and the outputs of low pass, high pass and band pass filters to a white noise input. These show the attenuation of high (low-pass) or low (high-pass) frequencies as described above.

The HALF POWER or CUT OFF frequencies, f_c, occur where the output power is reduced to 0.5 of the input power. Power ratios are usually given in decibels (dB)

$$dB = 10 \log_{10} P_{out}/P_{in}$$

For $P_{out}/P_{in}=0.5=-3.01$ dB. For this reason half power frequencies are known as -3 dB frequencies when specifying filter properties.

The ratio of output voltage to input voltage V_{out}/V_{in} is more useful than power ratios and filter characteristics are given as log-log plots of V_{out}/V_{in} against frequency. The relation between power and voltage is

$$P = V^2 /R$$

so the power ratio is equal to the square of the voltage ratio:

$$P_{out}/P_{in} = (V_{out}/V_{in})^2, \text{ hence } 1 \text{ dB} = 20 \log (V_{out}/V_{in})$$

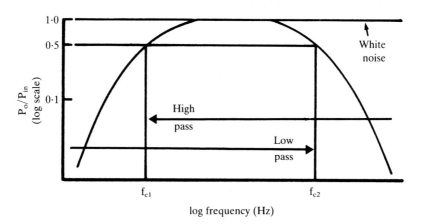

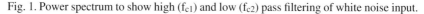

Fig. 1. Power spectrum to show high (f_{c1}) and low (f_{c2}) pass filtering of white noise input.

For

$$P_{out}/P_{in} = 0.5 \quad \text{then} \quad V_{out}/V_{in} = (0.5)^{0.5} = 0.7$$

The halfpower frequency, f_c, of the voltage ratio therefore occurs at $V_{out}/V_{in} = 0.7$, as shown in Fig. 3.

For a low pass filter, the BANDWIDTH is DC (zero frequency) to f_c. For a single stage resistor-capacitor (R-C) low pass filter (Fig. 2) the *power spectrum* at high frequencies has a final slope of -2 on a log-log plot, since the power declines with $1/f^2$. For *voltage* amplitude the log-log plot has a final slope of -1. Both correspond to -6 dB per octave (2 fold frequency change) or 20 dB per decade. A single stage filter of this kind is termed a *single pole* filter. More elaborate higher order filters contain several R-C stages arranged to optimize the roll-off and often have 2, 4, 8, 16 or 32 poles, giving corresponding slopes of -12, -24, -48, -96 or -192 db/octave. The properties of higher order filters are shown in Figs 3 and 4.

(2) The response of a low pass filter to a STEP INPUT of voltage is for most experiments the more important property. The sharp cut-off with frequency achieved with many higher order filters is at the expense of overshoot (or undershoot) of the amplitude following a step, as shown in Fig. 5, and will result in distortion of transients.

Ideal filter properties

(1) Sharp transition from conducting to non-conducting at $f=f_c$.
(2) Flat frequency response in the pass band ($f < f_c$ for lowpass; $f > f_c$ for high pass).
(3) No distortion of transient or step inputs.

There are 4 basic types of filter response commonly used. The good and bad characteristics of each are compared below and illustrated in Figs 3-5.

Simple R-C filters. Single stage R-C filters are encountered in oscilloscope and chart recorder amplifiers and are easily constructed. Roll-off is poor but can be improved by cascading R-C sections. However, attenuation at frequencies near $f=f_c$ is always poor.

Butterworth characteristic. Response in passband is flat and attenuation at $f=f_c$ is good. Suitable for noise analysis. Produce delay and overshoot in response to

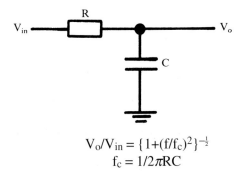

$$V_o/V_{in} = \{1+(f/f_c)^2\}^{-\frac{1}{2}}$$
$$f_c = 1/2\pi RC$$

Fig. 2. Single stage R-C low pass filter.

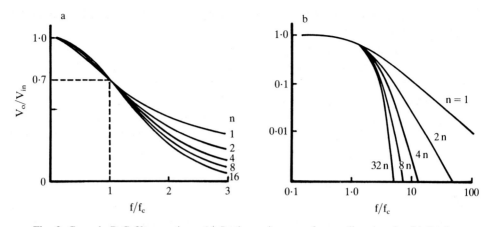

Fig. 3. Cascade R-C filter sections. (a) In the region near f_c on a linear scale. (b) On log coordinates. n is the number of cascaded sections (poles).

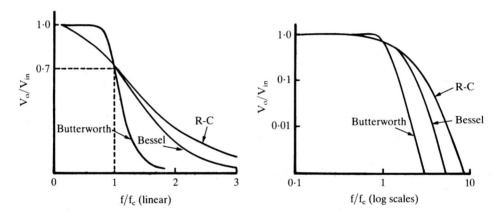

Fig. 4. Comparison of filter characteristics for n=6.

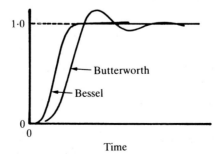

Fig. 5. Response of Butterworth and Bessel low pass filters (n=6) to a step input at t=0.

transient signals, therefore unsuitable for single channel currents, action potentials, synaptic potentials, voltage clamp currents etc.

Bessel characteristic. No delay and minimum overshoot with transient input. Suitable for single channel current etc. signals. Attenuation in region $f=f_c$ is poor and therefore not as good as Butterworth for spectral analysis of noise signals.

Tchebychev characteristic. Good attenuation at $f=f_c$ and a steep roll-off. However, the passband contains some degree of 'ripple' and transient inputs are distorted. This type of response is unsuitable for analysis but may be encountered in some equipment e.g. in FM tape recorders to remove the carrier frequency.

Data sampling and digitization

The main application of high order (4, 8, 16 pole) filters is in sampling a signal prior to digitization by a laboratory interface for computer display or analysis. During digitization the amplitude of the signal is measured at constant intervals determined by the sampling frequency, and stored as binary numbers.

Spectral (noise) analysis. The signal can be regarded as a sum of periodic waves of differing frequencies, phase and amplitude. The highest frequency that can be measured will be determined by the sampling frequency. The need to use high order low pass filters arises from the possibility of *aliasing* in the digitized record, i.e. the spurious addition of frequencies higher than 0.5 times the digitizing frequency to frequencies within the range sampled. If f_s is the digitizing frequency and $f \ll 0.5 f_s$, then the signal at frequencies of $nf_s \pm f$ (n is an integer) appears added to that at f. To avoid this, the maximum frequency that can be present in the record without producing aliasing is $f_N = 0.5\ f_s$ (the Nyquist frequency) so data are low-pass filtered at or below f_N during sampling. An illustration of aliasing is given in Fig. 6, which shows the periodic wave at the Nyquist frequency sampled twice every cycle, and waves of frequencies $(f_N - \Delta f)$ lower and $(f_N + \Delta f)$ higher. If both are present in a signal they are sampled and contribute to the total amplitude. However, their frequencies are indistinguishable at this sampling rate and the sum of their amplitudes would be attributed to $(f_N - \Delta f)$.

Maximum suppression of high frequencies is achieved with a high order Butterworth type response, which is suitable for noise analysis at frequencies up to $f_c = f_N$.

Transients. As mentioned above, the Butterworth response is unsuitable for transient signals. The Bessel response is OK but has poor suppression of frequencies in the region of the half power frequency. The bandwidth should be selected so as not to distort the rise and fall times of the transients. Generally for a Bessel type the 10-90% risetime for a step input is $0.34/f_c$. An action potential rises in less than 100 µs and a bandwidth of 5-10 kHz is needed to avoid distortion. Data should be sampled at 5-10 times the desired bandwidth to ensure good definition of the time course of the transient.

Sequential low pass filtering. Data may be filtered more than once before sampling or final display as a result of low pass filters in the several instruments used for recording, storing and playing back data. The net result is an approximate addition of

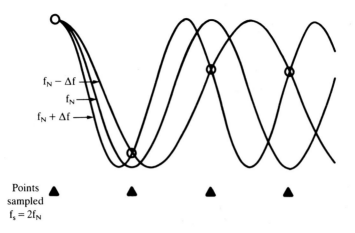

Fig. 6. Aliaising of (fN+Δf) with fN−Δf).

filters so that $1/f^2_c = 1/f^2_1 + {}^1/f^2_2...$ The most likely source of unexpected filtering is the high order filter of the playback amplifier of F.M. tape recorders.

7. Instruments

It is important to know (1) what is required of an instrument in order to make a particular measurement, (2) to be able to test and calibrate an instrument to see how well it satisfies the requirements, and (3) to build, modify or repair electronic circuits when necessary. These notes will concern voltage (microelectrode) amplifiers and current-to-voltage (patch clamp) amplifiers.

Microelectrode amplifier

It is useful to think of the amplifier as an 'ideal' voltmeter (i.e. with none of the following faults) connected to noise generators, input resistance and capacitance as indicated in Fig. 7.

The main requirements for microelectrode recording are as follows:

(1) The current flowing into the input, i.e. from the cell, should be small enough that appreciable redistribution of ions does not occur as a result of the transmembrane flux produced by this current. Also, a potential across the tip of the microelectrode will result, which varies with the electrode resistance. Values <10 pA are alright except for very small cells; values of 1 pA are obtainable with good JFET inputs, but lower values, e.g. for use with high resistance ion-sensitive electrodes, require special input amplifiers. Inputs with 'bridge' arrangements to balance out the contribution of the microelectrode resistance during current injection should be carefully checked for leakage at the zero current setting.

The procedure for measuring input current is simply from the change in voltage output on short cicuiting a high value resistor (100 MOhm) connected between input and amplifier ground.

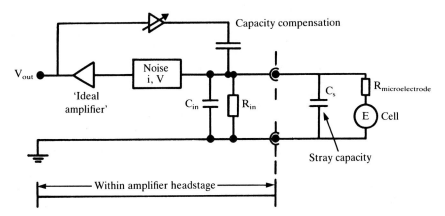

Fig. 7. Schematic diagram of microelectrode amplifier headstage and microelectrode.

(2) The amplifier should have high input resistance (impedance) compared with the microelectrode - the proportion of the membrane potential measured is $R_{in}/(R_{in}+R_{me})$. Most JFET inputs are 10^{12} Ω but varactor amplifiers or MOSFET inputs (10^{15} Ω) are needed for high resistance ion sensitive microelectrodes.

(3) Input noise should be small. Voltage noise with the input grounded is usually about 20 μV peak-peak at dc-10 kHz bandwidth. The major noise source is the microelectrode which contributes Johnson (resistance) noise plus an excess noise arising in the microelectrode, approx. 200 μV p-p for 10 MΩ electrode. Current noise in the amplifier input flows through the microelectrode and may be important for high resistance electrodes-1 pA r.m.s. current noise through a 100 MΩ electrode gives 100 μV r.m.s., about 400 μV p-p. As a result of contributions from these sources, noise increases more than proportionally with microelectrode resistance. This aspect of amplifier performance should be measured with both low and high values of test resistor for comparison with the Johnson noise, given by $V(rms)=(4kTf_cR)^{0.5}$ (k is Boltzmanns constant 1.36×10^{-9} Joule/degree, T temperature °Kelvin, R resistance, Ω, and f_c the bandwidth, Hz).

(4) Response time is usually limited by 'stray' capacitance to ground (C_s) arising from input transistors, connecting wires and across microelectrode walls to bath solution. The response to a rectangular input voltage rises exponentially with time constant $\tau=R_{me}.C_s$ - typical values of 50 MΩ and 10 pF give $\tau=500$ μs ($f_c=1/(2\pi\tau)=320$ Hz). Sources of capacitance are:

(1) between drain and gate of FET input transistors, 3-8 pF. Can be reduced by careful amplifier design (e.g. 'bootstrapping' or cascode input configuration) to 0.1-0.5 pF.

(2) Connecting wires, through proximity to screens. This can be reduced by minimizing length of wires and driving screens with the output voltage of the amplifier (see below).

(3) Capacitance across the microelectrode wall to the bath. This can be reduced by decreasing the depth of fluid in the bath. Painting the electrode with conductive paint,

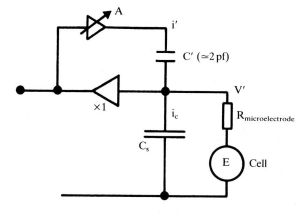

$$i_c = C_s dV'/dt \quad i' = A.C'dV'/dt$$
Gain A adjusted for $A.C' = C_s$ to give $i' = i_c$.

Fig. 8. Capacity compensation.

insulating with varnish and driving the paint screen with the ×1 output is very effective. Driven screens of this kind place the same voltage signal on the screen as that present on the input, thereby removing capacitative coupling between input and screen for signals of similar waveform, without loss of shielding from unwanted external scources. However, current noise applied to the screen from the output will be transmitted capacitatively into the input and so may make recordings with high resistance electrodes noisier.

Capacity compensation/neutralization: (see Fig. 8) The current through the stray capacity is compensated with current generated by a variable amplified output (1-10×) applied through a fixed capacitor to the input, generating a current proportional to dV/dt. Problems occur first in adjusting the compensation correctly so as not to distort the input signal, and also from the injection of current noise with the compensating signal. Capacity compensation works best when the stray capacity is initially small, so the precautions aimed at reducing stray capacity cited above are still worthwhile.

Procedures that may be used to test the response time of the microelectrode and amplifier (with or without compensation) are, (1) applying a rectangular voltage to the bath solution with just the tip of the electrode in the solution or (2) applying a triangular pulse through a small value ($\approx$ 1 pF) capacitor into the input. This latter procedure injects a rectangular pulse of current into the input and avoids spurious coupling through the capacitance of the microelectrode wall from the bath solution, which is present with the former method.

Patch clamp amplifier

The patch clamp amplifier is a current to voltage (I/V) amplifier with basic configuration

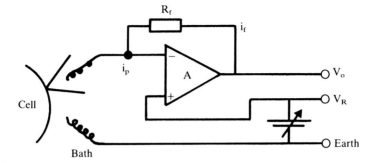

where the amplifier A maintains the (−) input at the same potential as (+) i.e. V_R, by negative feedback through resistor R_f. Thus, if the current flowing into (−) is neglected, $i_p + i_f = 0$ and $(V_O − V_R) = i_f R_f = −i_p R_f$.

Background noise in patch clamp recording depends critically on the impedence of pipette-cell seal at the input and on the feedback resistance, R_f. It is important to remember that *current* noise at the input matters in patch clamp recording. Sources of noise are (1) the seal and feedback resistances and (2) voltage noise in the amplifier input.

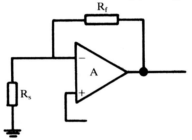

(1) The (−) input is connected to ground via R_s (seal and bath) and R_f (amplifier output). The source (seal) resistance R_s and feedback R_f give rise to voltage noise of approximately

$$V_{rms} = (4kTf_cR)^{\frac{1}{2}}$$

for ideal resistances, where k is Boltzmanns constant, T temperature (Kelvin) and f_c is the upper frequency limit (bandwidth, i.e. low pass filter setting). The current noise

$$i_{rms} = V_{rms}/R,$$

so

$$i_{rms} = (4kTf_c/R)^{\frac{1}{2}}$$

Thus, background current noise due to source and feedback resistances decreases as $1/R$. Values of R_f of 10-50 GΩ are used for this reason, and good seal resistances (5-50 GΩ) are necessary for low noise recording (see also next section).

(2) Voltage noise arises in the JFET transistors used for the input stage of the amplifier. This is due to (a) 'shot noise' of the input current, resulting from movement of discrete charge carriers (b) thermal variations in internal current flow through the JFET seen as input voltage noise when feedback is applied to the input. This voltage noise gives rise to currents flowing in the stray capacitance of the input. The currents increase greatly with recording frequency such that spectral density of input current

$$S_i = (2\pi f C)^2.S_V$$

where f is the frequency bandwidth, C the total capacitance and S_V the spectral density of input voltage noise.

Sources of input capacitance are (1) within the amplifier, mainly across FET junctions and from input lead to ground, approx. 10-20 pF. (2) Across the holder and pipette to adjacent grounded surfaces e.g. microscope and screens. (3) Across the pipette wall to the bath solution. Use of Sylgard resin or other treatments to coat the pipette exterior reduces capacitance considerably by decreasing creep of fluid along the outside of pipette glass and by increasing the thickness of the pipette wall. Minimising the bath level and drying the electrode holder when changing pipettes also reduces capacitance due to fluid films.

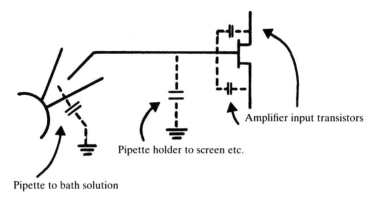

Sources of stray capacitance.

For a good amplifier, a coated pipette and a good seal, the contributions of the electronics, the pipette and the seal to the total noise are approximately equal, as indicated in Fig. 9.

Frequency response of patch clamp amplifiers. The large values of feedback resistor used in the patch amplifier result in an output time course, following a sudden or step input current, which is dominated by the parallel stray capacity associated with the resistor. The patch clamp thus has a low-pass filter charactaristic, so a 50 GΩ resistor with capacity across the terminals of only 0.1 pF gives a response time constant of τ=5 ms to a step input. This is too slow.

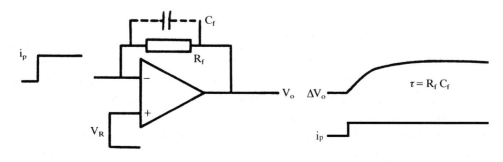

Provided parallel capacity is uniformly distributed over the resistor the response is approximately a single exponential. The output voltage for a step input of current i_p is

$$\Delta V = i_p R_f (1 - e^{-t/\tau})$$

A compensating circuit is employed to correct for this slow response by (a) differentiating the response,

$$\frac{dv}{dt} = \frac{-iR_f}{\tau} e^{-t/\tau}$$

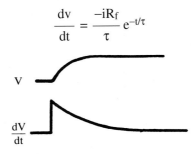

(b) scaling the differentiated response by τ and adding it to the response itself

$$V + \tau dV/dt = -iR_f$$

This procedure is valid for any input waveform and is not affected by pipette capacitance, providing compensation for C_f independent of recording conditions, and can give response time constants of 20-50 µs for a step input. A circuit that performs differentiation, scaling and addition which is often used for compensation is

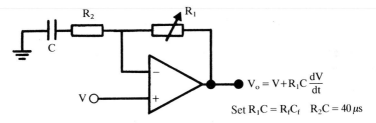

$$V_o = V + R_1 C \frac{dV}{dt}$$

Set $R_1 C = R_f C_f$ $R_2 C = 40\,\mu s$

This operates as a voltage follower at low frequencies but increases in gain at high frequencies. Time constant R_1C is adjusted to the same value as that produced by stray feedback capacitance in the I/V amplifier; R2 provides damping to prevent oscillation. If the patch clamp input stage amplifier has a more complicated response, as is the case with commercial switchable resistor designs, then additional waveform shaping circuits are used to compensate the response.

Setting up the compensation: this is most easily done with a square wave current injected into the input, adjusting R, (either a panel mounted or internal potentiometer) to give a square output, flat topped up to ~10 ms. A square input current is achieved by capacitively coupling a good triangular voltage waveform to the input, simply by clamping the open end of a screened cable in the vicinity of the input pin. The more complicated commercial patch clamps may require adjustments of 4 or more trimpots. The response should be checked with both high amplitude signals (often

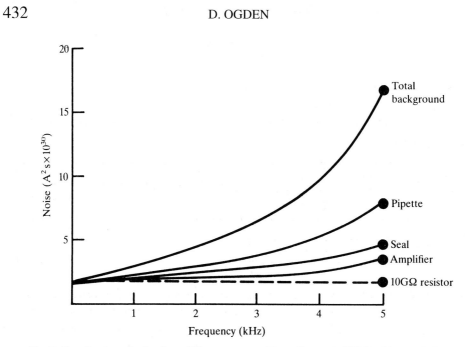

Fig. 9. Contribution of noise from different sources. From Sigworth (1983) with permission.

provided in commercial amplifiers) and low (<10 pA) amplitude as encountered in single channel recording.

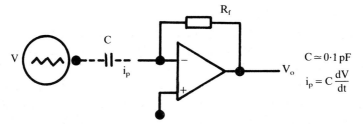

Setting the gain: this is done by connecting a precisely know resistance (about 100 MΩ) between input and ground, applying a voltage step V to the non-inverting (V_{ref}) or command input to give $i_p = V/R$ and looking at the output deflection V_o:

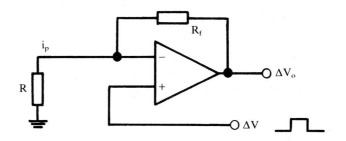

Thus the gain (V/A) is

$$\Delta V_o/i_p = \frac{\Delta V_o R}{\Delta V}$$

The gain is adjusted to give convenient units of V_o/i_p e.g. 10 mV/pA by an internal potentiometer at a later stage of amplification.

The current measurement circuitry of a patch clamp usually consists of the following stages:

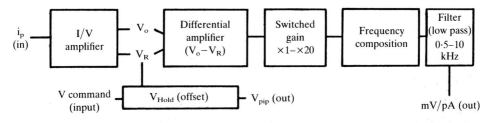

It should be noted that the polarity of the output of the initial I/V amplifier stage is retained (i.e. not inverted at a later stage) in most commercial instruments (one exception is the Biologic^{TM}), so $V_o=-i_pR_f$.Gain, with i_p positive for current into the amplifier input. Outward membrane currents are negative pipette current in whole cell clamp and outside out patch, and V_O changes in accord with the convention that outward movement of cations is positive. For cell attached recording and inside out patch inward current would deflect V_O positive. Data should be inverted when necessary to conform to the convention.

Capacitor feedback in patch clamp amplifiers

The noise due to the high value feedback resistors used in most patch clamp amplifiers can be avoided by using a capacitor as the feedback element, resulting in an output that is the integral of the pipette current and which is differentiated at the next stage of processing. As well as the absence of Johnson noise, capacitor feedback gives a wider range of current input, which is restricted by the rate of change of the amplifier output rather than the amplitude of the voltage applied across the feedback element. The I/V amplifier is in this case an integrator and suffers the restriction, discussed earlier, that standing offset currents (e.g. through the seal) are integrated along with the signal. The feedback capacitor therefore requires discharge as the potential across it approaches the maximum that can be supplied. This is done by automatic switching that result in reset transients of ≈ 50 μs in the record. The real advantage for patch clamp recording is an $\approx 30\%$ reduction in noise that can be achieved with careful single channel recording.

The wide range also has advantages in recording large synaptic currents and in lipid bilayer recording. The properties of the capacitative feedback amplifier are discussed by Finkel (1991).

Voltage command inputs

High resistance seals permit large changes of potential in the pipette without large current flow across the seal into the pipette from the bath. In response to a voltage step applied to V_{ref}, the I/V amplifier produces an output such that current flow through R_f causes the same potential change in the pipette. In order to improve the signal to noise, so as to reduce the contribution of noise on the command, these are divided by a factor, often 10, in the I/V amplifier i.e. $V_{ref}=0.1$ $V_{Command}$.

The current flowing into the pipette is the sum of currents (a) through the membrane patch (e.g. single channel currents), (b) through the seal resistance and (c) to charge the input capacity of the amplifier and pipette; (a) is the quantity measured, (b) produces a small, relatively constant offset with good seals, (c) produces large transients in response to rapid potential changes ($i=CdV/dt$) and may lead to saturation of the I/V amplifier or frequency compensation circuit. These transients obscure single channel currents following a voltage step, and if saturation of the amplifier occurs, may result in loss of voltage control in the pipette and, because of the long recovery times of amplifiers in the circuit, loss of data for several ms following a step.

To make recordings of voltage gated channel currents it is essential to compensate for the transient initial capacity current. Compensation is applied to the pipette input of the I/V amplifier *via* a capacitor (C_i, value ≈ 2 pF) so as to supply capacity current that would otherwise be provided by the output of the I/V amplifier. For this purpose, the voltage command applied to V_{ref} is modified by a separate parallel circuit, so that its amplitude and risetime can be varied, inverted and injected via the capacitor into the input. It is adjusted to remove the capacity transients to give a square, resistive initial step on the 'pipette current' output.

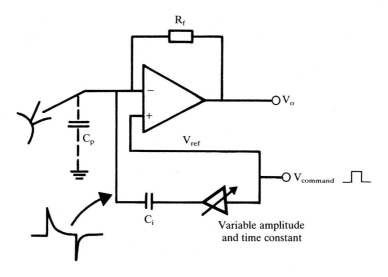

C_p=stray pipette capacity, C_i is the capacitor for injection of compensation current.

Whole cell recording. The recording of currents from whole cells, after breaking the membrane between pipette and cell, has electrical problems associated with the extra cell capacity to be charged through the series resistance of the pipette/cell junction:

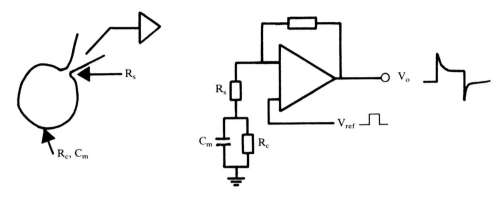

Voltage pulses to V_{ref} (and hence the pipette) cause current to flow into the cell through R_s to charge the cell capacitance, C_m. This gives rise to a current output with initial transients of time constant

$$\tau_c = \frac{R_s R_c C_m}{R_s + R_c} \quad \text{(N.B. for } R_c \gg R_s, \tau \simeq R_s C_m\text{)}$$

and an initial amplitude V/R_s. Thus, after compensation for fast transients, these much slower transients may be used to calculate R_s and C_m. The capacity current required to charge C_m may be compensated in the same way as for the pipette capacitance, i.e. by applying a current of variable magnitude and risetime to the input by injection through the capacitor C_i. This slow capacitance compensation is generated in parallel with the fast compensation and added to the signal injected through C_i. Calibration of the compensation circuit allows C_m and R_s to be read from dials on front panel calibrated potentiometers.

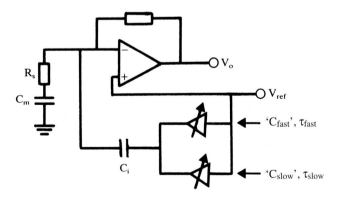

Unlike a microelectrode amplifier, the use of slow capacitance compensation does not improve the speed of clamping the cell membrane potential (unless the amplifiers have reached saturation) but simply removes the capacitative transient from the pipette current and amplifier output. The limiting frequency characteristics of the response of the cell for noise analysis or to a potential step has $\tau \approx R_s C_m$ as before.

The presence of the series resistance R_s gives rise to an error voltage between the clamped pipette potential and the true value of the cell membrane potential,

$$V_c - V_p = i_p R_s.$$

Thus, it is important to know the value of R_s so that corrections can be applied to current/voltage data for the error of V_c. Also, as mentioned above, the transient response of the system is attenuated by the $R_s C_m$ time constant, which for cells of e.g. 10-20 pF and R_s of 10-20 MΩ may produce low pass filtering at around 400 Hz.

The effect of R_s is to underestimate V_c by an amount proportional to the recorded current. Compensation for this effect ('series resistance compensation') may be made by feeding back a proportion of the current signal to V_{ref}. In commercial amplifiers the value of series resistance is taken from the whole cell transient cancellation, multiplied by i_p and a proportion, up to about 80%, and added to V_{ref}. However, this often results in instability and some adjustment to the phase of the compensation may be present. In practice the value of R_s often increases or fluctuates on a minute or shorter timescale during recording, often making series resistance compensation imprecise. As with other cases where compensation is applied, the best results are obtained with a low initial series resistance.

8. Grounding and screening

Ground lines or earths serve four distinct functions in electrophysiological equipment. These are:

(1) *Safety*. All instrument cases and other enclosures of apparatus with a mains supply must have a reliable connection to the mains earth through the plug.

(2) *Reference potential*. Provide the reference (zero) potential for measurements of cell potentials and also for each stage of signal processing that occurs within instruments. Any unwanted signal present on the reference ground of an amplifier will appear in the output and be passed on to the next stage. The reference point is the signal ground of the input amplifier or oscilloscope input. Leads connecting the signal ground should carry no current and run next to or twisted with the signal from the input amplifier, to reduce the area of loop susceptible to magnetic interference.

(3) *Current returns*. To return current to the common of power supplies from e.g. zener diodes, relays, lamps, decoupling circuits. These common returns should be run separately from reference grounds to a central grounding point.

(4) *Screening*. Electrostatic screening to prevent interference from mainly 50 (or 60) Hz is achieved with a Faraday cage and by ensuring that all conducting mechanical parts such as microscope objective, condenser and stage, the baseplate and micromanipulators have good, low resistance («1 Ω) connections to ground. The closer a component is to the pipette, the more important a good connection.

Magnetic interference from transformers or motors occurs by induction in circuit loops of large area oriented across the magnetic field generated. This form of interference is often of 100 or 150 Hz and can be prevented by re-routing ground lines to minimize the area presented to the field, or by moving the source of the interference.

It is usual to arrange grounding to a central point, often on the oscilloscope input, running separate lines for reference and screens to equipment in the cage or rack. Current returns go to the common of the power supply first, which is in turn connected by a single wire to the central ground point. The central point can be taken to the mains earth *via* the oscilloscope cable or by a separate lead.

9. Test equipment

Testing can mostly be done with electrophysiological apparatus already present i.e. an oscilloscope, a precisely timed square wave or pulse generator, such as a Digitimer, and an accurate voltage source or calibrator. Digital multimeters (or DMM) provide accurate measurement of steady potential, current and resistance. A good signal generator or other equipment for occasional use can often be shared or borrowed.

Electronics reading list

CLAYTON, G. B. (1979). *Operational Amplifiers*. 2nd Edition, London: Butterworths.
HOROWITZ, P. & HILL, W. (1989). *The Art of Electronics*. 2nd Edition, Cambridge: C.U.P.
PURVES, R. D. (1981). *Microelectrode Methods for Intracellular Recording and Iontophoresis*. London, New York: Academic Press.
FINKEL, A. S. (1991). Progress in instrumentation technology for recording from single channels and small cells. In *Molecular Neurobiology* (ed. Chad, J. & Wheal, H.V.). Oxford: IRL Press.

Specialist articles

MARSHALL, M. A. (1976). Earthing, Shielding and Filtering Problems. 4 articles in *Wireless World*, August, September, November and December, 1976.
PURVES, R. D. (1983). High impedence electronics. 2 articles in *Wireless World*, March, April, 1983.
SIGWORTH, F. J. (1983). Electronic design of the patch clamp: In *Single Channel Recording*. (ed. Sakmann & Neher). N.Y.: Plenum.
WILLIAMS, T. (1991). *The Circuit Designer Companion*. Butterworth-Heinemann. (Good chapter on earthing and shielding.)

Applications manuals from: Siliconix (FETs), National Semiconductor (FETs and 0p Amps), Burr Brown (0p Amps).

Index

Note: illustrations are represented by *italic page numbers*, tables by **bold numbers**.